NR Coregulators and Human Diseases

NR Coregulators and Human Diseases

Rakesh Kumar
The University of Texas M. D. Anderson Cancer Center, USA

Bert W. O'Malley
Baylor College of Medicine, USA

World Scientific

NEW JERSEY • LONDON • SINGAPORE • BEIJING • SHANGHAI • HONG KONG • TAIPEI • CHENNAI

Published by

World Scientific Publishing Co. Pte. Ltd.
5 Toh Tuck Link, Singapore 596224
USA office: 27 Warren Street, Suite 401-402, Hackensack, NJ 07601
UK office: 57 Shelton Street, Covent Garden, London WC2H 9HE

British Library Cataloguing-in-Publication Data
A catalogue record for this book is available from the British Library.

NUCLEAR RECEPTOR COREGULATORS AND HUMAN DISEASES

ISBN-13 978-981-270-536-5
ISBN-10 981-270-536-8
ISBN-13 978-981-270-537-2 (pbk)
ISBN-10 981-270-537-6 (pbk)

Typeset by Stallion Press
E-mail: enquiries@stallionpress.com

Printed in Singapore by Mainland Press Pte Ltd

Contents

About the Editors

Rakesh Kumar

Professor Rakesh Kumar is currently the John G. and Marie Stella Kenedy Memorial Foundation Chair at the M.D. Anderson Cancer Center, where he is Professor of Molecular and Cellular Oncology and of Biochemistry and Molecular Biology. Professor Kumar is also an adjunct Professor at the Baylor College of Medicine. Dr. Kumar's research is directed at defining the mechanisms of estrogen receptor action with a special focus on subcellular localization and master chromatin modifiers. He has discovered the novel targets and functions of the MTA family of nuclear receptor coregulators, and opened new avenues for research. He also was the first to recognize a mechanistic role of PAK1 in cancer cell invasiveness and hormone action, discovering its physiologic substrates, and identifying the nuclear localization and functions of PAK1. He serves on the editorial boards of major cancer journals as well as on peer-review grant panels. Professor Kumar has received several awards and honors for his research excellence.

Bert W. O'Malley

Professor Bert W. O'Malley is currently the Tom Thompson and Distinguished Service Professor at the Baylor College of Medicine. His laboratory discovered that steroid hormones and nuclear receptors act on genes to regulate the synthesis of messenger RNAs. He then went on to discover the "missing link coregulators" (coactivators/corepressors) that decipher all of the transcriptional instructions in the receptors. Coactivators are "master genes" that have immense regulatory influences on tissue development and physiology because they activate the subfamilies of genes in a manner designed to coordinately regulate cell physiology and metabolism. Of course, the dysfunctions in coactivators (or corepressors) lead to serious disease consequences but can serve as new markers for diagnosis and therapies. Professor O'Malley was a founder of the field of molecular endocrinology and is a member of the National Academy of Sciences and the Institute of Medicine. He has received many honorary doctorate degrees and numerous international awards and honors.

Preface

Nuclear receptor coregulators have experienced an explosive early development over their founding decade. The number of coactivators and corepressors has grown to over 300. The molecular biology of coactivators has informed us of a cadre of diverse and interesting mechanisms of transcriptional action, including chromatin modification and remodeling; initiation of transcription; elongation; alternative RNA splicing and finally, protein degradation. Over the past five years, researchers have demonstrated that coactivators have expanded their pleiotropic actions in multiple cell compartments where they shepherd functions of the numerous gene products required to regulate large physiologic processes such as growth, metabolism, and inflammation. The discovery of coactivators has also resulted in the production of over 90 mouse knockout models for the study of heritable diseases. Of the 300 currently discovered coregulators, about 165 already have been demonstrated to result in human pathologies and heritable dysfunction. They have been demonstrated to be causal in numerous instances of embryonic lethality; growth retardation; maturation; mental retardation; metabolic and endocrine disorders; inflammatory disorders; malignancies; reproductive; and cardiovascular abnormalities. The editors wish to thank the Nuclear Receptor Signaling Atlas Consortium for their continued support in the area of NR coregulator research. The editors also wish to point out that a great deal of primary and unpublished data in this field is summarized on the NURSA web site (www.NURSA.org).

Since these coregulator "master genes" are poised to pay big future dividends to the field of medicine, we felt it timely to compile the first book, written by the top experts in the field and dedicated to the physiologic and pathologic promises of coregulator research. The book will contribute to the foundation of Coregulator Biology as an emerging discipline in medical sciences.

Bert W. O'Malley, M.D. Rakesh Kumar, Ph.D.

protein (CBP), the related p300,[d] or the SRC coactivators, indicating that no simple protein signature exists for NR coregulators.

The nuclear receptors, on the other hand, share a high level of structural conservation in their central DNA binding domain, and, to a lesser extent, in the carboxyl terminal ligand binding domain (LBD). The intricacy of conserved α-helical structures in the LBD and their positional changes induced upon ligand binding provided the structural determinants for the identification of a conserved motif in proteins that bind to this domain. This so called "LXXLL-motif" (where L is leucine and X is any amino acid) or "NR box" was found to be necessary and sufficient to form a helical structure with which proteins bind to a "charge clamp" formed by the ligand-activated nuclear receptor.[14,15] Because proteins binding to this critical and hormone-sensitive region of the receptor inevitably alter the transactivation potency of nuclear receptors, the LXXLL motif was thought to serve as a signature motif for coregulators. This, however, is only partially valid, as many NR coregulators do not contain an NR box but still associate (through direct binding or tethered to another coregulator) with the nuclear receptors. Moreover, coregulator interactions also occur with other parts of the receptor that are not NR box-dependent. Thus, the LXXLL sequence is a signature motif for only a subset of coactivators or corepressors that bind in a ligand-dependent manner to the charge clamp of activated nuclear receptors.

Many coregulators indeed contain one or more LXXLL motifs. A compilation of nuclear receptor coregulators curated at NURSA (Nuclear Receptor Signaling Atlas; www.nursa.org) at the time of this writing includes 303 coregulator genes of which 149 have NR boxes in their amino acid sequence. While many (84) coregulators on this list have only one LXXLL motif, others have two (36) or several, with PELP1 (proline, glutamic acid and leucine rich protein 1) topping the list with eleven NR boxes.

Soon after the discovery of the SRC-1 coactivator, we realized that the same molecule could also be employed to repress gene function.[16] This unexpected "switch" in activity is realized now to be a frequent trait for both coactivators and coregulators. PELP1, separately cloned

[d] The SRC-family of coregulators as well as the related CBP and p300 are not specific for only the nuclear receptors, as they activate transcription mediated by non-nuclear receptor transcription factors too.

and termed MNAR (modulator of nongenomic activity of estrogen receptor), is a good example of a multi-tasking transcriptional coregulator that participates in both enhancement and repression of gene function. It participates in genomic and nongenomic responses to ER signaling.[17] It was postulated to play an important role in endocrine cancers as it increases E2-mediated cell proliferation and is involved in E2-mediated tumorigenesis and metastasis.[18] PELP1/MNAR coactivates ER- and RXR-mediated transcription but corepresses GR, NUR77, and non-nuclear receptor transcription factors such as AP-1 and NF-κB.[19] This interesting ambivalence towards acting as either positively as coactivator or negatively as corepressor is not uncommon for NR coregulators, nor is their ability to affect many transcription factors. The differential use of the many LXXLL helices may be a mechanism to modulate the efficacy of association with particular nuclear receptors,[20,21] and while a differential use of an activator and a repressor domain within a single coregulator is possible, it is uncommon, and the differential transcriptional activity is more readily achieved through interactions with other molecules carrying enzymatic activities tethered to the coregulator and brought into the transcriptional unit.

In the case of PELP1/MNAR, the N-terminal leucine-rich region was observed to interact with HDAC2 where it exhibits repressive activity when bound to GR. In addition, it was shown that the C-terminal acidic activation domain, which contains homopolymeric glutamic acids stretches, binds to the hypoacetylated histones H3 and H4 and prevents them from becoming substrates of histone acetyltransferases, promoting and maintaining the hypoacetylated state of histones at the target genomic site.[19] Hypoacetylated histones are associated with compact, transcriptionally inert, chromatin. When bound to ER, however, PELP1/MNAR recruits the p300/CBP histone acetyltransferases to the transcription unit,[17] thus reversing its role so as to maintain hyperacetylated histones. Acetylated histones are found in transcriptionally permissive, or "loose", chromatin. This example illustrates one of the hallmarks of NR coregulator function, namely the ability to provide or to recruit diverse enzymatic activities to nuclear receptors for the purpose of modulation of transcription. It is therefore not a surprise that the intense research being done on transactivation is revealing that a network of sequentially exchanged coregulator complexes contribute to transcription, rather than it being the work of individual proteins. These coregulator complexes have a heterogeneous composition and are not discrete, separable entities but share many molecules and subcomplexes.

1.2.2 *Diversity in enzymatic activities and complexes*

To better understand the unexpectedly diverse roles that the NR coregulators play in human diseases, one must first recognize the value of their extensive involvement in diverse cellular processes. The elucidation of many complexes identified to serve the roles in NR-mediated target gene expression excoriates transcription control from a solely promoter-proximal process and establishes it as a cell system-wide central event that is linked to epigenetic programs. Because coregulators act as part of multi-protein complexes, carrying along enzymatic activities such as acetylating and deacetylating entities, methylases and demethylases, protein kinases and phosphatases, ubiquitin and SUMO ligase activities, pseudouridylases, and ATP-dependent chromatin remodeling activities, they exert stratified actions in cellular processes as diverse as transcription (including transcriptional elongation, RNA splicing and RNA transport), translation, DNA replication, and cell cycle control. The targets for these enzymatic activities are components of the coregulator complexes, nuclear receptors and other transcription factors, components of the basal transcription machinery, and the chromatin adjacent to the genes they regulate.

To complicate matters, the eukaryotic cell makes use of a combination of these enzymatic activities, connecting diverse biological processes with transcription control, which, paradoxically at a first glance, also includes the cessation of transcription via the ubiquitin proteasome pathway.[22,23] Moreover, new evidence is emerging that suggests another level of abstraction in the control and integration of cellular signals by the differential post-translational modification of the NR coregulators themselves. While the histone modification code may have evolved to define the transcriptional state of polymerase II-driven gene expression, a coregulator modification code will define complex composition and its target preferences. All this, and possibly much more, as we are still discovering new NR coregulators and are learning more about their functional diversity, has to be considered in addition to the temporal and spatial aspects of coregulator and nuclear receptor expression, as well as the availability and type of ligands for nuclear receptor activation.

The focus of this essay (and book) is to discuss the involvement of selected NR coregulators in human diseases. We therefore must refer to other reviews for a more detailed account of the biochemical and cellular aspects of these molecules.[24–29] However, we conclude with a more precise definition of "NR coregulator", a prelude of which shall include that this interesting class of molecules lacks a defining signature motif and is unlikely described with available gene ontologies. NR coregulators comprise a heterogeneous and functionally distinct group of molecules that associate with nuclear receptors by direct binding or in a complex with other molecules and integrate cellular signals by tethering varied enzymatic activities to the transcription unit for the purpose of precisely regulating nuclear receptor-mediated target gene expression. The entirety of this regulation is coactivation or corepression.

1.2.3 *Steroid receptor RNA activator*

Although one is often tempted to think of coregulators solely being proteins, such a generalization cannot be made. The reason for this is that one NR coregulator has been described to function as an RNA molecule.[30] Albeit unique for its kind, steroid RNA activator (SRA) displays many properties intrinsic to NR coregulators. This RNA was shown to coactivate some, but not all nuclear receptors *in vitro* and *in vivo* through associations with other NR coregulators in distinct complexes. For example, SRA, which differentially coactivates ERα and ERβ[31,32] and enhances the transactivation of a recombinant AR mutant lacking the amino-terminal activation domain only in the presence of SRC-1,[30] is in a human ER-AF1 complex along with the DEAD-box RNA helicases p68 and p72, SRC-1 and splicing factor SF3A subunit 1.[33,34] This ribonucleoprotein complex interacts via SF3A1 with the ligand-activated ERα that is phosphorylated at Ser118, and via SRA and components of the SRC-1/CBP histone acetyltransferease complex and their LXXLL motifs with the LBD of the receptor. Thus, the ER-AF1 ribonucleoprotein appears to function as a communication link between different pathways such as MAPK-mediated growth factor and estrogen signaling, nuclear receptor transcription via histone acetylation, and ER-phosphorylation-dependent RNA splicing. The physical interaction of RNA helicases p68 and p72 with SRA is required for these proteins to function as ERα-specific coactivators. This suggests that SRA has a central structural role in the RNA-associated proteins in the complex.

While the DEAD-box RNA helicases function in all aspects of RNA biology,[35] including the processing of rRNA and microRNAs,[36] a number of NR coregulators that contain RNA recognition motifs (RRM) have been described that also play roles in mRNA splicing, such as CAPER,[37,38] CoAA/CoAM,[39] PGC-1[40] and p72.[41] Many RNAs, however, have not yet been identified for RRM-containing proteins, suggesting that uncharacterized RNA binding partners likely represent other potential RNA coregulators.

We have identified six secondary RNA structural motifs in SRA, each capable of participating individually in the coactivation function of SRA.[42] Further evidence for distinct associations of SRA RNA structures with discrete coregulator proteins was provided by the characterization of another ribonucleoprotein complex, in which the SRA-potentiated steroid receptor transcription activity is suppressed by binding to the RRMs of SHARP (SMRT/HDAC1 associated repressor protein), a potent transcriptional repressor that associates with the nuclear receptor corepressor SMRT (silencing mediator of retinoid and thyroid receptors) and at least five members of the NuRD repressor complex including histone deacetylase (HDAC) 1 and HDAC2.[43] This ribonucleoprotein complex thus integrates nuclear receptor activation and repression. SHARP was found to be an important regulator of Wnt signaling in cancers with β-catenin dysregulation.[44]

In yet another complex, a distinct stem-loop RNA structure of SRA is recognized by the single RRM of SLIRP (SRA stem-loop-interacting RNA-binding protein) to repress steroid receptor transactivation in a SRA- and RRM-dependent manner.[45] SLIRP also augments the effect of tamoxifen on ER transactivation and modulates the association of SRC-1 with SRA. SRA and the RNA helicases p68/p72 have also been found to coregulate MyoD and thus control skeletal muscle differentiation.[46,47] These publications broaden the biological activity of this non-coding RNA, and embolden us to suggest that all coregulator complexes may eventually be shown to involve functional non-coding RNAs.

Another attribute that SRA shares with protein coregulators is that it is subject to modification. SRA is modified post-transcriptionally by the pseudouridine synthases Pus1p and Pus3p, which are known to target certain transfer RNAs, but also associate with nuclear receptors and function as coactivators through pseudouridylation of SRA.[48,49] Because hyperpseudouridylated SRA acquires dominant-negative activity *in vivo*, the modification of SRA signifies a post-transcriptional mechanism that triggers a coactivator-corepressor switch in SRA, affording another means to regulate transcription.

SRA is implicated in carcinogenesis of hormone-responsive tissues via aberrant overstimulation of steroid receptor activity by ER and PR in breast and ovary tissues and by AR in the prostate.[30,50–52] The fact that overexpression of SRA in murine mammary glands lead to pre-neoplastic lesions but not to malignant tumors[52] highlights the additional complexity that is thought to arise from multiple SRA ribonucleoprotein complexes and also possibly due to coactivator-corepressor switching from its distinctly postranscriptionally modified species. Inactivating mutations in human Pus1p are a cause of myopathy, with lactic acidosis and sideroblastic anemia (MLASA), an autosomal recessive oxidative phosphorylation disorder specific to skeletal muscle and bone marrow.[53–55] It has been suggested that the facial dysmorphisms in MLASA patients may be due to the inability of SRA to function as a coactivator in this disease context.[54]

1.3 NR Coregulators in Human Diseases

Hormone action is highly complex and pleiotropic, largely as a consequence of coregulator biology which we have discussed above. This fact bodes for involvement in a wide spectrum of human diseases such as inherited genetic diseases, metabolic syndrome and cancer.[e] It is clear that in addition to its genetic fidelity, the activity of a gene product is determined by expression, modification, and turnover. A discussion of NR coregulators and human diseases must therefore consider the genetic heterogeneity of each coregulator gene, the intricate balance of their spatial and temporal expression, the modulation of coregulator activity via composite modification codes, and the cessation of function by protein/RNA degradation. Considering the fact that NR coregulators reside together with other molecules in multifaceted ribonucleoprotein complexes and act in a concerted manner to regulate nuclear receptor transcription, it is not always possible to assess the contribution of a single molecule in the involvement in human pathologies. In addition, while comprehensive data is available on inherited gene heterogeneity and NR coregulator expression due to numerous gene analyses and

---

[e] This also makes coregulators promising targets for effective therapeutics.

microarray- and Q-PCR-based efforts, respectively, much less is known as to how epigenetic mutations or post-translational modifications impact NR coregulator function in humans. We therefore will have to infer such knowledge from laboratory experiments and mouse knockout studies. Presently, we also suspect that the real extent and involvement of NR coregulators in human diseases is much larger than what our analyses indicate. New technologies and improved and wider applications of existing methodologies such as genome-wide chromatin immunoprecipitation assays, protein interaction networks and gene polymorphism-disease relationship maps will likely complement our current understanding and lead to the realization that NR coregulators play critical roles in a broad spectrum of human diseases.

1.3.1 *The NURSA list of NR coregulators*

The identification of NR coregulator-related diseases (Table 1.1) is based on a carefully curated list of genes for which the gene products meet the "restrictive" definition of NR coregulators presented above, and that an abstract for a peer-reviewed paper describing the characterization of a NR coreglulator function is available through PubMed. Excluded from the list are gene products that have all the hallmarks of functioning as coregulators but for which direct evidence of their ability to modulate nuclear receptor-mediated transcription is not clear (for examples, they only may have been reported to coactivate or repress AP-1 or NF-κB-mediated transcription).

While most gene products on the list are commonly recognized as NR coregulators, some entries are not primarily thought of as coregulators and deserve explanation. These include regulatory proteins such as p53, BRCA1, some cyclins, β-catenin and a few others. Tumor protein 53, which is primarily known for its essential role in the regulation of cell cycle and responses to genotoxic stress, was reported to repress androgen-induced transactivation of prostate-specific antigen by disrupting human AR amino-carboxyl terminal intramolecular interaction.[56] p53 is found mutated in a number of different tumors,[57,58] and germline mutations are frequently detected in cancer-prone families with Li-Fraumeni syndrome.[59,60]

Similarly, BRCA1 (breast cancer type 1 susceptibility protein), a nuclear phosphoprotein encoded by a gene in which inherited mutations are associated with susceptibility to breast and ovarian cancers, interacts with and regulates the activity of ER and AR.[61] BRCA1 has significant roles in different fundamental cellular processes, including the control of gene expression, chromatin remodeling, DNA repair, cell cycle checkpoint control, and has been linked to hormone-responsive cancers.[62–64]

Cyclins are a family of proteins involved in the progression of cells through the cell cycle. They form complexes with cyclin-dependent kinases (cdk), where different types of cyclins are active at distinct phases during the cell cycle where they phosphorylate different substrates. Cyclin D1 was also shown to function as a bifunctional NR coregulator for ER,[65] while Cyclin D3 has been reported to specifically enhance RAR α-mediated transcription through an interaction with the coactivator CRABPII.[66] Cellular retinoic acid binding proteins (CRABs) are implicated in the synthesis, degradation and control of the steady state levels of retinoic acids. CRABPII was identified as a ligand-independent coregulator of the RAR and RXR members of the NR superfamily.[67] Cyclin E, whose expression is frequently increased in human cancers, was identified as an AF-1 selective coactivator of androgen receptor mediated transcription.[68]

The "adherens junction" protein β-catenin is a multifunctional gene product that is activated by signaling through receptors for Wnt proteins, which are secreted signaling molecules that regulate developmental processes. As a NR coregulator, β-catenin was first shown to coactivate AR,[69] and subsequently found to modulate the transcriptional activity of several other members of the NR superfamily.[70–73] β-catenin uses several tandem repeats of the armadillo domain to form superhelical protein interaction interfaces which can bind to many different proteins. Overexpression of β-catenin was associated with basal cell carcinoma, leading to an increase in proliferation of related tumors. Consistent with this, somatic mutations at potential GSK3B phosphorylation sites result in the accumulation of this protein in the cytosol and are responsible for some colon cancers, ovarian and prostate carcinomas, hepatoblastomas and hepatocellular carcinomas. Many rare germ-line genetic variants of β-catenin also contribute to the inherited susceptibility to colorectal adenomas (see Table 1.1 for references).

Two interesting entries in our list of NR coregulators are DAX-1 (dosage-sensitive sex reversal adrenal hypoplasia congenita critical region on the X chromosome, gene 1, NR0B1) and SHP (Small heterodimer partner, NR0B2), which are more commonly thought of as atypical nuclear receptors. Both DAX-1 and SHP lack important domains characteristically found in members of the NR superfamily, and by forming heterodimers with many nuclear receptors, they act as dominant-negative regulators of transcription. SHP lacks a DNA binding domain and heterodimerizes with a plethora of receptors (AR, ER, GR, TR, CAR, RAR, RXR, HNF4, LRH-1, ERR, PXR and Nur77) to repress transactivation. SHP also functions as a NR corepressor through competition with coactivators by using its conserved LXXLL motif to bind to the AF2 of ligand-activated receptors.[74] Mutations in SHP lead to mild obesity and insulin resistance.[75] DAX-1 heterodimeric interactions with nuclear receptors include SF1, ERs, SHP, and DAX1A, an alternatively spliced DAX-1 that is more commonly expressed than DAX-1 in all tissues except the testis. Similar to SHP, DAX-1 also functions as a LXXLL-containing corepressor for activated ER.[76] Mutations in DAX-1 result in X-linked congenital adrenal hypoplasia and the associated disease, hypogonadotropic hypogonadism, the failure to undergo puberty,[77–79] and duplication of the DAX-1 gene causes dosage-sensitive sex reversal.[80–82] All considered, DAX-1 and SHP share more coregulator than nuclear receptor features and thus can more rightly be thought of as NR coregulators.

1.3.2 *An anthology of NR coregulators in human diseases*

Our attempt to associate human diseases with NR coregulators by extensively reviewing published literature resulted in 165 unique coregulators that were reported to be involved in at least one human pathology (Table 1.1). Because we have not considered defects in immortalized cell lines nor pathologies inferred by targeted gene deletions, we are tempted to conclude that our already impressive number of associations is a restrictive count. This number will almost certainly increase as clinical translational research in human coregulator pathological states continues.

Still, how good is this anthology? Knowing the limitations of searching publicly available information, we aimed to provide metrics for our own compilation and thus compared the list of coregulator-disease associations with information available from large gene-centered databases. We downloaded ftp-accessible human disease data from the University of California Santa Cruz (UCSC — spDisease), online mendelian inheritance in Man from the Johns Hopkins University (OMIM — Morbid

Table 1.1. NR Coregulators in Human Diseases

Familiar Symbol	Approved Symbol	Coregulator Function	LXXLL	Disease	Description	Reference
AFX	MLLT7	[1]	1	acute leukemia	translocation, fused to MLL	[2]
Akt	AKT1	[3, 4]	1	prostate cancer	amplification	[5]
				multiple cancers	review	[6]
				schizophrenia	UE	[7]
				oral cancers	Ser473 phosphorylation	[8]
ARA24	RAN	[9]	0	Kennedy's disease	androgen insensitivity	[9]
				prostate cancer	OE	[10]
ARA55	TGFB1I1	[11]	1	prostate cancer	UE	[12, 13]
ARA70	NCOA4	[14]	1	polycystic ovarian syndrome	OE	[15]
				prostate cancer	OE	[16]
				prostate cancer	UE	[10]
ARNIP	RCHY1	[17]	0	lung cancer	OE	[18]
				prostate cancer	OE	[19]
ART-27	UXT	[20]	0	prostate, breast cancer	UE	[21]
ASC-2	NCOA6	[22]	2	breast cancer	mutations	[23]
Atro	Gug	[24]	0	neuroblastoma	translocation t(1; 15) (p36.2; q24)	[25]
BCAS2	BCAS2	[26]	0	breast cancer	amplified	[27]
BCAS3	BCAS3	[28]	0	breast cancer	OE, amplified, translocated	[29–31]

(Continued)

Table 1.1. *(Continued)*

Familiar Symbol	Approved Symbol	Coregulator Function	LXXLL	Disease	Description	Reference
Bcl3	BCL3	[32]	1	leukemia, lymphoma	translocation, oncogene	[33]
				development	nonsyndromic cleft lip	[34]
BRCA1	BRCA1	[35]	0	breast cancer	mutations, genetic susceptibility	[23, 36, 37]
				ovarian cancer	mutations, genetic susceptibility	[38]
				breast, ovarian cancer		[39–41]
				multiple cancers	BRCA1 in hormone-responsive cancers	[42]
BRCA2	BRCA2	[43]	1	Fanconi anemia	mutations	[41, 44]
				Wilms tumor	truncations	[45]
				prostate cancer	mutations	[46]
				breast cancer	polymorphisms	[37, 40, 47, 48]
BRG1	SMARCA4	[49–51]	2	lung cancer	polymorphism	[52]
				breast cancer	LOH	[53]
				non-small cell lung cancer	UE	[54]
				gastric tumors	OE	[55]
				prostate cancer	OE	[56]
BRM	SMARCA2	[49–51]	2	non-small cell lung cancer	UE	[54]
Brn-3b	Pou4f2	[57]	0	breast cancer	OE	[58, 59]

(Continued)

Table 1.1. *(Continued)*

Familiar Symbol	Approved Symbol	Coregulator Function	LXXLL	Disease	Description	Reference
Brx	AKAP13	[60]	3	glucocorticoid resistance hypersensitivity syndromes	mutations	[61]
				breast cancer	Lys526Gln polymorphism	[37]
				endometriosis	OE	[62]
c-Jun	JUN	[63]	0	Hodgkin lymphoma	OE	[64]
CALR	CALR	[65]	1	multiple other diseases	ulcerative colitis and Crohn's disease	[66, 67]
				liver diseases	autoantigen, hepatic and coeliac disease	[68]
				bladder cancer	OE	[69]
				colon cancer	OE	[70]
CARM1	CARM1	[71]	0	prostate cancer	OE	[72]
				vascular diseases	homocysteine plasma regulation	[73]
CAV1	CAV1	[74]	0	prostate cancer	mutations, genetic susceptibility	[75]

(Continued)

Table 1.1. *(Continued)*

Familiar Symbol	Approved Symbol	Coregulator Function	LXXLL	Disease	Description	Reference
				cardiovascular symptoms	atherosclerosis, cardiac hypertrophy, cardiomyopathy, pulmonary hypertension, neointimal hyperplasia	[76]
				breast cancer	mutations	[77]
				breast cancer	OE & amplified	[78]
				lung cancer	UE in pulmonary fibrosis	[79]
				brain tumors (meningiomas)	OE	[80]
CBP	Crebbp	[81]	3	Rubinstein-Taybi Syndrome	heterozygous mutations	[82, 83]
				leukemia	translocation; MOZ fusion proteins	[84]
				Huntington's disease	mutant huntingtin binds to and inactivates/degrates CBP	[85]
				Spinal and bulbar muscular atrophy (SBMA)	CBP sequestration by polyglutamine tracts	[86]
Cdc25B	CDC25B	[87]	1	brain tumor	OE	[88]
				esophagus	esophageal squamous cell carcinomas	[89]
				breast tumor	OE	[90]

(Continued)

Table 1.1. *(Continued)*

Familiar Symbol	Approved Symbol	Coregulator Function	LXXLL	Disease	Description	Reference
CDK5	CDK5	[91]	2	Parkinson's disease	phosphorylation	[92, 93]
Cdk7	CDK7	[94]	1	Alzheimer's disease	aging of the hippocampus	[95]
CFL1	CFL1	[96]	0	breast cancer	OE	[97]
CITED1	CITED1	[98]	1	thyroid cancer	OE in papillary thyroid carcinoma	[99]
CoAA/ CoAM	RBM14	[100]	0	multiple cancers	OE and amplified	[101]
COBRA1	COBRA1	[102]	3	gastric cancer	OE	[103]
CR6	GADD45G	[104]	1	multiple cancers	UE	[105]
				liver cancer	UE	[106]
CTIP-1	Bcl11a	[107]	0	breast cancer	mutations	[23]
CTIP-2	BCL11B	[108]	0	acute myelocytic leukemia	translocation	[109]
Cyclin A2	CCNA2	[110]	0	multiple cancers	OE	[111–113]
				breast cancer	OE and amplified	[114, 115]
Cyclin D1	CCND1	[116]	0	multiple cancers	OE	[117, 118]
				breast cancer	A870G polymorphism	[119, 120]
				pituitary cancer	polymorphism	[121]
				oral cancer	amplified in squamous cell carcinoma	[122]

(Continued)

Table 1.1. *(Continued)*

Familiar Symbol	Approved Symbol	Coregulator Function	LXXLL	Disease	Description	Reference
				leukemia	large cell lymphoma with the t(11;14) translocation.	[123]
				colorectal cancer	A870G polymorphism	[124]
Cyclin D3	CCND3	[125]	2	bladder cancer	OE	[126]
				colorectal cancer		[127]
				multiple myelomas	OE	[128]
				laryngeal squamous cell carcinoma	OE	[129]
				large B-cell lymphoma	amplified	[130]
Cyclin E	CCNE1	[131]	0	breast cancer	OE	[132]
				bladder cancer	OE	[133]
				multiple cancers		[113]
DAP3	DAP3	[134]	0	thymoma	OE	[135]
				asthma	SNPs	[136]
DAX-1	NR0B1	[137]	1	adrenal hypoplasia congenita	mutations, Addison's disease included	[138–143]
				hypogonadotropic hypogonadism	mutations	[138, 144]
				dosage-sensitive sex reversal	duplication of the DAX1 gene	[145–147]

(Continued)

Table 1.1. (*Continued*)

Familiar Symbol	Approved Symbol	Coregulator Function	LXXLL	Disease	Description	Reference
				Ewing's sarcoma	DAX-1 OE by EWS/FLI translocation product	[148]
Daxx	Daxx	[149]	1	prostate cancer	OE	[150]
DJ-1	PARK7	[151]	0	Parkinson's Disease		[152–160]
				Dementia	haplo-insufficiency (disputed)	[161]
E6AP	UBE3A	[162]	2	Angelman syndrome	germline mutations	[163, 164]
				breast, prostate cancers	UE	[165]
EFP	TRIM25	[166]	0	breast cancer	OE	[167]
ELL	ELL	[168]	1	leukemias	OE	[168, 169]
				acute myeloid leukemia	translocation to MLL	[170]
ERR-10	LOH3CR2A	[171]	2	head and neck cancer	UE	[172]
Faf-1	FAF1	[173]	0	gastric cancer	UE	[174]
FEN-1	FEN1	[175]	1	Huntington's disease		[176]
				Werner syndrome (aging)		[177]
FHL2	FHL2	[178]	0	prostate cancer	nuclear expression	[179]
				breast cancer	OE	[180]

(*Continued*)

Table 1.1. *(Continued)*

Familiar Symbol	Approved Symbol	Coregulator Function	LXXLL	Disease	Description	Reference
FKHR	FOXO1A	[1]	1	alveolar rhabdomyosarcoma	PAX-FKHR fusion proteins in alveolar rhabdomyosarcomas	[181]
				Leukemia	constitutive phosphorylation in acute myeloid leukemia	[182]
Fli-1	Fliih	[183]	2	skin cancer		[184, 185]
				Ewing's sarcoma	translocation, EWS-FLI1 fusion	[186]
				Smith-Magenis syndrome	intrachromosomal rearrangement	[187]
FLNa	FLNA	[188]	1	frontometaphyseal dysplasia	mutations	[189]
				periventricular nodular heterotopia		[190]
				cardiac valvular dystrophy	mutations	[191]
				Ehlers-Danlos syndrome	mutations (C383T)	[192]
				otopalatodigital syndrome	Arg196Trp missense mutation	[193]
FoxG1	FOXG1	[194]	0	brain cancer	OE and amplified in medulloblastoma	[195]

(Continued)

Table 1.1. *(Continued)*

Familiar Symbol	Approved Symbol	Coregulator Function	LXXLL	Disease	Description	Reference
FOG-2	Zfpm2	[196]	0	heart disease, tetralogy of Fallot	mutations	[197]
				congenital diaphragmatic hernia	mutations	[198]
				Myelodysplastic syndrome	translocation, AML1-OG2/ZFPM2	[199]
G9a	EHMT2	[200]	0	Prader-Willi syndrome	CpG methylation	[201]
Gadd45	GADD45A	[104]	1	pancreatic cancer	OE	[202–204]
				breast cancer	UE, methylated	[205]
				lung cancer	UE	[206]
GRIP120/hn RNP U	HNRPU	[207]	1	breast cancer	OE	[208]
GSN	GSN	[209]	0	nephrotic syndrome	G654A mutation	[210]
				amyloidosis	D187Y, D187N mutations	[211, 212]
				skin disease	G654A or G654T mutation	[213]
				cerebral amyloid angiopathy	amyloid angiopathy	[214]
				breast cancer	mutations	[23]
HDAC1	HDAC1	[215]	0	gastric cancer	OE	[216]
				colorectal cancer	OE by RT-PCR	[217]

(Continued)

Table 1.1. *(Continued)*

Familiar Symbol	Approved Symbol	Coregulator Function	LXXLL	Disease	Description	Reference
HDAC2	HDAC2	[218]	0	colorectal cancer	frameshift mutation in hereditary nonpolyposis colorectal cancer	[219]
				Cushing disease	OE	[218]
				colorectal cancer	OE by RT-PCR, regulation	[217, 220]
HDAC3	HDAC3	[221, 222]	0	astrocyclic glial tumors	OE	[223]
HDAC4	HDAC4	[224]	0	oral cleft	mutations	[225]
				breast cancer	mutations	[23]
Hey1	HEY1	[226]	0	prostate cancer	UE, abnormal subcellular distribution	[226]
HMG-1	HMGB1	[227]	0	Sjogren's syndrome	OE	[228]
				systemic rheumatic diseases	OE in rheumatoid arthritis, systemic lupus erythematosus (SLE) and Sjogren's syndrome	[229]
HMG-2	HMGB2	[227]	0	systemic sclerosis	HMG autoantibodies	[230]
Hr	hr	[231]	2	papular atrichia	mutations	[232]
				androgenetic alopecia	mutations (significance disputed)	[233]
				alopecia universalis	linkage mapping	[234]

(Continued)

Table 1.1. *(Continued)*

Familiar Symbol	Approved Symbol	Coregulator Function	LXXLL	Disease	Description	Reference
JAB1	COPS5	[235]	1	other tumors	OE in oral squamous cell carcinomas	[236]
				liver cancer	chromosomal aberrations in hepatocellular carcinoma	[237]
				skin cancer	OE in matastatic melanomas	[238]
Ku70	XRCC6	[239]	1	breast cancer autoimmune disorders	mutations	[240]
Ku80	XRCC5	[239]	0	connective tissue disease	scleroderma-polymyositis overlap syndrome, autoantibodies	[241]
				aging	Werner syndrome, autoantigen	[242]
				oral cancers	UE in malignant melanomas of the oral cavity	[243]
				multiple cancers	polymorphism	[244, 245]
LATS2/ KPM	LATS2	[246]	0	gliomas	breast cancer and acute lymphoblastic leukemia	[247]
				acute lymphoblastic leukemia	downregulation	[248]
				breast cancer	UE, promoter hypermethylation	[249]

(Continued)

Table 1.1. *(Continued)*

Familiar Symbol	Approved Symbol	Coregulator Function	LXXLL	Disease	Description	Reference
Lsd1	Aof2	[250]	1	prostate cancer	OE	[179]
MLL2	MLL2	[251]	6	mixed-lineage leukemia		[252]
				solid tumors	amplified in solid tumors	[253]
MN1	MN1	[254]	0	acute myeloid leukemia	translocation t(12; 22) to TEL	[255, 256]
				meningioma		[257]
MPG	MPG	[258]	0	astrocytic tumors	OE	[259]
mPus1p	Pus1	[260]	1	myopathy, lactic acidosis and sideroblastic anemia	missense and stop mutations	[261–264]
MTA1	MTA1	[265]	0	multiple cancers	OE	[266]
				lung cancer	OE by RT-PCR	[267]
				liver cancer	OE by RT-PCR	[268]
				colorectal cancer	OE by RT-PCR	[217]
MTA2	MTA2	[269]	0	ovarian cancer	OE	[270]
MUC1	MUC1	[271]	1	gastric cancers	polymorphism in MUC1 tandem repeat	[272, 273]
				prostate cancer	OE	[274]
				breast cancer	aberrant glycosylation, splicing	[275–277]
				pancreatic cancer	OE	[278]

(Continued)

Table 1.1. *(Continued)*

Familiar Symbol	Approved Symbol	Coregulator Function	LXXLL	Disease	Description	Reference
MyD118	GADD45B	[104]	2	hypospadias	UE by MA	[279]
				liver cancer	UE	[280]
mZac1b		[281]	1	diabetes	transient neonatal diabetes mellitus	[282]
				ovarian, breast cancers	expression is lost	[283, 284]
				Beckwith-Wiedemann syndrome	polymorphism	[285]
				skin tumors	expression is lost in basal cell carcinomas	[286]
N-COR	Ncor1	[287]	1	colorectal carcinomas	OE by RT-PCR	[217]
				endometrial cancer	OE by RT-PCR	[288]
NM23-H2	NME2	[289, 290]	0	multiple cancers	UE, deletion	[291, 292]
				skin disease, psoriasis		[293]
				Alzheimer's disease, Down syndrome	UE	[294]
				erythroleukemia	leukemia cell line-specific expression	[295]
NSD1	Nsd1	[296]	2	familial gigantism	mutations	[297]
				Sotos syndrome	mutations	[298–300]
				Weaver syndrome	mutations	[301]
				Beckwith-Wiedemann syndrome	mutations	[302]

(Continued)

Table 1.1. *(Continued)*

Familiar Symbol	Approved Symbol	Coregulator Function	LXXLL	Disease	Description	Reference
ORCA	SQSTM1	[303]	1	Paget disease	P392L mutation	[304–308]
P-TEFb	CDK9	[309]	1	cardiac hypertrophy	OE	[310]
P/CAF	PCAF	[311]	1	colorectal cancer		[217]
				solid cancers	HATs, HDACs, and HMTs gene associations by Q-PCR	[312]
p300	EP300	[313]	3	leukemia	MOZ fusion	[84, 314]
				leukemia	polymorphism	[315]
				intestinal, gastric tumors	LOH by PCR	[316]
				Rubinstein-Taybi syndrome	heterozygous mutations in EP300 gene by Q-PCR	[83]
				multiple cancer	mutations	[317–319]
p53	TP53	[320]	1	Li-Fraumeni syndrome		[321, 322]
				breast, colorectal cancers	mutations	[23]
				multiple cancers		[323]
p54nrb	NONO	[324]	0	breast cancer	isoforms	[325]
				kidney cancer	fusion to the TFE3 DNA-binding domain in papillary renal cell carcinoma	[326]

(Continued)

Table 1.1. *(Continued)*

Familiar Symbol	Approved Symbol	Coregulator Function	LXXLL	Disease	Description	Reference
p57Kip2	Cdkn1c	[327]	0	Beckwith-Wiedemann syndrome	germline mutations	[328, 329]
				lung cancer	UE in nonsmall cell lung cancer	[330]
				squamous cell carcinoma	laryngeal, UE	[331]
				placenta defects	UE in tetraploid hydropic placentas	[332]
				pancreatic cancer	UE, correlation with cycline, PCNA OE	[333, 334]
				diabetes mellitus type II	mutations	[335]
p68	DDX5	[336]	1	breast cancer	OE	[337]
PAD4	PADI4	[338]	0	rheumatoid arthritis	OE	[339, 340]
				multiple cancers	OE	[341]
PARP-1	PARP1	[342]	3	multiple cancers		[343, 344]
				coeliac disease	haplotypes (promoter)	[345]
				Parkinson's disease	OE	[346]
				amyotrophic lateral sclerosis	protein stabilization	[347]
				leiomyomata	OE	[348]
				rheumatoid arthritis	promoter polymorphisms	[349]

(Continued)

Table 1.1. *(Continued)*

Familiar Symbol	Approved Symbol	Coregulator Function	LXXLL	Disease	Description	Reference
PBP/TRAP-220	PPARBP	[350, 351]	2	epilepsy	UE	[352]
PC2	PCQAP	[353]	0	schizophrenia	CAG trinucleotide repeats	[354]
PDEF	SPDEF	[355]	1	breast cancer	OE	[356]
				multiple cancers	OE	[357]
PDK1	PDPK1	[358]	0	kidney disease	TSC2/PKD1 contiguous gene syndrome	[359]
PELP1	PELP1	[360]	11	salivary duct carcinoma	OE by Western blotting and immunohistochemistry	[361]
				endometrial cancer	OE by Western blotting and immunohistochemistry	[362]
				breast cancer	Altered subcellular localization	[363]
PGC-1	Ppargc1a	[364]	1	diabetes	polymorphism, Gly482Ser variant	[365–369]
				gestational diabetes mellitus	polymorphism	[370]
				cardiac diseases	(Reviews)	[371, 372]
				gallstone disease	UE	[373]
				metabolic syndrome	Gly482Ser variant	[374–376]

(Continued)

Table 1.1. (*Continued*)

Familiar Symbol	Approved Symbol	Coregulator Function	LXXLL	Disease	Description	Reference
				diabetic retinopathy	Gly482Ser variant	[377]
				hypertension	Gly482Ser variant	[378, 379]
				breast cancer	correlation with PPARγ expression	[37, 380]
				Duchenne muscular dystrophy		[381]
				thyroid oncocytomas	OE	[382]
				obesity	UE	[383]
				insulin resistance	UE	[384]
				acute porphyria		[385]
PGC-1β/ PERC	PPARGC1B	[386, 387]	2	obesity, type 2 diabetes	genetic variations correlate with obesity	[388]
				breast cancer	polymorphism	[37]
				thyroid oncocytomas	OE	[382]
PIAS1	PIAS1	[389]	1	prostate cancer	expression (UE + OE)	[10, 390]
PIAS3	PIAS3	[391]	1	liver disease	OE in alcoholic and hepatitis C cirrhosis	[392]
				multiple cancers	OE	[393]
PIAS4	PIAS4	[394]	2	preleukemia	myelodysplasic syndrome, UE	[395]

(*Continued*)

Table 1.1. *(Continued)*

Familiar Symbol	Approved Symbol	Coregulator Function	LXXLL	Disease	Description	Reference
Pin1	PIN1	[396]	0	Alzheimer's disease	promoter polymorphisms, UE	[397–399]
				colorectal cancer	OE, β-catenin correlation	[400]
				liver cancer	OE, β-catenin correlation	[401]
				squamous cell carcinoma	OE, cyclinD1 correlation	[402]
				multiple cancers	OE	[403]
				salivary tumors	salivary adenoid cystic carcinomas	[404]
PKCdelta	PRKCD	[405]	0	MERRF syndrome	OE in mitochondrial myopathies and encephalomyopathies	[406]
				squamous cell carcinoma	UE	[407]
Pod-1	Tcf21	[408, 409]	0	lung cancer, squamous cell carcinomas	LOH, methylated	[410]
PPM1D	PPM1D	[411]	0	breast cancer	amplified	[412, 413]
				ovarian cancer	gene amplification	[414]
				brain tumors	medulloblastoma and neuroblastoma	[415, 416]
PRAME	PRAME	[417]	7	multiple cancers	OE	[418]
				leukemia	acute myeloid leukemia, OE	[419, 420]
PRC	PPRC1		2	thyroid oncocytoma	OE	[382]

(Continued)

Table 1.1. *(Continued)*

Familiar Symbol	Approved Symbol	Coregulator Function	LXXLL	Disease	Description	Reference
PRMT1	PRMT1	[421]	0	breast cancer	splice variants UE	[422]
Prox1	PROX1	[423]	2	liver cancer	UE	[424]
				pancreatic cancer	UE	[425]
PSF	SFPQ	[426]	0	kidney cancer	translocation to TFE3	[427]
PTα	PTMA	[428]	0	prostate cancer	OE	[429]
PTEN	PTEN	[430, 431]	0	Cowden disease		[432]
				Bannayan-Zonana syndrome	mutations	[433]
				Bannayan-Riley-Ruvalcaba syndrome	germline mutation, R335X	[434, 435]
				PTEN hamartoma tumor syndrome	mutations	[436]
				breast cancer	tumor inhibition by trastuzumab	[437, 438]
				brain tumors	mutations in anaplastic astrocytoma and glioblastoma multiforme	[439, 440]
				autism	mutations	[441]
				diabetes	polymorphisms in 5′-UTR	[442]
				T-cell acute lymphoblastic leukemia/lymphoma (T-ALL)	deleted	[443]

(Continued)

Table 1.1. *(Continued)*

Familiar Symbol	Approved Symbol	Coregulator Function	LXXLL	Disease	Description	Reference
R-MGMT	MGMT	[444]	1	lung cancer	OE in MGMT enhancer polymorphism (1099C→T)	[445]
				brain tumor	MGMT 84Phe polymorphism	[446]
				bladder cancer	exon 3 polymorphisms and haplotype	[447]
RACK1	GNB2L1	[448]	0	bipolar disorder	association with brain protein kinase C	[449]
				aging, Alzheimer's disease	age-related decline in leukocytes	[450, 451]
				alcoholism	alcohol addiction susceptibility due to polymorphism	[452]
RAF	IDE	[453, 454]	1	lung cancer	OE (disputed)	[455]
				Alzheimer's disease	genetic predisposition to disease (disputed)	[456]
RanBPM	RANBP9	[457]	0	inflammatory myofibroblastic tumors	ALK fusion	[458]
RAP46	BAG1	[459]	0	breast, lung cancer	differential expression in nucleus and cytoplasm	[460]
				brain tumor	OE of cytoplasmic BAG-1	[461]

(Continued)

Table 1.1. (*Continued*)

Familiar Symbol	Approved Symbol	Coregulator Function	LXXLL	Disease	Description	Reference
				prostate cancer	OE	[462]
				sleep disorder	narcolepsy, association by microsatellite	[463]
Rb	RB1	[464, 465]	0	multiple cancers		[466, 467]
				retinoblastoma	mutations	[467–472]
				osteosarcoma		[473]
				brain tumors	Rb deletion polymorphism in human gliomas	[474]
Rbp2	JARID1A	[475]	2	skin cancer	UE	[476]
REA	PHB2	[477]	1	breast cancer	UE by PCR	[478]
REGγ	PSME3	[479]	0	autoimmune diseases	OE	[480]
				connective tissue diseases	autoantigen	[481]
RelA	RELA	[482]	1	parathyroid tumors		[483]
				salivary gland tumors		[484]
				prostate cancer	nuclear localization	[485]
				kidney diseases	steroid-resistant nephrotic syndrome	[486, 487]
RIP140	NRIP1	[488]	9	infertility	G75G polymorphism, association with ESR1 g.938T>C	[489]
				endometriosis	mutations, R448G	[490]

(*Continued*)

Table 1.1. (*Continued*)

Familiar Symbol	Approved Symbol	Coregulator Function	LXXLL	Disease	Description	Reference
SAP30	Sap30	[491]	0	basal cell carcinoma	OE	[492]
SENP1	SENP1	[493]	0	prostate cancer	OE	[494]
SHARP	SPEN	[495]	1	colon, ovarian cancer	OE in cancers with disregulation of β-catenin/TCF-mediated transcription	[496]
SHP	NR0B2	[497]	1	obesity	genetic variations	[498]
				insulin resistance		[499]
Sirt1	Sirt1	[500]	1	multiple cancers	OE in chemo-treated cancers	[501]
				metabolic syndrome	polymorphism	[502]
Six3	SIX3	[503]	0	brain development	holoprosencephaly, mutations	[504–506]
Smad3	SMAD3	[507]	1	colorectal cancer	mutations	[23]
				multiple diseases	polymorphism	[508]
				gastric cancer	UE	[509]
Smad4	SMAD4	[510]	0	colorectal cancer	mutations, UE	[23, 511, 512]
				gastric cancer	UE, OE, polymorphism	[513, 514]
				head and neck cancer	mutations	[515]
				breast cancer	deletion	[516]

(*Continued*)

Table 1.1. *(Continued)*

Familiar Symbol	Approved Symbol	Coregulator Function	LXXLL	Disease	Description	Reference
				pancreatic cancer	mutations, UE, allelic loss of 18q	[517, 518]
				thyroid cancer	aberrant splicing	[519]
				cervical carcinomas	UE	[520]
				prostate cancer	UE	[521]
				Juvenile polyposis syndrome	mutations	[522, 523]
SMRT	NCOR2	[524]	0	endometrial cancer	OE by RT-PCR	[288]
				bipolar disorder	mutations (disputed)	[525]
SNURF	Rnf4	[526]	0	testicular cancer	UE	[527]
				Angelman, Prader-Willi Syndromes	imprinting defect	[528, 529]
SRA	SRA1	[530]	0	ovarian cancer	OE	[531]
				breast cancer	OE	[532]
				steroid-dependent tumors	OE in tumors of the breast, uterus, ovay	[533]
SRC-1	NCOA1	[534]	7	prostate cancer	OE in AR positive cancers, UE in hormone-refractory prostate tumors	[390, 535, 536]
				breast cancer	OE	[537]
				endometrial cancer	OE by RT-PCR	[288]
				rhabdomyosarcoma	NCOA1-PAX3 translocation	[538]

(Continued)

Table 1.1. *(Continued)*

Familiar Symbol	Approved Symbol	Coregulator Function	LXXLL	Disease	Description	Reference
SRC-2/ GRIP1	Ncoa2	[539, 540]	4	breast cancer	OE in intraductal carcinomas, correlation with ER expression	[541]
				brain cancer	correlation with PR expression	[542]
				prostate cancer		[543]
				leukemia	MOZ-TIF2 fusion in acute myeloid leukemia	[84, 544–549]
				endometrial cancer	OE by RT-PCR	[288]
SRC-3/ AIB1	NCOA3	[550–554]	5	endometrial cancer	OE	[288, 555]
				colorectal cancer	OE	[556]
				breast cancer	SNP	[557]
				oral cancers		[558, 559]
				liver cancer	amplified and correlation with c-myc	[560]
				liposarcomas	amplified	[561]
				prostate cancer	polymorphisms (poly-Q) in AIB1 and AR	[562]
				breast cancer	poly-Q polymorphisms and correlation with BRCA1/2 mutations	[557, 563–568]

(Continued)

Table 1.1. *(Continued)*

Familiar Symbol	Approved Symbol	Coregulator Function	LXXLL	Disease	Description	Reference
				breast cancer	OE, correlation with ER, PR, p53, Her2/neu, other CoRegs	[569–574]
				gastric cancers	amplified and OE	[572, 573]
				ovarian cancer	amplified and OE	[575, 576]
				different diseases	poly-Q polymorphisms	[577–582]
				pancreatic cancer	amplified and OE	[583, 584]
				breast cancer	amplified and OE	[550, 585–588]
				bone	poly-Q polymorphisms, ER status	[589]
				obesity	correlation with PR	[590]
				uterine cancer	OE	[591]
SRY	SRY	[592]	0	teratozoospermia	microdeletions	[593]
				disorders of sex development	mutations	[594]
				gonadal dysgenesis, Swyer syndrome	mutations	[595–599]
β-catenin	CTNNB1	[600]	5	skin cancer	OE in fibromatosis, Dupuytren's disease	[601]
				multiple cancers	OE	[602, 603]
				colorectal cancer	mutations	[604]

(Continued)

Table 1.1. *(Continued)*

Familiar Symbol	Approved Symbol	Coregulator Function	LXXLL	Disease	Description	Reference
				liver cancer	mutations in CTNNB1 genes in hepatocellular carcinoma and hepatoblastoma	[605]
				pilomatrixoma	skin tumor of the head and neck	[606]
				brain tumors, Wilm's tumor	loss of an entire copy of chromosome 6 in medulloblastoma	[607, 608]
STAT3	STAT3	[609]	1	squamous cell carcinomas	activation, nuclear translocation	[610]
				ovarian cancer	activation, nuclear translocation	[611, 612]
				gastric cancers		[613]
				multiple cancers	(review)	[614]
				asthma	polymorphisms	[615]
SUMO-1	SUMO1	[616]	0	type I diabetes		[617]
				neuronal intranuclear inclusion disease		[618]
				polyglutamine diseases		[619]

(Continued)

Table 1.1. (*Continued*)

Familiar Symbol	Approved Symbol	Coregulator Function	LXXLL	Disease	Description	Reference
SYT	SS18	[620]	0	lung cancer, synovial sarcomas	SYT-SSXZ fusion protein in synovial sarcomas	[621]
TAF-Iβ	SET	[622]	0	ovarian cancer	OE	[623]
				leukemia	fusion transcripts	[624]
TBL1	TBL1X	[625]	0	deafness	partial gene deletion	[626]
TBP	TBP	[627]	1	Creutzfeldt-Jakob disease	trinucleotide repeat expansion	[628]
				diabetes	trinucleotide repeat expansion	[629]
				spinocerebellar ataxia-17 & neurodegenerative diseases	trinucleotide repeat expansion	[630–636]
				Huntington's disease	trinucleotide repeat expansion	[637, 638]
				schizophrenia	trinucleotide repeat expansion	[639]
TDG	TDG	[640]	0	lung cancer	polymorphisms	[445]
				colorectal cancer		[641]
TGIF	TGIF	[642]	0	gastric cancer		[643]
				holoprosencephaly	mutations	[505, 644]

(*Continued*)

Table 1.1. *(Continued)*

Familiar Symbol	Approved Symbol	Coregulator Function	LXXLL	Disease	Description	Reference
TIF1α	TRIM24	[645, 646]	2	thyroid cancer	translocation to RET	[647]
				leukemia	OE	[648]
TIP27	JAZF1	[649]	0	endometrial neoplasms	JAZF1-JJAZ1 gene fusion	[650, 651]
Transgelin	TAGLN	[652]	0	colon cancer	UE	[653]
TRAP100	THRAP4	[654]	6	breast cancer	OE by microarray	[655]
TRAP230	MED12	[656]	2	schizophrenia	HOPA12bp polymorphism	[657, 658]
				hypothyroidism	HOPA12bp polymorphism	[659]
TRIP230	TRIP11	[660]	4	acute myelogenous leukemia	translocation to PDGFRB	[661]
TRRAP	TRRAP	[662]	9	pancreatic cancer	OE	[663]
TRX	TXN	[664, 665]	0	lung cancer	OE	[666]
				colorectal cancer	OE	[667]
				peripheral arterial disease	OE	[668]
TrxR1b	TXNRD1	[669]	0	arteriosclerosis	OE in atherosclerotic plaques	[670]
TSC2	TSC2	[671]	0	pancreatic islet-cell tumors		[672]
				tuberous sclerosis	mutations	[673–676]
				gangliogliomas	splice variant	[677]
				lung	lymphangioleiomyomatosis	[678]

(Continued)

Table 1.1. *(Continued)*

Familiar Symbol	Approved Symbol	Coregulator Function	LXXLL	Disease	Description	Reference
TSG101	TSG101	[679]	0	AIDS	polymorphisms	[680]
				gastric tumors	OE	[681]
				cervical cancer	aberrant transcript delta 154-1054	[682]
				acute myeloid leukaemia	aberrant transcript	[683]
				prostate cancer	deletions	[684]
				breast cancer	aberrant transcript	[685, 686]
TZF	ZNF318	[687]	0	breast cancer	mutations	[23]
Ubc9	UBE2I	[688, 689]	0	multiple cancers	OE	[690]
Vav3	VAV3	[691]	2	prostate cancer	OE	[691, 692]
WSTF	BAZ1B	[693]	3	Williams-Beuren syndrome	deletion	[694]
XAP2	Aip	[695]	0	brain tumors	mutations	[696, 697]
XBP-1	XBP1	[698]	0	liver cancer	OE	[699]
				breast cancer	OE	[700]
				bipolar disorder	polymorphism	[701–703]
				schizophrenia	polymorphism	[701, 704]
14-3-3 eta	YWHAH	[705]	0	schizophrenia	polymorphism	[706]

Legend: Symbols type fonts are in species-specific cases: A symbol in all upper case type font indicates that the initial characterization of the NR coregulator function was for a human gene product, symbols in regular case type font indicate non-human NR coregulators in the initial characterization. All disease associations are for human gene products.

OE; overexpression, UE; underexpression, LXXLL; number(s) of sequence motif(s) in the human protein sequence (see text for discussion).

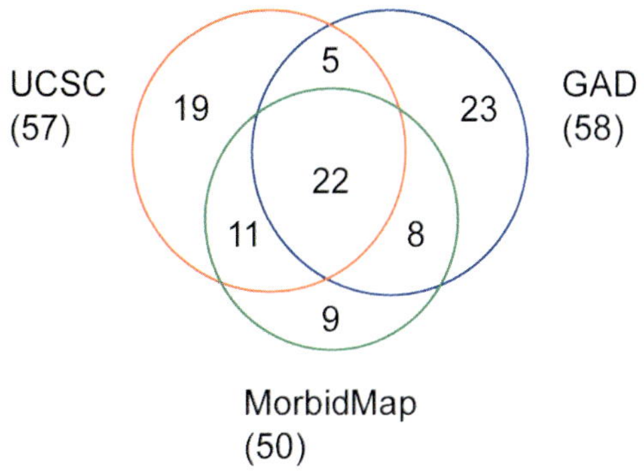

Fig. 1.1. NR Coregulator-associated human diseases in public databases: Venn-Diagram of three sets of public data on NR coregulator diseases. UCSC reports human diseases for 57 NR coregulators, GAD and OMIM/MorbidMap 58 and 50 associations respectively. In our literature search with 303 NR coregulators curated at NURSA (www.nursa.org), we found 165 NR coregulators that play critical roles in at least one human disease. UCSC; University of California Santa Cruz spDisease dataset, GAD; Gene Association Database genemap dataset, MorbidMap dataset from John Hopkins University Online Mendelian Inheritance in Man.

Map)[83] and from the genetic association database (GAD — genemap[84]). We then organized the information in a relational database. In the UCSC dataset, we found SwissProt/UniProt disease entries for 57 NR coregulators (19% coverage), while GAD and MorbidMap have annotations for 58 and 50 NR coregulator genes respectively (Figure 1.1). Our literature search contained all NR coregulator annotations provided by SwissProt/UniProt and OMIM/MorbidMap,[f] and significantly extended the coverage for this class of molecules by providing disease associations for 165 NR coregulators (55% coverage).

While this brief comparative analysis shows that our literature search was fruitful, it cannot provide conclusive evidence that NR coregulators play more important roles in human diseases than other gene products do. Similarly, the experimental assessment of the expression levels of NR coregulators in human cancers or other diseases may provide useful information with regards to the pathways through which they act, but due to the lack of knowledge on the absolute number of

[f] The GAD database also lists disputed findings.

genes involved in the biology that led to the altered expression, it can only add suggestive evidence. However, this is not the case if one considers gene mutations. A simple count of NR coregulator genes found to be mutated in a systematic and unbiased large cohort study, set against the total number of genes analyzed, could provide further hard evidence for the absolute importance of NR coregulators in human diseases. Currently, the most extensive data available on sequencing of DNA from human cancer tissues represents the analysis of 13,023 genes for somatic mutations in breast and colorectal tumors. In this dataset, the authors identified and verified 189 candidate cancer genes (CAN genes) that were mutated at a significant frequency.[64] We found nine different NR coregulators in this set of CAN genes, which is twice as many genes as one would expect for the total of 303 NR coregulators identified so far. There were 122 and 69 CAN genes reported in breast and colorectal cancers respectively. Seven NR coregulator genes reside in the breast cancer group, three in the colorectal cancer group, with TP53 being mutated to a significant extent in both cancer types as expected (Table 1.2).

Therefore, we must conclude that NR coregulators are more frequent targets in at least two major human cancers. Both, breast and colorectal cancers are clinically important worldwide, accounting for about 20% of total cancer diagnoses.[64] In the US, breast cancer is expected to strike more than 200,000 women and kill nearly 41,000 this year alone, and colorectal cancer is expected to strike nearly 150,000 Americans and kill more than 55,000 this year.[85]

Table 1.2. NR Coregulators in the CAN Dataset.[23]

Familiar Symbol	Approved Symbol	Coreglator Function	LXXLL	CAN Tissue
ASC-2	NCOA6	[22]	2	breast
BRCA1	BRCA1	[35]	0	breast
CTIP-1	Bcl11a	[107]	0	breast
GSN	GSN	[209]	0	breast
HDAC4	HDAC4	[224]	0	breast
p53	TP53	[320]	1	breast & colorectal
Smad3	SMAD3	[507]	1	colorectal
Smad4	SMAD4	[510]	0	colorectal
TZF	ZNF318	[687]	1	breast

Two important questions arise: Why do so many candidate cancer genes exist, and why are the coregulators overrepresented in the group of candidate cancer genes? It is generally accepted that genetic mutations provide the most reliable indicator for the importance of a gene in human neoplasia. Still, a single genetic alteration is hardly the cause for a disease,[g] since functional redundancy in the eukaryotic cell has evolved to protect it against the potential harm resulting from the disruption of a single gene. Mutated genes are thus "contributing to" rather than causing malignancies, which also may explain why so many genes not previously suspected to be involved in malignancies recently have been identified to play roles in the genesis of an oncogenic disease.

In the case of the NR coregulators, overlapping gene functions have been inferred based on the expression and functional studies carried out *in vitro* and *in vivo*, with targeted gene deletions in mice being especially helpful in elucidating the significance of individual NR coregulators in their entangling functions. Mouse gene knock-out models have been generated for about one third of the NR coregulators, and more than half of these targeted deletions result in embryonic lethality. This high level of necessity for many of the NR coregulators collectively underlines their importance in embryonic development, a period where networks of molecular pathways have to be established rather than maintained. Once established, however, these pathways are constantly subjected to functional tests because they have to adapt and respond to new or changing requirements imposed by genetic programs and epigenetic events. The different, and at times dramatic, responses to changing levels of sex steroid hormones (estrogens, progestins, testosterone), for example, are not only functionally dependent on the cognate nuclear receptors, but also have different impacts based on the "type of program" a particular cell is responsive to at the time of hormonal signaling. For example, it is well known that testosterone illicits a distinct response in the ovary that differs from the action of androgens in the testes, in the skeletal muscle, bone or in the brain, and that these diverse reactions are both sex and age dependent. The integrative entity that effects different cellular programs and responds to specific epigenetic programs for controlled gene expression is composed of the

[g] 'Single-gene' alterations that cause diseases — such as sickle-cell anaemia, cystic fibrosis or muscular dystrophy — exist but are exceptions to the rule.

transcriptional coregulators and the transcription factors to which they respond. Because of their integrative function, NR coregulators constitute a critical class of molecules that allow for distinct and appropriate genetic programs to be executed in different tissues during discrete stages of development or for cell homeostasis.

It is believed that most human cancers have genetic mutation rates that are similar to those observed in normal cells but that tumorigenic mutations lead to a self-reinforcing cascade that results in proliferation of the harmful mutation.[86] A mutation leads to a structural change in the encoding protein which is tested and re-evaluated immediately for its appropriateness for still serving cellular programs. The cellular programs themselves are subjected to adaptation, causing rapidly acquired structural and possibly functional diversity. Structural modifications also can lead to impaired function and pathological conditions. Cancer cells that survive the challenge (challenged genetically or through epigenetic events) have successfully altered their cellular programs and represent a changed genetic fingerprint of somatic mutations and different gene expression profiles.[h]

All considered, it is therefore not a surprise to us that a large number of CAN genes must exist, and that the NR coregulator genes are overrepresented in genes with mutations that may drive tumorigenesis. Such mutations have been shown to be directly responsible for tumor progression and are the only ones known to be useful as diagnostic and therapeutic targets.

1.3.3 *The bullet points*

The interpretative value of our collection would benefit if we could classify diseases according to accepted ontologies, separate cancer from non-cancer diseases, and provide a distinction for inherited genes that predispose an individual to cancer and for those genes that are mutated somatically, as spontaneous mutations that arise only in the cancer itself. Unfortunately, the nature of the data — publications with inconsistent, or misapplied, ontologies — does not allow a simple, universal

[h] It will be interesting to learn more about the dynamic changes of NR coregulator expression and posttranslational modifications that occur *during* tumorogenesis, and compare the rates of changes with the ones that occur in normal cell homeostasis. Currently, these types of data are sparse, but it is only a matter of time until studies in NR coregulators expression and modification are systematically applied and extended to all disease states.

classification scheme. Currently, the diseases in our list are specified according to the tissue types they affect, with limited additional information in some cases. We will further curate the list of NR coregulator-associated diseases and invite the reader to visit the NURSA web site to look at updated data in the future.

Broadly, the compilation of NR coregulator associations in human diseases includes gene products reported in the literature to be mutated, misexpressed, or to exist as a genetic heterogeneity that led to pathological situations. About 61% of these disease associations are for aberrant expression (over- or underexpressed, mostly in cancers), and the greater part of mutations are somatic mutations (23% polymorphisms, 15% translocations). The vast majority of human disease associations for NR coregulators were reported for cancers (75%), followed by aging, Alzheimer's or Parkinson's disease (collectively 7%), schizophrenia and bipolar disorder (5%), and diabetes (5%). This distribution reflects mostly the interests of the initial investigating scientists over the past five to six years, and would be expected to change over the next decade, for example, in favor of increasing metabolic disease relationships.

1.4 NR Coregulators in Human Cancers

When considering the plethora of functions that the nuclear receptors play in tissues such as breast, ovary, prostate and lungs, and the interest in these organs due to their relationship with cancer, the relatively large number of coregulators that appear to be involved in cancer is to be anticipated.

1.4.1 *Expression*

High throughput genomic analyses such as microarray-based mRNA profiling and genome-wide chromatin immunoprecipitation, along with highly sensitive and specific technologies such as quantitative real-time PCR, have together provided a comprehensive picture in how NR coregulator expression correlates with human pathologies. Most NR coregulators are expressed at the mRNA level in a rather constitutive manner in normal tissues (PGC-1α being the most noticeable exception, see below), but respond with dynamic changes in transcript and protein levels in chronically challenged cells.

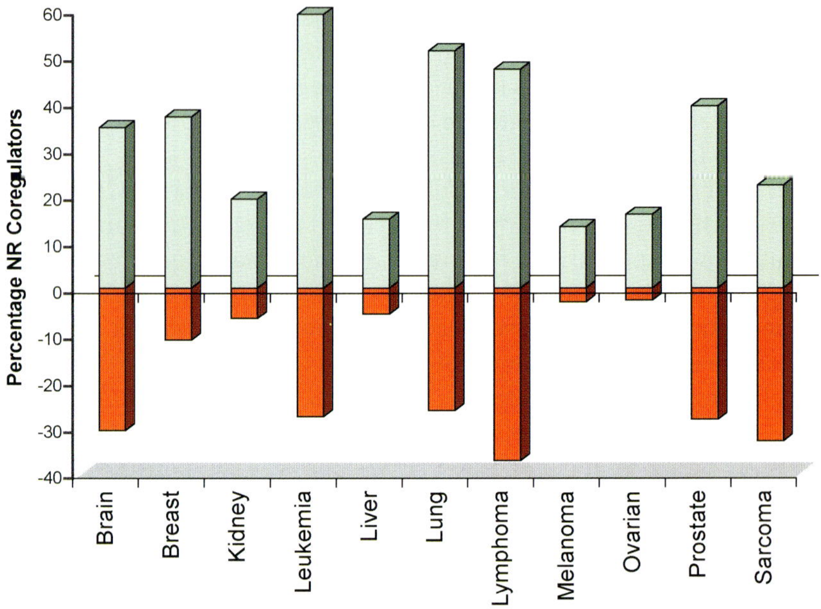

Fig. 1.2. NR coregulator-expression: Combined relative counts in percentage of total NR coregulators with disease associations reported to be over- or underexpressed in selected human tissues or cancer types.

NR coregulator misexpression represents the largest group of related pathologies in our list. We thus wanted to compare the expression data reported in the literature with array data from Oncomine (http://www.oncomine.org)[87] — a global collection of cancer profiling data. Knowing the limitations of plowing through pre-mined data, we assessed the levels of NR coregulator expression in various human cancer tissues.[88] Figure 1.2 illustrates that, broadly, more NR coregulators are over- than underexpressed, and that in some cancers, such as leukemia and lymphoma, the number of NR coregulators reported misexpressed are close to the total number of NR coregulator genes identified and included in our list.[i]

The question thus arises as to whether NR coregulator misexpression is a pervasive mediator in the etiology of human cancers. Although often reported in the literature as one obvious possible cause for

[i] Because some NR coregulators were reported to be over- as well as underexpressed in a tissue, the actual number of distinct coregulators misexpressed in this tissue is actually smaller than indicated in our counts.

endocrine-related cancers, the levels of gene expression are unreliable indicators for the contribution of the misexpressed gene to cancer, simply because a disturbance of any network inevitably leads to a multitude of such cellular changes. Over- or underexpression of NR coregulators are likely to reflect a global rearrangement of gene expression due to the adaptation in the cancer cell and thus should be considered as a consequence of a pathological situation rather than a causal role in the genesis of cancer. The reason for this lies in the ability of the NR coregulators and the associated enzymatic activities to serve as molecular sensors of diverse signaling inputs that enable them to integrate transcriptional responses rather than emanate primary cellular signals for their further amplification. Still, the misexpression of key regulatory factors in an intricate network of functions has the potential for amplification of the risk for diseases, including cancer, and also can alter the biological activities of therapeutic nuclear receptor ligands.[89]

Perturbations of the coregulator-transcription network due to misexpression restate the concept of coregulator competition as a regulatory strategy in the eukaryotic cell. For example, for β-catenin, it was shown that elevated levels of homodomain-containing proteins compete with TCF/LEF for nuclear β-catenin, thus switching transcriptional events to dictate cell-lineage determination.[90] Other potentially harmful examples of changes in expression levels include "feed-forward" mechanisms, whereby coregulator-stimulated nuclear receptors also control the expression of coregulator genes that are in the same transcription control loop, ensuring allosteric regulation (see BCAS3 and PGC-1α below for examples). Lastly, relative changes in the expression levels also can overwrite transcriptional events that are coregulated by proteins that are present in "limiting" concentrations such as due to haploinsufficiency (for example, Rubinstein-Taybi syndrome from haploinsufficiency for CBP or p300[91,92]).

1.4.2 *Post-translational modifications*

Equally important in accurately assessing the function of a protein in a cell is to include in the evaluation its regulation capacity at the protein level. In addition to its protein expression level, the activity of a NR coregulator can be controlled by post-translational modifications, but also by modifications of its target, by the activation state of the target nuclear receptor, and by degradation of both, coregulator and target.[22,93,94]

This need for obtaining a "complete" picture before appraising the contributing values of NR coregulators in human diseases (or any protein in any pathology, as a matter of fact) has yet to find widespread acceptance in the field of molecular biology. Clinical data on post-transcriptional and post-translational modifications and *in vivo* degradation is relatively sparse, and our knowledge in this matter is implied from *in vitro* experiments in tissue culture settings or genetically altered mice. Because our compilation of NR coregulator-disease association is based on human clinical studies only, the list is biased towards expression data.

The ability to make sense of contextual information along with the expression and mutation data undoubtedly will succeed as better tools for identifying modification sites for phoshorylation, mono/di/tri-methylation, acetylation, ubiquitylation, and sumoylation. In addition, new software is needed that integrates legacy data with new discoveries from large throughput mass spectrometry datasets to mine data by newer standards. Using this combinatorial approach, a better understanding of the molecular aspects of tumorigeneis is clearly feasible for the near future. Equally important is to study the expression-regulating elements of genes as they may contain functional polymorphisms with the potential for modulating the risk of various diseases, including cancer.

Many phosphorylation sites have been identified *in vitro* so far, but little is known about the precise cellular kinases that target the proteins for modification. Recently, a significant publication presented a new bioinformatics approach that predicts the kinases responsible for specific phosphorylation sites by exploiting both the inherent propensity of kinase catalytic domains to phosphorylate particular sequence motifs and contextual information such as the coexpression of kinases and their substrates, their co-occurrence in the genome, and network data.[95] The complete set of predictions, which was made available to the public, consists of 7,143 site-specific substrate interactions between 1,759 target proteins and 68 kinases. We mined this data for NR coregulator targets and found 107 substrate predictions for all NR coregulators (35% coverage), and 69 predictions for coregulators with disease associations (42%). This represents a vast increase in coverage considering the relatively low number of total predicted substrates (1,759) for all possible human proteins.[j] Thus, the NR coregulators distinguish themselves by being overly represented as targets for post-translational modifications by phosphorylation.

This is another quantitative demonstration that NR coregulators are key components of cell signaling processes. The complex nature of cell signaling would not be possible if protein-protein interactions were established only through static interactions between structurally defined domains. Post-transcriptional and post-translational modifications of coregulators (and other proteins) provide the dynamism and diversity necessary for these molecules to act as versatile mediators and integrators of cellular signaling.

1.4.3 *Somatic mutations*

Cancer is most likely the only generally accepted disease type in which somatic mutations are frequently pathogenic. Germline mutations clearly result in hereditary predispositions to cancer, while somatic mutations in an oncogene or tumor suppressor gene[k] will initiate the neoplastic process if the structural consequences of the mutation are not detrimental for the cell (acceptance by rejecting apoptosis). Tumor suppressor genes can lose their gene function through mutations such as missense mutations, deletions or insertions, or from epigenetic silencing. On the other hand, oncogenes are usually mutated in such a way that they become activated or to improve in activity (overall gain of gene function). This may occur through gene amplification, chromosomal translocations, or more subtly, by single point mutations.

Activating codon changes in β-catenin, for example, have oncogenic activity resulting in tumor development. Somatic mutations at potential GSK3 phosphorylation sites result in the accumulation of β-catenin in the cytosol, thus preventing the phosphorylation-dependent, ubiquitin-mediated degradation of the protein. Such activating mutations have

[j] The predictions are primarily based on mammalian data, and the paper nor the website (http://networkin.info) do not specify the species that constitutes the total predicted 1,759 substrates. If the predicted substrates were all unique human genes, the coverage would be about 7%, if they were all human proteins, the coverage would be less than 2%.

[k] Caretaker genes are a third class of cancer genes that promote tumorigenesis when their function of keeping genetic alterations at a minimum is impaired. DNA repair enzyme genes, for example, belong to this class of genes.

been found in various tumor types, including colon cancers, ovarian and prostate carcinomas, hair matrix cell tumors, hepatocellular carcinomas, and hepatoblastomas.[96–101] Other cases showing evidence of activation of the Wnt signaling pathway through activating β-catenin mutations were identified in a distinct molecular subgroup of medulloblastomas[102] found in Wilms' tumors patients that have an unusual germ line WT1 mutation.[103] Mutations of the WT1 gene on chromosome 11 are observed in approximately 20% of Wilms' tumors, and about half of the Wilms' tumors with mutations in WT1 also carry mutations the gene encoding the proto-oncogene beta-catenin.[104,105]

Some NR coregulator genes display a plethora of mutations that co-orchestrate different human pathologies. The cyclin D1 gene, for example, was found to be translocated to immunoglobulin gene regions [t(11;14)(q13;q32)] in certain B-lymphocytic malignancies, particularly mantle-cell lymphoma,[106] while a common A/G polymorphism at nucleotide 870 (codon 242) alters mRNA splicing to produce two transcripts, with the consequence of increased susceptibility to colorectal cancer[107] and breast cancer.[108,109] This A/G polymorphism was also associated with tumor grade in sporadic pituitary adenomas.[110] Amplification of the cyclin D1 gene was identified in oral squamous cell carcinoma cases,[111] and the gene is generally overexpressed in many different tumors.[112,113] Other NR coregulators with gene amplifications in human diseases are BCAS2 and BCAS3, Caveolin 1, CoAA/CoAM, MLL2, PPM1D, SRC-3/AIB1, and the cyclins A2, D1, D3.

Both breast cancer amplified sequence 2 (BCAS2, also termed DNA amplified in mammary carcinoma-1, DAM1) and 3 (BCAS3) were identified as overexpressed and amplified genes in some breast cancer cell lines and are subsequently shown to be coactivators for ERα.[114,115] Despite the similar terminology, these genes do not share any sequence similarity. The BCAS2 protein was found to be a component of the spliceosome and also to associate with other nuclear receptors in a ligand-independent fashion.[114] The BCAS3 gene can transpose to another breast carcinoma amplified sequence (BCAS4) with unknown function.[116] In addition to binding to nuclear receptors, the BCAS3 protein physically associates with histone H3 and the histone acetyltransferase complex protein P/CAF (p300/CBP-associated factor), while also possessing histone acetyltransferase activity itself. BCAS3 needs the NR coregulator PELP1/MNAR to function as an ERα coactivator. Interestingly, BCAS3 is also an ERα target gene, and PELP1/MNAR, together with metastasis-associated protein 1 (MTA1), another coregulator of ERα transactivation, acts as a transcriptional activator of BCAS3 expression, providing allosteric regulation whereby ER activation triggers a positive feedback loop through its ability to promote the expression of several of its own coregulators.

Caveolin-1 (CAV1), a principal component of plasma membrane vesicles (caveolae) and a platform protein for many signal transduction pathways, was shown to be a positive regulator of ERα signal transduction and a NR coactivator.[117] CAV1 links integrin subunits to the tyrosine kinase FYN, which is an initiating step in coupling integrins to the Ras-ERK pathway and promoting cell cycle progression. CAV1 is a negative regulator of the Ras-p42/44 MAP kinase cascade. The protein has been documented in several neoplasms with a controversial role in cell proliferation (anti-proliferative and pro-apoptotic properties), tumor development and progression (Table 1.1).

Coactivator activator A (CoAA, also termed SYT-interacting protein; SIP, or RNA binding motif protein 14; RBM14) is a RRM-containing coactivator that promotes transcription through synergistic interactions with other NR coregulators such as thyroid hormone receptor-binding protein (TRBP) and CBP.[39] An alternative splice form of CoAA, termed coactivator modulator CoAM, functions as a likely transcriptional repressor. Together, CoAA/CoAM regulate mRNA alternative splicing and affect both transcription and splicing in a promoter-preferential manner. With regard to cancer, the CoAA gene is overexpressed and amplified in some non-small cell lung carcinomas, squamous cell skin carcinomas and lymphomas.[118]

Protein phosphatase magnesium-dependent 1 delta (PPM1D) was initially characterized as a p53-regulated phosphatase responsible for the inactivation of p38 MAPK and consequent inactivation of p53. It was shown to be a coregulator through its stimulation of the transcriptional activity of several nuclear receptors, including PR.[119] Myeloid/lymphoid or mixed-lineage leukemia 2 (MLL2) is a component of a novel Set1-like complex in mammalian cells. The protein was shown to bind ERα via its two LXXLL motifs to coactivate estrogenic signal transduction pathways in breast cancer cells.[120] MLL2 also has histone H3 Lys4 methyltransferase activity that is dependent on menin, a protein mutated in multiple neoplasia type I (MEN1). MLL2 is amplified in some solid tumors[121] and was functionally linked to mixed-lineage leukemia.[122] The probably most recognizable NR coregulator to function as compelling oncogene, however, is AIB1, which we like to call SRC-3.

1.4.4 *SRC-3, a bona fide NR coactivator in tumorigenesis and a target for cellular programs*

AIB1/SRC-3 is the third member of the SRC/p160 coactivator family. This gene has redundantly been characterized as AIB1 (amplified in breast cancer-1[123]), ACTR (activator of thyroid hormone receptor[124]), p/CIP (p300/CBP-interacting protein[15]), RAC3 (receptor-associated coactivator 3[125]), TRAM-1 (thyroid hormone receptor activator molecule 1[126]) and also named NCOA3 (nuclear receptor coactivator 3). SRC-3 displays a plethora of somatic mutations and functions to justly classify it as a cancer gene. It was originally reported to be amplified in approximately 10% and overexpressed in 64% of primary breast tumors,[123] and was subsequently found to be amplified and overexpressed in many other human cancers as well (Table 1.1). Clinical studies in ER and HER2/neu positive breast cancer tissues show that the overexpression of SRC-3 correlates with tamoxifen resistance,[127] confirming earlier tissue culture experiments.[128] In some ER- and PR-negative breast cancers, the overexpression of SRC-3 was found to correlate with p53 detection and HER2/neu overexpression, indicating that abundant SRC-3 may impact on breast cancer by a mechanism not entirely dependent on steroid receptor coexpression and which may involve other oncogenic events, such as p53 protein stabilization and HER2/neu overexpression.[129]

The carboxyl-terminal interaction domain of the SRC-3 protein contains homopolymeric glutamines ranging from 26 to 32 residues. Somatic polymorphisms of polyglutamine tract length in SRC-3 have been identified in breast cancer, and now are also linked to cancers of the prostate, bone and other diseases (Table 1.1). Also, women with germ-line BRCA1/2 mutations are at significantly higher breast cancer risk if they carry SRC-3 alleles with an expanded poly-Q tract.[130]

Oncogenic potential for SRC-3 is substantiated by studies in transgenic mice that overexpress SRC-3. These mice suffer from mammary hypertrophy, hyperplasia, abnormal postweaning involution, and the development of malignant mammary tumors.[131] In contrast, SRC-3 knockout mice are resistant to chemical carcinogen-induced mammary tumorigenesis[132] and are resistant to v-Ha-ras-induced breast cancer initiation and progression.[133] SRC-3$^{-/-}$ mice also show the inhibition of spontaneous prostate cancer progression,[134] and TR$\beta^{-/-}$/SRC-3$^{-/-}$ mice show delayed thyroid carcinogenesis,[135] supporting the idea that SRC-3 is a powerful oncogenic gene. Surprisingly, in the lymphatic system, SRC-3 acts as a tumor suppressor by displaying B-cell lymphomas in gene-deleted mice.[136] These two faces of SRC-3 highlight the fact that SRC-3 is a versatile protein, allowing the cell to decide between proliferation or growth suppression in a context-dependent manner.

SRC-3 is a pleiotropic regulator in eukaryotic systems biology and is also implied in adipogenic and energy balance programs on the basis of targeted gene deletions in mice, which are lean and have reduced fat mass. It was recently shown that SRC-3 acts synergistically with CAAT/enhancer-binding protein C/EBP to control important aspects of adipose differentiation by targeting gene expression of PPARγ2.[137] Finally, SRC-3 is an important regulator of cytokine mRNA translation, and a critical determinant for prevention of toxic shock syndrome.[138] By activating the translational silencers TIA-1 and TIAR, SRC-3 inhibits the mRNA translation of pro-inflammatory cytokines. This physiological function of SRC-3 depends upon a nongenomic translational event, indicating that a coregulator protein can serve as both a transcriptional coactivator and a translational corepressor, depending on its target genes and cellular and signaling context.

SRC-3 is a broad-specificity transcriptional coregulator that mediates the activating functions of nuclear receptors and other transcription factors. Like other coregulators, SRC-3 forms multiple, distinct complexes which contain diverse enzymatic activities and functions to coactivate different classes of signal-dependent transcription factors. Together with SRC-1[1] and SRC-2, which also was characterized as Grip-1 (glucocorticoid receptor interacting protein1[139]) and TIF2 (transcriptional intermediary factor 2[140]), SRC-3 forms the structurally related p160 family of coactivator proteins, whose carboxy-terminal domains mediate interactions with the histone acetyltransferases and the NR coregulators CBP and p300, while the amino-terminal basic helix-loop-helix/PAS-containing domains form interactions with additional NR coregulators such as CoCoA (coiled-coil coactivator[141]), GAC63 (grip1-associated coactivator 63[142]) and CARM1 (coactivator-associated arginine methyltransferase 1[10]). These interactions are regulated by post-translational phosphorylation,[143,144] whereby different combinations of site-specific phosphorylations provide interaction specificity codes for different coregulators and transcription factors.[145] Phosphorylation is linked closely to the oncogenic-proliferative potential of SRC-3.[145,146]

Ubiquitinylation is another post-translational modification that is essential for the regulation of cellular signaling.[147,148] Its role in NR coregulator activation and degradation is now generally accepted.[22] It was recently demonstrated that transcriptional activation and turnover of SRC-3 are events controlled by cellular signaling using a phosphorylation-dependent ubiquitinylation code.[23] More specifically, GSK3 kinase-mediated phosphorylation of a specific peptide sequence in SRC-3 precedes non-proteolytic activation of SRC-3 through ubiquitinylation, whereby multiple mono-ubiquitinylation events promote increased SRC-3 coactivator potency and transcription factor specificity. Ultimately, the transition from mono-ubiquitinylation to long-chain polyubiquitinylation leads to SRC-3 degradation. Because the course of polyubiquitinylation is processive during the transcriptional activation of transcription factors, the phosphorylation-dependent ubiquitinylation functions as a transcriptional "time clock" to stimulate coactivator activation as well as to limit its lifetime.[23]

Finally, a post-translational modification of SRC-3 normally not thought of as an important functional event, is brought by yet another NR coregulator — Pin1. Peptidyl-prolyl isomerase 1 (Pin1) catalyzes the cis/trans isomerization of proline residues adjacent to phosphorylated serine/threonine residues to induce conformational changes in the SRC-3 protein, thereby modulating the interactions between SRC-3 and CBP/p300.[149] Experimental induced down-regulation of Pin1 in breast cancer cells was shown to reduce ligand-dependent transcription by nuclear receptors, suggesting that the peptidyl-prolyl cis/trans isomerase activity is an intrinsic part to signal integration at the transcription level.[149] Pin1 is frequently overexpressed in human cancers and has been implicated in oncogenesis.[150] A significant correlation was found between Pin1 and β-catenin protein expression in liver and colorectal cancers,[151,152] and with cyclin D1 expression in squamous cell carcinomas.[153] In addition to regulating the function of certain proteins after phosphorylation and to playing critical roles in cell cycle regulation and cancer development,[154] Pin-1 also is implicated in the protection against age-dependent neurodegeneration. Pin1 gene-deletion in mice shows pathologies resembling many aspects of human Alzheimer's disease.[155] Interestingly, Pin1 was found to be down-regulated or inhibited

pathologies. Cardiovascular disease, for example, is still the leading cause of death in the US as of 2007.[190] Metabolic diseases such as diabetes and obesity may predispose ultimately to the greatest eventual pathological impact. Different than in tumors, where mutations in cancer genes enhance clonal expansions that initiate neoplastic processes, non-cancer diseases are often idiopathic and of epigenetic origin. Genetic susceptibility for non-cancer diseases due to germline mutations are relatively rare compared to wider-spread conditions such as heart disease or diabetes. Some inherited diseases or syndromes[m] that highlight the essential biological actions that come from NR coregulators include Angelman and Rubinstein-Taybi syndromes.

1.5.1 *Eponymous syndromes*

Imprinted maternal mutations in the UBE3A (ubiquitin protein ligase E3A) allele, for example, are one cause for Angelman syndrome.[191,192] Angelman syndrome (AS) is characterized by features such as severe motor and intellectual retardation, microcephaly, ataxia, frequent jerky limb movements and arm flapping, hypotonia, hyperactivity, hypopigmentation, seizures, absence of speech, frequent smiling and laughter and abnormal gait. The prevalence of AS among children is about 1/15,000. The UBE3A gene, which is maternally expressed in the brain and biallelically expressed in other tissues, encodes an E3 ubiquitin-protein ligase termed E6-AP (E6-associated protein) and has been characterized as a NR coregulator.[193] It is a component of ER degradation via the ubiquitin-proteasome pathway in a calcium and calmodulin-regulated manner.[194]

Inherited mutations in the UBE3A gene, however, are a rare cause of AS.[195] About 70% of the AS patients show *de novo* maternal deletions at 15q11-q13. Interestingly, a clinically distinct disorder called Prader-Willi syndrome is caused by paternal deletion of the same region.[196] Patients clinically diagnosed for AS often show mutations in the methyl CpG binding protein 2 (MECP2) gene. MECP2 gene mutations are the cause of some cases of Rett syndrome,[197] a progressive neurological developmental disorder and a common cause of mental retardation. The MECP2 protein directly binds to the NR corepressor mSin3A, which

[m] Syndromes represent meaningful collections of several clinically recognizable features and may lack specific conditions as the underlying cause.

interacts with class I histone deacetylase, and to the SKI protein, recruiting them to methyl-CpG regions to suppress transcription. The c-ski proto-onocogene product is required for MECP2-mediated transcriptional repression,[198] but also binds to SMAD proteins for the repression of transforming growth factor-β signaling[199] through interactions with corepressor complexes containing the NR coregulators HDAC1, NCoR, SMRT and/or mSin3A. These examples demonstrate the fragility of transcription control and the multitude of diseases that can manifest if only a few key molecular players are deviant.

Another inherited autosomal dominant condition with disparate features and genetic heterogeneity is summarized as Rubinstein-Taybi syndrome, also known as "Broad Thumb-Hallux" syndrome. This condition is characterized by short stature, mental retardation, distinctive facial features, and broad thumbs and first toes, and is caused by mutations in the genes encoding the transcriptional coactivators CREB-binding proteins (CREBBP) and/or EP300.[91,92,200–202] Both CBP and the functionally and structurally related protein p300 are broad transcriptional coregulators that integrate cellular signals for multiple transcription factors, including members of the nuclear receptor superfamily,[203,204] in a wide variety of developmental and physiological processes. Both CBP and p300 are potent histone acetyltransferases, indicating that this disorder may be caused in part by aberrant chromatin regulation. The fact that the inactivation of one allele of CBP or p300 is sufficient to cause Rubinstein-Taybi syndrome points toward the absolute necessity of maintaining a critical level of HAT activity for normal development.

CBP and p300 are implicated in other diseases on the basis of expression, sequestration and somatic mutations. For example, the CBP protein was found to be incorporated into nuclear inclusions formed by polyglutamine-containing proteins in patients with spinal and bulbar muscular atrophy.[205] Similarly, CBP but not p300, was found depleted from its normal nuclear location and trapped and functionally repressed in polyglutamine aggregates in human postmortem brains and tissue cultures of patients with Huntington's disease.[206] Huntington's disease is a rare inherited neurological disorder that is caused by a trinucleotide repeat expansion in the Huntingtin (Htt) gene and is one of several polyglutamine-repeat diseases.

p300 gene alterations were identified in colon, breast, ovarian and gastric carcinomas.[207–209] Also, chromosome abnormalities caused by recurrent reciprocal translocations that fuse the MOZ (monocytic leukemia zinc finger protein, also known as ZNF220 or MYST3) gene to the NR coregulator and acetyltransferase genes CBP, p300 and SRC-2/TIF2 were reported in acute myeloid leukemia.[210] These fusion proteins are thought to account for leukemogenesis due to the aberrant regulation of histone acetylation. This example reiterates the importance of maintaining a critical level of histone acetyltransferase activity, which is surprising considering the relative abundance and widespread expression profiles of NR coregulators with HAT activity and other histone modifying enzymes in a cell.

1.5.2 *Metabolic syndrome*

Metabolic syndrome describes a cluster of conditions that collectively increase a person's risk for cardiovascular disease and/or diabetes. The term includes insulin resistance syndrome, syndrome X and Reaven's syndrome. Metabolic syndrome is becoming a major health liability in the western world, affecting about one quarter of the US population, and is the leading predisposition to death by a wide margin.[190] Metabolic syndrome (according to WHO) is associated with insulin resistance and/or impaired glucose tolerance and with the variable coexistence of hyperinsulinemia, central obesity, hypertension, dyslipidemia (high triglycerides and low HDL cholesterol) and microalbumineria. The etiology is complex and determined by genetic as well as environmental and behavioral factors. At the molecular level, the critical role of select nuclear receptors and their coregulators is particularly apparent by the relationship between alterations of their functions and the occurrence of major metabolic diseases.

1.5.2.1 *Nuclear receptors in metabolic syndrome*

The nuclear receptors with major roles in homeostasis are the "metabolic sensors" including the PPARs,[n] the LXRs, FXR, HNF4α which bind metabolic cofactors such as oxysterols and bile acids, or the xenobiotic-sensing PXR and CAR. Other members of the superfamily including the RXRs, GR and TR play secondary roles in metabolic syndrome

[n] Nuclear receptors in plural indicate gene isotypes; PPAR has three genes, denoted with the suffix α, β/∂, γ, LXR has two genes (α and β), and RXR has three genes (α, β, γ).

as well. Briefly, the PPARs are key regulators of adipogenesis by binding to lipid molecules to control the expression of a variety of genes in several pathways of lipid metabolism. In particular, all PPAR isotypes bind unsaturated fatty acids and some eicosanoids, derived mainly from arachidonic acid and linoleic acid, with varying affinities. PPARα also responds to fibrates that are hypolipidemic drugs, and PPARγ also bind thiazolidinedione-based insulin sensitizers. The expression and activity of the PPARs and their coactivator proteins of the PGC-1 family (below) is dynamically regulated in several cardiomyopathic and metabolic diseases.[211] A polymorphism in the PPARα protein (Lys162Val), alone or in interaction with dietary fat intake, was associated with components of the metabolic syndrome.[212] Lastly, PPARγ agonists[o] are a class of agents for the treatment of type 2 diabetes mellitus that act through their ability to improve insulin sensitivity.

By binding oxysterols and other derivatives of cholesterol metabolism, LXRα and LXRβ participate in the cholesterol sensing processes and regulate important aspects of cholesterol and fatty acid metabolism. Cholesterol homeostasis is also impacted by FXR, which binds bile acids, thus acting as a metabolite sensor, creating a negative-feedback loop in the enterohepatic circulation by moderating liver bile acid production when bound to its ligand. Because bile acids are an important means for the disposal of excess cholesterol, FXR forms a critical component of a cholesterol homeostatic network. HNF4α, a key regulator of hepatic lipid homeostasis and mainly expressed in the liver, intestine, and pancreatic β-cells, controls the expression of genes involved in glucose, fatty acid, and cholesterol metabolism. CAR and PXR regulate hepatic bile acid and bilirubin detoxification and elimination pathways through coordinating the induced expression of genes directing oxidative, conjugative and transport phases of endobiotic and xenobiotic metabolism.[213–215] RXRs participate in most of the fatty acid and cholesterol metabolic regulations as the common heterodimers nuclear receptor partner of PPARs, the LXRs and FXR. All three isotypes of RXR (α, β, γ) are activated *in vitro*[p] by 9-*cis*-retinoic acid, a synthetic isomer

[o] Rosiglitazone, marketed as "Avandia", is a selective ligand of PPARγ, and thus does not activate other PPAR isotypes. In addition to effects on insulin resistance, it has anti-inflammatory effects by lowering nuclear factor kappa-B (NFκB) levels.

[p] The nature of the major natural RXR ligand is debated, but RXR was found activated by the long-chain polyunsaturated fatty acid docohexaenoic acid in the adult mouse brain.

of all-*trans*-retinoic acid, and thus control the transcription of the heterodimer target genes in the absence of a ligand for their heterodimer partners. This in turn contributes to the "phantom ligand" effect of this retinoid.[216] Lastly, GR has been implicated in metabolic syndrome due to its maintenance of basal and stress-related homeostasis[217] and by adrenal secretion of cortisol and regeneration of cortisol in peripheral tissues,[218] while thyroid hormones reduce adiposity via increased metabolic rate.[q] TRα governs basal metabolic rate while TRβ mediates cholesterol and the TSH lowering effects of thyroid hormones.[219]

Overall, given the extent of participation by the nuclear receptors in various aspects of cell homeostasis and energy metabolism, it is reasonable to think that their coregulators will also play central roles here as well as, which we shall discuss below.

1.5.2.2 *PGC-1 coregulators in metabolic syndrome*

The PPARγ coactivator-1 (PGC-1) family of coregulators consists of PGC-1α, PGC-1β and the PGC-1-related coactivator (PRC).[220–223] PGC-1α (peroxisome proliferator-activated receptor γ, coactivator 1 α; PPARGC1A) was originally cloned as PGC-1 and shown to enhance the activity of PPARγ and TR on the uncoupling protein-1 promoter.[223] In tissues with high oxidative capacity, PGC-1α was found to be induced by cold exposure and by β-adrenergic signaling, indicating that PGC-1α is expressed when an organism is required to alter its metabolic program in response to cold temperatures or exercise.[224] PGC-1α stimulates its own expression in a cell type-specific manner. In skeletal muscle, it coactivates myocyte enhancer factor 2 (MEF2), which itself regulates PGC-1α mRNA transcription.[225] In white adipose tissues, PGC-1α coactivates PPARγ, which also controls transcription of the PGC-1α gene.[226] In addition, PGC-1α controls the nature of its own transcripts by directing the use of alternative promoters and differential internal splicing.[40] More recently, a rhythmical expression profile of PGC-1α was reported in mice, controlling the expression of other circadian genes through coactivation of the ROR isotypes, thus integrating circadian patterns with energy metabolism.[227]

PGC-1α interacts with several nuclear receptors in a ligand-dependent manner through its single LXXLL motif, but can also bind

[q] Unfortunately they also cause direct cardiac acceleration.

constitutively to other transcription factors. The targets of PGC-1α protein are determined by its phosphorylation, methylation and acetylation status.[29,221,228–230] Like other NR coregulators, PGC-1α shows remarkable integrative features through its ability to associate with different coregulator complexes, including the SRC-1 and CBP/p300-containing HAT complex and the mediator complex.

Due to its multiple interactions, dynamic expression and post-translational modification states, PGC-1α has the regulatory finesse to control cellular energy metabolism, systemic glucose metabolism and insulin sensitivity, and tissue-specific metabolic processes such as adipogenesis, mitochondrial biogenesis, fatty acid β-oxidation, FOXO1-mediated hepatic gluconeogenesis,[224,231] and in HNF4α-mediated regulation of lipoprotein metabolism.[232] Other work demonstrates its role in skeletal muscle homeostasis through the protection from atrophy (and thus guarding from Duchenne muscular dystrophy) by suppressing FOXO3 action and atrophy-specific gene transcription,[233,234] but also in the regulation and protection from reactive oxygen species.[235] Finally, PGC-1–deficient and overexpressing transgenic mouse models have bolstered the metabolic vital function of this coactivator.[236] These mice will likely be valuable tools for the design of therapeutic strategies and agents.

Two homologues to PGC-1α, PGC-1β (also termed PERC for PGC-1 related estrogen receptor coactivator; PPARGC1B), and PGC-1–related coactivator (PRC; with the official symbol PPRC1), were shown to coactivate several nuclear receptors including ERα.[221,222] Like PGC-1α, PGC-1β is strongly expressed in tissues with high oxidative capacity, such as heart, slow-twitch skeletal muscle, and heat-dissipating brown adipose tissue (BAT), where both coactivators serve critical roles in the regulation of mitochondrial functional capacity and cellular energy metabolism.[222] PRC, on the other hand, is expressed rather ubiquitously.[220] Different splice variants have been detected also for PGC-1β.[237]

PGC-1α and PGC-1β show considerable functional similarities by activating gene regulatory programs that drive increased capacity for cellular energy production, but tissue-specific differences exist. For example, in liver, PGC-1α enhances hepatic lipoprotein metabolism but has no impact on lipogenesis, while PGC-1β increases lipogenesis and lipoprotein transport, but does not control gluconeogenesis.[238] PGC-1β gene knock-out mouse models showed impaired mitochondrial functions and defects in adipose metabolism, confirming the overlapping functions with PGC-1α observed *in vivo*,[239] but also revealed new and non-redundant roles in controlling mitochondrial oxidative energy metabolism by modulating circadian activities.[240] Much less is known about the biological roles of PRC, but all three PRC-1 genes were found to be significantly overexpressed in thyroid oncocytomas.[241]

Defects in energy metabolism — aberrant mitochondrial functions, but also anomalous glucogenesis and adipogenesis — have been associated with many human diseases and are the central phenomena in metabolic syndrome. Because the PGC-1 coactivators function as master regulators of many essential metabolic pathways, dysregulation of the PGC-1 regulatory axis contributes to the pathogenesis of common disease states, and the PGC-1 proteins thus associate with various aspects of metabolic syndrome such as type 2 diabetes,[242,243] lipodystrophy and obesity,[244] hepatic insulin resistance,[245] porphyria (disorders of certain enzymes in the heme biosynthesis pathway),[246] cardiomyopathy,[247] and also with neurodegeneration. Moreover, gene mutations in the PGC-1α gene's promoter and a common polymorphism in the gene's coding region (Gly482Ser) are associated with type 2 diabetes,[r] gestational diabetes mellitus,[249] diabetic retinopathy (damage to the retina),[251] lipodystrophy and hypertension.[252] Decreased hepatic expression of PGC-1α due to a defect in inducible expression has been linked to cholesterol cholelithiasis (gallstones disease).[253]

Several neurodegenerative disorders due to mitochondrial dysfunction have been linked to Alzheimer's, Parkinson's and Huntington's diseases.[254] Neurodegeneration (spongiform lesions in the striatum) and behavioral deficiencies (anxiety and hyperactivity) were described in PGC-1α knockout mice[236] reminiscent of mouse models for Huntington's disease. Sporadic lesions in the substantia nigra and the hippocampus have parallels in mouse models for Parkinson's and Alzheimer's diseases.

Variation of PGC-1β may contribute to the pathogenesis of obesity, with a widespread Ala203Pro allele being a risk factor for the development of this common disorder.[255] PGC-1β expression is positively correlated with fat oxidation and nonoxidative glucose metabolism, whereas the expression of PGC-1α in muscle has been positively related to insulin-stimulated glucose uptake and oxidation. Finally, inherited variants in

[r] Reports exist of studies that dispute this association.[248–250]

both PGC-1α and PGC-1β genes have been implicated in familial breast cancer susceptibility due to the impact on estrogen signaling.[256] Overall, there is little doubt that the PGC-1 coregulators serve key regulatory functions in the dynamic transcriptional control of energy metabolic pathways in a variety of mammalian tissues, and that dysregulation of the PGC-1 regulatory circuit contributes to pathologic forms of metabolic syndrome.

1.5.2.3 *Other NR coregulators in metabolic syndrome*

Other NR coregulators such as the p160 coactivators and the corepressors Sirt-1, RIP140 and NCoR also have been shown to be important regulators of metabolism.[257] Targeted gene deletions in mice have indicated that opposing roles in energy homeostasis exist for the SRC-1 and SRC-2 coactivators. While SRC-1 deleted mice show decreased energy expenditure and thus are prone to obesity, SRC-2 ablated animals are leaner, possibly due to the blunted induction of PPARγ target genes involved in adipogenesis and mitochondrial uncoupling.[258–260] Recent metabolomics studies reveal SRC-1 to be a key metabolic regulator of the switch from glucose utilization to fatty acid oxidation during fasting (unpublished data). SRC-3 deficiency in knockout mice causes growth retardation and reduced fat mass. The leaner phenotype in these mice was attributed to the loss of CAAT enhancer binding protein-β (C/EBPβ) coactivation that is needed to fully drive PPARγ expression.[137]

Sirtuins (silent mating type information regulation 2 homolog) are nicotinamide adenine dinucleotide (NAD)-dependent enzymes that deacetylate lysine residues on various proteins. They have a wide range of cellular activities and are implicated in a number of human diseases.[261] Sirtuins are potential targets for metabolic syndrome on the basis of their involvement in caloric restriction.[262] Caloric restriction was shown to have beneficial effects on glucose metabolism and to reduce the incidence of age-related disorders such as diabetes, cancer and cardiovascular diseases. They can also protect against neurodegeneration as shown by the phenotypes that exist in animal models of Huntington's, Alzheimer's, and Parkinson's diseases.[263] Of the seven mammalian sirtuins, (SIRT1-SIRT7) only SIRT1 has been characterized as a NR coregulator so far. Interestingly, the NAD-dependent deacetylase SIRT1 is thought to link caloric restriction and energy homeostasis to increases in lifespan. SIRT1 attenuates adipogenesis by activating critical components of the response to caloric restriction in mammals through NCoR and SMRT-assisted repression of PPARγ transcription.[264] SIRT1 also inhibits dihydroxytestosterone-induced AR signaling by binding to the nuclear receptor to deacetylate a conserved lysine motif, which then prevents coactivator-induced intramolecular interactions between the AR amino and carboxyl termini.[265] The association of three Sirt1 single nucleotide polymorphisms with variations in energy homeostasis in Finnish subjects is supported by *in vivo* laboratory data. It implicates SIRT1 as a key coregulator for energy and metabolic homeostasis.[266] It is becoming clear that the sitruin group of transcriptional corepressors, which are structurally different from other histone deacetylases, has emerging pathogenetic roles in cancer, diabetes, muscle differentiation, heart diseases, neurodegeneration and aging. Thus, they represent a promising pharmacological target for the search and design of new drugs to treat components of metabolic syndrome.[267] In line with this, resveratrol, a natural component in grapes (and red wine), is a known activator of SIRT1 and was shown to improve metabolic functions in mice.[266,268]

Other corepressors implicated in the regulation of additional select aspects of metabolic syndrome involve the histone deacetylasess HDAC1 and HDAC3. The deacetylase inhibitor valproic acid was clinically shown to lead to weight gain,[269,270] most likely by allowing for adipocyte differentiation through suppression of HDAC1- and HDAC3-mediated repression of C/EBPβ- and PPARγ-mediated transactivation, respectively. Another important corepressor involved in metabolic regulation is RIP140 (receptor-interacting protein 140 or NRIP1 for nuclear receptor interacting protein 1). RIP140 is a widely expressed NR coregulator displaying critical roles in adipose biology by repressing transcription of many genes involved in adipogenesis and carbohydrate metabolism. RIP140 also functions in female reproductive physiology, specifically in the release of the ovum during ovulation.[271] RIP140 knockout mice are lean, resistant to obesity and show an increased sensitivity to insulin.[272] A common polymorphism located within the RIP140 gene coding for an Arg448Gly mutation has been associated with endometriosis,[273] and a silent polymorphism has been associated with azoospermia and severe oligozoospermia in Spanish men.[274] Finally, PTEN/MNAR is involved in metabolic syndrome on the basis of overexpression in some type 2 diabetic patients with a polymorphism in the 5′-untranslated region of the PTEN gene.[275]

1.6 Concluding Remarks

In this review we have broadly linked NR coregulators to a diverse variety of human diseases. While we might be surprised about the multitude of associations, the general picture that emerges is that cells have complex regulatory needs that can and do fail due to faults in NR coregulator function. The NR coregulators are involved in numerous diseases due to their fundamental function as signal integrators, which converge to regulate important pathways through their pleiotropic interactions with nuclear receptors and different enzymes in macromolecular ribonucleoprotein complexes. As such, coregulators are "master genes" that have become regulatory hubs for the coordinated control of broad transcriptional programs. We, as long lived organisms, must constantly respond to a broad range of complex, environmental conditions. This is a challenge that has been made possible by the proliferation of coregulator molecules.[276]

While we are still elucidating the fundamental science of nuclear receptor and coregulator-assisted regulation of transcription, it is important to note that genes do not work alone. A focus on pathways and system-wide integration will lead to a better understanding of the gestalt of cellular homeostasis, something that cannot be known from studying individual genes in isolation. Ultimately, this expanded understanding of nuclear receptors and their coregulators should lead to practical benefits for patients through the development of effective pharmaceuticals.

Acknowledgments

The authors thank Anna Malovannaya for reading the manuscript critically and Cécile Lanz for technical assistance.

This work was supported by grants from the National Institutes of Health and NURSA (NIDDK, NHLBI, NIEHS: U19 DK62434).

References

The following references are for the citations in the text. Tables 1.1 & 1.2 have a separate list of references.

1. Onate SA, *et al.*, Sequence and characterization of a coactivator for the steroid hormone receptor superfamily, *Science* **270**(5240):1354–1357, 1995.

2. Horlein AJ, *et al.*, Ligand-independent repression by the thyroid hormone receptor mediated by a nuclear receptor corepressor, *Nature* **377**(6548):397–404, 1995.
3. Seol W, *et al.*, Two receptor interacting domains in the nuclear hormone receptor corepressor RIP13/N-CoR, *Mol Endocrinol* **10**(12):1646–1655, 1996.
4. Chen JD, Evans RM, A transcriptional corepressor that interacts with nuclear hormone receptors, *Nature* **377**(6548):454–457, 1995.
5. Bannister AJ, Kouzarides T, The CBP coactivator is a histone acetyltransferase, *Nature* **384**(6610):641–643, 1996.
6. Blanco JC, *et al.*, The histone acetylase PCAF is a nuclear receptor coactivator, *Genes Dev* **12**(11):1638–1651, 1998.
7. Korzus E, *et al.*, Transcription factor-specific requirements for coactivators and their acetyltransferase functions, *Science* **279**(5351):703–707, 1998.
8. Ogryzko VV, *et al.*, The transcriptional coactivators p300 and CBP are histone acetyltransferases, *Cell* **87**(5):953–959, 1996.
9. Spencer TE, *et al.*, Steroid receptor coactivator-1 is a histone acetyltransferase, *Nature* **389**(6647):194–198, 1997.
10. Chen D, *et al.*, Regulation of transcription by a protein methyltransferase, *Science* **284**(5423):2174–2177, 1999.
11. Baniahmad A, *et al.*, The tau 4 activation domain of the thyroid hormone receptor is required for release of a putative corepressor(s) necessary for transcriptional silencing, *Mol Cell Biol* **15**(1):76–86, 1995.
12. McDonnell DP, Vegeto E, O'Malley BW, Identification of a negative regulatory function for steroid receptors, *Proc Natl Acad Sci USA* **89**(22):10563–10567, 1992.
13. NURSA, http://www.nursa.org/datasets.cfm?doi=10.1621/datasets.01001, 2007.
14. Heery DM, *et al.*, A signature motif in transcriptional coactivators mediates binding to nuclear receptors, *Nature* **387**(6634):733–736, 1997.
15. Torchia J, *et al.*, The transcriptional coactivator p/CIP binds CBP and mediates nuclear receptor function, *Nature* **387**(6634):677–684, 1997.
16. Weiss RE, *et al.*, Mice deficient in the steroid receptor coactivator 1 (SRC-1) are resistant to thyroid hormone, *EMBO J* **18**(7):1900–1904, 1999.
17. Vadlamudi RK, *et al.*, Molecular cloning and characterization of PELP1, a novel human coregulator of estrogen receptor alpha, *J Biol Chem* **276**(41):38272–38279, 2001.
18. Rajhans R, Vadlamudi RK, Comprehensive analysis of recent biochemical and biologic findings regarding a newly discovered protein-PELP1/MNAR, *Clin Exp Metastasis* **23**(1):1–7, 2006.
19. Choi YB, Ko JK, Shin J, The transcriptional corepressor, PELP1, recruits HDAC2 and masks histones using two separate domains, *J Biol Chem* **279**(49):50930–50941, 2004.

20. McInerney EM, *et al.*, Determinants of coactivator LXXLL motif specificity in nuclear receptor transcriptional activation, *Genes Dev* **12**(21):3357–3368, 1998.
21. Zhou G, *et al.*, Nuclear receptors have distinct affinities for coactivators: Characterization by fluorescence resonance energy transfer, *Mol Endocrinol* **12**(10):1594–1604, 1998.
22. Lonard DM, *et al.*, The 26S proteasome is required for estrogen receptor-alpha and coactivator turnover and for efficient estrogen receptor-alpha transactivation, *Mol Cell* **5**(6):939–948, 2000.
23. Wu RC, *et al.*, SRC-3 coactivator functional lifetime is regulated by a phospho-dependent ubiquitin time clock, *Cell* **129**(6):1125–1140, 2007.
24. Glass CK, Rosenfeld MG, The coregulator exchange in transcriptional functions of nuclear receptors, *Genes Dev* **14**(2):121–141, 2000.
25. Lonard DM, O'Malley BW, The expanding cosmos of nuclear receptor coactivators, *Cell* **125**(3):411–414, 2006.
26. Malik S, Roeder RG, Dynamic regulation of pol II transcription by the mammalian Mediator complex, *Trends Biochem Sci* **30**(5):256–263, 2005.
27. McKenna NJ, Lanz RB, O'Malley BW, Nuclear receptor coregulators: Cellular and molecular biology, *Endocr Rev* **20**(3):321–344, 1999.
28. Rosenfeld MG, Lunyak VV, Glass CK, Sensors and signals: A coactivator/corepressor/epigenetic code for integrating signal-dependent programs of transcriptional response, *Genes Dev* **20**(11):1405–1428, 2006.
29. Spiegelman BM, Heinrich R, Biological control through regulated transcriptional coactivators, *Cell* **119**(2):157–167, 2004.
30. Lanz RB, *et al.*, A steroid receptor coactivator, SRA, functions as an RNA and is present in an SRC-1 complex, *Cell* **97**(1):17–27, 1999.
31. Coleman KM, *et al.*, SRA coactivation of estrogen receptor-alpha is phosphorylation-independent, and enhances 4-hydroxytamoxifen agonist activity, *Biochem Biophys Res Commun* **323**(1):332–338, 2004.
32. Deblois G, Giguere V, Ligand-independent coactivation of ERalpha AF-1 by steroid receptor RNA activator (SRA) via MAPK activation, *J Steroid Biochem Mol Biol* **85**(2–5):123–131, 2003.
33. Masuhiro Y, *et al.*, Splicing potentiation by growth factor signals via estrogen receptor phosphorylation, *Proc Natl Acad Sci USA* **102**(23):8126–8131, 2005.
34. Watanabe M, *et al.*, A subfamily of RNA-binding DEAD-box proteins acts as an estrogen receptor alpha coactivator through the N-terminal activation domain (AF-1) with an RNA coactivator, SRA, *EMBO J* **20**(6):1341–1352, 2001.
35. Fuller-Pace FV, DExD/H box RNA helicases: Multifunctional proteins with important roles in transcriptional regulation, *Nucleic Acids Res* **34**(15):4206–4215, 2006.

36. Fukuda T, *et al.*, DEAD-box RNA helicase subunits of the Drosha complex are required for processing of rRNA and a subset of microRNAs, *Nat Cell Biol* **9**(5):604–611, 2007.
37. Dowhan DH, *et al.*, Steroid hormone receptor coactivation and alternative RNA splicing by U2AF65-related proteins CAPERalpha and CAPERbeta, *Mol Cell* **17**(3):429–439, 2005.
38. Jung DJ, *et al.*, Molecular cloning and characterization of CAPER, a novel coactivator of activating protein-1 and estrogen receptors, *J Biol Chem* **277**(2):1229–1234, 2002.
39. Iwasaki T, Chin WW, Ko L, Identification and characterization of RRM-containing coactivator activator (CoAA) as TRBP-interacting protein, and its splice variant as a coactivator modulator (CoAM), *J Biol Chem* **276**(36):33375–33383, 2001.
40. Monsalve M, *et al.*, Direct coupling of transcription and mRNA processing through the thermogenic coactivator PGC-1, *Mol Cell* **6**(2):307–316, 2000.
41. Honig A, *et al.*, Regulation of alternative splicing by the ATP-dependent DEAD-box RNA helicase p72, *Mol Cell Biol* **22**(16):5698–5707, 2002.
42. Lanz RB, *et al.*, Distinct RNA motifs are important for coactivation of steroid hormone receptors by steroid receptor RNA activator (SRA), *Proc Natl Acad Sci USA* **99**(25):16081–16086, 2002.
43. Shi Y, *et al.*, Sharp, an inducible cofactor that integrates nuclear receptor repression and activation, *Genes Dev* **15**(9):1140–1151, 2001.
44. Feng Y, *et al.*, Drosophila split ends homologue SHARP functions as a positive regulator of Wnt/beta-catenin/T-cell factor signaling in neoplastic transformation, *Cancer Res* **67**(2):482–491, 2007.
45. Hatchell EC, *et al.*, SLIRP, a small SRA binding protein, is a nuclear receptor corepressor, *Mol Cell* **22**(5):657–668, 2006.
46. Caretti G, Lei EP, Sartorelli V, The DEAD-box p68/p72 proteins and the noncoding RNA steroid receptor activator SRA: Eclectic regulators of disparate biological functions, *Cell Cycle* **6**(10):1172–1176, 2007.
47. Caretti G, *et al.*, The RNA helicases p68/p72 and the noncoding RNA SRA are coregulators of MyoD and skeletal muscle differentiation, *Dev Cell* **11**(4):547–560, 2006.
48. Zhao X, *et al.*, Regulation of nuclear receptor activity by a pseudouridine synthase through posttranscriptional modification of steroid receptor RNA activator, *Mol Cell* **15**(4):549–558, 2004.
49. Zhao X, *et al.*, Pus3p- and Pus1p-dependent pseudouridylation of steroid receptor RNA activator controls a functional switch that regulates nuclear receptor signaling, *Mol Endocrinol* **21**(3):686–699, 2007.
50. Leygue E, *et al.*, Expression of the steroid receptor RNA activator in human breast tumors, *Cancer Res* **59**(17):4190–4193, 1999.

51. Murphy LC, *et al.*, Altered expression of estrogen receptor coregulators during human breast tumorigenesis, *Cancer Res* **60**(22):6266–6271, 2000.
52. Lanz RB, *et al.*, Steroid receptor RNA activator stimulates proliferation as well as apoptosis *in vivo*, *Mol Cell Biol* **23**(20):7163–7176, 2003.
53. Bykhovskaya Y, *et al.*, Missense mutation in pseudouridine synthase 1 (PUS1) causes mitochondrial myopathy and sideroblastic anemia (MLASA), *Am J Hum Genet* **74**(6):1303–1308, 2004.
54. Fernandez-Vizarra E, *et al.*, Nonsense mutation in pseudouridylate synthase 1 (PUS1) in two brothers affected by myopathy, lactic acidosis and sideroblastic anaemia (MLASA), *J Med Genet* **44**(3):173–180, 2007.
55. Patton JR, *et al.*, Mitochondrial myopathy and sideroblastic anemia (MLASA): Missense mutation in the pseudouridine synthase 1 (PUS1) gene is associated with the loss of tRNA pseudouridylation, *J Biol Chem* **280**(20):19823–19828, 2005.
56. Shenk JL, *et al.*, p53 represses androgen-induced transactivation of prostate-specific antigen by disrupting hAR amino- to carboxyl-terminal interaction, *J Biol Chem* **276**(42):38472–38479, 2001.
57. Bartkova J, *et al.*, DNA damage response as a candidate anti-cancer barrier in early human tumorigenesis, *Nature* **434**(7035):864–870, 2005.
58. Petitjean A, *et al.*, TP53 mutations in human cancers: Functional selection and impact on cancer prognosis and outcomes, *Oncogene* **26**(15):2157–2165, 2007.
59. Iwakuma T, Lozano G, Flores ER, Li-Fraumeni syndrome: A p53 family affair, *Cell Cycle* **4**(7):865–867, 2005.
60. Oren M, Decision making by p53: Life, death and cancer, *Cell Death Differ* **10**(4):431–442, 2003.
61. Fan S, *et al.*, BRCA1 inhibition of estrogen receptor signaling in transfected cells, *Science* **284**(5418):1354–1356, 1999.
62. James CR, *et al.*, BRCA1, a potential predictive biomarker in the treatment of breast cancer, *Oncologist* **12**(2):142–150, 2007.
63. Rosen EM, *et al.*, BRCA1 in hormone-responsive cancers, *Trends Endocrinol Metab* **14**(8):378–385, 2003.
64. Sjoblom T, *et al.*, The consensus coding sequences of human breast and colorectal cancers, *Science* **314**(5797):268–274, 2006.
65. Zwijsen RM, *et al.*, CDK-independent activation of estrogen receptor by cyclin D1, *Cell* **88**(3):405–415, 1997.
66. Despouy G, *et al.*, Cyclin D3 is a cofactor of retinoic acid receptors, modulating their activity in the presence of cellular retinoic acid-binding protein II, *J Biol Chem* **278**(8):6355–6362, 2003.
67. Delva L, *et al.*, Physical and functional interactions between cellular retinoic acid binding protein II and the retinoic acid-dependent nuclear complex, *Mol Cell Biol* **19**(10):7158–7167, 1999.

68. Yamamoto A, *et al.*, Cyclin E as a coactivator of the androgen receptor, *J Cell Biol* **150**(4):873–880, 2000.
69. Truica CI, Byers S, Gelmann EP, Beta-catenin affects androgen receptor transcriptional activity and ligand specificity, *Cancer Res* **60**(17):4709–4713, 2000.
70. Botrugno OA, *et al.*, Synergy between LRH-1 and beta-catenin induces G1 cyclin-mediated cell proliferation, *Mol Cell* **15**(4):499–509, 2004.
71. Kouzmenko AP, *et al.*, Wnt/beta-catenin and estrogen signaling converge *in vivo*, *J Biol Chem* **279**(39):40255–40258, 2004.
72. Mizusaki H, *et al.*, Dax-1 (dosage-sensitive sex reversal-adrenal hypoplasia congenita critical region on the X chromosome, gene 1) gene transcription is regulated by wnt4 in the female developing gonad, *Mol Endocrinol* **17**(4):507–519, 2003.
73. Song LN, *et al.*, Beta-catenin binds to the activation function 2 region of the androgen receptor and modulates the effects of the N-terminal domain and TIF2 on ligand-dependent transcription, *Mol Cell Biol* **23**(5):1674–1687, 2003.
74. Johansson L, *et al.*, The orphan nuclear receptor SHP utilizes conserved LXXLL-related motifs for interactions with ligand-activated estrogen receptors, *Mol Cell Biol* **20**(4):1124–1133, 2000.
75. Nishigori H, *et al.*, Mutations in the small heterodimer partner gene are associated with mild obesity in Japanese subjects, *Proc Natl Acad Sci USA* **98**(2):575–580, 2001.
76. Zhang H, *et al.*, DAX-1 functions as an LXXLL-containing corepressor for activated estrogen receptors, *J Biol Chem* **275**(51):39855–39859, 2000.
77. Balsamo A, *et al.*, A new DAX1 gene mutation associated with congenital adrenal hypoplasia and hypogonadotropic hypogonadism, *Am J Med Genet A* **135**(3):292–296, 2005.
78. Mantovani G, *et al.*, DAX1 and X-linked adrenal hypoplasia congenita: Clinical and molecular analysis in five patients, *Eur J Endocrinol* **154**(5):685–689, 2006.
79. Reutens AT, *et al.*, Clinical and functional effects of mutations in the DAX-1 gene in patients with adrenal hypoplasia congenital, *J Clin Endocrinol Metab* **84**(2):504–511, 1999.
80. Bardoni B, *et al.*, A dosage sensitive locus at chromosome Xp21 is involved in male to female sex reversal, *Nat Genet* **7**(4):497–501, 1994.
81. Fechner PY, *et al.*, Report of a kindred with X-linked (or autosomal dominant sex-limited) 46,XY partial gonadal dysgenesis, *J Clin Endocrinol Metab* **76**(5):1248–1253, 1993.
82. German J, *et al.*, Genetically determined sex-reversal in 46,XY humans, *Science* **202**(4363):53–56, 1978.

83. Hamosh A, *et al.*, Online Mendelian Inheritance in Man (OMIM), a knowledgebase of human genes and genetic disorders, *Nucleic Acids Res* **33**(Database issue):D514–517, 2005.
84. Becker KG, *et al.*, The genetic association database, *Nat Genet* **36**(5):431–432, 2004.
85. American Cancer Society News Center, www.cancer.org/docroot/NWS/, 2007.
86. Lengauer C, Kinzler KW, Vogelstein B, Genetic instabilities in human cancers, *Nature* **396**(6712):643–649, 1998.
87. Rhodes DR, Chinnaiyan AM, Integrative analysis of the cancer transcriptome, *Nat Genet* **37 Suppl**:S31–37, 2005.
88. Lonard DM, Lanz RB, O'Malley BW, Nuclear Receptor Coregulators and Human Disease, *Endocr Rev*, 2007.
89. Shang Y, Brown M, Molecular determinants for the tissue specificity of SERMs, *Science* **295**(5564):2465–2468, 2002.
90. Olson LE, *et al.*, Homeodomain-mediated beta-catenin-dependent switching events dictate cell-lineage determination, *Cell* **125**(3):593–605, 2006.
91. Petrij F, *et al.*, Rubinstein-Taybi syndrome caused by mutations in the transcriptional coactivator CBP, *Nature* **376**(6538):348–351, 1995.
92. Zimmermann N, *et al.*, Confirmation of EP300 gene mutations as a rare cause of Rubinstein-Taybi syndrome, *Eur J Hum Genet*, 2007.
93. Lonard DM, Tsai SY, O'Malley BW, Selective estrogen receptor modulators 4-hydroxytamoxifen and raloxifene impact the stability and function of SRC-1 and SRC-3 coactivator proteins, *Mol Cell Biol* **24**(1):14–24, 2004.
94. Perissi V, *et al.*, A corepressor/coactivator exchange complex required for transcriptional activation by nuclear receptors and other regulated transcription factors, *Cell* **116**(4):511–526, 2004.
95. Linding R, *et al.*, Systematic discovery of *in vivo* phosphorylation networks, *Cell* **129**(7):1415–1426, 2007.
96. Barker N, Morin PJ, Clevers H, The Yin-Yang of TCF/beta-catenin signaling, *Adv Cancer Res* **77**:1–24, 2000.
97. Chan EF, *et al.*, A common human skin tumour is caused by activating mutations in beta-catenin, *Nat Genet* **21**(4):410–413, 1999.
98. Martensson A, *et al.*, Beta-catenin expression in relation to genetic instability and prognosis in colorectal cancer, *Oncol Rep* **17**(2):447–452, 2007.
99. Saldanha G, *et al.*, Nuclear beta-catenin in basal cell carcinoma correlates with increased proliferation, *Br J Dermatol* **151**(1):157–164, 2004.
100. Taniguchi K, *et al.*, Mutational spectrum of beta-catenin, AXIN1, and AXIN2 in hepatocellular carcinomas and hepatoblastomas, *Oncogene* **21**(31):4863–4871, 2002.

101. Varallo VM, *et al.*, Beta-catenin expression in Dupuytren's disease: Potential role for cell-matrix interactions in modulating beta-catenin levels *in vivo* and *in vitro*, *Oncogene* **22**(24):3680–3684, 2003.
102. Uschkereit C, *et al.*, Different CTNNB1 mutations as molecular genetic proof for the independent origin of four Wilms tumours in a patient with a novel germ line WT1 mutation, *J Med Genet* **44**(6):393–396, 2007.
103. Clifford SC, *et al.*, Wnt/Wingless pathway activation and chromosome 6 loss characterize a distinct molecular sub-group of medulloblastomas associated with a favorable prognosis, *Cell Cycle* **5**(22):2666–2670, 2006.
104. Call KM, *et al.*, Isolation and characterization of a zinc finger polypeptide gene at the human chromosome 11 Wilms' tumor locus, *Cell* **60**(3): 509–520, 1990.
105. Maiti S, *et al.*, Frequent association of beta-catenin and WT1 mutations in Wilms tumors, *Cancer Res* **60**(22):6288–6292, 2000.
106. Tsujimoto Y, *et al.*, Molecular cloning of the chromosomal breakpoint of B-cell lymphomas and leukemias with the t(11; 14) chromosome translocation, *Science* **224**(4656):1403–1406, 1984.
107. Le Marchand L, *et al.*, Association of the cyclin D1 A870G polymorphism with advanced colorectal cancer, *JAMA* **290**(21):2843–2848, 2003.
108. Caldon CE, *et al.*, Cell cycle control in breast cancer cells, *J Cell Biochem* **97**(2):261–274, 2006.
109. Shu XO, *et al.*, Association of cyclin D1 genotype with breast cancer risk and survival, *Cancer Epidemiol Biomarkers Prev* **14**(1):91–97, 2005.
110. Simpson DJ, *et al.*, Cyclin D1 (CCND1) genotype is associated with tumour grade in sporadic pituitary adenomas, *Carcinogenesis* **22**(11):1801–1807, 2001.
111. Wong YK, *et al.*, Cyclin D1 genotype in areca-associated oral squamous cell carcinoma, *J Oral Pathol Med* **32**(5):265–270, 2003.
112. Fu M, *et al.*, Minireview: Cyclin D1: Normal and abnormal functions, *Endocrinology* **145**(12):5439–5447, 2004.
113. Knudsen KE, *et al.*, Cyclin D1: Polymorphism, aberrant splicing and cancer risk, *Oncogene* **25**(11):1620–1628, 2006.
114. Qi C, *et al.*, Potentiation of estrogen receptor transcriptional activity by breast cancer amplified sequence 2, *Biochem Biophys Res Commun* **328**(2):393–398, 2005.
115. Gururaj AE, *et al.*, Estrogen induces expression of BCAS3, a novel estrogen receptor-alpha coactivator, through PELP1, *Mol Endocrinol*, 2007.
116. Barlund M, *et al.*, Cloning of BCAS3 (17q23) and BCAS4 (20q13) genes that undergo amplification, overexpression, and fusion in breast cancer, *Genes Chromosomes Cancer* **35**(4):311–317, 2002.
117. Schlegel A, *et al.*, Caveolin-1 potentiates estrogen receptor alpha (ERalpha) signaling. Caveolin-1 drives ligand-independent nuclear

translocation and activation of ERalpha, *J Biol Chem* **274**(47): 33551–33556, 1999.
118. Sui Y, *et al.*, Gene amplification and associated loss of 5′ regulatory sequences of CoAA in human cancers, *Oncogene* **26**(6):822–835, 2007.
119. Proia DA, *et al.*, Dual roles for the phosphatase PPM1D in regulating progesterone receptor function, *J Biol Chem* **281**(11):7089–7101, 2006.
120. Mo R, Rao SM, Zhu YJ, Identification of the MLL2 complex as a coactivator for estrogen receptor alpha, *J Biol Chem* **281**(23):15714–15720, 2006.
121. Huntsman DG, *et al.*, MLL2, the second human homolog of the Drosophila trithorax gene, maps to 19q13.1 and is amplified in solid tumor cell lines, *Oncogene* **18**(56):7975–7984, 1999.
122. Hess JL, Mechanisms of transformation by MLL, *Crit Rev Eukaryot Gene Expr* **14**(4):235–254, 2004.
123. Anzick SL, *et al.*, AIB1, a steroid receptor coactivator amplified in breast and ovarian cancer, *Science* **277**(5328):965–968, 1997.
124. Chen H, *et al.*, Nuclear receptor coactivator ACTR is a novel histone acetyltransferase and forms a multimeric activation complex with P/CAF and CBP/p300, *Cell* **90**(3):569–580, 1997.
125. Li H, Gomes PJ, Chen JD, RAC3, a steroid/nuclear receptor-associated coactivator that is related to SRC-1 and TIF2, *Proc Natl Acad Sci USA* **94**(16):8479–8484, 1997.
126. Takeshita A, *et al.*, TRAM-1, A novel 160-kDa thyroid hormone receptor activator molecule, exhibits distinct properties from steroid receptor coactivator-1, *J Biol Chem* **272**(44):27629–27634, 1997.
127. Kirkegaard T, *et al.*, Amplified in breast cancer 1 in human epidermal growth factor receptor — positive tumors of tamoxifen-treated breast cancer patients, *Clin Cancer Res* **13**(5):1405–1411, 2007.
128. Shou J, *et al.*, Mechanisms of tamoxifen resistance: Increased estrogen receptor-HER2/neu cross-talk in ER/HER2-positive breast cancer, *J Natl Cancer Inst* **96**(12):926–935, 2004.
129. Bouras T, Southey MC, Venter DJ, Overexpression of the steroid receptor coactivator AIB1 in breast cancer correlates with the absence of estrogen and progesterone receptors and positivity for p53 and HER2/neu, *Cancer Res* **61**(3):903–907, 2001.
130. Rebbeck TR, *et al.*, Modification of BRCA1- and BRCA2-associated breast cancer risk by AIB1 genotype and reproductive history, *Cancer Res* **61**(14):5420–5224, 2001.
131. Torres-Arzayus MI, *et al.*, High tumor incidence and activation of the PI3K/AKT pathway in transgenic mice define AIB1 as an oncogene, *Cancer Cell* **6**(3):263–274, 2004.

132. Kuang SQ, *et al.*, Mice lacking the amplified in breast cancer 1/steroid receptor coactivator-3 are resistant to chemical carcinogen-induced mammary tumorigenesis, *Cancer Res* **65**(17):7993–8002, 2005.
133. Kuang SQ, *et al.*, AIB1/SRC-3 deficiency affects insulin-like growth factor I signaling pathway and suppresses γ-Ha-ras-induced breast cancer initiation and progression in mice, *Cancer Res* **64**(5):1875–1885, 2004.
134. Chung AC, *et al.*, Genetic ablation of the amplified-in-breast cancer 1 inhibits spontaneous prostate cancer progression in mice, *Cancer Res* **67**(12):5965–5975, 2007.
135. Ying H, Willingham MC, Cheng SY, The steroid receptor coactivator-3 is a tumor promoter in a mouse model of thyroid cancer, *Oncogene*, 2007.
136. Coste A, *et al.*, Absence of the steroid receptor coactivator-3 induces B-cell lymphoma, *EMBO J* **25**(11):2453–2464, 2006.
137. Louet JF, *et al.*, Oncogenic steroid receptor coactivator-3 is a key regulator of the white adipogenic program, *Proc Natl Acad Sci USA* **103**(47):17868–17873, 2006.
138. Yu C, *et al.*, An essential function of the SRC-3 coactivator in suppression of cytokine mRNA translation and inflammatory response, *Mol Cell* **25**(5):765–778, 2007.
139. Hong H, *et al.*, GRIP1, a novel mouse protein that serves as a transcriptional coactivator in yeast for the hormone binding domains of steroid receptors, *Proc Natl Acad Sci USA* **93**(10):4948–4952, 1996.
140. Voegel JJ, *et al.*, TIF2, a 160 kDa transcriptional mediator for the ligand-dependent activation function AF-2 of nuclear receptors, *EMBO J* **15**(14):3667–3675, 1996.
141. Kim JH, Li H, Stallcup MR, CoCoA, a nuclear receptor coactivator which acts through an N-terminal activation domain of p160 coactivators, *Mol Cell* **12**(6):1537–1549, 2003.
142. Chen YH, Kim JH, Stallcup MR, GAC63, a GRIP1-dependent nuclear receptor coactivator, *Mol Cell Biol* **25**(14):5965–5972, 2005.
143. Font de Mora J, Brown M, AIB1 is a conduit for kinase-mediated growth factor signaling to the estrogen receptor, *Mol Cell Biol* **20**(14):5041–5047, 2000.
144. Wu RC, *et al.*, Regulation of SRC-3 (pCIP/ACTR/AIB-1/RAC-3/TRAM-1) coactivator activity by I kappa B kinase, *Mol Cell Biol* **22**(10):3549–3561, 2002.
145. Wu RC, *et al.*, Selective phosphorylations of the SRC-3/AIB1 coactivator integrate genomic reponses to multiple cellular signaling pathways, *Mol Cell* **15**(6):937–949, 2004.
146. Wu RC, Smith CL, O'Malley BW, Transcriptional regulation by steroid receptor coactivator phosphorylation, *Endocr Rev* **26**(3):393–399, 2005.

147. Varshavsky A, Regulated protein degradation, *Trends Biochem Sci* **30**(6):283–286, 2005.
148. Varshavsky A, The early history of the ubiquitin field, *Protein Sci* **15**(3):647–654, 2006.
149. Yi P, *et al.*, Peptidyl-prolyl isomerase 1 (Pin1) serves as a coactivator of steroid receptor by regulating the activity of phosphorylated steroid receptor coactivator 3 (SRC-3/AIB1), *Mol Cell Biol* **25**(21):9687–9699, 2005.
150. Bao L, *et al.*, Prevalent overexpression of prolyl isomerase Pin1 in human cancers, *Am J Pathol* **164**(5):1727–1737, 2004.
151. Kim CJ, *et al.*, Pin1 overexpression in colorectal cancer and its correlation with aberrant beta-catenin expression, *World J Gastroenterol* **11**(32):5006–5009, 2005.
152. Pang RW, *et al.*, PIN1 expression contributes to hepatic carcinogenesis, *J Pathol* **210**(1):19–25, 2006.
153. Fukuchi M, *et al.*, Prolyl isomerase Pin1 expression predicts prognosis in patients with esophageal squamous cell carcinoma and correlates with cyclinD1 expression, *Int J Oncol* **29**(2):329–334, 2006.
154. Yeh ES, Means AR, PIN1, the cell cycle and cancer, *Nat Rev Cancer* **7**(5):381–388, 2007.
155. Liou YC, *et al.*, Role of the prolyl isomerase Pin1 in protecting against age-dependent neurodegeneration, *Nature* **424**(6948):556–561, 2003.
156. Balastik M, *et al.*, Pin1 in Alzheimer's disease: Multiple substrates, one regulatory mechanism? *Biochim Biophys Acta* **1772**(4):422–429, 2007.
157. Lu KP, Pinning down cell signaling, cancer and Alzheimer's disease, *Trends Biochem Sci* **29**(4):200–209, 2004.
158. Segat L, *et al.*, PIN1 promoter polymorphisms are associated with Alzheimer's disease, *Neurobiol Aging* **28**(1):69–74, 2007.
159. Sultana R, *et al.*, Oxidative modification and down-regulation of Pin1 in Alzheimer's disease hippocampus: A redox proteomics analysis, *Neurobiol Aging* **27**(7):918–925, 2006.
160. Chan EM, *et al.*, AML1-FOG2 fusion protein in myelodysplasia, *Blood* **105**(11):4523–4526, 2005.
161. Pizzuti A, *et al.*, Mutations of ZFPM2/FOG2 gene in sporadic cases of tetralogy of Fallot, *Hum Mutat* **22**(5):372–377, 2003.
162. Ackerman KG, *et al.*, Fog2 is required for normal diaphragm and lung development in mice and humans, *PLoS Genet* **1**(1):58–65, 2005.
163. Vogelstein B, Kinzler KW, Cancer genes and the pathways they control, *Nat Med* **10**(8):789–799, 2004.
164. Friedberg EC, DNA damage and repair, *Nature* **421**(6921):436–440, 2003.
165. Rinaldo C, *et al.*, MDM2-regulated degradation of HIPK2 prevents p53Ser46 phosphorylation and DNA damage-induced apoptosis, *Mol Cell* **25**(5):739–750, 2007.

166. Trotman LC, *et al.*, Ubiquitination regulates PTEN nuclear import and tumor suppression, *Cell* **128**(1):141–156, 2007.
167. Celebi JT, *et al.*, Phenotypic findings of Cowden syndrome and Bannayan-Zonana syndrome in a family associated with a single germline mutation in PTEN, *J Med Genet* **36**(5):360–364, 1999.
168. Longy M, *et al.*, Mutations of PTEN in patients with Bannayan-Riley-Ruvalcaba phenotype, *J Med Genet* **35**(11):886–889, 1998.
169. Marsh DJ, *et al.*, Germline mutations in PTEN are present in Bannayan-Zonana syndrome, *Nat Genet* **16**(4):333–334, 1997.
170. Marsh DJ, *et al.*, PTEN mutation spectrum and genotype-phenotype correlations in Bannayan-Riley-Ruvalcaba syndrome suggest a single entity with Cowden syndrome, *Hum Mol Genet* **8**(8):1461–1472, 1999.
171. Maser RS, *et al.*, Chromosomally unstable mouse tumours have genomic alterations similar to diverse human cancers, *Nature* **447**(7147):966–971, 2007.
172. Li P, Nicosia SV, Bai W, Antagonism between PTEN/MMAC1/TEP-1 and androgen receptor in growth and apoptosis of prostatic cancer cells, *J Biol Chem* **276**(23):20444–20450, 2001.
173. Nan B, *et al.*, The PTEN tumor suppressor is a negative modulator of androgen receptor transcriptional activity, *J Mol Endocrinol* **31**(1):169–183, 2003.
174. Henry KW, *et al.*, Tuberous sclerosis gene 2 product modulates transcription mediated by steroid hormone receptor family members, *J Biol Chem* **273**(32):20535–20539, 1998.
175. Goncharova EA, Krymskaya VP, Pulmonary lymphangioleiomyomatosis (LAM): Progress and current challenges, *J Cell Biochem*, 2007.
176. Warner DR, *et al.*, Functional interaction between Smad, CREB binding protein, and p68 RNA helicase, *Biochem Biophys Res Commun* **324**(1):70–76, 2004.
177. Hayes SA, *et al.*, SMAD3 represses androgen receptor-mediated transcription, *Cancer Res* **61**(5):2112–2118, 2001.
178. Yanagisawa J, *et al.*, Convergence of transforming growth factor-beta and vitamin D signaling pathways on SMAD transcriptional coactivators, *Science* **283**(5406):1317–1321, 1999.
179. Kang HY, *et al.*, Differential modulation of androgen receptor-mediated transactivation by Smad3 and tumor suppressor Smad4, *J Biol Chem* **277**(46):43749–43756, 2002.
180. Friedl W, *et al.*, Juvenile polyposis: Massive gastric polyposis is more common in MADH4 mutation carriers than in BMPR1A mutation carriers, *Hum Genet* **111**(1):108–111, 2002.
181. Howe JR, *et al.*, The prevalence of MADH4 and BMPR1A mutations in juvenile polyposis and absence of BMPR2, BMPR1B, and ACVR1 mutations, *J Med Genet* **41**(7):484–491, 2004.

182. Embuscado EE, *et al.*, Immortalizing the complexity of cancer metastasis: Genetic features of lethal metastatic pancreatic cancer obtained from rapid autopsy, *Cancer Biol Ther* **4**(5):548–554, 2005.
183. Tang ZH, *et al.*, The relationship between loss expression of DPC4/Smad4 gene and carcinogenesis of pancreatobiliary carcinoma, *Hepatobiliary Pancreat Dis Int* **1**(4):624–629, 2002.
184. Wang W, *et al.*, Essential role of Smad3 in angiotensin II-induced vascular fibrosis, *Circ Res* **98**(8):1032–1039, 2006.
185. Dubrovska A, *et al.*, TGFbetal/Smad3 counteracts BRCA1-dependent repair of DNA damage, *Oncogene* **24**(14):2289–2297, 2005.
186. Shin S, Verma IM, BRCA2 cooperates with histone acetyltransferases in androgen receptor-mediated transcription, *Proc Natl Acad Sci USA* **100**(12):7201–7206, 2003.
187. Yeh S, *et al.*, Retinoblastoma, a tumor suppressor, is a coactivator for the androgen receptor in human prostate cancer DU145 cells, *Biochem Biophys Res Commun* **248**(2):361–367, 1998.
188. Singh P, Coe J, Hong W, A role for retinoblastoma protein in potentiating transcriptional activation by the glucocorticoid receptor, *Nature* **374**(6522):562–565, 1995.
189. Vierimaa O, *et al.*, Pituitary adenoma predisposition caused by germline mutations in the AIP gene, *Science* **312**(5777):1228–1230, 2006.
190. Jemal A, *et al.*, Cancer statistics 2007, CA *Cancer J Clin* **57**(1):43–66, 2007.
191. Kishino T, Lalande M, Wagstaff J, UBE3A/E6-AP mutations cause Angelman syndrome, *Nat Genet* **15**(1):70–73, 1997.
192. Rapakko K, Kokkonen H, Leisti J, UBE3A gene mutations in Finnish Angelman syndrome patients detected by conformation sensitive gel electrophoresis, *Am J Med Genet A* **126**(3):248–252.
193. Nawaz Z, *et al.*, The Angelman syndrome-associated protein, E6-AP, is a coactivator for the nuclear hormone receptor superfamily, *Mol Cell Biol* **19**(2):1182–1189, 1999.
194. Li L, *et al.*, E6AP and calmodulin reciprocally regulate estrogen receptor stability, *J Biol Chem* **281**(4):1978–1985, 2006.
195. Sahoo T, *et al.*, Identification of novel deletions of 15q11q13 in Angelman syndrome by array-CGH: Molecular characterization and genotype-phenotype correlations, *Eur J Hum Genet*, 2007.
196. Cassidy SB, Schwartz S, Prader-Willi and Angelman syndromes. Disorders of genomic imprinting, *Medicine (Baltimore)* **77**(2):140–151, 1998.
197. Amir RE, *et al.*, Rett syndrome is caused by mutations in X-linked MECP2, encoding methyl-CpG-binding protein 2, *Nat Genet* **23**(2):185–188, 1999.
198. Kokura K, *et al.*, The Ski protein family is required for MeCP2-mediated transcriptional repression, *J Biol Chem* **276**(36):34115–34121, 2001.

199. Ueki N, Hayman MJ, Direct interaction of Ski with either Smad3 or Smad4 is necessary and sufficient for Ski-mediated repression of transforming growth factor-beta signaling, *J Biol Chem* **278**(35):32489–32492, 2003.
200. Bartholdi D, *et al.*, Genetic heterogeneity in Rubinstein-Taybi syndrome: Delineation of the phenotype of the first patients carrying mutations in EP300, *J Med Genet* **44**(5):327–333, 2007.
201. Bartsch O, *et al.*, DNA sequencing of CREBBP demonstrates mutations in 56% of patients with Rubinstein-Taybi syndrome (RSTS) and in another patient with incomplete RSTS, *Hum Genet* **117**(5):485–493, 2005.
202. Roelfsema JH, *et al.*, Genetic heterogeneity in Rubinstein-Taybi syndrome: Mutations in both the CBP and EP300 genes cause disease, *Am J Hum Genet* **76**(4):572–580, 2005.
203. Chakravarti D, *et al.*, Role of CBP/P300 in nuclear receptor signaling, *Nature* **383**(6595):99–103, 1996.
204. Kamei Y, *et al.*, A CBP integrator complex mediates transcriptional activation and AP-1 inhibition by nuclear receptors, *Cell* **85**(3):403–414, 1996.
205. McCampbell A, *et al.*, CREB-binding protein sequestration by expanded polyglutamine, *Hum Mol Genet* **9**(14):2197–2202, 2000.
206. Nucifora FC, Jr, *et al.*, Interference by huntingtin and atrophin-1 with cbp-mediated transcription leading to cellular toxicity, *Science* **291**(5512): 2423–2428, 2001.
207. Bryan EJ, *et al.*, Mutation analysis of EP300 in colon, breast and ovarian carcinomas, *Int J Cancer* **102**(2):137–141, 2002.
208. Gayther SA, *et al.*, Mutations truncating the EP300 acetylase in human cancers, *Nat Genet* **24**(3):300–303, 2000.
209. Muraoka M, *et al.*, p300 gene alterations in colorectal and gastric carcinomas, *Oncogene* **12**(7):1565–1569, 1996.
210. Troke PJ, *et al.*, MOZ fusion proteins in acute myeloid leukaemia, *Biochem Soc Symp* **73**:23–39, 2006.
211. Finck BN, The PPAR regulatory system in cardiac physiology and disease, *Cardiovasc Res* **73**(2):269–277, 2007.
212. Robitaille J, *et al.*, Association between the PPARalpha-L162V polymorphism and components of the metabolic syndrome, *J Hum Genet* **49**(9):482–489, 2004.
213. Echchgadda I, *et al.*, The Xenobiotic-sensing Nuclear Receptors PXR, CAR and Orphan Nuclear Receptor HNF-4{alpha} in the Regulation of Human Steroid-Bile Acid-Sulfotransferase, *Mol Endocrinol*, 2007.
214. Tien ES, Negishi M, Nuclear receptors CAR and PXR in the regulation of hepatic metabolism, *Xenobiotica* **36**(10 11): 1152–1163, 2006.
215. Wagner M, *et al.*, CAR and PXR agonists stimulate hepatic bile acid and bilirubin detoxification and elimination pathways in mice, *Hepatology* **42**(2):420–430, 2005.

216. Shulman AI, Mangelsdorf DJ, Retinoid x receptor heterodimers in the metabolic syndrome, *N Engl J Med* **353**(6):604–615, 2005.
217. Chrousos GP, Kino T, Glucocorticoid action networks and complex psychiatric and/or somatic disorders, *Stress* **10**(2):213–219, 2007.
218. Walker BR, Andrew R, Tissue production of cortisol by 11beta-hydroxysteroid dehydrogenase type 1 and metabolic disease, *Ann N Y Acad Sci* **1083**:165–184, 2006.
219. Grover GJ, Mellstrom K, Malm J, Therapeutic potential for thyroid hormone receptor-beta selective agonists for treating obesity, hyperlipidemia and diabetes, *Curr Vasc Pharmacol* **5**(2):141–154, 2007.
220. Andersson U, Scarpulla RC, Pgc-1-related coactivator, a novel, serum-inducible coactivator of nuclear respiratory factor 1-dependent transcription in mammalian cells, *Mol Cell Biol* **21**(11):3738–3749, 2001.
221. Kressler D, *et al.*, The PGC-1-related protein PERC is a selective coactivator of estrogen receptor alpha, *J Biol Chem* **277**(16):13918–13925, 2002.
222. Lin J, *et al.*, Peroxisome proliferator-activated receptor gamma coactivator 1beta (PGC-1beta), a novel PGC-1-related transcription coactivator associated with host cell factor, *J Biol Chem* **277**(3):1645–1648, 2002.
223. Puigserver P, *et al.*, A cold-inducible coactivator of nuclear receptors linked to adaptive thermogenesis, *Cell* **92**(6):829–839, 1998.
224. Lin J, Handschin C, Spiegelman BM, Metabolic control through the PGC-1 family of transcription coactivators, *Cell Metab* **1**(6):361–370, 2005.
225. Czubryt MP, *et al.*, Regulation of peroxisome proliferator-activated receptor gamma coactivator 1 alpha (PGC-1 alpha) and mitochondrial function by MEF2 and HDAC5, *Proc Natl Acad Sci USA* **100**(4):1711–1716, 2003.
226. Hondares E, *et al.*, Thiazolidinediones and rexinoids induce peroxisome proliferator-activated receptor-coactivator (PGC)-1alpha gene transcription: An autoregulatory loop controls PGC-1alpha expression in adipocytes via peroxisome proliferator-activated receptor-gamma coactivation, *Endocrinology* **147**(6):2829–2838, 2006.
227. Liu C, *et al.*, Transcriptional coactivator PGC-1alpha integrates the mammalian clock and energy metabolism, *Nature* **447**(7143):477–481, 2007.
228. Knutti D, Kressler D, Kralli A, Regulation of the transcriptional coactivator PGC-1 via MAPK-sensitive interaction with a repressor, *Proc Natl Acad Sci USA* **98**(17):9713–9718, 2001.
229. Rodgers JT, *et al.*, Nutrient control of glucose homeostasis through a complex of PGC-1alpha and SIRT1, *Nature* **434**(7029):113–118, 2005.
230. Teyssier C, *et al.*, Activation of nuclear receptor coactivator PGC-1alpha by arginine methylation, *Genes Dev* **19**(12):1466–1473, 2005.
231. Puigserver P, Spiegelman BM, Peroxisome proliferator-activated receptor-gamma coactivator 1 alpha (PGC-1 alpha): Transcriptional coactivator and metabolic regulator, *Endocr Rev* **24**(1):78–90, 2003.

232. Rhee J, *et al.*, Partnership of PGC-1alpha and HNF4alpha in the regulation of lipoprotein metabolism, *J Biol Chem* **281**(21):14683–14690, 2006.
233. Sandri M, *et al.*, PGC-1alpha protects skeletal muscle from atrophy by suppressing FoxO3 action and atrophy-specific gene transcription, *Proc Natl Acad Sci USA* **103**(44):16260–16265, 2006.
234. Handschin C, *et al.*, PGC-1alpha regulates the neuromuscular junction program and ameliorates Duchenne muscular dystrophy, *Genes Dev* **21**(7):770–783, 2007.
235. St-Pierre J, *et al.*, Suppression of reactive oxygen species and neurodegeneration by the PGC-1 transcriptional coactivators, *Cell* **127**(2):397–408, 2006.
236. Lin J, *et al.*, Defects in adaptive energy metabolism with CNS-linked hyperactivity in PGC-1alpha null mice, *Cell* **119**(1):121–135, 2004.
237. Meirhaeghe A, *et al.*, Characterization of the human, mouse and rat PGC1 beta (peroxisome-proliferator-activated receptor-gamma coactivator 1 beta) gene *in vitro* and *in vivo*, *Biochem J* **373**(Pt 1):155–165, 2003.
238. Lin J, *et al.*, Hyperlipidemic effects of dietary saturated fats mediated through PGC-1beta coactivation of SREBP, *Cell* **120**(2):261–273, 2005.
239. Vianna CR, *et al.*, Hypomorphic mutation of PGC-1beta causes mitochondrial dysfunction and liver insulin resistance, *Cell Metab* **4**(6):453–464, 2006.
240. Sonoda J, *et al.*, PGC-1beta controls mitochondrial metabolism to modulate circadian activity, adaptive thermogenesis, and hepatic steatosis, *Proc Natl Acad Sci USA* **104**(12):5223–5228, 2007.
241. Savagner F, *et al.*, PGC-1-related coactivator and targets are upregulated in thyroid oncocytoma, *Biochem Biophys Res Commun* **310**(3):779–784, 2003.
242. Mootha VK, *et al.*, PGC-1alpha-responsive genes involved in oxidative phosphorylation are coordinately downregulated in human diabetes, *Nat Genet* **34**(3):267–273, 2003.
243. Patti ME, *et al.*, Coordinated reduction of genes of oxidative metabolism in humans with insulin resistance and diabetes: Potential role of PGC1 and NRF1, *Proc Natl Acad Sci USA* **100**(14):8466–8471, 2003.
244. Semple RK, *et al.*, Expression of the thermogenic nuclear hormone receptor coactivator PGC-1alpha is reduced in the adipose tissue of morbidly obese subjects, *Int J Obes Relat Metab Disord* **28**(1):176–179, 2004.
245. Yang X, Enerback S, Smith U, Reduced expression of FOXC2 and brown adipogenic genes in human subjects with insulin resistance, *Obes Res* **11**(10):1182–1191, 2003.
246. Thadani H, Deacon A, Peters T, Diagnosis and management of porphyria, *BMJ* **320**(7250):1647 1651, 2000.
247. Finck BN, Kelly DP, Peroxisome proliferator-activated receptor gamma coactivator-1 (PGC-1) regulatory cascade in cardiac physiology and disease, *Circulation* **115**(19):2540–2548, 2007.

248. Lacquemant C, *et al.*, No association between the G482S polymorphism of the proliferator-activated receptor-gamma coactivator-1 (PGC-1) gene and Type II diabetes in French Caucasians, *Diabetologia* **45**(4):602–603, 2002.
249. Leipold H, *et al.*, Peroxisome proliferator-activated receptor gamma coactivator-1alpha gene variations are not associated with gestational diabetes mellitus, *J Soc Gynecol Investig* **13**(2):104–107, 2006.
250. Stumvoll M, *et al.*, The Gly482Ser variant in the peroxisome proliferator-activated receptor gamma coactivator-1 is not associated with diabetes-related traits in non-diabetic German and Dutch populations, *Exp Clin Endocrinol Diabetes* **112**(5):253–257, 2004.
251. Petrovic MG, *et al.*, Gly482Ser polymorphism of the peroxisome proliferator-activated receptor-gamma coactivator-1 gene might be a risk factor for diabetic retinopathy in Slovene population (Caucasians) with type 2 diabetes and the Pro12Ala polymorphism of the PPARgamma gene is not, *Diabetes Metab Res Rev* **21**(5):470–474, 2005.
252. Andersen G, *et al.*, PGC-1alpha Gly482Ser polymorphism associates with hypertension among Danish whites, *Hypertension* **45**(4):565–570, 2005.
253. Bertolotti M, *et al.*, Decreased hepatic expression of PPAR-gamma coactivator-1 in cholesterol cholelithiasis, *Eur J Clin Invest* **36**(3):170–175, 2006.
254. Schon EA, Manfredi G, Neuronal degeneration and mitochondrial dysfunction, *J Clin Invest* **111**(3):303–312, 2003.
255. Andersen G, *et al.*, Evidence of an association between genetic variation of the coactivator PGC-1beta and obesity, *J Med Genet* **42**(5):402–407, 2005.
256. Wirtenberger M, *et al.*, Associations of genetic variants in the estrogen receptor coactivators PPARGC1A, PPARGC1B and EP300 with familial breast cancer, *Carcinogenesis* **27**(11):2201–2208, 2006.
257. Feige JN, Auwerx J, Transcriptional coregulators in the control of energy homeostasis, *Trends Cell Biol* **17**(6):292–301, 2007.
258. Jeong JW, *et al.*, The genomic analysis of the impact of steroid receptor coactivators ablation on hepatic metabolism, *Mol Endocrinol* **20**(5):1138–1152, 2006.
259. Picard F, *et al.*, SRC-1 and TIF2 control energy balance between white and brown adipose tissues, *Cell* **111**(7):931–941, 2002.
260. Wang Z, *et al.*, Critical roles of the p160 transcriptional coactivators p/CIP and SRC-1 in energy balance, *Cell Metab* **3**(2):111–122, 2006.
261. Yamamoto H, Schoonjans K, Auwerx J, Sirtuin functions in health and disease, *Mol Endocrinol* **21**(8):1745–1755, 2007.
262. Guarente L, Sirtuins as potential targets for metabolic syndrome, *Nature* **444**(7121):868–874, 2006.
263. Bordone L, Guarente L, Calorie restriction, SIRT1 and metabolism: Understanding longevity, *Nat Rev Mol Cell Biol* **6**(4):298–305, 2005.

264. Picard F, *et al.*, Sirt1 promotes fat mobilization in white adipocytes by repressing PPAR-gamma, *Nature* **429**(6993):771–776, 2004.
265. Fu M, *et al.*, Hormonal control of androgen receptor function through SIRT1, *Mol Cell Biol* **26**(21):8122–8135, 2006.
266. Lagouge M, *et al.*, Resveratrol improves mitochondrial function and protects against metabolic disease by activating SIRT1 and PGC-1alpha, *Cell* **127**(6):1109–1122, 2006.
267. Porcu M, Chiarugi A, The emerging therapeutic potential of sirtuin-interacting drugs: From cell death to lifespan extension, *Trends Pharmacol Sci* **26**(2):94–103, 2005.
268. Baur JA, *et al.*, Resveratrol improves health and survival of mice on a high-calorie diet, *Nature* **444**(7117):337–342, 2006.
269. Blaheta RA, *et al.*, Expression level of neural cell adhesion molecule (NCAM) inversely correlates with the ability of neuroblastoma cells to adhere to endothelium *in vitro*, *Cell Commun Adhes* **9**(3):131–147, 2002.
270. Verrotti A, *et al.*, Endocrine and metabolic changes in epileptic patients receiving valproic acid, *J Pediatr Endocrinol Metab* **18**(5):423–430, 2005.
271. Cavailles V, *et al.*, Nuclear factor RIP140 modulates transcriptional activation by the estrogen receptor, *EMBO J* **14**(15):3741–3751, 1995.
272. Leonardsson G, *et al.*, Nuclear receptor corepressor RIP140 regulates fat accumulation, *Proc Natl Acad Sci USA* **101**(22):8437–8442, 2004.
273. Caballero V, *et al.*, Preliminary molecular genetic analysis of the Receptor Interacting Protein 140 (RIP140) in women affected by endometriosis, *J Exp Clin Assist Reprod* **2**:11, 2005.
274. Galan JJ, *et al.*, Multilocus analyses of estrogen-related genes reveal involvement of the ESR1 gene in male infertility and the polygenic nature of the pathology, *Fertil Steril* **84**(4):910–918, 2005.
275. Ishihara H, *et al.*, Association of the polymorphisms in the 5′-untranslated region of PTEN gene with type 2 diabetes in a Japanese population, *FEBS Lett* **554**(3):450–454, 2003.
276. O'Malley BW, Molecular biology. Little molecules with big goals, *Science* **313**(5794):1749–1750, 2006.

References Tables 1.1 & 1.2

1. Schuur ER, *et al.*, Ligand-dependent interaction of estrogen receptor-alpha with members of the forkhead transcription factor family, *J Biol Chem* **276**(36):33554–33560, 2001.
2. Borkhardt A, *et al.*, Cloning and characterization of AFX, the gene that fuses to MLL in acute leukemias with a t(X; 11)(q13; q23), *Oncogene* **14**(2):195–202, 1997.

3. Lin HK, *et al.*, Akt suppresses androgen-induced apoptosis by phosphorylating and inhibiting androgen receptor, *Proc Natl Acad Sci USA* **98**(13):7200–7205, 2001.
4. Masuyama N, *et al.*, Akt inhibits the orphan nuclear receptor Nur77 and T-cell apoptosis, *J Biol Chem* **276**(35):32799–32805, 2001.
5. Edwards J, *et al.*, Gene amplifications associated with the development of hormone-resistant prostate cancer, *Clin Cancer Res* **9**(14):5271–5281, 2003.
6. Woodgett JR, Recent advances in the protein kinase B signaling pathway, *Curr Opin Cell Biol* **17**(2):150–157, 2005.
7. Emamian ES, *et al.*, Convergent evidence for impaired AKT1-GSK3beta signaling in schizophrenia, *Nat Genet* **36**(2):131–137, 2004.
8. Yu Z, *et al.*, Phosphorylation of Akt (Ser473) predicts poor clinical outcome in oropharyngeal squamous cell cancer, *Cancer Epidemiol Biomarkers Prev* **16**(3):553–558, 2007.
9. Hsiao PW, *et al.*, The linkage of Kennedy's neuron disease to ARA24, the first identified androgen receptor polyglutamine region-associated coactivator, *J Biol Chem* **274**(29):20229–20234, 1999.
10. Li P, *et al.*, Heterogeneous expression and functions of androgen receptor co-factors in primary prostate cancer, *Am J Pathol* **161**(4):1467–1474, 2002.
11. Fujimoto N, *et al.*, Cloning and characterization of androgen receptor coactivator, ARA55, in human prostate, *J Biol Chem* **274**(12):8316–8321, 1999.
12. Mestayer C, *et al.*, Expression of androgen receptor coactivators in normal and cancer prostate tissues and cultured cell lines, *Prostate* **56**(3):192–200, 2003.
13. Miyoshi Y, *et al.*, Expression of AR associated protein 55 (ARA55) and androgen receptor in prostate cancer, *Prostate* **56**(4):280–286, 2003.
14. Yeh S, Chang C, Cloning and characterization of a specific coactivator, ARA70, for the androgen receptor in human prostate cells, *Proc Natl Acad Sci USA* **93**(11):5517–5521, 1996.
15. Quezada S, *et al.*, Evaluation of steroid receptors, coregulators, and molecules associated with uterine receptivity in secretory endometria from untreated women with polycystic ovary syndrome, *Fertil Steril* **85**(4):1017–1026, 2006.
16. Hu YC, *et al.*, Functional domain and motif analyses of androgen receptor coregulator ARA70 and its differential expression in prostate cancer, *J Biol Chem* **279**(32):33438–33446, 2004.
17. Beitel LK, *et al.*, Cloning and characterization of an androgen receptor N-terminal-interacting protein with ubiquitin-protein ligase activity, *J Mol Endocrinol* **29**(1):41–60, 2002.
18. Duan W, *et al.*, Expression of Pirh2, a newly identified ubiquitin protein ligase, in lung cancer, *J Natl Cancer Inst* **96**(22):1718–1721, 2004.

19. Logan IR, *et al.*, Human PIRH2 enhances androgen receptor signaling through inhibition of histone deacetylase 1 and is overexpressed in prostate cancer, *Mol Cell Biol* **26**(17):6502–6510, 2006.
20. Markus SM, *et al.*, Identification and characterization of ART-27, a novel coactivator for the androgen receptor N terminus, *Mol Biol Cell* **13**(2):670–682, 2002.
21. Taneja SS, *et al.*, ART-27, an androgen receptor coactivator regulated in prostate development and cancer, *J Biol Chem* **279**(14):13944–13952, 2004.
22. Lee SK, *et al.*, A nuclear factor, ASC-2, as a cancer-amplified transcriptional coactivator essential for ligand-dependent transactivation by nuclear receptors *in vivo*, *J Biol Chem* **274**(48):34283–34293, 1999.
23. Sjoblom T, *et al.*, The consensus coding sequences of human breast and colorectal cancers, *Science* **314**(5797):268–274, 2006.
24. Wang L, *et al.*, Histone deacetylase-associating Atrophin proteins are nuclear receptor corepressors, *Genes Dev* **20**(5):525–530, 2006.
25. Amler LC, *et al.*, Identification and characterization of novel genes located at the t(1;15)(p36.2;q24) translocation breakpoint in the neuroblastoma cell line NGP, *Genomics* **64**(2):195–202, 2000.
26. Qi C, *et al.*, Potentiation of estrogen receptor transcriptional activity by breast cancer amplified sequence 2, *Biochem Biophys Res Commun* **328**(2):393–398, 2005.
27. Maass N, *et al.*, Amplification of the BCAS2 gene at chromosome 1p13.3-21 in human primary breast cancer, *Cancer Lett* **185**(2):219–223, 2002.
28. Gururaj AE, *et al.*, Estrogen Induces Expression of BCAS3, a Novel Estrogen Receptor-{alpha} Coactivator, through Proline-, Glutamic Acid-, and Leucine-Rich Protein-1 (PELP1), *Mol Endocrinol* **21**(8):1847–1860, 2007.
29. Barlund M, *et al.*, Cloning of BCAS3 (17q23) and BCAS4 (20q13) genes that undergo amplification, overexpression, and fusion in breast cancer, *Genes Chromosomes Cancer* **35**(4):311–317, 2002.
30. Monni O, *et al.*, Comprehensive copy number and gene expression profiling of the 17q23 amplicon in human breast cancer, *Proc Natl Acad Sci USA* **98**(10):5711–5716, 2001.
31. Sinclair CS, *et al.*, The 17q23 amplicon and breast cancer, *Breast Cancer Res Treat* **78**(3):313–322, 2003.
32. Na SY, *et al.*, Bcl3, an IkappaB protein, as a novel transcription coactivator of the retinoid X receptor, *J Biol Chem* **273**(47):30933–30938, 1998.
33. McKeithan TW, *et al.*, Cloning of the chromosome translocation breakpoint junction of the t(14;19) in chronic lymphocytic leukemia, *Proc Natl Acad Sci USA* **84**(24):9257–9260, 1987.
34. Stein J, *et al.*, Nonsyndromic cleft lip with or without cleft palate: Evidence of linkage to BCL3 in 17 multigenerational families, *Am J Hum Genet* **57**(2):257–272, 1995.

35. Fan S, *et al.*, BRCA1 inhibition of estrogen receptor signaling in transfected cells, *Science* **284**(5418):1354–1356, 1999.
36. James CR, *et al.*, BRCA1, a potential predictive biomarker in the treatment of breast cancer, *Oncologist* **12**(2):142–150, 2007.
37. Wirtenberger M, *et al.*, Associations of genetic variants in the estrogen receptor coactivators PPARGC1A, PPARGC1B and EP300 with familial breast cancer, *Carcinogenesis* **27**(11):2201–2208, 2006.
38. Sogaard M, Kjaer SK, Gayther S, Ovarian cancer and genetic susceptibility in relation to the BRCA1 and BRCA2 genes. Occurrence, clinical importance and intervention, *Acta Obstet Gynecol Scand* **85**(1):93–105, 2006.
39. Lux MP, Fasching PA, Beckmann MW, Hereditary breast and ovarian cancer: Review and future perspectives, *J Mol Med* **84**(1):16–28, 2006.
40. Smith KL and Robson ME, Update on hereditary breast cancer, *Curr Oncol Rep* **8**(1):14–21, 2006.
41. Walsh T, King MC, Ten genes for inherited breast cancer, *Cancer Cell* **11**(2):103–105, 2007.
42. Rosen EM, *et al.*, BRCA1 in hormone-responsive cancers, *Trends Endocrinol Metab* **14**(8):378–385, 2003.
43. Shin S, Verma IM, BRCA2 cooperates with histone acetyltransferases in androgen receptor-mediated transcription, *Proc Natl Acad Sci USA* **100**(12):7201–7206, 2003.
44. Neuhausen SL, *et al.*, Haplotype and phenotype analysis of nine recurrent BRCA2 mutations in 111 families: Results of an international study, *Am J Hum Genet* **62**(6):1381–1388, 1998.
45. Reid S, *et al.*, Biallelic BRCA2 mutations are associated with multiple malignancies in childhood including familial Wilms tumour, *J Med Genet* **42**(2):147–151, 2005.
46. Edwards SM, *et al.*, Two percent of men with early-onset prostate cancer harbor germline mutations in the BRCA2 gene, *Am J Hum Genet* **72**(1):1–12, 2003.
47. Healey CS, *et al.*, A common variant in BRCA2 is associated with both breast cancer risk and prenatal viability, *Nat Genet* **26**(3):362–364, 2000.
48. Sliwinski T, *et al.*, Polymorphisms of the BRCA2 and RAD51 Genes in Breast Cancer, *Breast Cancer Res Treat* **94**(2):105–109, 2005.
49. Chiba H, *et al.*, Two human homologues of Saccharomyces cerevisiae SWI2/SNF2 and Drosophila brahma are transcriptional coactivators cooperating with the estrogen receptor and the retinoic acid receptor, *Nucleic Acids Res* **22**(10):1815–1820, 1994.
50. Ichinose H, *et al.*, Ligand-dependent interaction between the estrogen receptor and the human homologues of SWI2/SNF2, *Gene* **188**(1):95–100, 1997.

51. Muchardt C, Yaniv M, A human homologue of Saccharomyces cerevisiae SNF2/SWI2 and Drosophila brm genes potentiates transcriptional activation by the glucocorticoid receptor, *EMBO J* **12**(11):4279–4290, 1993.
52. Medina PP, *et al.*, Genetic and epigenetic screening for gene alterations of the chromatin-remodeling factor, SMARCA4/BRG1, in lung tumors, *Genes Chromosomes Cancer* **41**(2):170–177, 2004.
53. Decristofaro MF, *et al.*, Characterization of SWI/SNF protein expression in human breast cancer cell lines and other malignancies, *J Cell Physiol* **186**(1):136–145, 2001.
54. Reisman DN, *et al.*, Loss of BRG1/BRM in human lung cancer cell lines and primary lung cancers: Correlation with poor prognosis, *Cancer Res* **63**(3):560–566, 2003.
55. Sentani K, *et al.*, Increased expression but not genetic alteration of BRG1, a component of the SWI/SNF complex, is associated with the advanced stage of human gastric carcinomas, *Pathobiology* **69**(6):315–320, 2001.
56. Sun A, *et al.*, Aberrant expression of SWI/SNF catalytic subunits BRG1/BRM is associated with tumor development and increased invasiveness in prostate cancers, *Prostate* **67**(2):203–213, 2007.
57. Budhram-Mahadeo V, Parker M, Latchman DS, POU transcription factors Brn-3a and Brn-3b interact with the estrogen receptor and differentially regulate transcriptional activity via an estrogen response element, *Mol Cell Biol* **18**(2):1029–1041, 1998.
58. Budhram-Mahadeo V, *et al.*, The Brn-3b POU family transcription factor represses expression of the BRCA-1 anti-oncogene in breast cancer cells, *Oncogene* **18**(48):6684–6691, 1999.
59. Lee SA, *et al.*, Expression of the Brn-3b transcription factor correlates with expression of HSP-27 in breast cancer biopsies and is required for maximal activation of the HSP-27 promoter, *Cancer Res* **65**(8):3072–3080, 2005.
60. Rubino D, *et al.*, Characterization of Brx, a novel Dbl family member that modulates estrogen receptor action, *Oncogene* **16**(19):2513–2526, 1998.
61. Kino T, Chrousos GP, Tissue-specific glucocorticoid resistance-hypersensitivity syndromes: Multifactorial states of clinical importance, *J Allergy Clin Immunol* **109**(4):609–613, 2002.
62. Hearns-Stokes R, *et al.*, Expression of the proto-oncoprotein breast cancer nuclear receptor auxiliary factor (Brx) is altered in eutopic endometrium of women with endometriosis, *Fertil Steril* **85**(1):63–70, 2006.
63. Bubulya A, *et al.*, c-Jun can mediate androgen receptor-induced transactivation, *J Biol Chem* **271**(40):24583–24589, 1996.
64. Mathas S, *et al.*, Aberrantly expressed c-Jun and JunB are a hallmark of Hodgkin lymphoma cells, stimulate proliferation and synergize with NF-kappa B, *EMBO J* **21**(15):4104–4113, 2002.

65. Dedhar S, *et al.*, Inhibition of nuclear hormone receptor activity by calreticulin, *Nature* **367**(6462):480–483, 1994.
66. Coppolino MG, Dedhar S, Calreticulin, *Int J Biochem Cell Biol* **30**(5):553–558, 1998.
67. Watanabe K, *et al.*, Anti-calreticulin antibodies in patients with inflammatory bowel disease, *Fukushima J Med Sci* **52**(1):1–11, 2006.
68. Sanchez D, *et al.*, Epitopes of calreticulin recognised by IgA autoantibodies from patients with hepatic and coeliac disease, *J Autoimmun* **21**(4):383–392, 2003.
69. Iwaki H, *et al.*, Diagnostic potential in bladder cancer of a panel of tumor markers (calreticulin, gamma-synuclein, and catechol-o-methyltransferase) identified by proteomic analysis, *Cancer Sci* **95**(12):955–961, 2004.
70. Brunagel G, *et al.*, Identification of calreticulin as a nuclear matrix protein associated with human colon cancer, *J Cell Biochem* **89**(2):238–243, 2003.
71. Chen D, *et al.*, Regulation of transcription by a protein methyltransferase, *Science* **284**(5423):2174–2177, 1999.
72. Hong H, *et al.*, Aberrant expression of CARM1, a transcriptional coactivator of androgen receptor, in the development of prostate carcinoma and androgen-independent status, *Cancer* **101**(1):83–89, 2004.
73. Kullo IJ, *et al.*, Novel genomic loci influencing plasma homocysteine levels, *Stroke* **37**(7):1703–1709, 2006.
74. Schlegel A, *et al.*, Caveolin-1 potentiates estrogen receptor alpha (ERalpha) signaling. caveolin-1 drives ligand-independent nuclear translocation and activation of ERalpha, *J Biol Chem* **274**(47):33551–33556, 1999.
75. Haeusler J, *et al.*, Association of a CAV-1 haplotype to familial aggressive prostate cancer, *Prostate* **65**(2):171–177, 2005.
76. Williams TM, Lisanti MP, The Caveolin genes: From cell biology to medicine, *Ann Med* **36**(8):584–595, 2004.
77. Li T, *et al.*, Caveolin-1 mutations in human breast cancer: Functional association with estrogen receptor alpha-positive status, *Am J Pathol* **168**(6): 1998–2013, 2006.
78. Savage K, *et al.*, Caveolin 1 is overexpressed and amplified in a subset of basal-like and metaplastic breast carcinomas: A morphologic, ultrastructural, immunohistochemical, and *in situ* hybridization analysis, *Clin Cancer Res* **13**(1):90–101, 2007.
79. Wang XM, *et al.*, Caveolin-1: A critical regulator of lung fibrosis in idiopathic pulmonary fibrosis, *J Exp Med* **203**(13):2895–2906, 2006.
80. Barresi V, *et al.*, Caveolin-1 in meningiomas: Expression and clinico-pathological correlations, *Acta Neuropathol (Berl)* **112**(5):617–626, 2006.
81. Kamei Y, *et al.*, A CBP integrator complex mediates transcriptional activation and AP-1 inhibition by nuclear receptors, *Cell* **85**(3):403–414, 1996.

82. Petrij F, *et al.*, Rubinstein-Taybi syndrome caused by mutations in the transcriptional coactivator CBP, *Nature* **376**(6538):348–351, 1995.
83. Zimmermann N, *et al.*, Confirmation of EP300 gene mutations as a rare cause of Rubinstein-Taybi syndrome, *Eur J Hum Genet* **15**(8):837–842, 2007.
84. Troke PJ, *et al.*, MOZ fusion proteins in acute myeloid leukaemia, *Biochem Soc Symp* (73):23–39, 2006.
85. Nucifora FC, Jr, *et al.*, Interference by huntingtin and atrophin-1 with cbp-mediated transcription leading to cellular toxicity, *Science* **291**(5512):2423–2428, 2001.
86. McCampbell A, *et al.*, CREB-binding protein sequestration by expanded polyglutamine, *Hum Mol Genet* **9**(14):2197–2202, 2000.
87. Ma ZQ, *et al.*, Cdc25B functions as a novel coactivator for the steroid receptors, *Mol Cell Biol* **21**(23):8056–8067, 2001.
88. Nakabayashi H, Hara M, Shimizu K, Prognostic significance of CDC25B expression in gliomas, *J Clin Pathol* **59**(7):725–728, 2006.
89. Sunada F, *et al.*, p53 negativity, CDC25B positivity, and metallothionein negativity are predictors of a response of esophageal squamous cell carcinoma to chemoradiotherapy, *World J Gastroenterol* **11**(36):5696–5700, 2005.
90. Ito Y, *et al.*, Expression of cdc25A and cdc25B phosphatase in breast carcinoma, *Breast Cancer* **11**(3):295–300, 2004.
91. Kino T, *et al.*, Cyclin-dependent kinase 5 differentially regulates the transcriptional activity of the glucocorticoid receptor through phosphorylation: Clinical implications for the nervous system response to glucocorticoids and stress, *Mol Endocrinol* **21**(7):1552–1568, 2007.
92. Maccioni RB, Munoz JP, Barbeito L, The molecular bases of Alzheimer's disease and other neurodegenerative disorders, *Arch Med Res* **32**(5):367–381, 2001.
93. Nakamura S, *et al.*, p35nck5a and cyclin-dependent kinase 5 colocalize in Lewy bodies of brains with Parkinson's disease, *Acta Neuropathol (Berl)* **94**(2):153–157, 1997.
94. Rochette-Egly C, *et al.*, Stimulation of RAR alpha activation function AF-1 through binding to the general transcription factor TFIIH and phosphorylation by CDK7, *Cell* **90**(1):97–107, 1997.
95. Zhu X, *et al.*, Neuronal CDK7 in hippocampus is related to aging and Alzheimer's disease, *Neurobiol Aging* **21**(6):807–813, 2000.
96. Ruegg J, *et al.*, Cofilin 1 is revealed as an inhibitor of glucocorticoid receptor by analysis of hormone-resistant cells, *Mol Cell Biol* **24**(21):9371–9382, 2004.
97. Wang W, *et al.*, The activity status of cofilin is directly related to invasion, intravasation, and metastasis of mammary tumors, *J Cell Biol* **173**(3):395–404, 2006.

98. Yahata T, *et al.*, Selective coactivation of estrogen-dependent transcription by CITED1 CBP/p300-binding protein, *Genes Dev* **15**(19):2598–2612, 2001.
99. Huang Y, *et al.*, Gene expression in papillary thyroid carcinoma reveals highly consistent profiles, *Proc Natl Acad Sci USA* **98**(26):15044–15049, 2001.
100. Iwasaki T, Chin WW, Ko L, Identification and characterization of RRM-containing coactivator activator (CoAA) as TRBP-interacting protein, and its splice variant as a coactivator modulator (CoAM), *J Biol Chem* **276**(36):33375–33383, 2001.
101. Sui Y, *et al.*, Gene amplification and associated loss of 5′ regulatory sequences of CoAA in human cancers, *Oncogene* **26**(6):822–835, 2007.
102. Aiyar SE, *et al.*, Attenuation of estrogen receptor alpha-mediated transcription through estrogen-stimulated recruitment of a negative elongation factor, *Genes Dev* **18**(17):2134–2146, 2004.
103. McChesney PA, *et al.*, Cofactor of BRCA1: A novel transcription factor regulator in upper gastrointestinal adenocarcinomas, *Cancer Res* **66**(3): 1346–1353, 2006.
104. Yi YW, *et al.*, Gadd45 family proteins are coactivators of nuclear hormone receptors, *Biochem Biophys Res Commun* **272**(1):193–198, 2000.
105. Ying J, *et al.*, The stress-responsive gene GADD45G is a functional tumor suppressor, with its response to environmental stresses frequently disrupted epigenetically in multiple tumors, *Clin Cancer Res* **11**(18): 6442–6449, 2005.
106. Sun L, *et al.*, GADD45gamma, down-regulated in 65% hepatocellular carcinoma (HCC) from 23 chinese patients, inhibits cell growth and induces cell cycle G2/M arrest for hepatoma Hep-G2 cell lines, *Mol Biol Rep* **30**(4):249–253, 2003.
107. Avram D, *et al.*, Isolation of a novel family of C(2)H(2) zinc finger proteins implicated in transcriptional repression mediated by chicken ovalbumin upstream promoter transcription factor (COUP-TF) orphan nuclear receptors, *J Biol Chem* **275**(14):10315–10322, 2000.
108. Senawong T, *et al.*, Involvement of the histone deacetylase SIRT1 in chicken ovalbumin upstream promoter transcription factor (COUP-TF)-interacting protein 2-mediated transcriptional repression, *J Biol Chem* **278**(44):43041–43050, 2003.
109. Bezrookove V, *et al.*, A novel t(6; 14)(q25-q27; q32) in acute myelocytic leukemia involves the BCL11B gene, *Cancer Genet Cytogenet* **149**(1): 72–76, 2004.
110. Narayanan R, *et al.*, Cyclin-dependent kinase activity is required for progesterone receptor function: Novel role for cyclin A/Cdk2 as a progesterone receptor coactivator, *Mol Cell Biol* **25**(1):264–277, 2005.

111. Mikami T, *et al.*, Bottom-up cell proliferation with cyclin A and p27Kip1 expression in ulcerative colitis-associated dysplasia, *Pathol Int* **56**(1):10–16, 2006.
112. Mrena J, *et al.*, Prognostic significance of cyclin A in gastric cancer, *Int J Cancer* **119**(8):1897–1901, 2006.
113. Yasmeen A, *et al.*, E- and A-type cyclins as markers for cancer diagnosis and prognosis, *Expert Rev Mol Diagn* **3**(5):617–633, 2003.
114. Aaltonen K, *et al.*, Reliability of cyclin A assessment on tissue microarrays in breast cancer compared to conventional histological slides, *Br J Cancer* **94**(11):1697–1702, 2006.
115. Husdal A, Bukholm G, Bukholm IR, The prognostic value and overexpression of cyclin A is correlated with gene amplification of both cyclin A and cyclin E in breast cancer patient, *Cell Oncol* **28**(3):107–116, 2006.
116. Zwijsen RM, *et al.*, CDK-independent activation of estrogen receptor by cyclin D1, *Cell* **88**(3):405–415, 1997.
117. Fu M, *et al.*, Minireview: Cyclin D1: Normal and abnormal functions, *Endocrinology* **145**(12):5439–5447, 2004.
118. Knudsen KE, *et al.*, Cyclin D1: Polymorphism, aberrant splicing and cancer risk, *Oncogene* **25**(11):1620–1628, 2006.
119. Caldon CE, *et al.*, Cell cycle control in breast cancer cells, *J Cell Biochem* **97**(2):261–274, 2006.
120. Shu XO, *et al.*, Association of cyclin D1 genotype with breast cancer risk and survival, *Cancer Epidemiol Biomarkers Prev* **14**(1):91–97, 2005.
121. Simpson DJ, *et al.*, Cyclin D1 (CCND1) genotype is associated with tumour grade in sporadic pituitary adenomas, *Carcinogenesis* **22**(11):1801–1807, 2001.
122. Wong YK, *et al.*, Cyclin D1 genotype in areca-associated oral squamous cell carcinoma, *J Oral Pathol Med* **32**(5):265–270, 2003.
123. Tsujimoto Y, *et al.*, Molecular cloning of the chromosomal breakpoint of B-cell lymphomas and leukemias with the t(11;14) chromosome translocation, *Science* **224**(4656):1403–1406, 1984.
124. Le Marchand L, *et al.*, Association of the cyclin D1 A870G polymorphism with advanced colorectal cancer, *JAMA* **290**(21):2843–2848, 2003.
125. Despouy G, *et al.*, Cyclin D3 is a cofactor of retinoic acid receptors, modulating their activity in the presence of cellular retinoic acid-binding protein II, *J Biol Chem* **278**(8):6355–6362, 2003.
126. Lopez-Beltran A, *et al.*, Cyclin D3 expression in primary Ta/T1 bladder cancer, *J Pathol* **209**(1):106–113, 2006.
127. Tanami H, *et al.*, Involvement of cyclin D3 in liver metastasis of colorectal cancer, revealed by genome-wide copy-number analysis, *Lab Invest* **85**(9):1118–1129, 2005.
128. Bergsagel PL, *et al.*, Cyclin D dysregulation: An early and unifying pathogenic event in multiple myeloma, *Blood* **106**(1):296–303, 2005.

129. Pruneri G, *et al.*, Cyclin D3 immunoreactivity is an independent predictor of survival in laryngeal squamous cell carcinoma, *Clin Cancer Res* **11**(1):242–248, 2005.
130. Kasugai Y, *et al.*, Identification of CCND3 and BYSL as candidate targets for the 6p21 amplification in diffuse large B-cell lymphoma, *Clin Cancer Res* **11**(23):8265–8272, 2005.
131. Yamamoto A, *et al.*, Cyclin E as a coactivator of the androgen receptor, *J Cell Biol* **150**(4):873–880, 2000.
132. Sieuwerts AM, *et al.*, Which cyclin E prevails as prognostic marker for breast cancer? Results from a retrospective study involving 635 lymph node-negative breast cancer patients, *Clin Cancer Res* **12**(11 Pt 1): 3319–3328, 2006.
133. Shariat SF, *et al.*, Correlation of cyclin D1 and E1 expression with bladder cancer presence, invasion, progression, and metastasis, *Hum Pathol* **37**(12):1568–1576, 2006.
134. Hulkko SM, Wakui H, Zilliacus J, The pro-apoptotic protein death-associated protein 3 (DAP3) interacts with the glucocorticoid receptor and affects the receptor function, *Biochem J* **349** (**Pt 3**):885–893, 2000.
135. Sasaki H, *et al.*, Arg and DAP3 expression was correlated with human thymoma stage, *Clin Exp Metastasis* **21**(6):507–513, 2004.
136. Hirota T, *et al.*, Association between genetic variation in the gene for death-associated protein-3 (DAP3) and adult asthma, *J Hum Genet* **49**(7):370–375, 2004.
137. Zhang H, *et al.*, DAX-1 functions as an LXXLL-containing corepressor for activated estrogen receptors, *J Biol Chem* **275**(51):39855–39859, 2000.
138. Balsamo A, *et al.*, A new DAX1 gene mutation associated with congenital adrenal hypoplasia and hypogonadotropic hypogonadism, *Am J Med Genet A* **135**(3):292–296, 2005.
139. Domenice S, *et al.*, Adrenocorticotropin-dependent precocious puberty of testicular origin in a boy with X-linked adrenal hypoplasia congenita due to a novel mutation in the DAX1 gene, *J Clin Endocrinol Metab* **86**(9):4068–4071, 2001.
140. Mantovani G, *et al.*, DAX1 and X-linked adrenal hypoplasia congenita: Clinical and molecular analysis in five patients, *Eur J Endocrinol* **154**(5):685–689, 2006.
141. Reutens AT, *et al.*, Clinical and functional effects of mutations in the DAX-1 gene in patients with adrenal hypoplasia congenital, *J Clin Endocrinol Metab* **84**(2):504–511, 1999.
142. Schwartz M, Blichfeldt S, Muller J, X-linked adrenal hypoplasia in a large Greenlandic family. Detection of a missense mutation (N4401) in the DAX-1 gene; implication for genetic counselling and carrier diagnosis, *Hum Genet* **99**(1):83–87, 1997.

143. Sekiguchi Y, *et al.*, Sibling cases of Addison's disease caused by DAX-1 gene mutations, *Intern Med* **46**(1):35–39, 2007.
144. Lin L, *et al.*, Analysis of DAX1 (NR0B1) and steroidogenic factor-1 (NR5A1) in children and adults with primary adrenal failure: Ten years' experience, *J Clin Endocrinol Metab* **91**(8):3048–3054, 2006.
145. Bardoni B, *et al.*, A dosage sensitive locus at chromosome Xp21 is involved in male to female sex reversal, *Nat Genet* **7**(4):497–501, 1994.
146. Fechner PY, *et al.*, Report of a kindred with X-linked (or autosomal dominant sex-limited) 46,XY partial gonadal dysgenesis, *J Clin Endocrinol Metab* **76**(5):1248–1253, 1993.
147. German J, *et al.*, Genetically determined sex-reversal in 46,XY humans, *Science* **202**(4363):53–56, 1978.
148. Kinsey M, Smith R, Lessnick SL, NR0B1 is required for the oncogenic phenotype mediated by EWS/FLI in Ewing's sarcoma, *Mol Cancer Res* **4**(11):851–859, 2006.
149. Mizuta H, Kuroda Y, Cloning and functional characterization of a rat Daxx that functions as a corepressor for the androgen receptor, *Cell Biol Int* **28**(8–9):609–614, 2004.
150. Waghray A, *et al.*, Identification of differentially expressed genes by serial analysis of gene expression in human prostate cancer, *Cancer Res* **61**(10):4283–4286, 2001.
151. Takahashi K, *et al.*, DJ-1 positively regulates the androgen receptor by impairing the binding of PIASx alpha to the receptor, *J Biol Chem* **276**(40):37556–37563, 2001.
152. Dekker MC, *et al.*, A clinical-genetic study of Parkinson's disease in a genetically isolated community, *J Neurol* **250**(9):1056–1062, 2003.
153. Hattori N, Mizuno Y, Pathogenetic mechanisms of parkin in Parkinson's disease, *Lancet* **364**(9435):722–724, 2004.
154. Hedrich K, *et al.*, DJ-1 (PARK7) mutations are less frequent than Parkin (PARK2) mutations in early-onset Parkinson's disease, *Neurology* **62**(3):389–394, 2004.
155. Klein C, *et al.*, PINK1, Parkin, and DJ-1 mutations in Italian patients with early-onset parkinsonism, *Eur J Hum Genet* **13**(9):1086–1093, 2005.
156. Lev N, *et al.*, Role of DJ-1 in Parkinson's disease, *J Mol Neurosci* **29**(3):215–225, 2006.
157. Lockhart PJ, *et al.*, Lack of mutations in DJ-1 in a cohort of Taiwanese ethnic Chinese with early-onset parkinsonism, *Mov Disord* **19**(9):1065–1069, 2004.
158. Macedo MG, *et al.*, The DJ-1L166P mutant protein associated with early onset Parkinson's disease is unstable and forms higher-order protein complexes, *Hum Mol Genet* **12**(21):2807–2816, 2003.
159. Maraganore DM, *et al.*, A limited role for DJ1 in Parkinson's disease susceptibility, *Neurology* **63**(3):550–553, 2004.

160. Morris HR, Genetics of Parkinson's disease, *Ann Med* **37**(2):86–96, 2005.
161. Vasquez AA, *et al.*, A deletion in DJ-1 and the risk of dementia — a population-based survey, *Neurosci Lett* **372**(3):196–199, 2004.
162. Nawaz Z, *et al.*, The Angelman syndrome-associated protein, E6-AP, is a coactivator for the nuclear hormone receptor superfamily, *Mol Cell Biol* **19**(2):1182–1189, 1999.
163. Jiang Y, *et al.*, Imprinting in Angelman and Prader-Willi syndromes, *Curr Opin Genet Dev* **8**(3):334–342, 1998.
164. Rapakko K, Kokkonen H, Leisti J, UBE3A gene mutations in Finnish Angelman syndrome patients detected by conformation sensitive gel electrophoresis, *Am J Med Genet A* **126**(3):248–252, 2004.
165. Gao X, *et al.*, Decreased expression of e6-associated protein in breast and prostate carcinomas, *Endocrinology* **146**(4):1707–1712, 2005.
166. Nakajima A, *et al.*, Ligand-dependent transcription of estrogen receptor alpha is mediated by the ubiquitin ligase EFP, *Biochem Biophys Res Commun* **357**(1):245–251, 2007.
167. Suzuki T, *et al.*, Estrogen-responsive finger protein as a new potential biomarker for breast cancer, *Clin Cancer Res* **11**(17):6148–6154, 2005.
168. Pascual-Le Tallec L, *et al.*, The elongation factor ELL (eleven-nineteen lysine-rich leukemia) is a selective coregulator for steroid receptor functions, *Mol Endocrinol* **19**(5):1158–1169, 2005.
169. Conaway JW, Conaway RC, Transcription elongation and human disease, *Annu Rev Biochem* **68**:301–319, 1999.
170. Rabbitts TH, Chromosomal translocations in human cancer, *Nature* **372**(6502):143–149, 1994.
171. Meng Q, *et al.*, ERR-10: A new repressor in transcriptional signaling activation of estrogen receptor-alpha, *FEBS Lett* **576**(1–2):190–200, 2004.
172. Xie Y, *et al.*, Molecular cloning and characterization of NAG-7: A novel gene downregulated in human nasopharyngeal carcinoma, *Chin Med J (Engl)* **114**(5):530–534, 2001.
173. Obradovic D, *et al.*, DAXX, FLASH, and FAF-1 modulate mineralocorticoid and glucocorticoid receptor-mediated transcription in hippocampal cells — toward a basis for the opposite actions elicited by two nuclear receptors? *Mol Pharmacol* **65**(3):761–769, 2004.
174. Bjorling-Poulsen M, *et al.*, The pro-apoptotic FAS-associated factor 1 is specifically reduced in human gastric carcinomas, *Int J Oncol* **23**(4): 1015–1023, 2003.
175. Schultz-Norton JR, *et al.*, The deoxyribonucleic acid repair protein flap endonuclease-1 modulates estrogen-responsive gene expression, *Mol Endocrinol* **21**(7):1569–1580, 2007.

176. Otto CJ, *et al.*, The "flap" endonuclease gene FEN1 is excluded as a candidate gene implicated in the CAG repeat expansion underlying Huntington disease, *Clin Genet* **59**(2):122–127, 2001.
177. Sharma S, Sommers JA, Brosh RM, Jr, *In vivo* function of the conserved non-catalytic domain of Werner syndrome helicase in DNA replication, *Hum Mol Genet* **13**(19):2247–2261, 2004.
178. Muller JM, *et al.*, FHL2, a novel tissue-specific coactivator of the androgen receptor, *EMBO J* **19**(3):359–369, 2000.
179. Kahl P, *et al.*, Androgen receptor coactivators lysine-specific histone demethylase 1 and four and a half LIM domain protein 2 predict risk of prostate cancer recurrence, *Cancer Res* **66**(23):11341–11347, 2006.
180. Gabriel B, *et al.*, Expression of the transcriptional coregulator FHL2 in human breast cancer: A clinicopathologic study, *J Soc Gynecol Investig* **13**(1):69–75, 2006.
181. Davicioni E, *et al.*, Identification of a PAX-FKHR gene expression signature that defines molecular classes and determines the prognosis of alveolar rhabdomyosarcomas, *Cancer Res* **66**(14):6936–6946, 2006.
182. Cheong JW, *et al.*, Constitutive phosphorylation of FKHR transcription factor as a prognostic variable in acute myeloid leukemia, *Leuk Res* **27**(12):1159–1162, 2003.
183. Lee YH, Campbell HD, Stallcup MR, Developmentally essential protein flightless I is a nuclear receptor coactivator with actin binding activity, *Mol Cell Biol* **24**(5):2103–2117, 2004.
184. Kubo M, *et al.*, Persistent down-regulation of Fli1, a suppressor of collagen transcription, in fibrotic scleroderma skin, *Am J Pathol* **163**(2):571–581, 2003.
185. Wang Y, Fan PS, Kahaleh B, Association between enhanced type I collagen expression and epigenetic repression of the FLI1 gene in scleroderma fibroblasts, *Arthritis Rheum* **54**(7):2271–2279, 2006.
186. Honsei N, *et al.*, Participation of nuclear localization signal 2 in the 3′-ETS domain of FLI1 in nuclear translocation of various chimeric EWS-FLI1 oncoproteins in Ewing tumor, *Int J Oncol* **29**(3):689–693, 2006.
187. Park JP, *et al.*, Smith-Magenis syndrome resulting from a *de novo* direct insertion of proximal 17q into 17p11.2, *Am J Med Genet* **77**(1):23–27, 1998.
188. Loy CJ, Sim KS, Yong EL, Filamin-A fragment localizes to the nucleus to regulate androgen receptor and coactivator functions, *Proc Natl Acad Sci USA* **100**(8):4562–4567, 2003.
189. Robertson SP, *et al.*, Frontometaphyseal dysplasia: Mutations in FLNA and phenotypic diversity, *Am J Med Genet A* **140**(16):1726–1736, 2006.
190. Sheen VL, Walsh CA, Periventricular heterotopia: New insights into Ehlers-Danlos syndrome, *Clin Med Res* **3**(4):229–233, 2005.

191. Kyndt F, *et al.*, Mutations in the gene encoding filamin A as a cause for familial cardiac valvular dystrophy, *Circulation* **115**(1):40–49, 2007.
192. Gomez-Garre P, *et al.*, Ehlers-Danlos syndrome and periventricular nodular heterotopia in a Spanish family with a single FLNA mutation, *J Med Genet* **43**(3):232–237, 2006.
193. Kondoh T, *et al.*, A Japanese case of oto-palato-digital syndrome type II: An apparent lack of phenotype-genotype correlation, *J Hum Genet* **52**(4):370–373, 2007.
194. Obendorf M, *et al.*, FoxG1, a member of the forkhead family, is a corepressor of the androgen receptor, *J Steroid Biochem Mol Biol* **104**(3–5):195–207, 2007.
195. Adesina AM, *et al.*, FOXG1 dysregulation is a frequent event in medulloblastoma, *J Neurooncol*, 2007.
196. Huggins GS, *et al.*, Friend of GATA 2 physically interacts with chicken ovalbumin upstream promoter-TF2 (COUP-TF2) and COUP-TF3 and represses COUP-TF2-dependent activation of the atrial natriuretic factor promoter, *J Biol Chem* **276**(30):28029–28036, 2001.
197. Pizzuti A, *et al.*, Mutations of ZFPM2/FOG2 gene in sporadic cases of tetralogy of Fallot, *Hum Mutat* **22**(5):372–377, 2003.
198. Ackerman KG, *et al.*, Fog2 is required for normal diaphragm and lung development in mice and humans, *PLoS Genet* **1**(1):58–65, 2005.
199. Chan EM, *et al.*, AML1-FOG2 fusion protein in myelodysplasia, *Blood* **105**(11):4523–4526, 2005.
200. Lee DY, *et al.*, Histone H3 lysine 9 methyltransferase G9a is a transcriptional coactivator for nuclear receptors, *J Biol Chem* **281**(13):8476–8485, 2006.
201. Xin Z, *et al.*, Role of histone methyltransferase G9a in CpG methylation of the Prader-Willi syndrome imprinting center, *J Biol Chem* **278**(17):14996–15000, 2003.
202. Dong M, *et al.*, Expression of Gadd45a and p53 proteins in human pancreatic cancer: Potential effects on clinical outcomes, *J Surg Oncol* **95**(4):332–336, 2007.
203. Schneider G, *et al.*, GADD45alpha is highly expressed in pancreatic ductal adenocarcinoma cells and required for tumor cell viability, *Int J Cancer* **118**(10):2405–2411, 2006.
204. Yamasawa K, *et al.*, Clinicopathological significance of abnormalities in Gadd45 expression and its relationship to p53 in human pancreatic cancer, *Clin Cancer Res* **8**(8):2563–2569, 2002.
205. Wang W, *et al.*, Analysis of methylation-sensitive transcriptome identifies GADD45a as a frequently methylated gene in breast cancer, *Oncogene* **24**(16):2705–2714, 2005.
206. Higashi H, *et al.*, Down-regulation of Gadd45 expression is associated with tumor differentiation in non-small cell lung cancer, *Anticancer Res* **26**(3A):2143–2147, 2006.

207. Eggert M, *et al.*, The glucocorticoid receptor is associated with the RNA-binding nuclear matrix protein hnRNP U, *J Biol Chem* **272**(45): 28471–28478, 1997.
208. Galande S, Chromatin (dis)organization and cancer: BUR-binding proteins as biomarkers for cancer, *Curr Cancer Drug Targets* **2**(2):157–190, 2002.
209. Nishimura K, *et al.*, Modulation of androgen receptor transactivation by gelsolin: A newly identified androgen receptor coregulator, *Cancer Res* **63**(16):4888–4894, 2003.
210. Ardalan MR, Shoja MM, Kiuru-Enari S, Amyloidosis-related nephrotic syndrome due to a G654A gelsolin mutation: The first report from the Middle East, *Nephrol Dial Transplant* **22**(1):272–275, 2007.
211. Chastan N, *et al.*, Cardiac conduction alterations in a French family with amyloidosis of the Finnish type with the p.Asp187Tyr mutation in the GSN gene, *Muscle Nerve* **33**(1):113–119, 2006.
212. Sipila K, Aula P, Database for the mutations of the Finnish disease heritage, *Hum Mutat* **19**(1):16–22, 2002.
213. Kiuru-Enari S, Keski-Oja J, Haltia M, Cutis laxa in hereditary gelsolin amyloidosis, *Br J Dermatol* **152**(2):250–257, 2005.
214. Yamada M, Cerebral amyloid angiopathy and gene polymorphisms, *J Neurol Sci* **226**(1–2):41–44, 2004.
215. Nagy L, *et al.*, Nuclear receptor repression mediated by a complex containing SMRT, mSin3A, and histone deacetylase, *Cell* **89**(3):373–380, 1997.
216. Choi JH, *et al.*, Expression profile of histone deacetylase 1 in gastric cancer tissues, *Jpn J Cancer Res* **92**(12):1300–1304, 2001.
217. Giannini R, Cavallini A, Expression analysis of a subset of coregulators and three nuclear receptors in human colorectal carcinoma, *Anticancer Res* **25**(6B):4287–4292, 2005.
218. Bilodeau S, *et al.*, Role of Brg1 and HDAC2 in GR trans-repression of the pituitary POMC gene and misexpression in Cushing disease, *Genes Dev* **20**(20):2871–2886, 2006.
219. Ropero S, *et al.*, A truncating mutation of HDAC2 in human cancers confers resistance to histone deacetylase inhibition, *Nat Genet* **38**(5):566–569, 2006.
220. Huang BH, *et al.*, Inhibition of histone deacetylase 2 increases apoptosis and p21Cip1/WAF1 expression, independent of histone deacetylase 1, *Cell Death Differ* **12**(4):395–404, 2005.
221. Fajas L, *et al.*, The retinoblastoma-histone deacetylase 3 complex inhibits PPARgamma and adipocyte differentiation, *Dev Cell* **3**(6):903–910, 2002.
222. Guenther MG, Barak O, Lazar MA, The SMRT and N-CoR corepressors are activating cofactors for histone deacetylase 3, *Mol Cell Biol* **21**(18):6091–6101, 2001.

223. Liby P, *et al.*, Elevated and deregulated expression of HDAC3 in human astrocytic glial tumours, *Folia Biol (Praha)* **52**(1–2):21–33, 2006.
224. Leong H, *et al.*, Recruitment of histone deacetylase 4 to the N-terminal region of estrogen receptor alpha, *Mol Endocrinol* **19**(12):2930–2942, 2005.
225. Park JW, *et al.*, High throughput SNP and expression analyses of candidate genes for non-syndromic oral clefts, *J Med Genet* **43**(7):598–608, 2006.
226. Belandia B, *et al.*, Hey1, a mediator of notch signaling, is an androgen receptor corepressor, *Mol Cell Biol* **25**(4):1425–1436, 2005.
227. Boonyaratanakornkit V, *et al.*, High-mobility group chromatin proteins 1 and 2 functionally interact with steroid hormone receptors to enhance their DNA binding *in vitro* and transcriptional activity in mammalian cells, *Mol Cell Biol* **18**(8):4471–4487, 1998.
228. Ek M, *et al.*, Increased extracellular levels of the novel proinflammatory cytokine high mobility group box chromosomal protein 1 in minor salivary glands of patients with Sjogren's syndrome, *Arthritis Rheum* **54**(7): 2289–2294, 2006.
229. Uesugi H, *et al.*, Prevalence and characterization of novel pANCA, antibodies to the high mobility group non-histone chromosomal proteins HMG1 and HMG2, in systemic rheumatic diseases, *J Rheumatol* **25**(4): 703–709, 1998.
230. Ayer LM, *et al.*, Antibodies to high mobility group proteins in systemic sclerosis, *J Rheumatol* **21**(11):2071–2075, 1994.
231. Thompson CC, Bottcher MC, The product of a thyroid hormone-responsive gene interacts with thyroid hormone receptors, *Proc Natl Acad Sci USA* **94**(16):8527–8532, 1997.
232. Kruse R, *et al.*, Novel hairless mutations in two kindreds with autosomal recessive papular atrichia, *J Invest Dermatol* **113**(6):954–959, 1999.
233. Hillmer AM, *et al.*, The hairless gene in androgenetic alopecia: Results of a systematic mutation screening and a family-based association approach, *Br J Dermatol* **146**(4):601–608, 2002.
234. Ahmad W, *et al.*, Alopecia universalis associated with a mutation in the human hairless gene, *Science* **279**(5351):720–724, 1998.
235. Chauchereau A, *et al.*, JAB1 interacts with both the progesterone receptor and SRC-1, *J Biol Chem* **275**(12):8540–8548, 2000.
236. Harada K, *et al.*, High expression of Jun activation domain-binding protein 1 (Jab1) is a strong prognostic marker in oral squamous cell carcinoma patients treated by UFT in combination with radiation, *Anticancer Res* **26**(2B):1615–1619, 2006.
237. Patil MA, *et al.*, Array-based comparative genomic hybridization reveals recurrent chromosomal aberrations and Jab1 as a potential target for 8q gain in hepatocellular carcinoma, *Carcinogenesis* **26**(12):2050–2057, 2005.

238. Ivan D, *et al.*, Expression of cell cycle inhibitor p27Kip1 and its inactivator Jab1 in melanocytic lesions, *Mod Pathol* **17**(7):811–818, 2004.
239. Mayeur GL, *et al.*, Ku is a novel transcriptional recycling coactivator of the androgen receptor in prostate cancer cells, *J Biol Chem* **280**(11):10827–10833, 2005.
240. Fu YP, *et al.*, Breast cancer risk associated with genotypic polymorphism of the nonhomologous end-joining genes: A multigenic study on cancer susceptibility, *Cancer Res* **63**(10):2440–2446, 2003.
241. Mimori T, *et al.*, Isolation and characterization of cDNA encoding the 80-kDa subunit protein of the human autoantigen Ku (p70/p80) recognized by autoantibodies from patients with scleroderma-polymyositis overlap syndrome, *Proc Natl Acad Sci USA* **87**(5):1777–1781, 1990.
242. Comai L, Li B, The Werner syndrome protein at the crossroads of DNA repair and apoptosis, *Mech Ageing Dev* **125**(8):521–528, 2004.
243. Korabiowska M, *et al.*, Decreased expression of Ku70/Ku80 proteins in malignant melanomas of the oral cavity, *Anticancer Res* **22**(1A):193–196, 2002.
244. De Ruyck K, *et al.*, Microsatellite polymorphisms in DNA repair genes XRCC1, XRCC3 and XRCC5 in patients with gynecological tumors: Association with late clinical radiosensitivity and cancer incidence, *Radiat Res* **164**(3):237–244, 2005.
245. Price EA, *et al.*, Rare microsatellite polymorphisms in the DNA repair genes XRCC1, XRCC3 and XRCC5 associated with cancer in patients of varying radiosensitivity, *Somat Cell Mol Genet* **23**(4):237–247, 1997.
246. Powzaniuk M, *et al.*, The LATS2/KPM tumor suppressor is a negative regulator of the androgen receptor, *Mol Endocrinol* **18**(8):2011–2023, 2004.
247. Eoli M, *et al.*, Molecular markers of gliomas: A clinical approach, *Neurol Res* **28**(5):538–541, 2006.
248. Jimenez-Velasco A, *et al.*, Downregulation of the large tumor suppressor 2 (LATS2/KPM) gene is associated with poor prognosis in acute lymphoblastic leukemia, *Leukemia* **19**(12):2347–2350, 2005.
249. Takahashi Y, *et al.*, Down-regulation of LATS1 and LATS2 mRNA expression by promoter hypermethylation and its association with biologically aggressive phenotype in human breast cancers, *Clin Cancer Res* **11**(4):1380–1385, 2005.
250. Metzger E, *et al.*, LSD1 demethylates repressive histone marks to promote androgen-receptor-dependent transcription, *Nature* **437**(7057):436–439, 2005.
251. Mo R, Rao SM, Zhu YJ, Identification of the MLL2 complex as a coactivator for estrogen receptor alpha, *J Biol Chem* **281**(23):15714–15720, 2006.
252. Hess JL, Mechanisms of transformation by MLL, *Crit Rev Eukaryot Gene Expr* **14**(4):235–254, 2004.

253. Huntsman DG, *et al.*, MLL2, the second human homolog of the Drosophila trithorax gene, maps to 19q13.1 and is amplified in solid tumor cell lines, *Oncogene* **18**(56):7975–7984, 1999.
254. Sutton AL, *et al.*, The 1,25(OH)2D3-regulated transcription factor MN1 stimulates vitamin D receptor-mediated transcription and inhibits osteoblastic cell proliferation, *Mol Endocrinol* **19**(9):2234–2244, 2005.
255. Buijs A, *et al.*, The MN1-TEL fusion protein, encoded by the translocation (12;22)(p13;q11) in myeloid leukemia, is a transcription factor with transforming activity, *Mol Cell Biol* **20**(24):9281–9293, 2000.
256. Heuser M, *et al.*, High meningioma 1 (MN1) expression as a predictor for poor outcome in acute myeloid leukemia with normal cytogenetics, *Blood* **108**(12):3898–3905, 2006.
257. Ragel BT, Jensen RL, Molecular genetics of meningiomas, *Neurosurg Focus* **19**(5):E9, 2005.
258. Likhite VS, *et al.*, Interaction of estrogen receptor alpha with 3-methyladenine DNA glycosylase modulates transcription and DNA repair, *J Biol Chem* **279**(16):16875–16882, 2004.
259. Kim NK, *et al.*, Expression of the DNA repair enzyme, N-methylpurine-DNA glycosylase (MPG) in astrocytic tumors, *Anticancer Res* **23**(2B):1417–1423, 2003.
260. Zhao X, *et al.*, Regulation of nuclear receptor activity by a pseudouridine synthase through posttranscriptional modification of steroid receptor RNA activator, *Mol Cell* **15**(4):549–558, 2004.
261. Fernandez-Vizarra E, *et al.*, Nonsense mutation in pseudouridylate synthase 1 (PUS1) in two brothers affected by myopathy, lactic acidosis and sideroblastic anaemia (MLASA), *J Med Genet* **44**(3):173–180, 2007.
262. Zeharia A, *et al.*, Mitochondrial myopathy, sideroblastic anemia, and lactic acidosis: An autosomal recessive syndrome in Persian Jews caused by a mutation in the PUS1 gene, *J Child Neurol* **20**(5):449–452, 2005.
263. Patton JR, *et al.*, Mitochondrial myopathy and sideroblastic anemia (MLASA): Missense mutation in the pseudouridine synthase 1 (PUS1) gene is associated with the loss of tRNA pseudouridylation, *J Biol Chem* **280**(20):19823–19828, 2005.
264. Bykhovskaya Y, *et al.*, Missense mutation in pseudouridine synthase 1 (PUS1) causes mitochondrial myopathy and sideroblastic anemia (MLASA), *Am J Hum Genet* **74**(6):1303–1308, 2004.
265. Mazumdar A, *et al.*, Transcriptional repression of oestrogen receptor by metastasis-associated protein 1 corepressor, *Nat Cell Biol* **3**(1):30–37, 2001.
266. Kumar R, Wang RA, Bagheri-Yarmand R, Emerging roles of MTA family members in human cancers, *Semin Oncol* **30**(5 Suppl 16):30–37, 2003.
267. Sasaki H, *et al.*, Expression of the MTA1 mRNA in advanced lung cancer, *Lung Cancer* **35**(2):149–154, 2002.

268. Hamatsu T, *et al.*, The role of MTA1 gene expression in human hepatocellular carcinoma, *Oncol Rep* **10**(3):599–604, 2003.
269. Cui Y, *et al.*, Metastasis-associated protein 2 is a repressor of estrogen receptor alpha whose overexpression leads to estrogen-independent growth of human breast cancer cells, *Mol Endocrinol* **20**(9):2020–2035, 2006.
270. Ji Y, *et al.*, Expression of MTA2 gene in ovarian epithelial cancer and its clinical implication, *J Huazhong Univ Sci Technolog Med Sci* **26**(3):359–362, 2006.
271. Wei X, Xu H, Kufe D, MUC1 oncoprotein stabilizes and activates estrogen receptor alpha, *Mol Cell* **21**(2):295–305, 2006.
272. Santos-Silva F, *et al.*, Thomsen-Friedenreich antigen expression in gastric carcinomas is associated with MUC1 mucin VNTR polymorphism, *Glycobiology* **15**(5):511–517, 2005.
273. Silva F, *et al.*, MUC1 gene polymorphism in the gastric carcinogenesis pathway, *Eur J Hum Genet* **9**(7):548–552, 2001.
274. Cozzi PJ, *et al.*, MUC1, MUC2, MUC4, MUC5AC and MUC6 expression in the progression of prostate cancer, *Clin Exp Metastasis* **22**(7):565–573, 2005.
275. Kawaguchi T, *et al.*, Lack of polymorphism in MUC1 tandem repeats in cancer cells is related to breast cancer progression in Japanese women, *Breast Cancer Res Treat* **92**(3):223–230, 2005.
276. Croce MV, *et al.*, Tissue and serum MUC1 mucin detection in breast cancer patients, *Breast Cancer Res Treat* **81**(3):195–207, 2003.
277. Schmid BC, *et al.*, Evaluation of MUC1 splice variants as prognostic markers in patients with ductal carcinoma *in situ* of the breast, *Oncol Rep* **10**(6):1981–1985, 2003.
278. Qu CF, *et al.*, MUC1 expression in primary and metastatic pancreatic cancer cells for *in vitro* treatment by (213)Bi-C595 radioimmunoconjugate, *Br J Cancer* **91**(12):2086–2093, 2004.
279. Wang Z, *et al.*, Up-regulation of estrogen responsive genes in hypospadias: Microarray analysis, *J Urol* **177**(5):1939–1946, 2007.
280. Qiu W, *et al.*, Down-regulation of growth arrest DNA damage-inducible gene 45beta expression is associated with human hepatocellular carcinoma, *Am J Pathol* **162**(6):1961–1974, 2003.
281. Huang SM, Stallcup MR, Mouse Zac1, a transcriptional coactivator and repressor for nuclear receptors, *Mol Cell Biol* **20**(5):1855–1867, 2000.
282. Kamiya M, *et al.*, The cell cycle control gene ZAC/PLAGL1 is imprinted — A strong candidate gene for transient neonatal diabetes, *Hum Mol Genet* **9**(3):453–460, 2000.
283. Cvetkovic D, *et al.*, Altered expression and loss of heterozygosity of the LOT1 gene in ovarian cancer, *Gynecol Oncol* **95**(3):449–455, 2004.

284. Kamikihara T, *et al.*, Epigenetic silencing of the imprinted gene ZAC by DNA methylation is an early event in the progression of human ovarian cancer, *Int J Cancer* **115**(5):690–700, 2005.
285. Arima T, *et al.*, ZAC, LIT1 (KCNQ1OT1) and p57KIP2 (CDKN1C) are in an imprinted gene network that may play a role in Beckwith-Wiedemann syndrome, *Nucleic Acids Res* **33**(8):2650–2660, 2005.
286. Basyuk E, *et al.*, The candidate tumor suppressor gene ZAC is involved in keratinocyte differentiation and its expression is lost in basal cell carcinomas, *Mol Cancer Res* **3**(9):483–492, 2005.
287. Horlein AJ, *et al.*, Ligand-independent repression by the thyroid hormone receptor mediated by a nuclear receptor corepressor, *Nature* **377**(6548): 397–404, 1995.
288. Kershah SM, *et al.*, Expression of estrogen receptor coregulators in normal and malignant human endometrium, *Gynecol Oncol* **92**(1):304–313, 2004.
289. Paravicini G, *et al.*, The metastasis suppressor candidate nucleotide diphosphate kinase NM23 specifically interacts with members of the ROR/RZR nuclear orphan receptor subfamily, *Biochem Biophys Res Commun* **227**(1):82–87, 1996.
290. Rayner K, *et al.*, NM23-H2, an estrogen receptor beta-associated protein, shows diminished expression with progression of atherosclerosis, *Am J Physiol Regul Integr Comp Physiol* **292**(2):R743–750, 2007.
291. Leone A, *et al.*, Somatic allelic deletion of nm23 in human cancer, *Cancer Res* **51**(9):2490–2493, 1991.
292. Stahl JA, *et al.*, Identification of a second human nm23 gene, nm23-H2, *Cancer Res* **51**(1):445–449, 1991.
293. Braun S, *et al.*, Novel roles of NM23 proteins in skin homeostasis, repair and disease, *Oncogene* **26**(4):532–542, 2007.
294. Kim SH, *et al.*, Human brain nucleoside diphosphate kinase activity is decreased in Alzheimer's disease and Down syndrome, *Biochem Biophys Res Commun* **296**(4):970–975, 2002.
295. Okabe-Kado J, Kasukabe T, Honma Y, Expression of cell surface NM23 proteins of human leukemia cell lines of various cellular lineage and differentiation stages, *Leuk Res* **26**(6):569–576, 2002.
296. Huang N, *et al.*, Two distinct nuclear receptor interaction domains in NSD1, a novel SET protein that exhibits characteristics of both corepressors and coactivators, *EMBO J* **17**(12):3398–3412, 1998.
297. van Haelst MM, *et al.*, Familial gigantism caused by an NSD1 mutation, *Am J Med Genet A* **139**(1):40–44, 2005.
298. Kurotaki N, *et al.*, Haploinsufficiency of NSD1 causes Sotos syndrome, *Nat Genet* **30**(4):365–366, 2002.

299. Nagai T, *et al.*, Sotos syndrome and haploinsufficiency of NSD1: Clinical features of intragenic mutations and submicroscopic deletions, *J Med Genet* **40**(4):285–289, 2003.
300. Tatton-Brown K, Rahman N, Sotos syndrome, *Eur J Hum Genet* **15**(3): 264–271, 2007.
301. Douglas J, *et al.*, NSD1 mutations are the major cause of Sotos syndrome and occur in some cases of Weaver syndrome but are rare in other overgrowth phenotypes, *Am J Hum Genet* **72**(1):132–143, 2003.
302. Baujat G, *et al.*, Paradoxical NSD1 mutations in Beckwith-Wiedemann syndrome and 11p15 anomalies in Sotos syndrome, *Am J Hum Genet* **74**(4):715–720, 2004.
303. Marcus SL, *et al.*, A p56(lck) ligand serves as a coactivator of an orphan nuclear hormone receptor, *J Biol Chem* **271**(44):27197–27200, 1996.
304. Beyens G, *et al.*, Evaluation of the role of the SQSTM1 gene in sporadic Belgian patients with Paget's disease, *Calcif Tissue Int* **75**(2):144–152, 2004.
305. Falchetti A, *et al.*, Two novel mutations at exon 8 of the Sequestosome 1 (SQSTM1) gene in an Italian series of patients affected by Paget's disease of bone (PDB), *J Bone Miner Res* **19**(6):1013–1017, 2004.
306. Hocking LJ, *et al.*, Domain-specific mutations in sequestosome 1 (SQSTM1) cause familial and sporadic Paget's disease, *Hum Mol Genet* **11**(22):2735–2739, 2002.
307. Laurin N, *et al.*, Recurrent mutation of the gene encoding sequestosome 1 (SQSTM1/p62) in Paget disease of bone, *Am J Hum Genet* **70**(6):1582–1588, 2002.
308. Michou L, *et al.*, Genetics of Paget's disease of bone, *Joint Bone Spine* **73**(3):243–248, 2006.
309. Lee DK, Duan HO, Chang C, Androgen receptor interacts with the positive elongation factor P-TEFb and enhances the efficiency of transcriptional elongation, *J Biol Chem* **276**(13):9978–9984, 2001.
310. Sano M, *et al.*, Activation of cardiac Cdk9 represses PGC-1 and confers a predisposition to heart failure, *EMBO J* **23**(17):3559–3569, 2004.
311. Blanco JC, *et al.*, The histone acetylase PCAF is a nuclear receptor coactivator, *Genes Dev* **12**(11):1638–1651, 1998.
312. Ozdag H, *et al.*, Differential expression of selected histone modifier genes in human solid cancers, *BMC Genomics* **7**:90, 2006.
313. Chakravarti D, *et al.*, Role of CBP/P300 in nuclear receptor signalling, *Nature* **383**(6595):99–103, 1996.
314. Kitabayashi I, *et al.*, Fusion of MOZ and p300 histone acetyltransferases in acute monocytic leukemia with a t(8;22)(p11;q13) chromosome translocation, *Leukemia* **15**(1):89–94, 2001.

315. Shigeno K, *et al.*, Disease-related potential of mutations in transcriptional cofactors CREB-binding protein and p300 in leukemias, *Cancer Lett* **213**(1):11–20, 2004.
316. Koshiishi N, *et al.*, p300 gene alterations in intestinal and diffuse types of gastric carcinoma, *Gastric Cancer* **7**(2):85–90, 2004.
317. Bryan EJ, *et al.*, Mutation analysis of EP300 in colon, breast and ovarian carcinomas, *Int J Cancer* **102**(2):137–141, 2002.
318. Gayther SA, *et al.*, Mutations truncating the EP300 acetylase in human cancers, *Nat Genet* **24**(3):300–303, 2000.
319. Muraoka M, *et al.*, p300 gene alterations in colorectal and gastric carcinomas, *Oncogene* **12**(7):1565–1569, 1996.
320. Shenk JL, *et al.*, p53 represses androgen-induced transactivation of prostate-specific antigen by disrupting hAR amino- to carboxyl-terminal interaction, *J Biol Chem* **276**(42):38472–38479, 2001.
321. Iwakuma T, Lozano G, Flores ER, Li-Fraumeni syndrome: A p53 family affair, *Cell Cycle* **4**(7):865–867, 2005.
322. Oren M, Decision making by p53: Life, death and cancer, *Cell Death Differ* **10**(4):431–442, 2003.
323. Bartkova J, *et al.*, DNA damage response as a candidate anti-cancer barrier in early human tumorigenesis, *Nature* **434**(7035):864–870, 2005.
324. Ishitani K, *et al.*, p54nrb acts as a transcriptional coactivator for activation function 1 of the human androgen receptor, *Biochem Biophys Res Commun* **306**(3):660–665, 2003.
325. Pavao M, *et al.*, Immunodetection of nmt55/p54nrb isoforms in human breast cancer, *BMC Cancer* **1**:15, 2001.
326. Clark J, *et al.*, Fusion of splicing factor genes PSF and NonO (p54nrb) to the TFE3 gene in papillary renal cell carcinoma, *Oncogene* **15**(18): 2233–2239, 1997.
327. Joseph B, *et al.*, p57(Kip2) cooperates with Nurr1 in developing dopamine cells, *Proc Natl Acad Sci USA* **100**(26):15619–15624, 2003.
328. Higashimoto K, *et al.*, Imprinting disruption of the CDKN1C/KCNQ1OT1 domain: The molecular mechanisms causing Beckwith-Wiedemann syndrome and cancer, *Cytogenet Genome Res* **113**(1–4):306–312, 2006.
329. Lam WW, *et al.*, Analysis of germline CDKN1C (p57KIP2) mutations in familial and sporadic Beckwith-Wiedemann syndrome (BWS) provides a novel genotype-phenotype correlation, *J Med Genet* **36**(7):518–523, 1999.
330. Pateras IS, *et al.*, Downregulation of the KIP family members p27(KIP1) and p57(KIP2) by SKP2 and the role of methylation in p57(KIP2) inactivation in nonsmall cell lung cancer, *Int J Cancer* **119**(11):2546–2556, 2006.
331. Fan GK, *et al.*, p57(kip2) expression is related to carcinogenesis and tumor progression in laryngeal tissues, *Acta Otolaryngol* **126**(3):301–305, 2006.

332. Fukunaga M, Immunohistochemical characterization of p57Kip2 expression in tetraploid hydropic placentas, *Arch Pathol Lab Med* **128**(8):897–900, 2004.
333. Yue H, Jiang HY, Expression of cell cycle regulator p57kip2, cyclinE protein and proliferating cell nuclear antigen in human pancreatic cancer: an immunohistochemical study, *World J Gastroenterol* **11**(32):5057–5060, 2005.
334. Yue H, *et al.*, Expression of p57kip2, Rb protein and PCNA and their relationships with clinicopathology in human pancreatic cancer, *World J Gastroenterol* **9**(2):377–380, 2003.
335. Nielsen EM, *et al.*, Studies of variations of the cyclin-dependent kinase inhibitor 1C and the cyclin-dependent kinase 4 genes in relation to type 2 diabetes mellitus and related quantitative traits, *J Mol Med* **83**(5):353–361, 2005.
336. Endoh H, *et al.*, Purification and identification of p68 RNA helicase acting as a transcriptional coactivator specific for the activation function 1 of human estrogen receptor alpha, *Mol Cell Biol* **19**(8):5363–5372, 1999.
337. Haines GK, *et al.*, Expression of the double-stranded RNA-dependent protein kinase (p68) in human breast tissues, *Tumour Biol* **17**(1):5–12, 1996.
338. Wang Y, *et al.*, Human PAD4 regulates histone arginine methylation levels via demethylimination, *Science* **306**(5694):279–283, 2004.
339. Ikari K, *et al.*, Association between PADI4 and rheumatoid arthritis: A replication study, *Arthritis Rheum* **52**(10):3054–3057, 2005.
340. Kearney PL, *et al.*, Kinetic characterization of protein arginine deiminase 4: A transcriptional corepressor implicated in the onset and progression of rheumatoid arthritis, *Biochemistry* **44**(31):10570–10582, 2005.
341. Chang X, Han J, Expression of peptidylarginine deiminase type 4 (PAD4) in various tumors, *Mol Carcinog* **45**(3):183–196, 2006.
342. Pavri R, *et al.*, PARP-1 determines specificity in a retinoid signaling pathway via direct modulation of mediator, *Mol Cell* **18**(1):83–96, 2005.
343. Masutani M, Nakagama H, Sugimura T, Poly(ADP-ribosyl)ation in relation to cancer and autoimmune disease, *Cell Mol Life Sci* **62**(7–8):769–783, 2005.
344. Woon EC, Threadgill MD, Poly(ADP-ribose)polymerase inhibition — where now? *Curr Med Chem* **12**(20):2373–2392, 2005.
345. Rueda B, *et al.*, Poly (ADP-ribose) polymerase-1 haplotypes are associated with coeliac disease, *Int J Immunogenet* **32**(4):245–248, 2005.
346. Soos J, *et al.*, The expression of PARP, NF-kappa B and parvalbumin is increased in Parkinson's disease, *Neuroreport* **15**(11):1715–1718, 2004.
347. Kim SH, *et al.*, PARP expression is increased in astrocytes but decreased in motor neurons in the spinal cord of sporadic ALS patients, *J Neuropathol Exp Neurol* **62**(1):88–103, 2003.

348. Huang SC, *et al.*, Enhanced polyadenosine diphosphate-ribosylation in gonadotropin-releasing hormone agonist-treated uterine leiomyoma, *J Clin Endocrinol Metab* **88**(10):5009–5016, 2003.
349. Pascual M, *et al.*, A poly(ADP-ribose) polymerase haplotype spanning the promoter region confers susceptibility to rheumatoid arthritis, *Arthritis Rheum* **48**(3):638–641, 2003.
350. Yuan CX, *et al.*, The TRAP220 component of a thyroid hormone receptor-associated protein (TRAP) coactivator complex interacts directly with nuclear receptors in a ligand-dependent fashion, *Proc Natl Acad Sci USA* **95**(14):7939–7944, 1998.
351. Zhu Y, *et al.*, Isolation and characterization of PBP, a protein that interacts with peroxisome proliferator-activated receptor, *J Biol Chem* **272**(41):25500–25506, 1997.
352. Li JM, *et al.*, Decreased expression of thyroid receptor-associated protein 220 in temporal lobe tissue of patients with refractory epilepsy, *Biochem Biophys Res Commun* **348**(4):1389–1397, 2006.
353. Fondell JD, *et al.*, Thyroid hormone receptor-associated proteins and general positive cofactors mediate thyroid hormone receptor function in the absence of the TATA box-binding protein-associated factors of TFIID, *Proc Natl Acad Sci USA* **96**(5):1959–1964, 1999.
354. De Luca A, *et al.*, Association study between CAG trinucleotide repeats in the PCQAP gene (PC2 glutamine/Q-rich-associated protein) and schizophrenia, *Am J Med Genet B Neuropsychiatr Genet* **116**(1):32–35, 2003.
355. Oettgen P, *et al.*, PDEF, a novel prostate epithelium-specific ets transcription factor, interacts with the androgen receptor and activates prostate-specific antigen gene expression, *J Biol Chem* **275**(2):1216–1225, 2000.
356. Feldman RJ, *et al.*, Pdef expression in human breast cancer is correlated with invasive potential and altered gene expression, *Cancer Res* **63**(15):4626–4631, 2003.
357. Ghadersohi A, *et al.*, Prostate derived Ets transcription factor shows better tumor-association than other cancer-associated molecules, *Oncol Rep* **11**(2):453–458, 2004.
358. Yin Y, *et al.*, 3-phosphoinositide-dependent protein kinase-1 activates the peroxisome proliferator-activated receptor-gamma and promotes adipocyte differentiation, *Mol Endocrinol* **20**(2):268–278, 2006.
359. Martignoni G, *et al.*, Renal disease in adults with TSC2/PKD1 contiguous gene syndrome, *Am J Surg Pathol* **26**(2):198–205, 2002.
360. Vadlamudi RK, *et al.*, Molecular cloning and characterization of PELP1, a novel human coregulator of estrogen receptor alpha, *J Biol Chem* **276**(41):38272–38279, 2001.
361. Vadlamudi RK, *et al.*, Novel estrogen receptor coactivator PELP1/MNAR gene and ERbeta expression in salivary duct adenocarcinoma: Potential therapeutic targets, *Hum Pathol* **36**(6):670–675, 2005.

362. Vadlamudi RK, *et al.*, Deregulation of estrogen receptor coactivator proline-, glutamic acid-, and leucine-rich protein-1/modulator of nongenomic activity of estrogen receptor in human endometrial tumors, *J Clin Endocrinol Metab* **89**(12):6130–6138, 2004.
363. Vadlamudi RK, *et al.*, Functional implications of altered subcellular localization of PELP1 in breast cancer cells, *Cancer Res* **65**(17):7724–7732, 2005.
364. Puigserver P, *et al.*, A cold-inducible coactivator of nuclear receptors linked to adaptive thermogenesis, *Cell* **92**(6):829–839, 1998.
365. Hara K, *et al.*, A genetic variation in the PGC-1 gene could confer insulin resistance and susceptibility to Type II diabetes, *Diabetologia* **45**(5):740–743, 2002.
366. Kim JH, *et al.*, Peroxisome proliferator-activated receptor gamma coactivator 1 alpha promoter polymorphisms are associated with early-onset type 2 diabetes mellitus in the Korean population, *Diabetologia* **48**(7):1323–1330, 2005.
367. Kunej T, *et al.*, A Gly482Ser polymorphism of the peroxisome proliferator-activated receptor-gamma coactivator-1 (PGC-1) gene is associated with type 2 diabetes in Caucasians, *Folia Biol (Praha)* **50**(5):157–158, 2004.
368. Muller YL, *et al.*, A Gly482Ser missense mutation in the peroxisome proliferator-activated receptor gamma coactivator-1 is associated with altered lipid oxidation and early insulin secretion in Pima Indians, *Diabetes* **52**(3):895–898, 2003.
369. Vimaleswaran KS, *et al.*, Peroxisome proliferator-activated receptor-gamma coactivator-1alpha (PGC-1alpha) gene polymorphisms and their relationship to Type 2 diabetes in Asian Indians, *Diabet Med* **22**(11):1516–1521, 2005.
370. Leipold H, *et al.*, Peroxisome proliferator-activated receptor gamma coactivator-1alpha gene variations are not associated with gestational diabetes mellitus, *J Soc Gynecol Investig* **13**(2):104–107, 2006.
371. Finck BN, Kelly DP, PGC-1 coactivators: Inducible regulators of energy metabolism in health and disease, *J Clin Invest* **116**(3):615–622, 2006.
372. Finck BN, Kelly DP, Peroxisome proliferator-activated receptor gamma coactivator-1 (PGC-1) regulatory cascade in cardiac physiology and disease, *Circulation* **115**(19):2540–2548, 2007.
373. Bertolotti M, *et al.*, Decreased hepatic expression of PPAR-gamma coactivator-1 in cholesterol cholelithiasis, *Eur J Clin Invest* **36**(3):170–175, 2006.
374. Ambye L, *et al.*, Studies of the Gly482Ser polymorphism of the peroxisome proliferator-activated receptor gamma coactivator-1alpha (PGC-1alpha) gene in Danish subjects with the metabolic syndrome, *Diabetes Res Clin Pract* **67**(2):175–179, 2005.
375. Lucia A, *et al.*, PPARGC1A genotype (Gly482Ser) predicts exceptional endurance capacity in European men, *J Appl Physiol* **99**(1):344–348, 2005.

376. Sookoian S, *et al.*, Peroxisome proliferator-activated receptor gamma and its coactivator-1alpha may be associated with features of the metabolic syndrome in adolescents, *J Mol Endocrinol* **35**(2):373–380, 2005.
377. Petrovic MG, *et al.*, Gly482Ser polymorphism of the peroxisome proliferator-activated receptor-gamma coactivator-1 gene might be a risk factor for diabetic retinopathy in Slovene population (Caucasians) with type 2 diabetes and the Pro12Ala polymorphism of the PPARgamma gene is not, *Diabetes Metab Res Rev* **21**(5):470–474, 2005.
378. Andersen G, *et al.*, PGC-1alpha Gly482Ser polymorphism associates with hypertension among Danish whites, *Hypertension* **45**(4):565–570, 2005.
379. Cheurfa N, *et al.*, The Gly482Ser polymorphism in the peroxisome proliferator-activated receptor-gamma coactivator-1 gene is associated with hypertension in type 2 diabetic men, *Diabetologia* **47**(11):1980–1983, 2004.
380. Jiang WG, Douglas-Jones A, Mansel RE, Expression of peroxisome-proliferator activated receptor-gamma (PPARgamma) and the PPARgamma co-activator, PGC-1, in human breast cancer correlates with clinical outcomes, *Int J Cancer* **106**(5):752–757, 2003.
381. Handschin C, *et al.*, PGC-1alpha regulates the neuromuscular junction program and ameliorates Duchenne muscular dystrophy, *Genes Dev* **21**(7):770–783, 2007.
382. Savagner F, *et al.*, PGC-1-related coactivator and targets are upregulated in thyroid oncocytoma, *Biochem Biophys Res Commun* **310**(3):779–784, 2003.
383. Semple RK, *et al.*, Expression of the thermogenic nuclear hormone receptor coactivator PGC-1alpha is reduced in the adipose tissue of morbidly obese subjects, *Int J Obes Relat Metab Disord* **28**(1):176–179, 2004.
384. Yang X, Enerback S, Smith U, Reduced expression of FOXC2 and brown adipogenic genes in human subjects with insulin resistance, *Obes Res* **11**(10):1182–1191, 2003.
385. Thadani H, Deacon A, Peters T, Diagnosis and management of porphyria, *BMJ* **320**(7250):1647–1651, 2000.
386. Kressler D, *et al.*, The PGC-1-related protein PERC is a selective coactivator of estrogen receptor alpha, *J Biol Chem* **277**(16):13918–13925, 2002.
387. Lin J, *et al.*, Peroxisome proliferator-activated receptor gamma coactivator 1beta (PGC-1beta), a novel PGC-1-related transcription coactivator associated with host cell factor, *J Biol Chem* **277**(3):1645–1648, 2002.
388. Andersen G, *et al.*, Evidence of an association between genetic variation of the coactivator PGC-1beta and obesity, *J Med Genet* **42**(5):402–407, 2005.
389. Tan J, *et al.*, Protein inhibitor of activated STAT-1 (signal transducer and activator of transcription-1) is a nuclear receptor coregulator expressed in human testis, *Mol Endocrinol* **14**(1):14–26, 2000.
390. Linja MJ, *et al.*, Expression of androgen receptor coregulators in prostate cancer, *Clin Cancer Res* **10**(3):1032–1040, 2004.

391. Junicho A, *et al.*, Protein inhibitor of activated STAT3 regulates androgen receptor signaling in prostate carcinoma cells, *Biochem Biophys Res Commun* **278**(1):9–13, 2000.
392. Starkel P, *et al.*, Deficient Stat3 DNA-binding is associated with high Pias3 expression and a positive anti-apoptotic balance in human end-stage alcoholic and hepatitis C cirrhosis, *J Hepatol* **43**(4):687–695, 2005.
393. Wang L and Banerjee S, Differential PIAS3 expression in human malignancy, *Oncol Rep* **11**(6):1319–1324, 2004.
394. Gross M, *et al.*, Distinct effects of PIAS proteins on androgen-mediated gene activation in prostate cancer cells, *Oncogene* **20**(29):3880–3887, 2001.
395. Ueda M, *et al.*, DNA microarray analysis of stage progression mechanism in myelodysplastic syndrome, *Br J Haematol* **123**(2):288–296, 2003.
396. Yi P, *et al.*, Peptidyl-prolyl isomerase 1 (Pin1) serves as a coactivator of steroid receptor by regulating the activity of phosphorylated steroid receptor coactivator 3 (SRC-3/AIB1), *Mol Cell Biol* **25**(21):9687–9699, 2005.
397. Lu KP, Pinning down cell signaling, cancer and Alzheimer's disease, *Trends Biochem Sci* **29**(4):200–209, 2004.
398. Segat L, *et al.*, PIN1 promoter polymorphisms are associated with Alzheimer's disease, *Neurobiol Aging* **28**(1):69–74, 2007.
399. Sultana R, *et al.*, Oxidative modification and down-regulation of Pin1 in Alzheimer's disease hippocampus: A redox proteomics analysis, *Neurobiol Aging* **27**(7):918–925, 2006.
400. Kim CJ, *et al.*, Pin1 overexpression in colorectal cancer and its correlation with aberrant beta-catenin expression, *World J Gastroenterol* **11**(32): 5006–5009, 2005.
401. Pang RW, *et al.*, PIN1 expression contributes to hepatic carcinogenesis, *J Pathol* **210**(1):19–25, 2006.
402. Fukuchi M, *et al.*, Prolyl isomerase Pin1 expression predicts prognosis in patients with esophageal squamous cell carcinoma and correlates with cyclinD1 expression, *Int J Oncol* **29**(2):329–334, 2006.
403. Bao L, *et al.*, Prevalent overexpression of prolyl isomerase Pin1 in human cancers, *Am J Pathol* **164**(5):1727–1737, 2004.
404. Zhou CX, Gao Y, Aberrant expression of beta-catenin, Pin1 and cylin D1 in salivary adenoid cystic carcinoma: Relation to tumor proliferation and metastasis, *Oncol Rep* **16**(3):505–511, 2006.
405. Kambhampati S, *et al.*, Activation of protein kinase C delta by all-trans-retinoic acid, *J Biol Chem* **278**(35):32544–32551, 2003.
406. Lee CF, *et al.*, Involvement of protein kinase C delta in the alteration of mitochondrial mass in human cells under oxidative stress, *Free Radic Biol Med* **40**(12):2136–2146, 2006.
407. D'Costa AM, *et al.*, The proapoptotic tumor suppressor protein kinase C-delta is lost in human squamous cell carcinomas, *Oncogene* **25**(3):378–386, 2006.

408. Cui S, *et al.*, Disrupted gonadogenesis and male-to-female sex reversal in Pod1 knockout mice, *Development* **131**(16):4095–4105, 2004.
409. Hong CY, *et al.*, Modulation of the expression and transactivation of androgen receptor by the basic helix-loop-helix transcription factor Pod-1 through recruitment of histone deacetylase 1, *Mol Endocrinol* **19**(9):2245–2257, 2005.
410. Smith LT, *et al.*, Epigenetic regulation of the tumor suppressor gene TCF21 on 6q23-q24 in lung and head and neck cancer, *Proc Natl Acad Sci USA* **103**(4):982–987, 2006.
411. Proia DA, *et al.*, Dual roles for the phosphatase PPM1D in regulating progesterone receptor function, *J Biol Chem* **281**(11):7089–7101, 2006.
412. Bulavin DV, *et al.*, Amplification of PPM1D in human tumors abrogates p53 tumor-suppressor activity, *Nat Genet* **31**(2):210–215, 2002.
413. Rauta J, *et al.*, The serine-threonine protein phosphatase PPM1D is frequently activated through amplification in aggressive primary breast tumours, *Breast Cancer Res Treat* **95**(3):257–263, 2006.
414. Hirasawa A, *et al.*, Association of 17q21-q24 gain in ovarian clear cell adenocarcinomas with poor prognosis and identification of PPM1D and APPBP2 as likely amplification targets, *Clin Cancer Res* **9**(6):1995–2004, 2003.
415. Mendrzyk F, *et al.*, Genomic and protein expression profiling identifies CDK6 as novel independent prognostic marker in medulloblastoma, *J Clin Oncol* **23**(34):8853–8862, 2005.
416. Saito-Ohara F, *et al.*, PPM1D is a potential target for 17q gain in neuroblastoma, *Cancer Res* **63**(8):1876–1883, 2003.
417. Epping MT, *et al.*, The human tumor antigen PRAME is a dominant repressor of retinoic acid receptor signaling, *Cell* **122**(6):835–847, 2005.
418. Epping MT, Bernards R, A causal role for the human tumor antigen preferentially expressed antigen of melanoma in cancer, *Cancer Res* **66**(22): 10639–10642, 2006.
419. Matsushita M, *et al.*, Preferentially expressed antigen of melanoma (PRAME) in the development of diagnostic and therapeutic methods for hematological malignancies, *Leuk Lymphoma* **44**(3):439–444, 2003.
420. Steinbach D, *et al.*, Clinical implications of PRAME gene expression in childhood acute myeloid leukemia, *Cancer Genet Cytogenet* **133**(2):118–123, 2002.
421. Koh SS, *et al.*, Synergistic enhancement of nuclear receptor function by p160 coactivators and two coactivators with protein methyltransferase activities, *J Biol Chem* **276**(2):1089–1098, 2001.
422. Scorilas A, *et al.*, Genomic organization, physical mapping, and expression analysis of the human protein arginine methyltransferase 1 gene, *Biochem Biophys Res Commun* **278**(2):349–359, 2000.

423. Qin J, *et al.*, Prospero-related homeobox (Prox1) is a corepressor of human liver receptor homolog-1 and suppresses the transcription of the cholesterol 7-alpha-hydroxylase gene, *Mol Endocrinol* **18**(10):2424–2439, 2004.
424. Shimoda M, *et al.*, A homeobox protein, prox1, is involved in the differentiation, proliferation, and prognosis in hepatocellular carcinoma, *Clin Cancer Res* **12**(20 Pt 1):6005–6011, 2006.
425. Schneider M, *et al.*, Role of lymphangiogenesis and lymphangiogenic factors during pancreatic cancer progression and lymphatic spread, *Int J Oncol* **28**(4):883–890, 2006.
426. Mathur M, Tucker PW, Samuels HH, PSF is a novel corepressor that mediates its effect through Sin3A and the DNA binding domain of nuclear hormone receptors, *Mol Cell Biol* **21**(7):2298–2311, 2001.
427. Mathur M, Samuels HH, Role of PSF-TFE3 oncoprotein in the development of papillary renal cell carcinomas, *Oncogene* **26**(2):277–283, 2007.
428. Martini PG, *et al.*, Prothymosin alpha selectively enhances estrogen receptor transcriptional activity by interacting with a repressor of estrogen receptor activity, *Mol Cell Biol* **20**(17):6224–6232, 2000.
429. Suzuki S, *et al.*, Expression of prothymosin alpha is correlated with development and progression in human prostate cancers, *Prostate* **66**(5): 463–469, 2006.
430. Li P, Nicosia SV, Bai W, Antagonism between PTEN/MMAC1/TEP-1 and androgen receptor in growth and apoptosis of prostatic cancer cells, *J Biol Chem* **276**(23):20444–20450, 2001.
431. Nan B, *et al.*, The PTEN tumor suppressor is a negative modulator of androgen receptor transcriptional activity, *J Mol Endocrinol* **31**(1):169–183, 2003.
432. Trotman LC, *et al.*, Ubiquitination regulates PTEN nuclear import and tumor suppression, *Cell* **128**(1):141–156, 2007.
433. Marsh DJ, *et al.*, Germline mutations in PTEN are present in Bannayan-Zonana syndrome, *Nat Genet* **16**(4):333–334, 1997.
434. Celebi JT, *et al.*, Phenotypic findings of Cowden syndrome and Bannayan-Zonana syndrome in a family associated with a single germline mutation in PTEN, *J Med Genet* **36**(5):360–364, 1999.
435. Longy M, *et al.*, Mutations of PTEN in patients with Bannayan-Riley-Ruvalcaba phenotype, *J Med Genet* **35**(11):886–889, 1998.
436. Marsh DJ, *et al.*, PTEN mutation spectrum and genotype-phenotype correlations in Bannayan-Riley-Ruvalcaba syndrome suggest a single entity with Cowden syndrome, *Hum Mol Genet* **8**(8):1461–1472, 1999.
437. Nagata Y, *et al.*, PTEN activation contributes to tumor inhibition by trastuzumab, and loss of PTEN predicts trastuzumab resistance in patients, *Cancer Cell* **6**(2):117–127, 2004.

438. Zhou XP, *et al.*, A 39-bp deletion polymorphism in PTEN in African American individuals: Implications for molecular diagnostic testing, *J Mol Diagn* **4**(2):114–117, 2002.
439. Kato H, *et al.*, PTEN gene mutation and high MIB-1 labeling index may contribute to dissemination in patients with glioblastoma, *J Clin Neurosci* **11**(1):37–41, 2004.
440. Smith JS, *et al.*, PTEN mutation, EGFR amplification, and outcome in patients with anaplastic astrocytoma and glioblastoma multiforme, *J Natl Cancer Inst* **93**(16):1246–1256, 2001.
441. Butler MG, *et al.*, Subset of individuals with autism spectrum disorders and extreme macrocephaly associated with germline PTEN tumour suppressor gene mutations, *J Med Genet* **42**(4):318–321, 2005.
442. Ishihara H, *et al.*, Association of the polymorphisms in the 5′-untranslated region of PTEN gene with type 2 diabetes in a Japanese population, *FEBS Lett* **554**(3):450–454, 2003.
443. Maser RS, *et al.*, Chromosomally unstable mouse tumours have genomic alterations similar to diverse human cancers, *Nature* **447**(7147):966–971, 2007.
444. Teo AK, *et al.*, The modified human DNA repair enzyme O(6)-methylguanine-DNA methyltransferase is a negative regulator of estrogen receptor-mediated transcription upon alkylation DNA damage, *Mol Cell Biol* **21**(20):7105–7114, 2001.
445. Krzesniak M, *et al.*, Polymorphisms in TDG and MGMT genes — epidemiological and functional study in lung cancer patients from Poland, *Ann Hum Genet* **68**(Pt 4):300–312, 2004.
446. Wiencke JK, *et al.*, Molecular features of adult glioma associated with patient race/ethnicity, age, and a polymorphism in O6-methylguanine-DNA-methyltransferase, *Cancer Epidemiol Biomarkers Prev* **14**(7):1774–1783, 2005.
447. Li C, *et al.*, Exon 3 polymorphisms and haplotypes of O6-methylguanine-DNA methyltransferase and risk of bladder cancer in southern China: A case-control analysis, *Cancer Lett* **227**(1):49–57, 2005.
448. Rigas AC, *et al.*, The scaffolding protein RACK1 interacts with androgen receptor and promotes cross-talk through a protein kinase C signaling pathway, *J Biol Chem* **278**(46):46087–46093, 2003.
449. Ron D, Signaling cascades regulating NMDA receptor sensitivity to ethanol, *Neuroscientist* **10**(4):325–336, 2004.
450. Corsini E, *et al.*, Age-related decline in RACK-1 expression in human leukocytes is correlated to plasma levels of dehydroepiandrosterone, *J Leukoc Biol* **77**(2):247–256, 2005.
451. Racchi M, *et al.*, RACK-1 expression and cytokine production in leukocytes obtained from AD patients, *Aging Clin Exp Res* **18**(2):153–157, 2006.

452. Wang H, Friedman E, Increased association of brain protein kinase C with the receptor for activated C kinase-1 (RACK1) in bipolar affective disorder, *Biol Psychiatry* **50**(5):364–370, 2001.
453. Kupfer SR, *et al.*, Receptor accessory factor enhances specific DNA binding of androgen and glucocorticoid receptors, *J Biol Chem* **268**(23):17519–17527, 1993.
454. Kupfer SR, Wilson EM, French FS, Androgen and glucocorticoid receptors interact with insulin degrading enzyme, *J Biol Chem* **269**(32):20622–20628, 1994.
455. Miwa W, *et al.*, Absence of activating mutations of the RAF1 protooncogene in human lung cancer, *Biol Chem Hoppe Seyler* **375**(10):705–709, 1994.
456. Nowotny P, *et al.*, Association studies between risk for late-onset Alzheimer's disease and variants in insulin degrading enzyme, *Am J Med Genet B Neuropsychiatr Genet* **136**(1):62–68, 2005.
457. Rao MA, *et al.*, RanBPM, a nuclear protein that interacts with and regulates transcriptional activity of androgen receptor and glucocorticoid receptor, *J Biol Chem* **277**(50):48020–48027, 2002.
458. Ma Z, *et al.*, Fusion of ALK to the Ran-binding protein 2 (RANBP2) gene in inflammatory myofibroblastic tumor, *Genes Chromosomes Cancer* **37**(1):98–105, 2003.
459. Kullmann M, *et al.*, RAP46 is a negative regulator of glucocorticoid receptor action and hormone-induced apoptosis, *J Biol Chem* **273**(23):14620–14625, 1998.
460. Tang SC, BAG-1, an anti-apoptotic tumour marker, *IUBMB Life* **53**(2):99–105, 2002.
461. Batistatou A, *et al.*, Estrogen receptor beta (ERbeta) protein expression correlates with BAG-1 and prognosis in brain glial tumours, *J Neurooncol* **77**(1):17–23, 2006.
462. Krajewska M, *et al.*, Expression of BAG-1 protein correlates with aggressive behavior of prostate cancers, *Prostate* **66**(8):801–810, 2006.
463. Wieczorek S, *et al.*, Screening for candidate gene regions in narcolepsy using a microsatellite based approach and pooled DNA, *J Mol Med* **82**(10): 696–705, 2004.
464. Singh P, Coe J, Hong W, A role for retinoblastoma protein in potentiating transcriptional activation by the glucocorticoid receptor, *Nature* **374**(6522):562–565, 1995.
465. Yeh S, *et al.*, Retinoblastoma, a tumor suppressor, is a coactivator for the androgen receptor in human prostate cancer DU145 cells, *Biochem Biophys Res Commun* **248**(2):361–367, 1998.
466. Khidr L, Chen PL, RB, the conductor that orchestrates life, death and differentiation, *Oncogene* **25**(38):5210–5219, 2006.

467. Yandell DW, *et al.*, Oncogenic point mutations in the human retinoblastoma gene: Their application to genetic counseling, *N Engl J Med* **321**(25): 1689–1695, 1989.
468. Dunn JM, *et al.*, Mutations in the RB1 gene and their effects on transcription, *Mol Cell Biol* **9**(11):4596–4604, 1989.
469. Friend SH, *et al.*, A human DNA segment with properties of the gene that predisposes to retinoblastoma and osteosarcoma, *Nature* **323**(6089):643–646, 1986.
470. Lee WH, *et al.*, Human retinoblastoma susceptibility gene: Cloning, identification, and sequence, *Science* **235**(4794):1394–1399, 1987.
471. Sakai T, *et al.*, Oncogenic germ-line mutations in Sp1 and ATF sites in the human retinoblastoma gene, *Nature* **353**(6339):83–86, 1991.
472. Sanchez-Sanchez F, *et al.*, Attenuation of disease phenotype through alternative translation initiation in low-penetrance retinoblastoma, *Hum Mutat* **28**(2):159–167, 2007.
473. Issing WJ, *et al.*, An association of the RB gene with osteosarcoma: Molecular genetic evaluation of a case of hereditary retinoblastoma, *Eur Arch Otorhinolaryngol* **250**(5):277–280, 1993.
474. Rothberg PG, *et al.*, A deletion polymorphism due to Alu-Alu recombination in intron 2 of the retinoblastoma gene: Association with human gliomas, *Mol Carcinog* **19**(2):69–73, 1997.
475. Chan SW, Hong W, Retinoblastoma-binding protein 2 (Rbp2) potentiates nuclear hormone receptor-mediated transcription, *J Biol Chem* **276**(30): 28402–28412, 2001.
476. Vogt T, *et al.*, Deficiency of a novel retinoblastoma binding protein 2-homolog is a consistent feature of sporadic human melanoma skin cancer, *Lab Invest* **79**(12):1615–1627, 1999.
477. Montano MM, *et al.*, An estrogen receptor-selective coregulator that potentiates the effectiveness of antiestrogens and represses the activity of estrogens, *Proc Natl Acad Sci USA* **96**(12):6947–6952, 1999.
478. Simon SL, *et al.*, Expression of a repressor of estrogen receptor activity in human breast tumors: Relationship to some known prognostic markers, *Cancer Res* **60**(11):2796–2799, 2000.
479. Li X, *et al.*, The SRC-3/AIB1 coactivator is degraded in a ubiquitin- and ATP-independent manner by the REGgamma proteasome, *Cell* **124**(2):381–392, 2006.
480. Cavazzana I, *et al.*, Clinical and serological features of 35 patients with anti-Ki autoantibodies, *Lupus* **14**(10):837–841, 2005.
481. Matsushita M, *et al.*, Autoimmune response to proteasome activator 28alpha in patients with connective tissue diseases, *J Rheumatol* **31**(2):252–259, 2004.
482. Palvimo JJ, *et al.*, Mutual transcriptional interference between RelA and androgen receptor, *J Biol Chem* **271**(39):24151–24156, 1996.

483. Corbetta S, *et al.*, Activity and function of the nuclear factor kappaB pathway in human parathyroid tumors, *Endocr Relat Cancer* **12**(4):929–937, 2005.
484. Zhang J, Peng B, Chen X, Expressions of nuclear factor kappaB, inducible nitric oxide synthase, and vascular endothelial growth factor in adenoid cystic carcinoma of salivary glands: Correlations with the angiogenesis and clinical outcome, *Clin Cancer Res* **11**(20):7334–7343, 2005.
485. Shukla S, *et al.*, Nuclear factor-kappaB/p65 (Rel A) is constitutively activated in human prostate adenocarcinoma and correlates with disease progression, *Neoplasia* **6**(4):390–400, 2004.
486. Aviles DH, *et al.*, Decreased expression of T-cell NF-kappaB p65 subunit in steroid-resistant nephrotic syndrome, *Kidney Int* **66**(1):60–67, 2004.
487. Valanciute A, *et al.*, NF-kappa B p65 antagonizes IL-4 induction by c-maf in minimal change nephrotic syndrome, *J Immunol* **172**(1):688–698, 2004.
488. Cavailles V, *et al.*, Nuclear factor RIP140 modulates transcriptional activation by the estrogen receptor, *EMBO J* **14**(15):3741–3751, 1995.
489. Galan JJ, *et al.*, Multilocus analyses of estrogen-related genes reveal involvement of the ESR1 gene in male infertility and the polygenic nature of the pathology, *Fertil Steril* **84**(4):910–918, 2005.
490. Caballero V, *et al.*, Preliminary molecular genetic analysis of the Receptor Interacting Protein 140 (RIP140) in women affected by endometriosis, *J Exp Clin Assist Reprod* **2**:11, 2005.
491. Laherty CD, *et al.*, SAP30, a component of the mSin3 corepressor complex involved in N-CoR-mediated repression by specific transcription factors, *Mol Cell* **2**(1):33–42, 1998.
492. Sironi E, *et al.*, Loss of heterozygosity on chromosome 4q32-35 in sporadic basal cell carcinomas: Evidence for the involvement of p33ING2/ING1L and SAP30 genes, *J Cutan Pathol* **31**(4):318–322, 2004.
493. Cheng J, *et al.*, SENP1 enhances androgen receptor-dependent transcription through desumoylation of histone deacetylase 1, *Mol Cell Biol* **24**(13):6021–6028, 2004.
494. Cheng J, *et al.*, Role of desumoylation in the development of prostate cancer, *Neoplasia* **8**(8):667–676, 2006.
495. Shi Y, *et al.*, Sharp, an inducible cofactor that integrates nuclear receptor repression and activation, *Genes Dev* **15**(9):1140–1151, 2001.
496. Feng Y, *et al.*, Drosophila split ends homologue SHARP functions as a positive regulator of Wnt/beta-catenin/T-cell factor signaling in neoplastic transformation, *Cancer Res* **67**(2):482–491, 2007.
497. Johansson L, *et al.*, The orphan nuclear receptor SHP utilizes conserved LXXLL-related motifs for interactions with ligand-activated estrogen receptors, *Mol Cell Biol* **20**(4):1124–1133, 2000.

498. Nishigori H, *et al.*, Mutations in the small heterodimer partner gene are associated with mild obesity in Japanese subjects, *Proc Natl Acad Sci USA* **98**(2):575–580, 2001.
499. Wang L, *et al.*, Orphan receptor small heterodimer partner is an important mediator of glucose homeostasis, *Mol Endocrinol* **20**(11):2671–2681, 2006.
500. Fu M, *et al.*, Hormonal control of androgen receptor function through SIRT1, *Mol Cell Biol* **26**(21):8122–8135, 2006.
501. Chu F, *et al.*, Control of multidrug resistance gene mdr1 and cancer resistance to chemotherapy by the longevity gene sirt1, *Cancer Res* **65**(22):10183–10187, 2005.
502. Lagouge M, *et al.*, Resveratrol improves mitochondrial function and protects against metabolic disease by activating SIRT1 and PGC-1alpha, *Cell* **127**(6):1109–1122, 2006.
503. Laflamme C, *et al.*, The homeotic protein Six3 is a coactivator of the nuclear receptor NOR-1 and a corepressor of the fusion protein EWS/NOR-1 in human extraskeletal myxoid chondrosarcomas, *Cancer Res* **63**(2):449–454, 2003.
504. Pasquier L, *et al.*, A new mutation in the six-domain of SIX3 gene causes holoprosencephaly, *Eur J Hum Genet* **8**(10):797–800, 2000.
505. Wallis D, Muenke M, Mutations in holoprosencephaly, *Hum Mutat* **16**(2):99–108, 2000.
506. Wallis DE, *et al.*, Mutations in the homeodomain of the human SIX3 gene cause holoprosencephaly, *Nat Genet* **22**(2):196–198, 1999.
507. Yanagisawa J, *et al.*, Convergence of transforming growth factor-beta and vitamin D signaling pathways on SMAD transcriptional coactivators, *Science* **283**(5406):1317–1321, 1999.
508. Watanabe Y, *et al.*, A catalog of 106 single-nucleotide polymorphisms (SNPs) and 11 other types of variations in genes for transforming growth factor-beta1 (TGF-beta1) and its signaling pathway, *J Hum Genet* **47**(9):478–483, 2002.
509. Han SU, *et al.*, Loss of the Smad3 expression increases susceptibility to tumorigenicity in human gastric cancer, *Oncogene* **23**(7):1333–1341, 2004.
510. Kang HY, *et al.*, Differential modulation of androgen receptor-mediated transactivation by Smad3 and tumor suppressor Smad4, *J Biol Chem* **277**(46):43749–43756, 2002.
511. Seshimo I, *et al.*, Expression and mutation of SMAD4 in poorly differentiated carcinoma and signet-ring cell carcinoma of the colorectum, *J Exp Clin Cancer Res* **25**(3):433–442, 2006.
512. Gallione CJ, *et al.*, SMAD4 mutations found in unselected HHT patients, *J Med Genet* **43**(10):793–797, 2006.

513. Kim JY, *et al.*, Smad4 expression in gastric adenoma and adenocarcinoma: Frequent loss of expression in diffuse type of gastric adenocarcinoma, *Histol Histopathol* **20**(2):543–549, 2005.
514. Wang LH, *et al.*, Inactivation of SMAD4 tumor suppressor gene during gastric carcinoma progression, *Clin Cancer Res* **13**(1):102–110, 2007.
515. Qiu W, *et al.*, Disruption of transforming growth factor beta-Smad signaling pathway in head and neck squamous cell carcinoma as evidenced by mutations of SMAD2 and SMAD4, *Cancer Lett* **245**(1–2):163–170, 2007.
516. Zhong D, *et al.*, Homozygous deletion of SMAD4 in breast cancer cell lines and invasive ductal carcinomas, *Cancer Biol Ther* **5**(6):601–607, 2006.
517. Embuscado EE, *et al.*, Immortalizing the complexity of cancer metastasis: Genetic features of lethal metastatic pancreatic cancer obtained from rapid autopsy, *Cancer Biol Ther* **4**(5):548–554, 2005.
518. Tang ZH, *et al.*, The relationship between loss expression of DPC4/Smad4 gene and carcinogenesis of pancreatobiliary carcinoma, *Hepatobiliary Pancreat Dis Int* **1**(4):624–629, 2002.
519. Lazzereschi D, *et al.*, A complex pattern of mutations and abnormal splicing of Smad4 is present in thyroid tumours, *Oncogene* **24**(34):5344–5354, 2005.
520. Baldus SE, *et al.*, Smad4 deficiency in cervical carcinoma cells, *Oncogene* **24**(5):810–819, 2005.
521. Horvath LG, *et al.*, Loss of BMP2, Smad8, and Smad4 expression in prostate cancer progression, *Prostate* **59**(3):234–242, 2004.
522. Friedl W, *et al.*, Juvenile polyposis: Massive gastric polyposis is more common in MADH4 mutation carriers than in BMPR1A mutation carriers, *Hum Genet* **111**(1):108–111, 2002.
523. Howe JR, *et al.*, The prevalence of MADH4 and BMPR1A mutations in juvenile polyposis and absence of BMPR2, BMPR1B, and ACVR1 mutations, *J Med Genet* **41**(7):484–491, 2004.
524. Chen JD, Evans RM, A transcriptional corepressor that interacts with nuclear hormone receptors, *Nature* **377**(6548):454–457, 1995.
525. Shink E, *et al.*, Exclusion of non-synonymous SNPs and a polyglutamine tract in SMRT/N-CoR2 as common deleterious mutation for bipolar disorder in the Sagnenay-Lac-St-Jean population, *Am J Med Genet B Neuropsychiatr Genet* **134**(1):10–12, 2005.
526. Moilanen AM, *et al.*, Identification of a novel RING finger protein as a coregulator in steroid receptor-mediated gene transcription, *Mol Cell Biol* **18**(9):5128–5139, 1998.
527. Hirvonen-Santti SJ, *et al.*, Down-regulation of estrogen receptor beta and transcriptional coregulator SNURF/RNF4 in testicular germ cell cancer, *Eur Urol* **44**(6):742–747, 2003.

528. Horsthemke B, Buiting K, Imprinting defects on human chromosome 15, *Cytogenet Genome Res* **113**(1–4):292–299, 2006.
529. Rodriguez-Jato S, *et al.*, Characterization of cis- and trans-acting elements in the imprinted human SNURF-SNRPN locus, *Nucleic Acids Res* **33**(15):4740–4753, 2005.
530. Lanz RB, *et al.*, A steroid receptor coactivator, SRA, functions as an RNA and is present in an SRC-1 complex, *Cell* **97**(1):17–27, 1999.
531. Hussein-Fikret S, Fuller PJ, Expression of nuclear receptor coregulators in ovarian stromal and epithelial tumours, *Mol Cell Endocrinol* **229**(1–2): 149–160, 2005.
532. Leygue E, *et al.*, Expression of the steroid receptor RNA activator in human breast tumors, *Cancer Res* **59**(17):4190–4193, 1999.
533. Lanz RB, *et al.*, Steroid receptor RNA activator stimulates proliferation as well as apoptosis *in vivo*, *Mol Cell Biol* **23**(20):7163–7176, 2003.
534. Onate SA, *et al.*, Sequence and characterization of a coactivator for the steroid hormone receptor superfamily, *Science* **270**(5240):1354–1357, 1995.
535. Agoulnik IU, *et al.*, Role of SRC-1 in the promotion of prostate cancer cell growth and tumor progression, *Cancer Res* **65**(17):7959–7967, 2005.
536. Maki HE, *et al.*, Screening of genetic and expression alterations of SRC1 gene in prostate cancer, *Prostate* **66**(13):1391–1398, 2006.
537. Myers E, *et al.*, Inverse relationship between ER-beta and SRC-1 predicts outcome in endocrine-resistant breast cancer, *Br J Cancer* **91**(9): 1687–1693, 2004.
538. Wachtel M, *et al.*, Gene expression signatures identify rhabdomyosarcoma subtypes and detect a novel t(2; 2)(q35; p23) translocation fusing PAX3 to NCOA1, *Cancer Res* **64**(16):5539–5545, 2004.
539. Hong H, *et al.*, GRIP1, a novel mouse protein that serves as a transcriptional coactivator in yeast for the hormone binding domains of steroid receptors, *Proc Natl Acad Sci USA* **93**(10):4948–4952, 1996.
540. Voegel JJ, *et al.*, TIF2, a 160 kDa transcriptional mediator for the ligand-dependent activation function AF-2 of nuclear receptors, *EMBO J* **15**(14):3667–3675, 1996.
541. Kurebayashi J, *et al.*, Expression levels of estrogen receptor-alpha, estrogen receptor-beta, coactivators, and corepressors in breast cancer, *Clin Cancer Res* **6**(2):512–518, 2000.
542. Carroll RS, *et al.*, Expression of a subset of steroid receptor cofactors is associated with progesterone receptor expression in meningiomas, *Clin Cancer Res* **6**(9):3570–3575, 2000.
543. Culig Z, *et al.*, Mechanisms of endocrine therapy-responsive and -unresponsive prostate tumours, *Endocr Relat Cancer* **12**(2):229–244, 2005.
544. Carapeti M, *et al.*, Consistent fusion of MOZ and TIF2 in AML with inv(8)(p11q13), *Cancer Genet Cytogenet* **113**(1):70–72, 1999.

545. Collins HM, *et al.*, MOZ-TIF2 alters cofactor recruitment and histone modification at the RARbeta2 promoter: Differential effects of MOZ fusion proteins on CBP- and MOZ-dependent activators, *J Biol Chem* **281**(25):17124–17133, 2006.
546. Deguchi K, *et al.*, MOZ-TIF2-induced acute myeloid leukemia requires the MOZ nucleosome binding motif and TIF2-mediated recruitment of CBP, *Cancer Cell* **3**(3):259–271, 2003.
547. Imamura T, *et al.*, Rearrangement of the MOZ gene in pediatric therapy-related myelodysplastic syndrome with a novel chromosomal translocation t(2; 8)(p23; p11), *Genes Chromosomes Cancer* **36**(4):413–419, 2003.
548. Kindle KB, *et al.*, MOZ-TIF2 inhibits transcription by nuclear receptors and p53 by impairment of CBP function, *Mol Cell Biol* **25**(3):988–1002, 2005.
549. Liang J, *et al.*, Acute mixed lineage leukemia with an inv(8)(p11q13) resulting in fusion of the genes for MOZ and TIF2, *Blood* **92**(6):2118–2122, 1998.
550. Anzick SL, *et al.*, AIB1, a steroid receptor coactivator amplified in breast and ovarian cancer, *Science* **277**(5328):965–968, 1997.
551. Chen H, *et al.*, Nuclear receptor coactivator ACTR is a novel histone acetyltransferase and forms a multimeric activation complex with P/CAF and CBP/p300, *Cell* **90**(3):569–580, 1997.
552. Li H, Gomes PJ, Chen JD, RAC3, a steroid/nuclear receptor-associated coactivator that is related to SRC-1 and TIF2, *Proc Natl Acad Sci USA* **94**(16):8479–8484, 1997.
553. Takeshita A, *et al.*, TRAM-1, A novel 160-kDa thyroid hormone receptor activator molecule, exhibits distinct properties from steroid receptor coactivator-1, *J Biol Chem* **272**(44):27629–27634, 1997.
554. Torchia J, *et al.*, The transcriptional coactivator p/CIP binds CBP and mediates nuclear-receptor function, *Nature* **387**(6634):677–684, 1997.
555. Balmer NN, *et al.*, Steroid receptor coactivator AIB1 in endometrial carcinoma, hyperplasia and normal endometrium: Correlation with clinicopathologic parameters and biomarkers, *Mod Pathol* **19**(12):1593–1605, 2006.
556. Xie D, *et al.*, Correlation of AIB1 overexpression with advanced clinical stage of human colorectal carcinoma, *Hum Pathol* **36**(7):777–783, 2005.
557. Burwinkel B, *et al.*, Association of NCOA3 polymorphisms with breast cancer risk, *Clin Cancer Res* **11**(6):2169–2174, 2005.
558. Chen YJ, *et al.*, Genome-wide profiling of oral squamous cell carcinoma, *J Pathol* **204**(3):326–332, 2004.
559. Liu CJ, *et al.*, Array-comparative genomic hybridization to detect genomewide changes in microdissected primary and metastatic oral squamous cell carcinomas, *Mol Carcinog* **45**(10):721–731, 2006.

560. Wang Y, *et al.*, Prognostic significance of c-myc and AIB1 amplification in hepatocellular carcinoma. A broad survey using high-throughput tissue microarray, *Cancer* **95**(11):2346–2352, 2002.
561. Fritz B, *et al.*, Microarray-based copy number and expression profiling in dedifferentiated and pleomorphic liposarcoma, *Cancer Res* **62**(11):2993–2998, 2002.
562. Hsing AW, *et al.*, Polymorphic CAG/CAA repeat length in the AIB1/SRC-3 gene and prostate cancer risk: A population-based case-control study, *Cancer Epidemiol Biomarkers Prev* **11**(4):337–341, 2002.
563. Colilla S, *et al.*, The joint effect of smoking and AIB1 on breast cancer risk in BRCA1 mutation carriers, *Carcinogenesis* **27**(3):599–605, 2006.
564. Hughes DJ, *et al.*, Breast cancer risk in BRCA1 and BRCA2 mutation carriers and polyglutamine repeat length in the AIB1 gene, *Int J Cancer* **117**(2):230–233, 2005.
565. Kadouri L, *et al.*, Polyglutamine repeat length in the AIB1 gene modifies breast cancer susceptibility in BRCA1 carriers, *Int J Cancer* **108**(3):399–403, 2004.
566. Montgomery KG, *et al.*, The AIB1 glutamine repeat polymorphism is not associated with risk of breast cancer before age 40 years in Australian women, *Breast Cancer Res* **7**(3):R353–356, 2005.
567. Rebbeck TR, *et al.*, Modification of BRCA1- and BRCA2-associated breast cancer risk by AIB1 genotype and reproductive history, *Cancer Res* **61**(14):5420–5424, 2001.
568. Spurdle AB, *et al.*, The AIB1 polyglutamine repeat does not modify breast cancer risk in BRCA1 and BRCA2 mutation carriers, *Cancer Epidemiol Biomarkers Prev* **15**(1):76–79, 2006.
569. Bouras T, Southey MC, Venter DJ, Overexpression of the steroid receptor coactivator AIB1 in breast cancer correlates with the absence of estrogen and progesterone receptors and positivity for p53 and HER2/neu, *Cancer Res* **61**(3):903–907, 2001.
570. Hudelist G, *et al.*, Expression of sex steroid receptors and their co-factors in normal and malignant breast tissue: AIB1 is a carcinoma-specific co-activator, *Breast Cancer Res Treat* **78**(2):193–204, 2003.
571. Jernstrom H, *et al.*, Genetic factors related to racial variation in plasma levels of insulin-like growth factor-1: Implications for premenopausal breast cancer risk, *Mol Genet Metab* **72**(2):144–154, 2001.
572. Sakakura C, *et al.*, Amplification and over-expression of the AIB1 nuclear receptor coactivator gene in primary gastric cancers, *Int J Cancer* **89**(3):217–223, 2000.
573. Xu FP, *et al.*, SRC-3/AIB1 protein and gene amplification levels in human esophageal squamous cell carcinomas, *Cancer Lett* **245**(1–2):69–74, 2007.

574. Zhao C, *et al.*, Elevated expression levels of NCOA3, TOP1, and TFAP2C in breast tumors as predictors of poor prognosis, *Cancer* **98**(1):18–23, 2003.
575. Li AJ, *et al.*, AIB1 polymorphisms predict aggressive ovarian cancer phenotype, *Cancer Epidemiol Biomarkers Prev* **14**(12):2919–2922, 2005.
576. Tanner MM, *et al.*, Frequent amplification of chromosomal region 20q12-q13 in ovarian cancer, *Clin Cancer Res* **6**(5):1833–1839, 2000.
577. Dai P, Wong LJ, Somatic instability of the DNA sequences encoding the polymorphic polyglutamine tract of the AIB1 gene, *J Med Genet* **40**(12):885–890, 2003.
578. Haiman CA, *et al.*, Polymorphisms in steroid hormone pathway genes and mammographic density, *Breast Cancer Res Treat* **77**(1):27–36, 2003.
579. Haiman CA, *et al.*, Polymorphic repeat in AIB1 does not alter breast cancer risk, *Breast Cancer Res* **2**(5):378–385, 2000.
580. Hayashi Y, *et al.*, Polymorphism of homopolymeric glutamines in coactivators for nuclear hormone receptors, *Endocr J* **46**(2):279–284, 1999.
581. Shibata A, *et al.*, Somatic gene alteration of AIB1 gene in patients with breast cancer, *Endocr J* **48**(2):199–204, 2001.
582. Shirazi SK, Bober MA, Coetzee GA, Polymorphic exonic CAG microsatellites in the gene amplified in breast cancer (AIB1 gene), *Clin Genet* **54**(1):102–103, 1998.
583. Astrow AB, When I can't make you live, *Hastings Cent Rep* **22**(1):45–46, 1992.
584. Ghadimi BM, *et al.*, Specific chromosomal aberrations and amplification of the AIB1 nuclear receptor coactivator gene in pancreatic carcinomas, *Am J Pathol* **154**(2):525–536, 1999.
585. Guan XY, *et al.*, Hybrid selection of transcribed sequences from microdissected DNA: Isolation of genes within amplified region at 20q11-q13.2 in breast cancer, *Cancer Res* **56**(15):3446–3450, 1996.
586. Iwase H, *et al.*, Clinical significance of AIB1 expression in human breast cancer, *Breast Cancer Res Treat* **80**(3):339–345, 2003.
587. List HJ, *et al.*, Expression of the nuclear coactivator AIB1 in normal and malignant breast tissue, *Breast Cancer Res Treat* **68**(1):21–28, 2001.
588. Murphy LC, *et al.*, Altered expression of estrogen receptor coregulators during human breast tumorigenesis, *Cancer Res* **60**(22):6266–6271, 2000.
589. Patel MS, *et al.*, Alleles of the estrogen receptor alpha-gene and an estrogen receptor cotranscriptional activator gene, amplified in breast cancer-1 (AIB1), are associated with quantitative calcaneal ultrasound, *J Bone Miner Res* **15**(11):2231–2239, 2000.
590. Wasserman L, *et al.*, Correlates of obesity in postmenopausal women with breast cancer: Comparison of genetic, demographic, disease-related, life history and dietary factors, *Int J Obes Relat Metab Disord* **28**(1):49–56, 2004.

591. Sakaguchi H, *et al.*, Clinical implications of steroid receptor coactivator (SRC)-3 in uterine endometrial cancers, *J Steroid Biochem Mol Biol* **104**(3–5):237–240, 2007.
592. Yuan X, *et al.*, SRY interacts with and negatively regulates androgen receptor transcriptional activity, *J Biol Chem* **276**(49):46647–46654, 2001.
593. Hellani A, *et al.*, Y chromosome microdeletions: Are they implicated in teratozoospermia? *Hum Reprod* **20**(12):3505–3509, 2005.
594. Nikolova G, Vilain E, Mechanisms of disease: Transcription factors in sex determination — Relevance to human disorders of sex development, *Nat Clin Pract Endocrinol Metab* **2**(4):231–238, 2006.
595. Battiloro E, *et al.*, A novel double nucleotide substitution in the HMG box of the SRY gene associated with Swyer syndrome, *Hum Genet* **100**(5–6):585–587, 1997.
596. Canto P, *et al.*, A mutation in the 5′ non-high mobility group box region of the SRY gene in patients with Turner syndrome and Y mosaicism, *J Clin Endocrinol Metab* **85**(5):1908–1911, 2000.
597. Domenice S, *et al.*, A novel missense mutation (S18N) in the 5′ non-HMG box region of the SRY gene in a patient with partial gonadal dysgenesis and his normal male relatives, *Hum Genet* **102**(2):213–215, 1998.
598. Jordan BK, *et al.*, Familial mutation in the testis-determining gene SRY shared by an XY female and her normal father, *J Clin Endocrinol Metab* **87**(7):3428–3432, 2002.
599. Schaffler A, *et al.*, Identification of a new missense mutation (Gly95Glu) in a highly conserved codon within the high-mobility group box of the sex-determining region Y gene: Report on a 46,XY female with gonadal dysgenesis and yolk-sac tumor, *J Clin Endocrinol Metab* **85**(6):2287–2292, 2000.
600. Truica CI, Byers S, Gelmann EP, Beta-catenin affects androgen receptor transcriptional activity and ligand specificity, *Cancer Res* **60**(17):4709–4713, 2000.
601. Varallo VM, *et al.*, Beta-catenin expression in Dupuytren's disease: Potential role for cell-matrix interactions in modulating beta-catenin levels *in vivo* and *in vitro*, *Oncogene* **22**(24):3680–3684, 2003.
602. Barker N, Morin PJ, Clevers H, The Yin-Yang of TCF/beta-catenin signaling, *Adv Cancer Res* **77**:1–24, 2000.
603. Saldanha G, *et al.*, Nuclear beta-catenin in basal cell carcinoma correlates with increased proliferation, *Br J Dermatol* **151**(1):157–164, 2004.
604. Martensson A, *et al.*, Beta-catenin expression in relation to genetic instability and prognosis in colorectal cancer, *Oncol Rep* **17**(2):447–452, 2007.
605. Taniguchi K, *et al.*, Mutational spectrum of beta-catenin, AXIN1, and AXIN2 in hepatocellular carcinomas and hepatoblastomas, *Oncogene* **21**(31):4863–4871, 2002.

606. Chan EF, *et al.*, A common human skin tumour is caused by activating mutations in beta-catenin, *Nat Genet* **21**(4):410–413, 1999.
607. Clifford SC, *et al.*, Wnt/Wingless pathway activation and chromosome 6 loss characterize a distinct molecular sub-group of medulloblastomas associated with a favorable prognosis, *Cell Cycle* **5**(22):2666–2670, 2006.
608. Uschkereit C, *et al.*, Different CTNNB1 mutations as molecular genetic proof for the independent origin of four Wilms tumours in a patient with a novel germ line WT1 mutation, *J Med Genet* **44**(6):393–396, 2007.
609. De Miguel F, *et al.*, Stat3 enhances transactivation of steroid hormone receptors, *Nucl Recept* **1**(1):3, 2003.
610. Shah NG, *et al.*, Stat3 expression in oral squamous cell carcinoma: Association with clinicopathological parameters and survival, *Int J Biol Markers* **21**(3):175–183, 2006.
611. Duan Z, *et al.*, Signal transducers and activators of transcription 3 pathway activation in drug-resistant ovarian cancer, *Clin Cancer Res* **12**(17):5055–5063, 2006.
612. Rosen DG, *et al.*, The role of constitutively active signal transducer and activator of transcription 3 in ovarian tumorigenesis and prognosis, *Cancer* **107**(11):2730–2740, 2006.
613. Choi JH, *et al.*, Phospho-Stat3 expression and correlation with VEGF, p53, and Bcl-2 in gastric carcinoma using tissue microarray, *Apmis* **114**(9):619–625, 2006.
614. Bowman T, *et al.*, STATs in oncogenesis, *Oncogene* **19**(21):2474–2488, 2000.
615. Litonjua AA, *et al.*, Polymorphisms in signal transducer and activator of transcription 3 and lung function in asthma, *Respir Res* **6**:52, 2005.
616. Poukka H, *et al.*, Covalent modification of the androgen receptor by small ubiquitin-like modifier 1 (SUMO-1), *Proc Natl Acad Sci USA* **97**(26):14145–14150, 2000.
617. Li M, *et al.*, SUMO wrestling with type 1 diabetes, *J Mol Med* **83**(7):504–513, 2005.
618. Pountney DL, *et al.*, SUMO-1 marks the nuclear inclusions in familial neuronal intranuclear inclusion disease, *Exp Neurol* **184**(1):436–446, 2003.
619. Ueda H, *et al.*, Enhanced SUMOylation in polyglutamine diseases, *Biochem Biophys Res Commun* **293**(1):307–313, 2002.
620. Perani M, *et al.*, The proto-oncoprotein SYT interacts with SYT-interacting protein/coactivator activator (SIP/CoAA), a human nuclear receptor coactivator with similarity to EWS and TLS/FUS family of proteins, *J Biol Chem* **280**(52):42863–42876, 2005.
621. Begueret H, *et al.*, Primary intrathoracic synovial sarcoma: A clinicopathologic study of 40 t(X; 18)-positive cases from the French Sarcoma Group and the Mesopath Group, *Am J Surg Pathol* **29**(3):339–346, 2005.

622. Loven MA, *et al.*, A novel estrogen receptor alpha-associated protein, template-activating factor Ibeta, inhibits acetylation and transactivation, *Mol Endocrinol* **17**(1):67–78, 2003.
623. Ouellet V, *et al.*, SET complex in serous epithelial ovarian cancer, *Int J Cancer* **119**(9):2119–2126, 2006.
624. Rosati R, *et al.*, Cryptic chromosome 9q34 deletion generates TAF-Ialpha/CAN and TAF-Ibeta/CAN fusion transcripts in acute myeloid leukemia, *Haematologica* **92**(2):232–235, 2007.
625. Guenther MG, *et al.*, A core SMRT corepressor complex containing HDAC3 and TBL1, a WD40-repeat protein linked to deafness, *Genes Dev* **14**(9):1048–1057, 2000.
626. Bassi MT, *et al.*, X-linked late-onset sensorineural deafness caused by a deletion involving OA1 and a novel gene containing WD-40 repeats, *Am J Hum Genet* **64**(6):1604–1616, 1999.
627. Saville B, *et al.*, Cooperative coactivation of estrogen receptor alpha in ZR-75 human breast cancer cells by SNURF and TATA-binding protein, *J Biol Chem* **277**(4):2485–2497, 2002.
628. Shatunov A, *et al.*, Small *de novo* duplication in the repeat region of the TATA-box-binding protein gene manifest with a phenotype similar to variant Creutzfeldt-Jakob disease, *Clin Genet* **66**(6):496–501, 2004.
629. Owerbach D, Pina L, Gabbay KH, Association of a CAG/CAA repeat sequence in the TBP gene with type I diabetes, *Biochem Biophys Res Commun* **323**(3):865–869, 2004.
630. Bruni AC, *et al.*, Behavioral disorder, dementia, ataxia, and rigidity in a large family with TATA box-binding protein mutation, *Arch Neurol* **61**(8):1314–1320, 2004.
631. Koide R, *et al.*, A neurological disease caused by an expanded CAG trinucleotide repeat in the TATA-binding protein gene: A new polyglutamine disease? *Hum Mol Genet* **8**(11):2047–2053, 1999.
632. Salvatore E, *et al.*, Characterization of nigrostriatal dysfunction in spinocerebellar ataxia 17, *Mov Disord* **21**(6):872–875, 2006.
633. Silveira I, *et al.*, Trinucleotide repeats in 202 families with ataxia: A small expanded (CAG)n allele at the SCA17 locus, *Arch Neurol* **59**(4):623–629, 2002.
634. van Roon-Mom WM, *et al.*, TATA-binding protein in neurodegenerative disease, *Neuroscience* **133**(4):863–872, 2005.
635. Wu YR, *et al.*, Analysis of polyglutamine-coding repeats in the TATA-binding protein in different neurodegenerative diseases, *J Neural Transm* **112**(4):539–546, 2005.
636. Zuhlke C, *et al.*, Different types of repeat expansion in the TATA-binding protein gene are associated with a new form of inherited ataxia, *Eur J Hum Genet* **9**(3):160–164, 2001.

637. Bauer P, *et al.*, Trinucleotide repeat expansion in SCA17/TBP in white patients with Huntington's disease-like phenotype, *J Med Genet* **41**(3):230–232, 2004.
638. Stevanin G, *et al.*, Huntington's disease-like phenotype due to trinucleotide repeat expansions in the TBP and JPH3 genes, *Brain* **126**(Pt 7): 1599–1603, 2003.
639. Chen CM, *et al.*, Expanded trinucleotide repeats in the TBP/SCA17 gene mapped to chromosome 6q27 are associated with schizophrenia, *Schizophr Res* **78**(2–3):131–136, 2005.
640. Chen D, *et al.*, T:G mismatch-specific thymine-DNA glycosylase potentiates transcription of estrogen-regulated genes through direct interaction with estrogen receptor alpha, *J Biol Chem* **278**(40):38586–38592, 2003.
641. Broderick P, *et al.*, Evaluation of NTHL1, NEIL1, NEIL2, MPG, TDG, UNG and SMUG1 genes in familial colorectal cancer predisposition, *BMC Cancer* **6**:243, 2006.
642. Sharma M, Sun Z, 5′TG3′ interacting factor interacts with Sin3A and represses AR-mediated transcription, *Mol Endocrinol* **15**(11):1918–1928, 2001.
643. Hu ZL, *et al.*, Expressions of TGIF, MMP9 and VEGF proteins and their clinicopathological relationship in gastric cancer, *Zhong Nan Da Xue Xue Bao Yi Xue Ban* **31**(1):70–74, 2006.
644. Gripp KW, *et al.*, Mutations in TGIF cause holoprosencephaly and link NODAL signalling to human neural axis determination, *Nat Genet* **25**(2):205–208, 2000.
645. Le Douarin B, *et al.*, A possible involvement of TIF1 alpha and TIF1 beta in the epigenetic control of transcription by nuclear receptors, *EMBO J* **15**(23):6701–6715, 1996.
646. Thenot S, *et al.*, Differential interaction of nuclear receptors with the putative human transcriptional coactivator hTIF1, *J Biol Chem* **272**(18): 12062–12068, 1997.
647. Klugbauer S, Rabes HM The transcription coactivator HTIF1 and a related protein are fused to the RET receptor tyrosine kinase in childhood papillary thyroid carcinomas, *Oncogene* **18**(30):4388–4393, 1999.
648. Gandini D, *et al.*, Preferential expression of the transcription coactivator HTIF1alpha gene in acute myeloid leukemia and MDS-related AML, *Leukemia* **16**(5):886–893, 2002.
649. Nakajima T, *et al.*, TIP27: A novel repressor of the nuclear orphan receptor TAK1/TR4, *Nucleic Acids Res* **32**(14):4194–4204, 2004.
650. Huang HY, Ladanyi M, Soslow RA, Molecular detection of JAZF1-JJAZ1 gene fusion in endometrial stromal neoplasms with classic and variant histology: Evidence for genetic heterogeneity, *Am J Surg Pathol* **28**(2): 224–232, 2004.

651. Koontz JI, *et al.*, Frequent fusion of the JAZF1 and JJAZ1 genes in endometrial stromal tumors, *Proc Natl Acad Sci USA* **98**(11):6348–6353, 2001.
652. Yang Z, *et al.*, Transgelin functions as a suppressor via inhibition of ARA54-enhanced androgen receptor transactivation and prostate cancer cell growth, *Mol Endocrinol* **21**(2):343–358, 2007.
653. Shields JM, Rogers-Graham K, Der CJ, Loss of transgelin in breast and colon tumors and in RIE-1 cells by Ras deregulation of gene expression through Raf-independent pathways, *J Biol Chem* **277**(12):9790–9799, 2002.
654. Wang Q, *et al.*, A coregulatory role for the TRAP-mediator complex in androgen receptor-mediated gene expression, *J Biol Chem* **277**(45):42852–42858, 2002.
655. Clark J, *et al.*, Identification of amplified and expressed genes in breast cancer by comparative hybridization onto microarrays of randomly selected cDNA clones, *Genes Chromosomes Cancer* **34**(1):104–114, 2002.
656. Ito M, *et al.*, Identity between TRAP and SMCC complexes indicates novel pathways for the function of nuclear receptors and diverse mammalian activators, *Mol Cell* **3**(3):361–370, 1999.
657. Kirov G, *et al.*, Association analysis of the HOPA12bp polymorphism in schizophrenia and manic depressive illness, *Am J Med Genet B Neuropsychiatr Genet* **118**(1):16–19, 2003.
658. Spinks R, *et al.*, Association of the HOPA12bp allele with a large X-chromosome haplotype and positive symptom schizophrenia, *Am J Med Genet B Neuropsychiatr Genet* **127**(1):20–27, 2004.
659. Philibert RA, *et al.*, Population-based association analyses of the HOPA12bp polymorphism for schizophrenia and hypothyroidism, *Am J Med Genet* **105**(1):130–134, 2001.
660. Chang KH, *et al.*, A thyroid hormone receptor coactivator negatively regulated by the retinoblastoma protein, *Proc Natl Acad Sci USA* **94**(17):9040–9045, 1997.
661. Abe A, *et al.*, Fusion of the platelet-derived growth factor receptor beta to a novel gene CEV14 in acute myelogenous leukemia after clonal evolution, *Blood* **90**(11):4271–4277, 1997.
662. Yanagisawa J, *et al.*, Nuclear receptor function requires a TFTC-type histone acetyl transferase complex, *Mol Cell* **9**(3):553–562, 2002.
663. Bashyam MD, *et al.*, Array-based comparative genomic hybridization identifies localized DNA amplifications and homozygous deletions in pancreatic cancer, *Neoplasia* **7**(6):556–562, 2005.
664. Liu GH, Qu J, Shen X, Thioredoxin-mediated negative autoregulation of peroxisome proliferator-activated receptor alpha transcriptional activity, *Mol Biol Cell* **17**(4):1822–1833, 2006.

665. Makino Y, *et al.*, Direct association with thioredoxin allows redox regulation of glucocorticoid receptor function, *J Biol Chem* **274**(5):3182–3188, 1999.
666. Kim HJ, *et al.*, Preferential elevation of Prx I and Trx expression in lung cancer cells following hypoxia and in human lung cancer tissues, *Cell Biol Toxicol* **19**(5):285–298, 2003.
667. Raffel J, *et al.*, Increased expression of thioredoxin-1 in human colorectal cancer is associated with decreased patient survival, *J Lab Clin Med* **142**(1):46–51, 2003.
668. Wahlgren CM, Pekkari K, Elevated thioredoxin after angioplasty in peripheral arterial disease, *Eur J Vasc Endovasc Surg* **29**(3):281–286, 2005.
669. Damdimopoulos AE, *et al.*, An alternative splicing variant of the selenoprotein thioredoxin reductase is a modulator of estrogen signaling, *J Biol Chem* **279**(37):38721–38729, 2004.
670. Furman C, *et al.*, Thioredoxin reductase 1 is upregulated in atherosclerotic plaques: Specific induction of the promoter in human macrophages by oxidized low-density lipoproteins, *Free Radic Biol Med* **37**(1):71–85, 2004.
671. Henry KW, *et al.*, Tuberous sclerosis gene 2 product modulates transcription mediated by steroid hormone receptor family members, *J Biol Chem* **273**(32):20535–20539, 1998.
672. Merritt JL, *et al.*, Extensive acrochordons and pancreatic islet-cell tumors in tuberous sclerosis associated with TSC2 mutations, *Am J Med Genet A* **140**(15):1669–1672, 2006.
673. Crino PB, Nathanson KL, Henske EP, The tuberous sclerosis complex, *N Engl J Med* **355**(13):1345–1356, 2006.
674. Jobert S, *et al.*, Deletion of 11 amino acids in tuberin associated with severe tuberous sclerosis phenotypes: Evidence for a new essential domain in the first third of the protein, *Eur J Hum Genet* **5**(5):280–287, 1997.
675. Sancak O, *et al.*, Mutational analysis of the TSC1 and TSC2 genes in a diagnostic setting: Genotype — phenotype correlations and comparison of diagnostic DNA techniques in Tuberous Sclerosis Complex, *Eur J Hum Genet* **13**(6):731–741, 2005.
676. York B, Lou D, Noonan DJ, Tuberin nuclear localization can be regulated by phosphorylation of its carboxyl terminus, *Mol Cancer Res* **4**(11):885–897, 2006.
677. Platten M, *et al.*, A novel splice site associated polymorphism in the tuberous sclerosis 2 (TSC2) gene may predispose to the development of sporadic gangliogliomas, *J Neuropathol Exp Neurol* **56**(7):806–810, 1997.
678. Goncharova EA, Krymskaya VP, Pulmonary lymphangioleiomyomatosis (LAM): Progress and current challenges, *J Cell Biochem*, 2007.
679. Watanabe M, *et al.*, A putative tumor suppressor, TSG101, acts as a transcriptional suppressor through its coiled-coil domain, *Biochem Biophys Res Commun* **245**(3):900–905, 1998.

680. Bashirova AA, *et al.*, Consistent effects of TSG101 genetic variability on multiple outcomes of exposure to human immunodeficiency virus type 1, *J Virol* **80**(14):6757–6763, 2006.
681. Koon N, *et al.*, Molecular targets for tumour progression in gastrointestinal stromal tumours, *Gut* **53**(2):235–240, 2004.
682. Klaes R, *et al.*, Significant increase of a specific variant TSG101 transcript during the progression of cervical neoplasia, *Eur J Cancer* **35**(5):733–737, 1999.
683. Lin PM, *et al.*, Aberrant TSG101 transcripts in acute myeloid leukaemia, *Br J Haematol* **102**(3):753–758, 1998.
684. Sun Z, *et al.*, Frequent abnormalities of TSG101 transcripts in human prostate cancer, *Oncogene* **15**(25):3121–3125, 1997.
685. Balz V, *et al.*, Analysis of BRCA1, TP53, and TSG101 germline mutations in German breast and/or ovarian cancer families, *Cancer Genet Cytogenet* **138**(2):120–127, 2002.
686. Li L, *et al.*, The TSG101 tumor susceptibility gene is located in chromosome 11 band p15 and is mutated in human breast cancer, *Cell* **88**(1):143–154, 1997.
687. Ishizuka M, *et al.*, A zinc finger protein TZF is a novel corepressor of androgen receptor, *Biochem Biophys Res Commun* **331**(4):1025–1031, 2005.
688. Gottlicher M, *et al.*, Interaction of the Ubc9 human homologue with c-Jun and with the glucocorticoid receptor, *Steroids* **61**(4):257–262, 1996.
689. Poukka H, *et al.*, Ubc9 interacts with the androgen receptor and activates receptor-dependent transcription, *J Biol Chem* **274**(27):19441–19446, 1999.
690. Mo YY, Moschos SJ, Targeting Ubc9 for cancer therapy, *Expert Opin Ther Targets* **9**(6):1203–1216, 2005.
691. Lyons LS, Burnstein KL, Vav3, a Rho GTPase guanine nucleotide exchange factor, increases during progression to androgen independence in prostate cancer cells and potentiates androgen receptor transcriptional activity, *Mol Endocrinol* **20**(5):1061–1072, 2006.
692. Dong Z, *et al.*, Vav3 oncogene is overexpressed and regulates cell growth and androgen receptor activity in human prostate cancer, *Mol Endocrinol* **20**(10):2315–2325, 2006.
693. Kitagawa H, *et al.*, The chromatin-remodeling complex WINAC targets a nuclear receptor to promoters and is impaired in Williams syndrome, *Cell* **113**(7):905–917, 2003.
694. Lu X, *et al.*, A novel human gene, WSTF, is deleted in Williams syndrome, *Genomics* **54**(2):241–249, 1998.
695. Froidevaux MS, *et al.*, The co-chaperone XAP2 is required for activation of hypothalamic thyrotropin-releasing hormone transcription *in vivo*, *EMBO Rep* **7**(10):1035–1039, 2006.

696. Georgitsi M, *et al.*, Molecular diagnosis of pituitary adenoma predisposition caused by aryl hydrocarbon receptor-interacting protein gene mutations, *Proc Natl Acad Sci USA* **104**(10):4101–4105, 2007.
697. Vierimaa O, *et al.*, Pituitary adenoma predisposition caused by germline mutations in the AIP gene, *Science* **312**(5777):1228–1230, 2006.
698. Ding L, *et al.*, Ligand-independent activation of estrogen receptor alpha by XBP-1, *Nucleic Acids Res* **31**(18):5266–5274, 2003.
699. Kishimoto T, *et al.*, Enhanced expression of a new class of liver-enriched b-Zip transcription factors, hepatocarcinogenesis-related transcription factor, in hepatocellular carcinomas of rats and humans, *Cell Growth Differ* **9**(4):337–344, 1998.
700. Perou CM, *et al.*, Molecular portraits of human breast tumours, *Nature* **406**(6797):747–752, 2000.
701. Chen W, *et al.*, A case-control study provides evidence of association for a functional polymorphism — 197C/G in XBP1 to schizophrenia and suggests a sex-dependent effect, *Biochem Biophys Res Commun* **319**(3): 866–870, 2004.
702. Kakiuchi C, *et al.*, Impaired feedback regulation of XBP1 as a genetic risk factor for bipolar disorder, *Nat Genet* **35**(2):171–175, 2003.
703. Masui T, *et al.*, A possible association between the — 116C/G single nucleotide polymorphism of the XBP1 gene and lithium prophylaxis in bipolar disorder, *Int J Neuropsychopharmacol* **9**(1):83–88, 2006.
704. Kakiuchi C, *et al.*, Association of the XBP1-116C/G polymorphism with schizophrenia in the Japanese population, *Psychiatry Clin Neurosci* **58**(4):438–440, 2004.
705. Kim YS, *et al.*, Role of 14-3-3 eta as a positive regulator of the glucocorticoid receptor transcriptional activation, *Endocrinology* **146**(7):3133–3140, 2005.
706. Toyooka K, *et al.*, 14-3-3 protein eta chain gene (YWHAH) polymorphism and its genetic association with schizophrenia, *Am J Med Genet* **88**(2):164–167, 1999.

Chapter 2

p160 Coactivators: Critical Mediators of Transcriptional Activation by Nuclear Receptors

Jeong Hoon Kim and Michael R. Stallcup

Nuclear receptors activate transcription by binding to specific enhancer elements associated with their target genes and recruiting coactivators, which remodel chromatin structure and recruit and activate RNA polymerase II and the basal transcription factors. A very large number of coactivators have been identified, suggesting that the above processes are extremely complex and highly regulated. Among the many coactivators involved, the p160 or steroid receptor coactivators (SRCs) were the first identified nuclear receptor coactivators and have been shown to play crucial roles. They are apparently among the first coactivators recruited to the promoter by hormone-activated nuclear receptors, and they appear to be required for many of the subsequent steps leading to chromatin remodeling and assembly of a transcription initiation complex. By associating with DNA-bound NRs, SRCs serve as large scaffolds to assemble other coactivators on the promoter, including histone acetyltransferases, histone methyltransferases, and other coactivators that help to assemble the transcription initiation complex. Inappropriate expression of SRCs has also been strongly implicated in a number of diseases, particularly hormone-dependent cancers. Thus, the study of this three-member coactivator family is helping to unlock the secrets of transcriptional activation, and at the same time, elucidate new potential diagnostic and therapeutic strategies for human diseases.

2.1 Introduction

The genetic information of higher eukaryotic organisms is embedded in the context of chromatin. Since expression of genes depends partly on their accessibility, regulation of chromatin architecture plays a critical role. The nucleosome, the basic building block of chromatin, is formed by the association of a 147-base-pair stretch of DNA with a histone octamer formed by two molecules of each of the four core histones, H2A, H2B, H3, and H4.[1,2] These nucleosomes can be further packaged with the aid of histone H1 and other non-histone proteins, resulting in condensed chromatin structures which restrict access to transcription factors and basal transcription machinery. Chromatin structure not only serves to compact the genome but also plays a major role in the regulation of gene expression *in vivo*. A variety of posttranslational modifications of the histone N-terminal tails can increase or decrease the DNA accessibility by inducing open or closed chromatin structures.[1,2] Such modifications are proposed to constitute a "histone code" that represents an epigenetic marking mechanism for controlling gene transcription and other chromatin-regulated processes.[2] Posttranslational modifications occur not only on histones but also on DNA-binding transcription factors and components of coactivator complexes at target gene promoters;[3–5] these modifications thus play major roles in all aspects of transcription and its regulation.

The nuclear receptors (NR) are ligand-regulated transcription factors including receptors for steroid and thyroid hormones, retinoic acid and vitamin D, as well as orphan receptors.[6–8] NRs share a structural organization that consists of a variable N-terminal region, a conserved DNA binding domain (DBD) containing two zinc fingers, and a conserved ligand-binding domain (LBD). NRs have two distinct transcriptional activation functions (AF), AF-1 in the N-terminal region and the hormone dependent AF-2 in the LBD. The overall three-dimensional structures of NR LBDs are similar, and ligand binding induces a conformational change which is necessary for binding of transcriptional coactivators.[3,9]

The observation of transcriptional interference or squelching between NRs without any direct interaction or any overlapping DNA binding suggested the existence of limiting amounts of common transcriptional coactivators that mediate the NR transcriptional function.[6] The observation that the transcriptional activities of NR are manifested in a tissue-specific manner also suggested that NRs do not function by themselves, but rather, require specific cellular cofactors for maximal responses. After the identification of the first NR coactivator, steroid receptor coactivator 1 (SRC-1),[10] more than 200 NR coactivators have been identified and characterized (http://www.nursa.org). Primary coactivators are defined here as proteins that interact directly with activation functions of NRs, whereas secondary coactivators associate indirectly with NRs through primary coactivators. Many coactivators bind NRs in a hormone-dependent manner. The role of coactivators is to mediate transcriptional activation. Coactivators do not bind directly to DNA but are recruited to specific promoters by ligand-activated and DNA-bound NRs or other classes of transcription factors. The action of promoter-associated coactivators on the process of transcription may involve enzymatic activities necessary for histone modification and chromatin remodeling. In addition, coactivators also function via protein-protein interaction with the components of the basal transcription machinery and thus serve as a bridge between NRs and the basal transcription machinery. Thus coactivators contribute to transcriptional activation by helping to remodel chromatin structure and recruit RNA polymerase II and its associated basal transcriptional machinery.

The large number of coactivators apparently involved in the process of transcriptional activation indicates that the process is extremely complex and involves many steps and sources of regulatory input of which we are currently unaware. Among the many known coactivators, three large complexes (Fig. 2.1) have received considerable attention, although we still understand their functions only partially. The p160 steroid receptor coactivators (SRC) and their associated coactivators[7,11] serve as platforms for the recruitment of histone modifying enzymes including histone acetyltransferases (HAT) and histone methyltransferases (HMT). SRCs also recruit a variety of non-enzymatic secondary coactivators which apparently contribute to transcriptional activation through protein-protein interactions. The ATP-dependent chromatin remodeling complexes, such as the SWI/SNF (switch defective/sucrose non-fermenter) complex, use energy from ATPase subunits such as BRM1 (brahma) and BRG1 (brahma-related gene 1) to modify the histone-DNA interface and introduce topological change in closed circular nucleosomal arrays; this apparently causes nucleosome sliding and increases DNA accessibility within nucleosomes.[8] The TRAP/DRIP/mediator complexes contain subunits that bind to NRs, recruit RNA polymerase II, and interact with the basal transcription apparatus.[12,13]

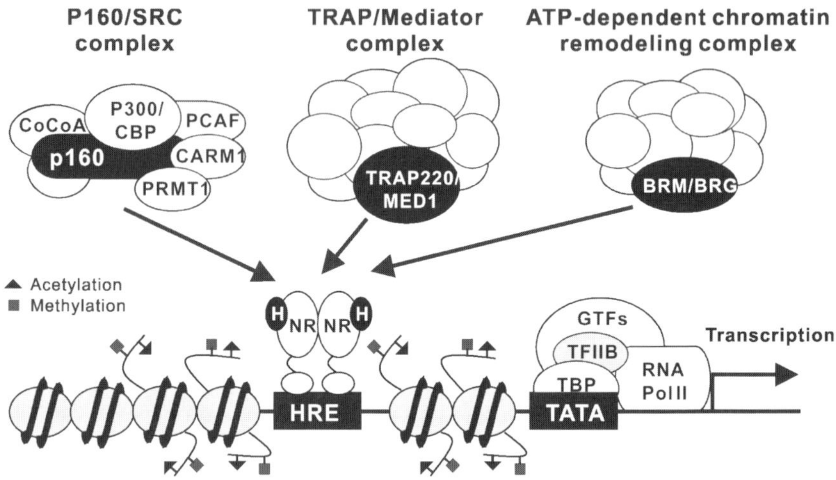

Fig. 2.1. Multiple coactivator complexes participate in NR-mediated transcription. The p160/SRC coactivator complex possesses histone acetyltransferase and histone methyltransferase activities, which cause local acetylation and methylation of nucleosomal histones. The SWI/SNF complex possesses ATP-dependent chromatin remodeling activities. The Mediator complex facilitates recruitment of RNA polymerase II (RNA Pol II) and general transcription factors (GTFs). These coactivator complexes are recruited by hormone-activated NRs to their target gene promoters and act in concert to remodel the chromatin structure, recruit the general transcription machinery, and enhance NR target gene transcription. H, hormone; HRE, hormone response element.

Presumably, the actions of these three coactivator complexes and many additional coactivators must be somehow choreographed or coordinated to integrate their actions toward the regulation of a single event — transcription initiation. Among many contributing NR coactivators, the SRCs appear to play a pivotal and perhaps seminal role in the complex process of transcriptional activation.

2.2 p160/SRC Coactivators

2.2.1 *Identification of SRC family members as NR coactivators*

SRC-1 was initially identified in a yeast two-hybrid screen as a ligand-dependent binding protein for the progesterone receptor (PR) ligand binding domain (LBD)[10,11] and was subsequently shown to interact with other NRs and potentiate NR activity in a ligand-dependent manner.[6,11] The observation that SRC-1 could reverse squelching of PR transcriptional activity by estrogen receptor (ER) suggested that it was one of the limiting cofactors required by both receptors. SRC-2/GRIP1 (glucocorticoid receptor interacting protein)/TIF-2 (transcriptional intermediary factor-2) was identified through its interaction with LBDs of glucocorticoid receptor (GR) and ER.[11] Overexpression of SRC-2, like SRC-1, is capable of relieving squelching by ER. SRC-3 was identified by several laboratories using different cloning strategies and has a variety of names, including p/CIP (p300/CBP cointegrator associated protein); RAC3 (RAR-associated coactivator 3); ACTR (activator of thyroid and retinoic acid receptor); AIB1 (amplified in breast cancer 1); and TRAM1 (thyroid receptor activator molecule 1).[11,14–16] All SRC family members are able to interact with and enhance transcriptional activity of NRs including ER; androgen receptor (AR); GR, thyroid hormone receptor (TR); retinoid X receptor (RXR); and many orphan nuclear receptors.[3,6,11] Although SRC members were initially cloned as NR coactivators, they also can function as coactivators for a wide variety of other DNA binding transcription factors including activator protein 1 (AP1); p53; myocyte enhancer factor 2C (MEF-2C); nuclear factor κB (NF-κB); aryl hydrocarbon receptor (AHR); AHR nuclear translocator (ARNT); signal transducer and activator of transcription (STATs); and transcription enhancer factor 4 (TEF-4)[7,17] This suggests that SRCs are important components for multiple regulatory pathways and, like p300/CBP, act as general coactivators for a wide range of DNA-binding transcription factors.

2.2.2 *Domain structure of p160/SRC coactivators*

p160/SRC coactivator proteins are about 160 kDa in size, have 50%–55% sequence similarity, and also share similar structural and functional domains [Fig. 2.2(A)]. The central NR interaction domain (NID) consisting of three NR box motifs (LXXLL, where L is leucine and X is any amino acid) interacts with a hydrophobic cleft in the NR LBD formed as a result of ligand-induced conformational changes.[3,9] Three LXXLL motifs are conserved within SRC family members. Interestingly, distinct LXXLL motifs exhibit differential binding affinity for different NRs, and the sequences flanking the LXXLL motifs are

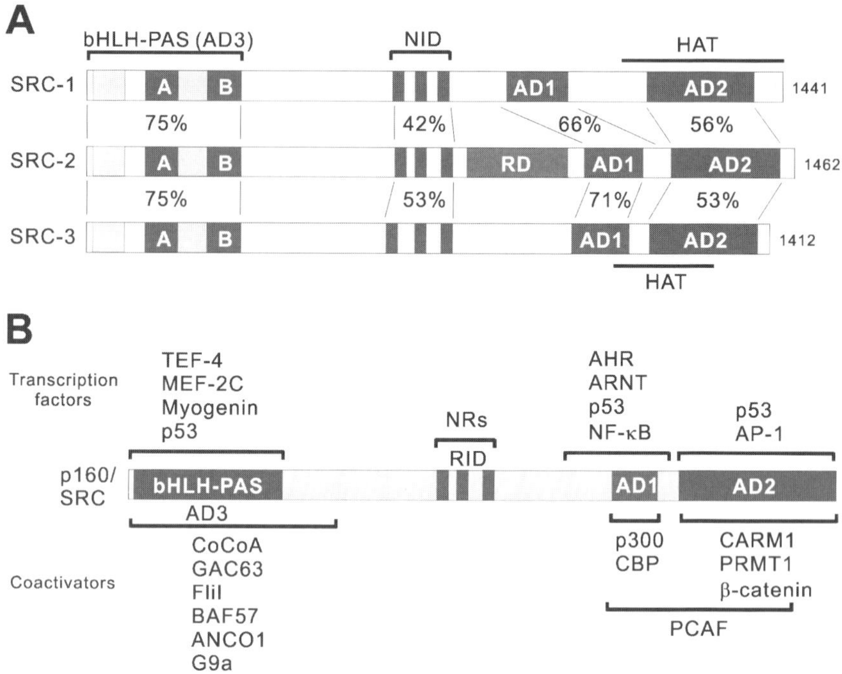

Fig. 2.2. Functional domains and interacting proteins of p160/SRC family members. (A) The domain structure of p160/SRC coactivators: The bHLH-PAS domain (AD3), NR interaction domain (NID), AD1, AD2, HAT domain, and GRIP1/SRC-2 repression domain (RD) are diagrammed. The central NID contains three NR box (LXXLL) motifs. SRC-1 and SRC-3 have a HAT domain near their C terminus. The percentage numbers indicate the amino acid similarity of SRC-1 and SRC-3 to SRC-2 in selected domains. The size of each protein, in number of amino acid residues, is indicated on the right. (B) p160/SRC-interacting proteins: Association of specific p160/SRC domains with DNA-binding transcription factors (top) and coactivators (bottom) is shown.

key determinants of the binding affinity and specificity of coactivators for their NR partners. The C-terminal region of SRCs can interact with the N-terminal AF-1 activation domains of some NRs.[7] The interactions with NRs are responsible for the association of SRCs with specific NR target gene promoters. Three intrinsic transcriptional activation domains (ADs), defined by their ability to activate transcription of reporter genes when the ADs are fused to Gal4 DBD, serve as binding sites for secondary coactivators and are thus responsible for transmitting the activating signal from the hormone-activated, DNA-bound NR to the transcription machinery[7] [Fig. 2.2(B)]. The C-terminal ADs of SRCs contribute to chromatin remodeling: AD1 recruits histone acetyltransferases p300; CREB binding protein (CBP); and PCAF.[7,14,15] and AD2 recruits histone methyltransferases CARM1 (coactivator-associated arginine methyltransferase 1) and PRMT1 (protein arginine methyltransferase 1).[7,18,19] These histone-modifying enzymes act synergistically with SRC coactivators to enhance NR function.[7,19] The C-terminal domains of SRC-1 and SRC-3 (ACTR) possess weak HAT activities [Fig. 2.2(A)].[15,20] The physiological substrates and functions of these HAT activities in NR-mediated transcription remain unclear. The N-terminal bHLH-PAS domain is the most conserved domain (75% similarity and 60% identity) among the SRC family members. It functions as a third AD (AD3).[21] PAS domains are multifunctional domains found on proteins from prokaryotes to humans. They serve as DNA-binding, protein-protein interaction, or ligand-binding surfaces in various bHLH-PAS transcription factors. The bHLH-PAS domain of SRC coactivators contains a bipartite nuclear localization signal (NLS) and mediates proteasome-dependent turnover.[22] It also serves as a binding site for DNA binding transcription factors including TEF-4, Tat, MEF-2C, myogenin, and p53 [Fig. 2.2(B)]. In addition, the bHLH-PAS domain can function as an AD (AD3) by recruiting other coactivators such as CoCoA, GAC63, FliI, G9a, hMMS19, cyclin T1, ANCO1, and BAF57.[21,23] Thus, SRCs are versatile proteins containing many protein-protein interaction surfaces. Different surfaces are used to interact with different DNA-binding transcription factors, and the specific activation domain used also varies, depending on the specific DNA-binding transcription factor to which the SRC is bound and the regulatory context of the target gene promoter.

2.2.3 *In vivo function: Redundancy and specificity of p160/SRC coactivator action*

p160/SRC coactivators play important roles in animal development and physiology. One important question for continuing research concerns *in vivo* functional redundancy versus specificity among the three members of the p160 family. Although widely expressed in many tissues and cell types, relative expression levels are tissue specific. The viable phenotypes of SRC-1, SRC-2, or SRC-3 knockout mice and the elevated expression of SRC-2 in the SRC-1 knockout mice suggest that SRC family members may be able to compensate partially for each other *in vivo*.[11] This functional redundancy may be due to their similar properties in terms of their interactions with NRs and secondary coactivators as well as their partially overlapping expression patterns in certain tissues. Partial redundancy among SRC family members could serve as a safety mechanism in the regulation of important biological processes in case the function of one or two of the family members is compromised. However, several reports indicate that their activities are not completely overlapping and that each member of the family has preferential functions in certain tissues with certain NRs.[11] SRC-1 knockout mice exhibit normal growth and fertility, but are partially resistant to several hormones including estrogen, androgen, progestin, and thyroid hormones. SRC-2 knockout mice grow normally, but have significantly reduced fertility in both male and female and enhanced adaptive thermogenesis and protection against obesity, whereas SRC-1 knockout mice were prone to obesity. SRC-3 knockout mice display growth retardation, abnormal development and function of the female reproductive system, and resistance to growth hormones. However, they do not exhibit the generalized resistance to steroid hormones in most target tissues that characterized the phenotype of SRC-1 null mice. These distinct phenotypic defects by knockout of each individual SRC suggest the presence of specific functions for each of these family members. In addition, plasmid microinjection experiments showed that either SRC-1 or SRC-2, but not SRC-3, could rescue transactivation of a retinoic acid response element (RARE)-linked reporter gene in SRC-1 immunodepleted cells, indicating a functional distinction between the SRC-1/SRC-2 and SRC-3 subfamilies.[11]

Cell type-specific expression and promoter-specific recruitment of SRCs also plays a role in the actions of selective estrogen receptor modulators (SERM). The SERM tamoxifen, which has estrogen-like activity in the uterus but antiestrogenic activity in the breast, is used widely in the treatment of breast cancer. However, due to its estrogenic activity in uterus, tamoxifen treatment for breast cancer is associated with proliferation of the endometrium, as well as increasing the risk of endometrial cancer. While expression levels of ER, SRC-2, and SRC-3 are similar in endometrial carcinoma Ishikawa and breast cancer MCF-7 cell lines, SRC-1 expression is much greater in the Ishikawa cell line, and this correlates with the increased estrogenic activity of tamoxifen in this cell line.[24] In Ishikawa cells, tamoxifen-bound ER recruits SRC-1 instead of corepressor to ER target genes that do not contain a classical ERE such as c-Myc and IGF-1 and stimulates their expression. Moreover, increasing SRC-1, but not SRC-2 and SRC-3, level in MCF-7 cells conferred the agonistic activity of tamoxifen, whereas knockdown of SRC-1 expression inhibited the estrogenic activity of tamoxifen in Ishikawa cells. The reason for the enhanced activity of SRC-1 (versus the other SRCs) or preferential recruitment of SRC-1 by ER at these promoters is currently not known.

In spite of the extensive sequence homology among SRC family members, there is sufficient sequence divergence to account for functional specificity. GRIP1/SRC-2, but not SRC-1 or SRC-3, can act as a GR corepressor at certain glucocorticoid response elements (GREs).[25] Hormone-activated GR binds to DNA-bound AP-1 protein (formed by a Jun-Fos heterodimer). In that context, GRIP1 potentiates GR-mediated repression of collagenase-3 gene expression in a GR- and hormone-dependent manner. This GRIP1-selective corepressor activity maps to a previously uncharacterized region (residues 765-1007) of GRIP1, which lacks evident sequence similarity to other SRC family members [Fig. 2.2(A)]. Indeed, this region of GRIP1 acted as a repression domain when expressed as a Gal4 fusion protein in a mammalian one-hybrid assay. Another example of GRIP1-selective corepressor activity is the requirement of GRIP1 for repression of TNFα (tumor necrosis factor α) gene expression by ER.[26] Unliganded ER is recruited to the TNFα gene promoter through protein-protein interaction in a TNFα-dependent manner and acts as a coactivator. Interestingly, estradiol represses TNFα gene expression by reversing the ligand-independent activation by ER; this involves recruitment of GRIP1 to the TNFα promoter, and recruited GRIP1 acts as a corepressor. In spite of some redundancy, each SRC family member also has selective functions, which are based on both tissue-specific expression patterns and different inherent properties dictated by their amino acid sequence differences.

2.2.4 *p160/SRC coactivators and diseases*

SRC family members play roles in the pathogenesis of several disease states, especially human malignancies.[27–30] AIB1/SRC-3 was initially identified from the amplified 20q12 chromosomal region of breast tumor cells. Indeed, this gene is amplified in 5%–10% of breast and ovarian tumors, and in an even higher percentage of cases, AIB1 protein is expressed significantly higher in tumor tissue than in the corresponding normal breast tissue. This was the first indication that alterations in coactivator expression may be associated with certain cancers. A number of other studies have subsequently observed increased expression of selected SRCs associated with tumorigenesis and progression of hormone-dependent cancers such as breast, prostate, ovarian, and endometrial cancers. The expression of SRC-1 and SRC-2 is upregulated in tissue specimens obtained from patients who failed prostate cancer endocrine therapy.[28] Increased expression of these coactivators is associated with enhanced activation of AR. Similar association between expression and high prostate cancer grade and stage was reported for SRC-3.[27] Elevated SRC-3 in combination with high expression of the receptor tyrosine kinase HER2/neu is frequently associated with substantially aggressive tamoxifen resistance.[4,29] It is proposed that enhanced signaling from the HER2/neu activates downstream kinases, which in turn phosphorylate and activate both ER and SRC-3, leading to enhanced transcriptional activation and tamoxifen resistance. This suggests the existence of a deleterious synergy between ER, SRC-3, and the HER2/neu signaling pathway. SRC-1, when overexpressed with HER2/neu, also correlates with tamoxifen resistance and a higher recurrence rate in breast cancer. Elevated expression of SRC-2 and SRC-3 is also associated with endometrial adenocarcinoma. A fusion between the TIF-2/SRC2 and MOZ genes at 8q13 and 8p11 has been identified in human acute myeloid leukemia.[30] The MOZ-TIF-2 fusion protein contains the C-terminal sequence of TIF-2, including the AD1 domain, thus facilitating its interaction with CBP/p300. As a consequence of this interaction, CBP is mislocalized from promyelocytic leukemia bodies, and cellular levels of CBP are depleted, leading to a reduced transcriptional activity of CBP-dependent activators such as NRs and p53. Consistent with this, AD1 integrity was found to be essential for the induction of acute myeloid leukemias by MOZ-TIF2 in mice.

2.3 p160/SRC-Associated Secondary Coactivators

In their roles as NR coactivators, the major functions delineated for p160/SRC coactivators are to use their LXXLL motifs to bind hormone-activated, promoter-bound NRs and from there, to use their ADs to assemble various secondary coactivators on the promoter.

Although there is a long list of secondary coactivators that bind SRCs [Fig. 2.2(B)], several critical secondary coactivators recruited by SRCs are discussed below.

2.3.1 *p300 and CBP bind to AD1 of SRCs*

p300 and CBP were originally identified as proteins that bind to the adenoviral E1A protein and the cAMP-response-element binding protein (CREB), respectively. They bind to (either directly or through other coactivators) and function as transcriptional coactivators for many different classes of DNA-binding transcription factors[5,31] [Fig. 2.3(A)]. p300 and CBP enhance hormone-dependent transcriptional activation in synergy with SRC coactivators and histone methyltransferase CARM1. Although p300/CBP directly interacts with NRs, its participation in transcriptional activation by NRs appears to be mediated through interaction with SRC family coactivators. Deletion of the NR-interacting LXXLL motif located at the N-terminus of p300/CBP did not significantly affect transactivation by NRs, whereas the p300/CBP region that interacts with SRCs and the AD1 region of SRCs (which binds p300/CBP) were absolutely essential.[7,32] p300/CBP and SRC coactivators also interact independently with PCAF leading to the formation of a large HAT complex.

p300/CBP possesses intrinsic HAT activity, and rates of gene transcription roughly correlate with the degree of histone acetylation, with hyperacetylated regions of the genome appearing to be more actively transcribed than hypoacetylated regions.[1,2,5,31] One of the roles of histone acetylation is simply to decrease charge-dependent interaction between basic histone N-terminal tails and polyanionic DNA through the neutralization of positive charges of histone tails. Acetylated histones can also mediate acetylation-dependent protein-protein interaction; for example, bromo domains are found on many proteins associated with transcription, and some bromo domains bind preferentially to acetylated (versus unmodified) histones. This is consistent with the histone code hypothesis that specific histone tail modifications provide docking sites to recruit proteins that effectively read the histone code and act in response to it.[2] Thus, HAT activity can increase the accessibility of DNA to transcription factors and coactivators by lowering the positive charge; decondensing local chromatin conformation; disrupting internucleosomal interactions made via the histone

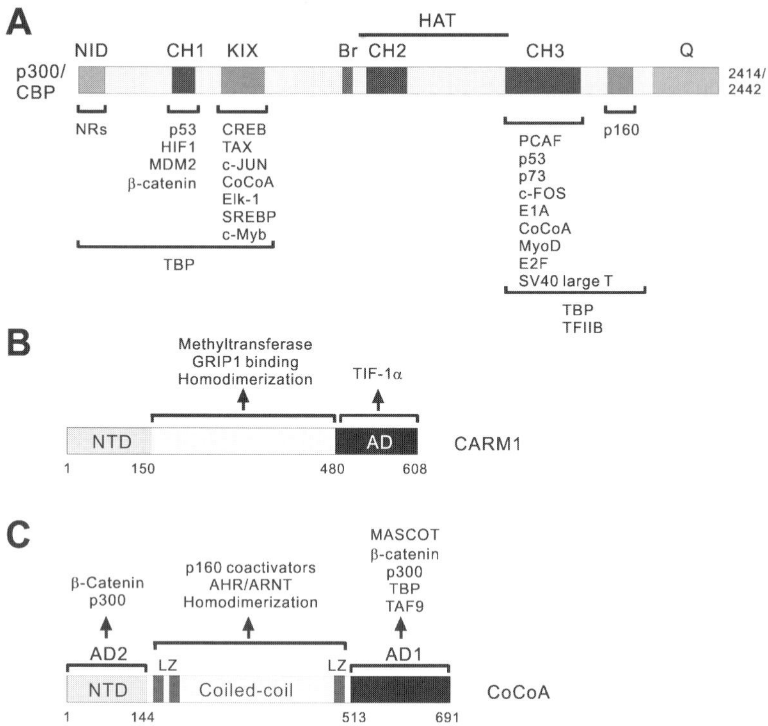

Fig. 2.3. Structural and functional domains of p160/SRC-associated secondary coactivators p300/CBP, CARM1, and CoCoA. (A) Domain structure of p300/CBP and their binding proteins: Labeled above the diagram are structural and functional domains, including NID, three CH domains, KIX, bromodomain (Br), glutamine (Q)-rich domain, and HAT domain. Below are indicated some proteins that interact with specific regions of p300/CBP. (B) Functional domains of CARM1: The methyltransferase, homodimerization, and GRIP1 binding domains are located in the central region of CARM1, which is highly conserved among PRMT family members. The C-terminal domain of CARM1 acts as an AD. (C) Structural and functional domains of CoCoA: The central coiled-coil domain of CoCoA interacts with the bHLH-PAS domains of p160/SRC coactivators and AHR/ARNT. In addition, there are two ADs that serve as protein-protein interaction surfaces for other coactivators.

tails; and/or recruiting additional cofactors through the histone code mechanism. Furthermore, p300/CBP can interact with RNA polymerase II and basal transcription factors such as TATA binding protein (TBP), and transcription factor IIB (TFIIB), and thus may also help to recruit RNA polymerase II and basal transcription machinery to the promoter.[5,31]

p300 and CBP share several conserved protein-protein interaction domains, including the bromodomain which recognizes acetylated lysines, three cysteine-histidine (CH)-rich domains, a KIX domain, and a glutamine (Q)-rich domain [Fig. 2.3(A)]. The N- and C-terminal regions of p300/CBP contain ADs, and the HAT activity resides in the central region of the protein. This modular organization may allow p300/CBP to provide a scaffold for assembly of multicomponent transcription coactivator complexes, which may vary according to the transcription factor and the context of the promoter on which p300 or CBP is bound.

Although p300 and CBP share extensive homology, genetic and molecular analyses suggest that they perform not only overlapping but also unique functions.[5,31] Homozygous knockouts in p300 and CBP both show similar, embryonic lethal phenotypes, together with similar defects in growth and neural tube closure. Moreover, p300/CBP double heterozygotes are invariably lethal, suggesting that p300 and CBP have overlapping as well as distinct functions during embryonic development. *cbp* heterozygous mice, but not *p300* heterozygous mice, show certain hematological defects and a predisposition to cancer. Homozygous p300 knockout fibroblasts have specific defects in retinoic acid-dependent transcription but retain normal CREB activity. Similarly, ribozyme-mediated ablation of p300, but not CBP, blocks retinoic acid-induced differentiation of F9 embryonal teratocarcinoma cells. Interestingly, ribozymes directed against either p300 or CBP can block retinoic acid-induced apoptosis of F9 cells. The observation that p300, but not CBP, is recruited to the pS2 promoter during the initial unproductive cycle of binding by activated ER also reflects nonredundant roles of p300 and CBP.[33]

2.3.2 *Protein Arginine Methyltransferases CARM1 and PRMT1 bind to AD2 of SRCs*

Methylation of arginine residues is a common posttranslational modification of proteins in eukaryotes. Arginine methylation has been implicated in the regulation of many cellular processes, including transcription, RNA processing, signal transduction, protein trafficking, and DNA repair.[19] This posttranslational modification is catalyzed by a family of PRMTs, of which there are at least eight members.[19] The PRMTs are classified into two groups: type I PRMTs (PRMT1, 3, 4, and 6) catalyze the formation of monomethylarginine and asymmetric dimethylarginine, whereas type II PRMTs (PRMT5 and 7) catalyze the formation of monomethylarginine and symmetric dimethylarginine. The catalytic domain of PRMTs resides in a highly conserved central region [Fig. 2.3(B)] composed of two parts: a highly conserved S-adenosylmethionine binding region formed by a Rossman fold and two α-helices; and a less conserved β-barrel structure that folds against the S-adenosylmethionine binding region to form the protein substrate binding cleft. In addition, each PRMT member has a unique N-terminal region, which in some PRMTs may cooperate with the methyltransferase domain to dictate substrate specificity.

2.3.2.1 *CARM1/PRMT4*

CARM1 (also referred to as PRMT4) was identified in a yeast two-hybrid screen using the AD2 region of GRIP1.[18] CARM1 interacts with all three SRCs and enhances transcriptional activation by NRs, but apparently only when SRC coactivators are present, suggesting that CARM1 functions as a secondary coactivator through its association with SRC coactivators (Fig. 2.4, top). CARM1 is a histone H3-specific arginine (R) methyltransferase and is able to methylate histone H3 at R2, R17, and R26. This methyltransferase activity of CARM1 is essential for its ability to act as a coactivator. This was the first clue that the methylation of histones at arginine may be a stimulating event for transcription. The conserved central catalytic region of CARM1 is also important for SRC coactivator binding and homodimerization activities [Fig. 2.3(B)].[7,19] In addition to this catalytic domain, the unique C-terminal AD of CARM1 is also required for the coactivation function. Chromatin immunoprecipitation studies indicate that CARM1 is specifically recruited to NR-regulated promoters *in vivo* in response to the hormone, and histone H3 methylated at R17 and R26 is indeed found to be enriched on hormone-activated promoters.[34,35] Arginine methylation thus represents a histone modification that correlates with the active state of transcription, much like acetylation. CARM1 acts synergistically with the acetyltransferases p300 and CBP to stimulate transcription by NRs, suggesting that histone methylation and histone acetylation work together to stimulate transcription to the full extent.[7] Indeed, histone modifications at the induced pS2 promoter are highly interdependent and formed in a stepwise manner.[35,36] For example, upon estrogen stimulation, CBP is recruited to the pS2 promoter and acetylates histone H3 lysine (K) 18. Following these events, CARM1 is bound to the promoter and methylates R17, therefore suggesting that the acetylation of histone H3 K18 is required for histone H3 R17 methylation by CARM1. This dependency might reflect the increased substrate affinity of CARM1 for the acetylated versus the unmodified substrate,[35] thus raising the possibility of cross-talk between methylation and acetylation. In addition to histone H3, additional CARM1 substrates have been identified by candidate approaches and enzyme reactions on protein arrays. CARM1 methylates p300/CBP and SRCs, and modulates their transcriptional activity.[19,37–40] CARM1 also methylates several RNA binding proteins including HuR, HuD, PABP1, TARPP, and several splicing and transcriptional elongation factors, suggesting its possible role in RNA processing.[19,41] CARM1 knockout embryos are small in size and die during late embryonic development or immediately after birth. CARM1 knockout embryos and CARM1 knockout mouse embryonic fibroblasts (MEFs) are deficient in some aspects of ER target gene expression, suggesting the essential role of CARM1 in ER-mediated transcriptional activation.[42] Because of its key role in various aspects of transcription, CARM1 may be a good target for the development of specific small molecule inhibitors.

2.3.2.2 *PRMT1*

PRMT1 has different roles in diverse biological processes. Like CARM1, PRMT1 interacts with AD2 of SRCs and serves as an NR coactivator.[19] PRMT1 is recruited to the ER target pS2 promoter in an E2-dependent manner,[36] and PRMT1 and CARM1 can act in a synergistic manner to enhance NR function.[7,19] The substrate specificity of CARM1 and PRMT1 is different, indicating distinct functions. PRMT1 specifically methylates R3 of histone H4 and is the major H4 R3 methyltransferase, as indicated by the loss of H4 R3 methylation in PRMT1 knockout cell lines.[19,43] Methylation of R3 on histone H4 promotes acetylation of H4 by p300/CBP that is a prerequisite for assembling a functional promoter complex, indicating once again, the cross-talk between two stimulating modifications. PRMT1 is required

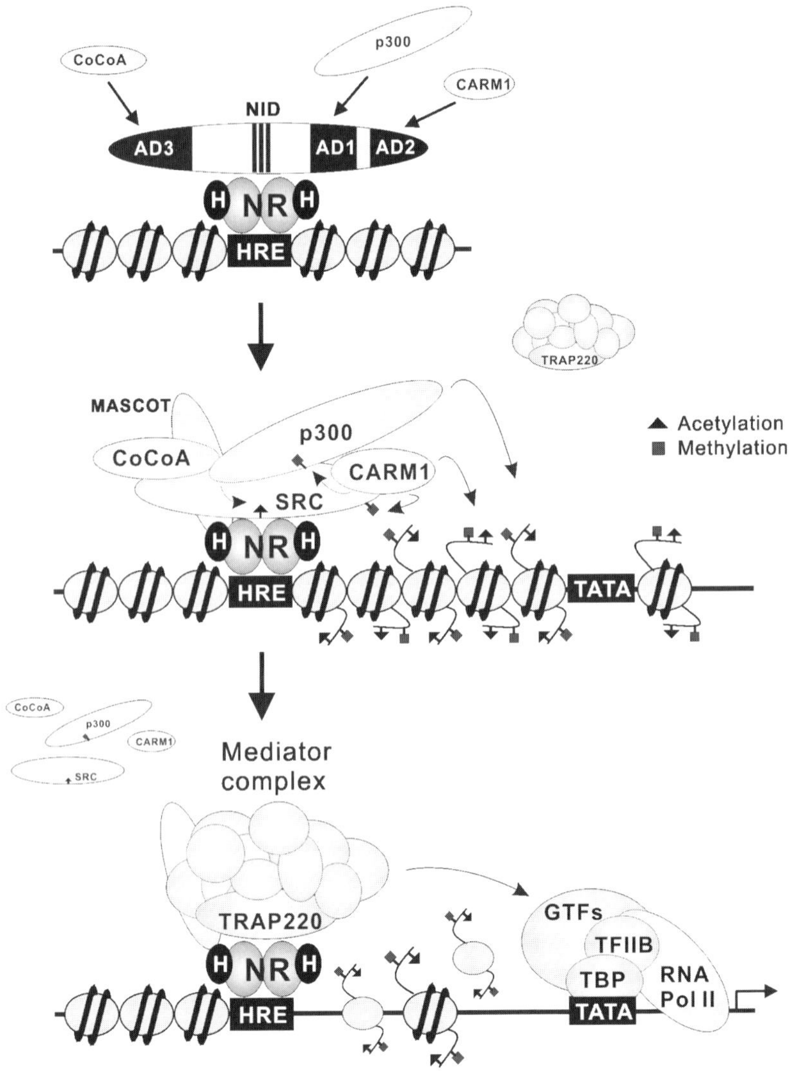

Fig. 2.4. Mechanism of action of p160/SRC coactivator complex. Hormone-activated NR binds to the HRE and recruits p160/SRC coactivators to the promoter by directly binding to the NID of p160/SRC coactivators. AD1 and AD2 in the C-terminal region of p160/SRC coactivators transmit the transcriptional activation signals by recruiting downstream secondary coactivators p300/CBP and CARM1, respectively. AD3 in the N-terminal bHLH-PAS region of p160/SRC coactivators also transmits the transcriptional activation signal by recruiting CoCoA, which interacts with p300/CBP and recruits MASCOT via

for differentiation during embryogenesis. Mice lacking a functional PRMT1 gene die at an early stage of embryogenesis as the head and nervous system fail to form.[44]

2.3.3 *CoCoA binds to AD3 of SRCs*

CoCoA, the product of the *calcoco1* gene, was isolated as a SRC binding protein in a yeast two-hybrid screen using the N-terminal bHLH-PAS (AD3) domain of GRIP1 as a bait.[21] CoCoA binds to SRCs but not directly to NRs, and acts synergistically in combination with p300 and CARM1 in transient transfection assays; this coactivator synergy is entirely dependent on the presence of a SRC with an intact N-terminus. CoCoA consists of three functional domains: the central coiled-coil domain, which interacts with the bHLH-PAS domains of SRCs and AHR/ARNT; the C-terminal activation domain (AD1); and the N-terminal activation domain (AD2) [Fig. 2.3(C)]. The C-terminal AD of CoCoA

---

protein-protein interactions. Enzymatic activities associated with p160/SRC coactivator complex results in nucleosomal remodeling and modifications of histone tails, such as histone methylation by CARM1 and PRMT1 and histone acetylation by p300/CBP and PCAF. Such modifications contribute to chromatin remodeling, making the promoter region more accessible. p300/CBP and CARM1 also modify components within the p160/SRC complex. Acetylation of p160/SRC coactivators by p300/CBP destabilizes NR-SRC interaction, and methylation of p300/CBP and p160/SRC coactivators by CARM1 destabilizes p300/CBP-SRC interaction and SRC-CARM1 interaction; these modifications thus cause dissociation of the p160/SRC complex from the promoter. The departure of p160/SRC complex (and presumably the NR as well) creates the opportunity for binding by additional NR dimers carrying a different coactivator complex, e.g. Mediator complex (shown in lower diagram) or SWI/SNF (not shown). Specific components of the p160/SRC complex may also facilitate subsequent recruitment of SWI/SNF (not shown), which contributes further to chromatin remodeling in an ATP-dependent manner. Action of the chromatin remodeling complexes caused dissociation of histones or loosening of the histone-DNA interaction, such that the promoter is more accessible for binding of the general transcription machinery. MASCOT, recruited initially as part of the p160/SRC complex, may help to facilitate subsequent recruitment of mediator complex to the promoter. Mediator recruits RNA Polymerase II and basal transcription machinery.

is essential for the coactivator function of CoCoA with NRs, GRIP1, and AHR/ARNT.[21,45] The ADs of CoCoA apparently contribute to transcriptional activation through multiple protein-protein interactions. Both ADs interact with p300/CBP and thus might help stabilize the SRC coactivator complex through the simultaneous interaction with SRC and p300.[46,47] CoCoA AD1 also interacts with components of TFIID including TBP and TBP associated factor 9 (TAF9) and thus may contribute to bridging between the SRC coactivator complex and basal transcription factors. CoCoA AD1 recruits MASCOT (Mediator associated SAP coactivator) which facilitates the recruitment of Mediator complex (Kim JH and Stallcup MR, unpublished results).

In addition to NRs, CoCoA binds to and cooperates synergistically with β-catenin as a secondary coactivator for AR and TCF/LEF,[48] and it also binds directly to and serves as a primary coactivator for AHR/ARNT and p53.[45] Endogenous CoCoA binds to the native promoters of target genes for NR, AHR/ARNT, p53, and TCF/LEF transcription factors and is required for efficient ER, GRIP1, AHR/ARNT, p53, TCF/LEF, and β-catenin function. Thus, CoCoA is a physiologically relevant coactivator for a diverse group of DNA-binding transcription factors.

TAX1 binding protein 1 (TAX1BP1), a CoCoA homologue encoded by the *calcoco3* gene, can also function as a NR coactivator.[49] TAX1BP1 and CoCoA have an overall sequence similarity of 45%, and both contain the central coiled-coil domain. The highest degree of conservation is in their N-terminal ADs, which are also homologous to nuclear dot protein 52 (NDP52, the product of the *calcoco2* gene). Interestingly, NR-mediated transcription was not significantly diminished in *tax1bp1* knockout cells. This observation suggests that CoCoA might fulfill a redundant function in the absence of TAX1BP1.

2.4 Mechanism of Action of p160/SRC Coactivators

2.4.1 *Assembly of SRC coactivator complex on NR target promoters*

The assembly of an SRC coactivator complex on the promoters of NR target genes is an early step in hormone-induced transcriptional activation and is mediated by numerous protein-protein interactions.[7,36]

After binding of activated NRs to their response elements within the regulatory region of hormone responsive genes, SRC coactivators appear to be among the first coactivators recruited (Fig. 2.4). The LBD of ER and other NRs consists of twelve α-helices, designated H1-H12. In the presence of agonist, the coactivator binding surface is formed by the folding of H12 against helices 3, 5, 6, and 11.[9] The LXXLL motifs form amphipathic α-helices in which the leucines create a hydrophobic surface that fits into the hydrophobic groove of the NR LBD. In contrast, H12 occupies the coactivator binding surface in the antagonist-bound NRs, thus blocking coactivator recruitment. These observations provide a structural basis for understanding the molecular mechanisms responsible for NR activation by its agonists and inhibition by its antagonists. Once bound to NRs, SRCs serve as a scaffold for the assembly of secondary coactivators (Fig. 2.4), using the three transcriptional ADs as docking sites. AD1 recruits HATs p300/CBP,[14,15] and AD2 recruits HMTs CARM1 and PRMT1.[18,19] The resulting histone modifications (acetylation of all four core histones in nearby nucleosomes and methylation of selective arginines on histones H3 and H4) contribute to local remodeling of chromatin structure and thereby increase promoter accessibility to downstream coactivators and the basal transcription machinery.[7,19] AD3 may recruit many downstream coactivators including CoCoA.[7,21] CoCoA ADs interact with p300/CBP, basal transcription factors TBP and TAF9, and other proteins which are still under investigation (Kim JH and Stallcup MR, unpublished results).

It is important to note that various cellular signal transduction pathways add another layer of regulation to the assembly and function of SRC coactivator complexes. In particular, phosphorylation of SRC proteins may result in increased affinity with target NRs.[3,4] For example, mitogen-activated protein kinase (MAPK)-induced phosphorylation of SRC-1 enhances its ability to function as a NR coactivator. SRC-1 can also be phosphorylated by protein kinase A in the absence of hormone, and this phosphorylation facilitates interaction of SRC-1 with p300/CBP and enhances its synergy with p300/CBP. Phosphorylation of several residues of SRC3 is also required for its effective interaction with CBP. In addition to phosphorylation, SRC proteins can also be modified by the small ubiquitin-like modifier SUMO.[50] Sumoylation of SRC-1 and SRC-2 in the NID enhances the interaction with NRs and enhances their coactivator function.

2.4.2 *Disassembly of SRC coactivator complex*

Hormone-activated NRs and their coactivators are not stably bound to promoters for long periods, but rather, they appear to associate and dissociate repeatedly with half times of a few seconds to one minute.[51] This cycle of binding to and dissociation from the promoter appears to be important for efficient transcriptional activation. Although the exact mechanism is not well understood, increasing evidence suggests that posttranslational modifications may promote SRC complex disassembly (Fig. 2.4). p300/CBP can acetylate SRC proteins at lysine residues adjacent to their NR interaction domains.[52] These acetylations destabilize the SRC and NR interaction and presumably facilitate coactivator release from hormone-bound NR. CARM1 can methylate an arginine residue within the C-terminal GRIP1 binding domain of p300.[39] This methylation inhibits the interaction between p300 and SRC and may also contribute to coactivator complex disassembly. CARM1 also methylates arginine residues in the CARM1 binding region of SRC-3 in response to estradiol.[37,38] Together, the above modifications induce dissociation of SRCs from CARM1, p300, and NRs, consequently, contributing to SRC complex dissociation. Furthermore, methylation regulates the stability of SRC-3 by causing increased degradation and enhanced cellular turnover.[38] In addition, NR and coactivator ubiquitylation is also important for clearing them from the promoter.[3] Ubiquitin proteasome-mediated degradation of NRs and coactivators provides an efficient clearance mechanism; this allows for the cyclical recruitment of NRs and coactivators to the promoter, and thus plays a positive role in NR-mediated transcription.

2.4.3 *Exchange and coordination between different coactivator complexes*

The fact that chromatinized transcription units are repressed compared with naked DNA indicates that nucleosome and chromatin remodeling are a critical part of gene regulation, prior to recruitment of the basal transcription machinery. Transcriptional activation from chromatin templates by NRs requires multiple coactivator complexes, including a SRC complex, a SWI/SNF-type ATPase complex, and a mediator complex. Because all of these are large protein complexes that interact with NR LBDs, their association with a given NR molecule may be mutually exclusive and occur in an ordered manner. In fact, chromatin immunoprecipitation analyses found that these and other coactivator complexes appear to be sequentially recruited to NR target promoters after ligand treatment.[32,36]

One possible mechanism for the ordered recruitment of distinct coactivator complexes could be that HAT activities of SRC complexes are required for the recruitment of mediator and SWI/SNF components. Indeed, the recruitment of both the SWI/SNF and mediator multi-protein complexes to chromatin was stabilized by histone acetylation by p300/CBP, which itself is recruited through interaction with SRC coactivators.[53] This model may explain why mediator complex supports transcription *in vitro* from naked DNA templates more efficiently than from chromatized templates. It may also explain the functional synergies between SRC coactivator complexes and mediator or SWI/SNF complexes. Another appealing possibility is that additional coactivators interact physically and functionally with one or more of the large coactivator complexes and help to coordinate their recruitment and/or activities. In fact, NR coactivator PGC-1α interacts with and stimulates the function of both p300 and mediator in NR-mediated transcription, suggesting that PGC-1α may coordinate the transition between NR/p300-dependent chromatin remodeling and NR/mediator-dependent transcription.[13,54]

Several lines of evidence indicate a close functional link between SRC and SWI/SNF complexes. The bHLH-PAS domain of SRC coactivators directly interacts with BAF57 found in all SWI/SNF remodeling complexes. This interaction appears to be important for the ability of the SRC proteins to potentiate transcription by ER.[8] Interestingly, CARM1 was found as a component of the SWI/SNF-like nucleosomal methylation activator complex (NUMAC), which includes at least eight components of SWI/SNF, including the critical ATPase subunit BRG1.[19] CARM1 in NUMAC preferentially methylates nucleosomal histone H3 rather than free histone H3, whereas free CARM1 preferentially methylates free histone H3. Reciprocally, CARM1 binds to and stimulates the ATPase activity of BRG1. CARM1 and BRG1 are both recruited to an ER target gene, and they cooperatively activate ER-dependent transcription. These findings suggest that the physical association of CARM1 and BRG1 mutually enhances their enzymatic activities to promote ER signaling. In addition, the CARM1- and GRIP1/SRC-2-interacting protein, Flightless I, also interacts with two components of the SWI/SNF complex, BRG1 and BAF53.[19] Furthermore, CoCoA also can interact with BAF170 (Kim JH and Stallcup MR, unpublished results). Thus, it appears that there are multiple protein-protein interactions between SRC and SWI/SNF complexes, and these interactions may contribute to the coordination of the recruitment and/or activities of these two coactivator complexes.

A possible link between the SRC and mediator complexes was suggested by the recent finding that MASCOT can associate with both mediator and SRC complexes (Kim JH and Stallcup MR, unpublished results). MASCOT (also called CCAR1) binds to the C-terminal AD1 of CoCoA and serves as a coactivator for NRs in collaboration with GRIP1/SRC-2, p300, CARM1, and CoCoA. Interestingly, MASCOT is required for the optimal recruitment of the mediator to an endogenous target gene of ER. It is proposed that MASCOT coordinates the activities of mediator and SRC complexes by bridging these coactivator complexes.

2.5 Conclusions and Future Perspectives

The identification of hundreds of potential or confirmed coregulators over the past two decades has revealed an unanticipated level of complexity in transcriptional regulation mechanisms in eukaryotic organisms. Transcriptional activation is a dynamic process regulated by post-translational modifications and protein-protein interactions that assemble, disassemble, and modulate the activities of a variety of DNA-binding proteins; coactivators; chromatin-associated proteins; basal transcription factors; and RNA polymerase complexes. Although much more work clearly remains to be done, we have begun to gain a deeper insight into the mechanism of transcriptional regulation of NRs and their coactivators; the role played by post-translational modifications and chromatin conformation; and the contribution of coactivators in the development of pathological conditions.

In many ways, recent advances in molecular biology methods, technology, and informatics have played and will continue to play critical roles in the identification and functional characterization of transcriptionally important coactivators. For example, the use of yeast two hybrid system as a genetic tool and the use of mass spectrometry analysis of biochemically isolated protein complexes have dramatically advanced efforts to identify new coregulators for NRs. These and other methods will undoubtedly continue to evolve. Chromatin immunoprecipitation has provided a wealth of information about *in vivo* coregulator localization on specific genes at specific time points. For a broader functional view, whole-genome analyses with DNA microarray technologies have been and will be used to identify genes that are bound and regulated by NRs and NR coregulators. They are also used to assess the impact of coactivator deficiencies and mutations on the expression of all genes for a given organism or cell type. In addition, the availability of transgenic, knock-out, and knock-down cell and animal models will facilitate the examination of the physiological roles of individual coregulators and the dissection of the hormone-dependent molecular genetic pathways influenced by specific coregulators.

Current information, although limited, suggests that differential recruitment of coactivators and corepressors will be an important contributing mechanism in gene regulation; this expands the regulatory potential of NRs by allowing for the selective recruitment and exchange of multiple coregulator complexes. In addition, while similar sets of coactivators may be recruited to many target genes, the actions of individual coactivators may be more critical on some target genes than on others. Therefore, the regulation of coactivator levels and activities is well suited for the rapid activation or repression of sets of genes in response to extracellular signals.

There is growing evidence that dysregulation of expression and function of SRC coactivators and their associated coactivators play different roles in disorders such as hormone-dependent cancers. Alterations in the level of a coregulator can change the fine balance between coactivators and corepressors and hence modulate signaling by NR bound by agonist and antagonist.[55] For this reason, antagonistic properties of anti-hormone drugs may be substantially reduced when coactivators are overexpressed or when their function is altered by mutations. Thus, it would not be surprising to find additional human diseases caused by coactivator overexpression or mutation. It may be informative to check NR coactivator levels or mutations in cancer patients and patients with anti-hormone resistance and other hormone-related diseases such as steroid resistance syndromes and reproductive dysfunction. Because of their involvement in so many important biological processes and also because of the rapid improvement in our understanding of their mechanisms of action, coactivators may become inviting targets for the design of pharmacological interventions for treatment of these disorders. Towards, that end, continued efforts to identify ligands that act as selective nuclear receptor modulators are particularly exciting. Each ligand that binds to an NR LBD apparently induces a different conformation, resulting in a ligand-specific pattern of coactivator recruitment and thus ligand-specific activation and/or inhibition of a subset of the full repertoire of biological responses elicited by the natural ligand. Thus, a clearer understanding of the mechanism of action of NR coactivators should assist in the development of clinically relevant pharmacological agents for hormone therapy.

On the basis of recent progress, it appears certain that additional coactivators and corepressors with diverse physiological functions will continue to be discovered, adding further complexity and enlightenment to our understanding of transcriptional regulation by NRs. Future research will need to address many fundamental questions that still remain to be answered. What are the specific contributions of each coactivator or each component of a coactivator complex to the molecular mechanism of transcriptional activation? With so many coactivators participating, how are their activities coordinated and integrated to regulate a single event, i.e. the initiation of transcription? How does the context of chromatin influence the binding and activities of coactivators? How are coactivators and the transcription machinery in general organized or compartmentalized in the nucleus, and how does nuclear architecture and organization contribute to transcriptional regulation? How does the regulatory context of different promoters cause different coactivator requirements for transcriptional activation? Since many so called coactivators can also function as corepressors in specific situations, what molecular interactions dictate whether the coregulator acts as coactivator or corepressor in a given situation? How do coactivators contribute to the etiology of hormone-dependent cancers and the disorders of the endocrine system? The field of NR coregulator research thus still offers tremendous opportunities for seminal discoveries and is perhaps ideally situated to unlock the secrets of the fundamental mechanisms of transcriptional regulation. Answering these exciting but daunting questions will require continued pursuit of current strategies as well as novel technologies and strategies developed by the next generation of scientists.

Acknowledgments

This work was supported by grant number DK43093 to M.R.S. from the National Institutes of Health. J.H.K. was supported in part by a postdoctoral traineeship from grant number T32 CA009320 from the National Institutes of Health.

References

1. Kouzarides T, Chromatin modifications and their function, *Cell* **128**:693–705, 2007.
2. Jenuwein T, Allis CD, Translating the histone code, *Science* **293**:1074–1080, 2001.
3. Rosenfeld MG, Lunyak VV, Glass CK, Sensors and signals: A coactivator/corepressor/epigenetic code for integrating signal-dependent programs of transcriptional response, *Genes Dev* **20**:1405–1428, 2006.
4. Wu RC, Smith CL, O'Malley BW, Transcriptional regulation by steroid receptor coactivator phosphorylation, *Endocr Rev* **26**:393–399, 2005.
5. Vo N, Goodman RH, CREB-binding protein and p300 in transcriptional regulation, *J Biol Chem* **276**:13505–13508, 2001.
6. McKenna NJ, O'Malley BW, Combinatorial control of gene expression by nuclear receptors and coregulators, *Cell* **108**:465–474, 2002.
7. Stallcup MR, Kim JH, Teyssier C, *et al.*, The roles of protein-protein interactions and protein methylation in transcriptional activation by nuclear receptors and their coactivators, *J Steroid Biochem Mol Biol* **85**:139–145, 2003.
8. Belandia B, Parker MG, Nuclear receptors: A rendezvous for chromatin remodeling factors, *Cell* **114**:277–280, 2003.
9. Nettles KW, Greene GL, Ligand control of coregulator recruitment to nuclear receptors, *Annu Rev Physiol* **67**:309–333, 2005.
10. Onate SA, Tsai SY, Tsai MJ, *et al.*, Sequence and characterization of a coactivator for the steroid hormone receptor superfamily, *Science* **270**:1354–1357, 1995.
11. Xu J, Li Q, Review of the *in vivo* functions of the p160 steroid receptor coactivator family, *Mol Endocrinol* **17**:1681–1692, 2003.
12. Malik S, Roeder RG, Dynamic regulation of pol II transcription by the mammalian Mediator complex, *Trends Biochem Sci* **30**:256–263, 2005.
13. Roeder RG, Transcriptional regulation and the role of diverse coactivators in animal cells, *FEBS Lett* **579**:909–915, 2005.
14. Torchia J, Rose DW, Inostroza J, *et al.*, The transcriptional co-activator p/CIP binds CBP and mediates nuclear-receptor function, *Nature* **387**:677–684, 1997.

15. Chen H, Lin RJ, Schiltz RL, *et al.*, Nuclear receptor coactivator ACTR is a novel histone acetyltransferase and forms a multimeric activation complex with P/CAF and CBP/p300, *Cell* **90**:569–580, 1997.
16. Anzick SL, Kononen J, Walker RL, *et al.*, AIB1, a steroid receptor coactivator amplified in breast and ovarian cancer, *Science* **277**:965–968, 1997.
17. Shang Y, Molecular mechanisms of oestrogen and SERMs in endometrial carcinogenesis, *Nat Rev Cancer* **6**:360–368, 2006.
18. Chen D, Ma H, Hong H, *et al.*, Regulation of transcription by a protein methyltransferase, *Science* **284**:2174–2177, 1999.
19. Lee DY, Teyssier C, Strahl BD, *et al.*, Role of protein methylation in regulation of transcription, *Endocr Rev* **26**:147–170, 2005.
20. Spencer TE, Jenster G, Burcin MM, *et al.*, Steroid receptor coactivator-1 is a histone acetyltransferase, *Nature* **389**:194–198, 1997.
21. Kim JH, Li H, Stallcup MR, CoCoA, a nuclear receptor coactivator which acts through an N-terminal activation domain of p160 coactivators, *Mol Cell* **12**:1537–1549, 2003.
22. Li C, Wu RC, Amazit L, *et al.*, Specific amino acid residues in the basic helix-loop-helix domain of SRC-3 are essential for its nuclear localization and proteasome-dependent turnover, *Mol Cell Biol* **27**:1296–1308, 2007.
23. Lee DY, Northrop JP, Kuo MH, *et al.*, Histone H3 lysine 9 methyltransferase G9a is a transcriptional coactivator for nuclear receptors, *J Biol Chem* **281**:8476–8485, 2006.
24. Shang Y, Brown M, Molecular determinants for the tissue specificity of SERMs, *Science* **295**:2465–2468, 2002.
25. Rogatsky I, Luecke HF, Leitman DC, *et al.*, Alternate surfaces of transcriptional coregulator GRIP1 function in different glucocorticoid receptor activation and repression contexts, *Proc Natl Acad Sci USA* **99**:16701–16706, 2002.
26. Cvoro A, Tzagarakis-Foster C, Tatomer D, *et al.*, Distinct roles of unliganded and liganded estrogen receptors in transcriptional repression, *Mol Cell* **21**:555–564, 2006.
27. Liao L, Kuang SQ, Yuan Y, *et al.*, Molecular structure and biological function of the cancer-amplified nuclear receptor coactivator SRC-3/AIB1, *J Steroid Biochem Mol Biol* **83**:3–14, 2002.
28. Culig Z, Comuzzi B, Steiner H, *et al.*, Expression and function of androgen receptor coactivators in prostate cancer, *J Steroid Biochem Mol Biol* **92**:265–271, 2004.
29. Yan J, Tsai SY, Tsai MJ, SRC-3/AIB1: Transcriptional coactivator in oncogenesis. *Acta Pharmacol Sin* **27**:387–394, 2006.
30. Troke PJ, Kindle KB, Collins HM, *et al.*, MOZ fusion proteins in acute myeloid leukaemia, *Biochem Soc Symp* **73**:23–39, 2006.

31. Chan HM, La Thangue NB, p300/CBP proteins: HATs for transcriptional bridges and scaffolds, *J Cell Sci* **114**:2363–2373, 2001.
32. Dilworth FJ, Chambon P, Nuclear receptors coordinate the activities of chromatin remodeling complexes and coactivators to facilitate initiation of transcription, *Oncogene* **20**:3047–3054, 2001.
33. Shang Y, Hu X, DiRenzo J, *et al.*, Cofactor dynamics and sufficiency in estrogen receptor-regulated transcription, *Cell* **103**:843–852, 2000.
34. Ma H, Baumann CT, Li H, *et al.*, Hormone-dependent, CARM1-directed, arginine-specific methylation of histone H3 on a steroid-regulated promoter, *Curr Biol* **11**:1981–1985, 2001.
35. Daujat S, Bauer UM, Shah V, *et al.*, Crosstalk between CARM1 methylation and CBP acetylation on histone H3, *Curr Biol* **12**:2090–2097, 2002.
36. Metivier R, Penot G, Hubner MR, *et al.*, Estrogen receptor-alpha directs ordered, cyclical, and combinatorial recruitment of cofactors on a natural target promoter, *Cell* **115**:751–763, 2003.
37. Feng Q, Yi P, Wong J, *et al.*, Signaling within a coactivator complex: Methylation of SRC-3/AIB1 is a molecular switch for complex disassembly, *Mol Cell Biol* **26**:7846–7857, 2006.
38. Naeem H, Cheng D, Zhao Q, *et al.*, The activity and stability of the transcriptional coactivator p/CIP/SRC-3 are regulated by CARM1-dependent methylation, *Mol Cell Biol* **27**:120–134, 2007.
39. Lee YH, Coonrod SA, Kraus WL, *et al.*, Regulation of coactivator complex assembly and function by protein arginine methylation and demethylimination, *Proc Natl Acad Sci USA* **102**:3611–3616, 2005.
40. Xu W, Chen H, Du K, *et al.*, A transcriptional switch mediated by cofactor methylation, *Science* **294**:2507–2511, 2001.
41. Cheng D, Cote J, Shaaban S, *et al.*, The arginine methyltransferase CARM1 regulates the coupling of transcription and mRNA processing, *Mol Cell* **25**:71–83, 2007.
42. Yadav N, Lee J, Kim J, *et al.*, Specific protein methylation defects and gene expression perturbations in coactivator-associated arginine methyltransferase 1-deficient mice, *Proc Natl Acad Sci USA* **100**:6464–6468, 2003.
43. Wang H, Huang ZQ, Xia L, *et al.*, Methylation of histone H4 at arginine 3 facilitating transcriptional activation by nuclear hormone receptor, *Science* **293**:853–857, 2001.
44. Pawlak MR, Scherer CA, Chen J, *et al.*, Arginine N-methyltransferase 1 is required for early postimplantation mouse development, but cells deficient in the enzyme are viable, *Mol Cell Biol* **20**:4859–4869, 2000.
45. Kim JH, Stallcup MR, Role of the coiled-coil coactivator (CoCoA) in aryl hydrocarbon receptor-mediated transcription, *J Biol Chem* **279**:49842–49848, 2004.

46. Kim JH, Yang CK, Stallcup MR, Downstream signaling mechanism of the C-terminal activation domain of transcriptional coactivator CoCoA, *Nucleic Acids Res* **34**:2736–2750, 2006.
47. Yang CK, Kim JH, Stallcup MR, Role of the N-terminal activation domain of the coiled-coil coactivator in mediating transcriptional activation by beta-catenin, *Mol Endocrinol* **20**:3251–3262, 2006.
48. Yang CK, Kim JH, Li H, *et al.*, Differential use of functional domains by coiled-coil coactivator in its synergistic coactivator function with beta-catenin or GRIP1, *J Biol Chem* **281**:3389–3397, 2006.
49. Chin KT, Chun AC, Ching YP, *et al.*, Human T-cell leukemia virus oncoprotein tax represses nuclear receptor-dependent transcription by targeting coactivator TAX1BP1, *Cancer Res* **67**:1072–1081, 2007.
50. Hilgarth RS, Murphy LA, Skaggs HS, *et al.*, Regulation and function of SUMO modification, *J Biol Chem* **279**:53899–53902, 2004.
51. Hager GL, Elbi C, Johnson TA, *et al.*, Chromatin dynamics and the evolution of alternate promoter states, *Chromosome Res* **14**:107–116, 2006.
52. Chen H, Lin RJ, Xie W, *et al.*, Regulation of hormone-induced histone hyperacetylation and gene activation via acetylation of an acetylase, *Cell* **98**:675–686, 1999.
53. Huang ZQ, Li J, Sachs LM, *et al.*, A role for cofactor-cofactor and cofactor-histone interactions in targeting p300, SWI/SNF and Mediator for transcription, *EMBO J* **22**:2146–2155, 2003.
54. Wallberg AE, Yamamura S, Malik S, *et al.*, Coordination of p300-mediated chromatin remodeling and TRAP/mediator function through coactivator PGC-1alpha, *Mol Cell* **12**:1137–1149, 2003.
55. Wang Q, Blackford JA, Jr, Song LN, *et al.*, Equilibrium interactions of corepressors and coactivators with agonist and antagonist complexes of glucocorticoid receptors, *Mol Endocrinol* **18**:1376–1395, 2004.

Chapter 3

Regulation of Nuclear Hormone Receptor Functions by Ubiquitin-Proteasome Pathway

Ayesha Ismail, Heath Catoe, Sarath Dhananjayan and Zafar Nawaz

The ubiquitin-proteasome pathway (UPP) constitutes the major pathway for degradation of nuclear and cytoplasmic proteins in the eukaryotic cell. The role of the UPP in cell cycle progression and signal transduction pathways is well known, but its role in nuclear hormone receptor (NHR) functions is a surprise to many scientists. The involvement of a number of UPP components in NHR-mediated transcription suggested a possible link between these two diverse pathways. Both NHRs and their coregulators (coactivators and corepressors) are ubiquitinated and degraded by the UPP. Interestingly, the UPP enzymes and components of both the 19S and 20S proteasomes are recruited to NHR target gene promoters suggesting that the UPP could be involved in NHR transcription regulation at multiple steps. In recent years, it has been demonstrated that the role of ubiquitin can be expanded well beyond its function of acting as a death-tag for the degradation of proteins by the 26S proteasome. Keeping in mind the involvement of the UPP in a variety of cellular processes, its deregulation or malfunction has been associated with a number of human diseases including cancers. Targeting the UPP as a potential for drug discovery in the treatment of cancer and neurodegenerative disorders is just beginning. In this chapter, we will discuss the probable role of the UPP in NHR regulated gene transcription and also the possibility of targeting the UPP for drug discovery for the treatment of endocrine cancers like breast and prostate cancers.

3.1 Introduction

Maintenance of protein homeostasis is essential for regulating normal cellular processes. Almost 25 years ago, the ubiqutin-proteasome pathway (UPP) was discovered and was shown to be an integral part of the degradation of misfolded, damaged or mutated cellular proteins. The UPP is a highly conserved system in eukaryotes that occurs in a multistep process. The beginning of the UPP degradation process starts with the shuttling of ubiquitin molecules through a cascade of enzymes leading to the covalent addition of ubiquitin to the protein targeted to be degraded. The covalent addition of ubiquitin is termed ubiquitination or ubiquitylation. The formation of a chain of ubiquitins on a substrate is termed polyubiquitination. The UPP ends with the proteolysis of the substrate protein, which is now marked with polyubiquitination, by the multicatalytic protein complex termed the 26S proteasome.[1]

In addition to targeting proteins for degradation, ubiquitination also plays vital roles in other cellular functions such as protein sorting, endocytosis, transcriptional regulation, cell cycle regulation, signal transduction and cell differentiation.[2,3] More than a decade ago, it was demonstrated that nuclear hormone receptors (NHRs) and their coregulators are targets of the UPP, and the UPP is involved in the regulation of NHR-dependent transactivation.[4] Transactivation is the term used to describe an increase in the rate of transcription through a trans-acting factor. The ubiquitin pathway enzymes and components of the 26S proteasome complex have been reported to be recruited onto the NHR target gene promoters and act as NHR coactivators, suggesting that the UPP is involved in NHR transactivation functions.[5–8] However, the link between NHR degradation and transactivation function is quite complex and not well understood. The effect of degradation of the receptor via the UPP on its transcriptional function can be receptor and promoter specific. The involvement of the UPP in NHR function was unexpected, and it is now being speculated that the proteasome components could also have non-proteolytic functions at different stages of NHR-mediated transcriptional activation. However, the precise mechanism by which the UPP modulates NHR transactivation is far from clear and needs further in depth investigation. The need for further research of the UPP is substantiated by its implication in many human diseases promoting the UPP as an attractive target for developing new drugs to treat disorders resulting from UPP malfunction.[9] In this chapter, we will give a general overview of the UPP's function, discuss its role in NHR function, describe disorders linked to the UPP, and discuss the UPP as a potential target for drug discovery.

3.2 Ubiquitin-Proteasome Pathway

The ubiquitin-proteasome pathway (UPP) is the major pathway in eukaryotic cells involved in the degradation of short-lived regulatory, abnormal or misfolded proteins as well as antigen presentation.[1] Drs Ciechanover, Hershko, and Rose were awarded the 2004 Nobel Prize in Chemistry for the discovery of the UPP. The degradation of proteins by the UPP is a highly specific, tightly regulated and complex process. As mentioned in the introduction, the UPP uses a 76 amino acid (aa) polypeptide, ubiquitin to mark protein substrates. The successive steps involved in the conjugation of ubiquitin to substrate proteins require an enzyme cascade, which will be further explained in the following section.

3.2.1 *Enzymatic cascade of the ubiquitin-proteasome pathway*

Proteins destined for destruction by the UPP must be tagged with ubiquitin in order to be recognized and degraded by the 26S proteasome. In the degradation process, the initial step is the activation of the glycine found on the carboxyl-terminus of ubiquitin. In an ATP-dependent mechanism, the glycine residue is covalently linked by a thioester bond to a cysteine residue in the active site on the ubiquitin-activating enzyme termed E1. This initial step is the only ATP-dependent mechanism in the UPP. In the next step of the cascade, activated ubiquitin is transferred to the ubiquitin-conjugating enzyme, termed E2, while retaining the high energy thioester bond linkage through the glycine on ubiquitin and an active site cysteine on the E2. The last step of the UPP enzyme cascade results in the conjugation of ubiquitin to the substrate protein. The ubiquitin-ligase enzyme, termed E3, coordinates the amide-isopeptide linkage between ubiquitin and substrate proteins. For the majority of protein substrates, ubiquitin is linked by its carboxy-terminus to the ε-amino group of the substrate protein's lysine (Lys) residues[10] (Fig. 3.1). However, some protein substrates have been shown to be ubiquitinated on their amino-terminus and through amino acids other than lysine.

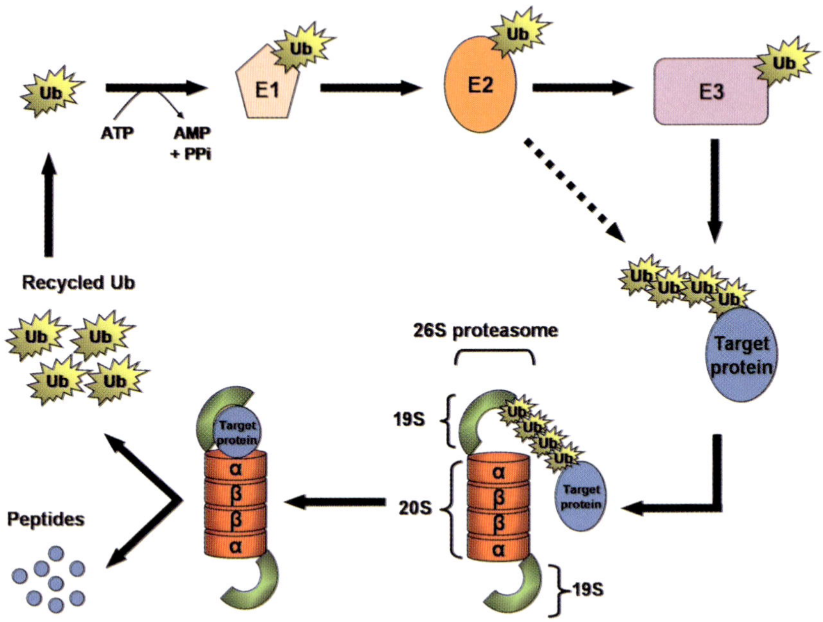

Fig. 3.1. Schematic representation of the ubiquitin-proteasome pathway. The ubiquitin molecule is attached to substrate proteins targeted for degradation via a cascade of three enzymes namely E1, E2 and E3. Initially the ubiquitin (Ub) is activated in the presence of ATP and covalently linked to the cysteine residue of the E1, ubiquitin-activiating enzyme. This is followed by the transfer of the activated ubiquitin from E1 to the cysteine residue of an E2, ubiquitin-conjugating enzyme. The activated ubiquitin from the E2 is either transferred directly to the substrate protein or transferred to the cysteine residue of an E3, ubiquitin-ligase enzyme. The E3 enzyme then transfers the ubiquitin to the lysine residue of the substrate protein. This series of enzymatic reactions results in the addition of the first ubiquitin moiety to the substrate protein. The E3 then aids in addition of a polyubiquitin chain to the protein. The target protein, tagged with a minimum of 4 ubiquitin molecules, is recognized by the 19S cap of the proteasome wherein, the deubiquitylating enzymes remove the ubiquitins, releasing free and reusable ubiquitin. The unfolded protein is then fed into the 20S catalytic core, which then degrades the protein, releasing peptides.

After the linkage of the first ubiquitin to the substrate protein, the E3 ligase catalyzes the formation of the polyubiquitin chain, in which the carboxy-terminus of each ubiquitin unit is linked to a Lys-48 residue of the previous ubiquitin. Proteins can either be tagged with a single ubiquitin (monoubiquitinated) or with multiple ubiquitin's (polyubiquitinated). It has been demonstrated that a protein targeted for degradation by the UPP must have a chain of at least four ubiquitin proteins linked to it. Other Lys residues on ubiquitin can be used to create chains not involved in signaling proteolysis, but this topic will be covered in proteasome-independent functions of ubiquitin. Till date, only a single known E1 with several E2 and E3 enzymes have been defined. Data from the human genome project have revealed over 40 potential E2s and over 500 potential E3s.[11] It is the E3s that are thought to give the UPP its specificity for targeting individual proteins for degradation, so the rest of this section will focus on their characteristics.

E3s can be broadly divided into two major families, namely the *HECT* (Homologous to the E6-Associated Protein carboxy terminus) E3s and the Really Interesting New Gene (RING) finger E3s. Depending on the type of E3 ligase, the transfer of ubiquitin to the substrate is either direct or indirect, and the E3 alone or in combination with the E2 determines the specificity for discreetly targeting proteins to the UPP.

In *HECT* domain E3 ligases, the E2 first transfers the activated ubiquitin to the E3 creating an ubiquitin-linked E3. Subsequently, the *HECT*-E3 transfers the ubiquitin to the target substrate protein. The E6-associated protein (E6-AP) was the founding member of this class of E3 ligases. E6-AP was discovered through the study of the human papillomavirus (HPV). HPV encodes a protein called E6, which specifically targets and inactivates the tumor suppressor protein p53. Later it was discovered that E6 protein acts as an adaptor between E6-AP and p53, facilitating the ubiquitination of p53 by E6-AP. The characterization of E6-AP led to the identification of a whole family of proteins that share a common ~350aa long carboxy-terminal *HECT* domain.[10]

The second group of E3 ligase family contains RING finger E3 ligases that catalyze a direct transfer of the ubiquitin moiety from E2 to the E3 bound substrate. Unlike the *HECT* E3s, the RING E3 ligases do not form a thioester intermediate with ubiquitin. These ligases simply serve as a docking site for the E2 enzyme and the substrate or they may be involved in the allosteric activation of the E2-bound substrate, which enhance the transfer of ubiquitin to the target substrate. However, till date, it is not clear which is the predominant function of RING finger E3 ligases.

The RING finger E3 ligases are further classified into two distinct families: single- or multi-subunit complexes. Single subunit E3s contain the substrate recognition element and the RING finger on the same polypeptide. Among the well understood single subunit RING E3s are the oncoprotein murine double minute 2 (MDM2) which ubiquitinates the p53 protein, the proto-oncoprotein c-Cbl which ubiquitylates growth factor receptors and inhibitors of apoptosis. The multi-subunit RING E3s include a small ring finger protein, a member of the cullin family of proteins as well as other subunits, some of which are involved in substrate recognition. Among the multi-subunit RING finger complexes are the Skp1-Cullin/Cdc53-F-box (SCF) complex, Anaphase promoting complex (APC) and the Von-Hippel Lindau-Cul2/Elongin B and C (VHL-CBC) complex. The SCF complexes are involved in the degradation of cell cycle-induced proteins as well as proteins involved in signal transduction. SCF E3s recognize and ubiquitinate diverse groups of phosphoproteins, with substrate specificity conferred by their F-box protein partners. The VHL-CBC complex E3 ligases are involved in the degradation of hypoxia inducible transcription factor 1α (HIF1α). The most complex of the multisubunit RING E3s is the APC that is involved in the degradation of cell cycle regulators.[10]

It is not clear how the E3s mediate both transfer of ubiquitin to the substrate Lys and adds ubiquitin moieties to the growing chain. In recent years, a ubiquitin chain elongation factor or E4 has also been discovered. Using yeast as a model system, it has been demonstrated that E4 catalyzes ubiquitin chain elongation in conjunction with E1, E2 and E3. E4 has also been shown to ubiquitinate substrates in the absence of *HECT* or RING E3 ligases, leading to speculations that it may be a subfamily of E3 ligases.[12] More studies are needed to define the precise role of E4 in the process of ubiquitination of proteins. The UPP enzyme cascade concludes with a ubiquitinated protein. This ubiquitinated protein is then degraded by the 26S proteasome. This step will be discussed in the next section.

3.2.2 *The 26S proteasome*

The mammalian 26S proteasome is a ~2.5MDa multi-subunit protein degradation complex composed of two major sub-complexes that are highly conserved among all eukaryotes: a 20S proteolytic core particle and a 19S regulatory particle. The 20S core is a barrel shaped structure made up of four rings of seven subunits each, which based on their sequence similarity, can be divided into an α-type and a β-type group. The two β-type member rings are on the inside and two α-type member rings are on the outside, thus giving the sub-complex a general structure of α1-7 β1-7 β1-7 α1-7. Within the 20S core complex, only three β-subunits (six subunits total in the complex) possess tryptic, chymotryptic and peptidylglutamyl-like proteolytic activity. The α-type subunits appear to serve as a template on which the β-type subunits assemble, and protect the cellular milieu from the β-type protease activity. The 19S regulatory cap, a multi-subunit assembly bound to either end of the cylindrical 20S proteolytic core, can be split into two parts: a lid and a base. An approximately 12 member non-ATPase lid and a six member AAA (ATPases associated with a variety of cellular activity) ATPase base forms a hexameric ring that contacts the α-type member ring of the 20S core complex.[13] Once the polyubiquitinated substrate is recognized by the 26S proteasome, certain deubiquitinating enzymes (DUBs) in the 19S lid complex cleave the ubiquitin chain and allow recycling of ubiquitin pool (Fig. 3.1). Two classes of DUBs are known, ubiquitin-specific proteases and ubiquitin carboxy-terminal hydrolases. This is followed by the unfolding of the substrate by the chaperone-like activity of the ATPases in the 19S base complex, which then feeds the substrate into the inner catalytic 20S core complex. The functions of most of the non-ATPase subunits are still largely unknown although S5a has been shown to bind polyubiquitin chains, and S13 and Poh1 (RPN11) have recently been shown to contain deubiquitylating-protease activity.[14] A critical issue that is yet to be resolved is the entry of protein substrates and the exit of proteolytic products from the 26S proteasome. Some clues about these processes are beginning to come to light. The opening into the 20S catalytic chamber is small (approximately 1.3 nm), and significant unfolding of the substrate is required. A molecular gate (amino-terminal tail of the α3-subunit) also guards the opening, but it is constitutively open when the 19S regulatory units are bound to the 20S proteasome[15] (Fig. 3.1).

An additional complex that associates with the 20S core complex is PA28 or 11S or REG. Unlike assembly of the 19S-20S-19S, complex formation with 11S-20S-11S is ATP-independent and digests only peptides but not ubiquitin-conjugated intact proteins. This complex is thought to function as an “immunoproteasome” to process antigens to be recognized by the class I major histocompatibility complex (MHC) and T-cell receptors. The MHC is responsible for discerning between self and non-self or infected cells. Also, the existence of an asymmetrical 19S-20S-11S complex has been demonstrated that has the potential to carry out two-step proteolysis, with the initial cleavage resulting in large peptides followed by trimming to yield smaller antigenic peptides.[16]

3.3 Ubiquitin-Proteasome Pathway's Involvement in NHR Function

In this section, we will discuss the role of the UPP components and NHR degradation via the UPP on NHR transactivation functions. Additionally, the different signals that trigger the degradation of NHRs by the UPP and the degradation of coregulators themselves by the UPP will also be discussed.

3.3.1 *Nuclear hormone receptors and their coregulators*

NHRs comprise one of the largest superfamily of ligand-regulated transcription factors among eukaryotes. Their ligands range from steroids such as estrogens for estrogen receptor (ER) or progestins for progesterone receptor (PR) to mineralocorticoids for mineralocorticoid receptor (MR) etc. In some cases, the ligand is not known. Therefore, these receptors are referred to as the "orphan receptors".[17]

NHRs are known to regulate gene expression by binding to specific DNA enhancer elements, called hormone response elements, in the promoters of their target genes. The primary function of NHRs is to upregulate or downregulate the transcription of their target genes, thereby playing a major role in diverse cellular processes including reproduction, metabolism, and development. The association of NHRs with general transcription factors (GTFs) and RNA polymerase II (RNA Pol II) leading to transcription is well documented. In the past two decades, it has become clear that transcriptional regulation of target genes by NHRs requires a cohort of accessory factors known as coregulators or cofactors.[18] Coregulators can be broadly divided into two classes: coactivators that enhance the transcriptional activity of NHRs and corepressors that repress NHR activity. Coactivators are recruited to agonist-bound receptors, whereas corepressors are bound either to unliganded receptors or receptors bound to antagonists. Depending on the type of coregulator recruited to the target gene promoter, their binding to the receptor leads to either a transcriptionally permissive or repressive environment at the promoter. Coregulators are known to have two main functions: they act either as bridging factors which provide a scaffold for protein–protein interactions or they facilitate relaxation of the highly compact chromatin structure by bringing in enzymatic activities such as acetyltransferase, methyltransferase, ATPase, ubiquitin-conjugation and ubiquitin-ligation to the target gene promoter.[19] The existence of coregulator proteins came to light from transcriptional interference or squelching experiments initially from yeast followed by experiments in mammalian cells.[20] These experiments demonstrated a decrease in the activation of receptor A in the presence of receptor B, suggesting the existence of a limited pool of activator proteins in the cell. These experiments suggested that different receptors compete for a common pool of coactivators that are essential for their transactivation functions. Till date, there are ~200 coactivators that are thought to regulate the activities of ~46 NHRs.[21] A detailed description of the different classes of coactivators discovered so far and their specific functions is beyond the scope of this chapter. This chapter will focus on NHR coregulators which belong to the UPP, and the mechanism of action by which they regulate NHR functions.

3.3.2 *Ubiquitin-proteasome pathway enzymes and components as coregulators of NHRs*

A number of ubiquitin proteasome pathway enzymes and components have been shown to act as coactivators of NHRs. The first UPP enzymes shown to exhibit coactivation activity were the yeast E3 ligase RSP5 (reversion of spt phenotype 5) and its human homologue RPF1/NEDD4 (receptor potentiating factor1/neural precursor cell expressed, developmentally downregulated 4).[22] This was followed by the discovery that another E3 ligase, E6-AP also modulates NHR functions.[6] Surprisingly, the ubiquitin ligase activity of both these enzymes was dispensable for their coactivation function, suggesting that these two activities were separable. Another E3 ligase, MDM2 has also been reported to enhance the transactivation functions of NHRs. Another example of a ubiquitin ligase/acetyltransferase enzyme is p300 which acts as a coactivator of NHRs and has been shown to cooperate with MDM2 and ubiquitylate p53. Since the E3 ligase enzymes worked in concert with their partner E2s, it was important to test if their compatible E2 enzymes were also involved in NHR transcriptional activity. In fact, it has been shown that the E2 enzymes, UBCH7 and UBCH5 were essential for ER, PR, androgen receptor (AR), glucocorticoid receptor (GR) and thyroid receptor (TR)-dependent transactivation of target genes.[8,23] The involvement of both the E3 and E2 enzymes in the transcriptional activation of NHR target genes was further substantiated by the recruitment of these enzymes onto the promoters of NHR target genes.[7,23]

In addition to the E2s and E3s, the components of the 19S regulatory particle and the 20S catalytic core particle are involved in NHR-dependent transactivation. The 19S proteasome activator's ATPase subunit SUG1/Rpt6/TRIP1/PSMC3 (suppressor of Gal1/Thyroid-receptor interacting protein), S1 and PSMC5/TBP1 (Tat-binding protein) have been shown to functionally interact with retinoid X receptor (RXR), vitamin D3 receptor (VDR), TR and ER in a ligand-dependent fashion.[7,24–26] The F-box proteins TBL1 and TBLR1 that are components of the SCF E3 ligase complex have been reported to facilitate ligand-dependent transactivation of a number of NHRs by mediating corepressor/coactivators exchange via the recruitment of the 19S regulatory subunit to the promoter.[23] Recently, it has been demonstrated that LMP2 (low molecular mass polypeptide2) subunit of the 20S catalytic core was recruited to the promoters of ER target genes. Furthermore, LMP2 protein was found to be present not just at the promoter, but along the entire sequence of the gene.[27] These findings showed that the UPP was necessary for ER-mediated gene transcription placing the UPP in an important role in not only NHR transactivation but also in general transcription activities as well. The UPP components and ubiquitin-like pathway enzymes that act as NHR coregulators have been summarized in Table 3.1.

3.3.3 *Ubiquitin-proteasome pathway's role in NHR-mediated transactivation*

As discussed above, the 26S proteasome is involved in the degradation of the majority of short-lived and newly synthesized proteins in the cell.

Table 3.1. Ubiquitin-proteasome Pathway Enzymes and Components as Coregulators of NHRs

Enzyme or Component	Function	Target Receptor
UbcH2	Ubiquitin Conjugation	TR
UbcH5	Ubiquitin Conjugation	ER, RAR, TR, PPARγ
UbcH7	Ubiquitin Conjugation	PR, GR, AR, RAR
E6-AP	Ubiquitin Ligation	AR, PR, GR, ER, TR
RPF1/NEDD4	Ubiquitin Ligation	PR, GR
HDM2/MDM2	Ubiquitin Ligation	ER, AR
BRCA1	Ubiquitin Ligation	ER
TBL1	Ubiquitin Ligation	ER, AR, TR, PPARγ
TBLR1	Ubiquitin Ligation	ER, RAR, PPARγ
S1/Rpn2	Non-ATPase	ER, TR, RAR, PPARγ
SUG1/Rpt6 [19S]	ATPase	ER, AR, VDR, RARγ
SUG2/Rpt4 [19S]	ATPase	PPARγ
LMP2/PSMB9 [20S]	Peptidase	ER
TBP1/PSMC3	ATPase	TR
MIP224/TBP7	ATPase	Orphan receptor
p300	Acetyltransferase/ Ubiquitin Ligation	ER, AR, GR, TR, RAR, VDR
Ubc9	SUMO Conjugation	AR, GR, ER, PPARγ
Ubc12	NEDD8 Conjugation	ER
Uba3	NEDD8 Activation	ER, AR, PR

Further review of the research shows a well established role for the UPP in regulating cell cycle and signal transduction. In the past decade, it has been demonstrated that nuclear hormone receptors (NHRs) are ubiquitinated and targeted to the proteasome for degradation in the course of their nuclear activities. In the following section, we will discuss the probable role of the UPP in the regulation of NHR functions.

The involvement of the different UPP enzymes and components in NHR transactivation function suggested that the UPP had an essential role in regulating transcription of NHR target genes. Since it is well known that the primary role of the 26S proteasome is to act as a garbage bin for cellular proteins, the obvious function of the UPP could be to attenuate NHR transcription by regulating the protein levels of NHRs. This possibility was supported by the observation that ER is targeted to the 26S proteasome with half-lives greater than three hours in absence of hormone and less than one hour in the presence of estradiol.[28] This was followed by findings that the degradation of ER is required for its transcriptional activity.[7,29] The link between receptor degradation and transcription was both fascinating and puzzling to scientists in the steroid receptor field. It was questioned why should two diverse processes that utilize energy be linked and what could be the possible outcome.

There are a large number of coactivator complexes that are known to associate with target gene promoters, and there has to be a sequential and orderly recruitment and release of coactivators onto and off the promoters of NHR target genes for proper transcription to ensue. This idea was supported by the recent observations which suggest that coactivators and receptor cycle on and off hormone-responsive gene promoters in a ligand-dependent fashion.[5,7,30] It is conceivable that the UPP enzymes and components are required for the exchange/removal of coactivators and receptors thereby regulating transcription of genes (Fig. 3.2). The concept that receptor degradation and transactivation are linked has been challenged recently and it has been shown that the processes of degradation and transactivation are separable. It is possible that for a subset of NHR target genes, UPP-dependent degradation of receptor is essential for its transcriptional activity whereas for other target genes, receptor degradation is not essential. This possibility was supported by a recent study which suggests that degradation of ER is essential for one target gene whereas it is not required for the transcription of another target gene.[31] Another possibility is that NHRs degradation via the UPP contributes differentially to receptor-dependent gene transcription in different cell types. In line with the findings with ER, other NHRs such as GR, AR, VDR, TR, RXR, and retinoic acid receptor (RAR) either unliganded or liganded are also ubiquitinated and targeted for degradation by the proteasome.[5,7,26,32] However, the observation of increased receptor degradation leading to increased transcription is not a universal phenomenon.

In the case of PR and GR, both receptors are degraded like ER upon binding to their cognate ligands, but proteasome inhibition lead to a decrease in transactivation of PR and increase in GR-mediated activity. On the other hand, both AR and VDR are stabilized in presence of their ligands. In addition, stabilization of AR and VDR by proteasome inhibition had opposing effects on their transcriptional activity. All these studies indicate that proteasomal degradation is receptor-specific and is far more complex than previously thought. The AR and VDR are

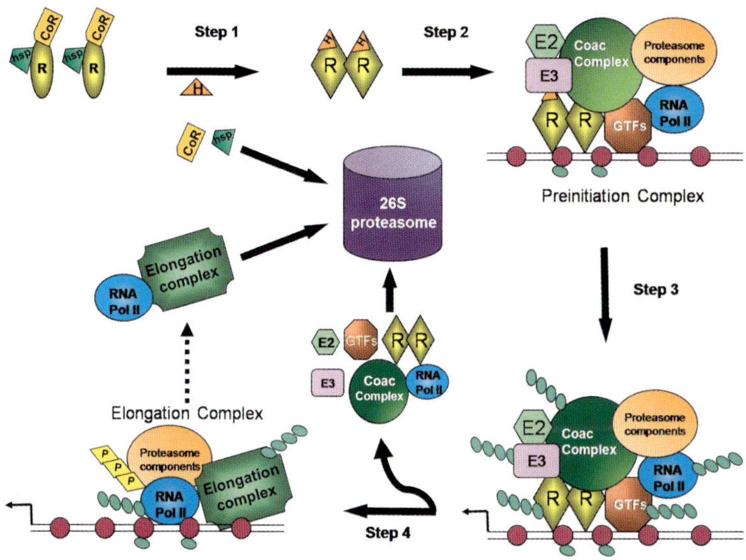

Fig. 3.2. A model representing the possible functions of the proteasome and its components in NHR-mediated transcription: NHRs (R) are found in an inactive state bound to negative coregulators such as corepressors (CoR) and heat shock proteins (hsp). In *Step 1*, upon binding to its cognate ligand (H), NHRs, dissociate from negative coregulators (these are degraded by the 26S proteasome), undergo conformational change, dimerize, undergo posttranslational modification and bind to their response elements in the chromatin. In *Step 2*, the receptor, bound to its response elements, recruits coactivators which include ubiquitin pathway enzymes and components (E2 ubiquitin-conjugating, E3 ubiquitin-ligase) that remodel and de-repress the compact chromatin, aid in exchange of coactivator complexes (Coac complex), and leads to recruitment of general transcription factors (GTFs) and RNA polymerase II (RNA Pol II) thereby facilitating the formation of preinitiation complex. These complex series of events lead to the initiation of transcription from the target gene promoter (*Step 3*). Many of the proteins in the preinitiation complex are thought to be ubiquitinated, possibly facilitating disassembly of the initiation complex and degradation of receptor, coactivator(s), GTFs and ubiquitin enzymes by the 26S proteasome (*Step 4*). This step is followed by phosphorylation of RNA Pol II at the carboxy-terminus, and once it is hyperphosphorylated, it is ubiquitinated which promotes recruitment of elongation factors. Based on experiments done in yeast and the fact that components of both 19S and 20S proteasomes were found associated with NHR-target gene promoters, it is possible that 19S and 20S components are also involved in restructuring and remodeling the elongation complex, facilitating the elongation phase of transcription. The association of these components and not the full 26S proteasome with the elongation complex indicates the components are also playing a non-proteolytic role in NHR-dependent transcription.

exceptions in the group of NHRs, as they are not degraded after ligand-binding, and on the contrary, their protein levels are upregulated in presence of the hormone.[33] Recently, it has been demonstrated that AR is ubiquitinated upon treatment with the androgen, and it does get degraded after approximately eight hours of androgen treatment. These findings suggest that the up-regulation of the receptor observed with hormone-bound receptor is not due to lack of degradation but rather from increased protein synthesis.[34] As in the case with ER, it has been demonstrated that AR responsive target genes are also associated with the 19S proteasome components which have been shown to cycle dynamically with AR on and off target gene's promoters. Hence, it can be implied that the UPP is also involved in clearing the promoter to facilitate reinitiation of transcription. As for the VDR, which like AR, is stabilized upon binding to its ligand, there are conflicting reports on the role of proteasomal degradation on its transcriptional activity.

On the other hand, the case of GR is both interesting and perplexing. GR is destabilized by its cognate ligand and at the same time, stabilization of the receptor by proteasome inhibition leads to increase in its transcriptional potency.[35] Both PR and GR can bind and activate the same target gene, but both of these receptors may recruit different coactivators to the target promoter to activate transcription. Hence, it is possible that one receptor such as PR recruits cofactors that should be removed by the proteasome for transcription to proceed, whereas for other receptors like GR the mechanism might be opposite, such that it recruits cofactors that need to be degraded by the proteasome for transcription to cease.[14] It has also been shown that the large pool of GR resulting from proteasome inhibition significantly increases the rate of transcription of target genes without affecting chromatin remodeling. Building on this hypothesis, it was observed that the GR antagonist 8-bromo-cyclic AMP could reduce target gene transcription in the absence of any change in chromatin remodeling. The precise mechanism for the differences observed with these two receptors is still not clear. To add to the observations of ubiquitination and the degradation of NHRs by the proteasome, it has been reported that some proteins like p21CIP/WAF1, ornithine decarboxylase, c-Jun, calmodulin and IκBα are degraded in an ubiquitin-independent manner, thus giving a clue of a similar mechanism operating with members of the NHR family.[36]

Apart from NHRs and their coregulators, proteins of the general transcription machinery such as RNA Pol II, TAFII 250 subunit of TFIID, TBP and TAF(II)135 have been shown to be targets of the UPP. RNA Pol II has been reported to be ubiquitinated and degraded by the proteasome pathway in response to stress conditions such as UV-light radiation, as a mechanism to inhibit transcription elongation.[14] RNA Pol II is a substrate for both RPF1 and SCF E3 ligases leading to its degradation. Taken together, these findings indicate that proteins involved in the transcription process in general are targets susceptible to regulation by the 26S proteasome.

3.3.4 *Signals for NHR degradation via the 26S proteasome*

As discussed above, there are a number of different scenarios for degradation and subsequent transcriptional outcome of the different NHRs. The question lingers as to what are the signals that trigger the degradation of NHRs by the proteasome machinery. The first signal seems to be the ligand itself which is receptor specific for inducing degradation or stabilization. Pure agonists like estradiol induce ER degradation, whereas partial agonists like tamoxifen and RU486 have been shown to stabilize ER and PR respectively. But this is not true for all receptors, ligand binding does not act as a degradation signal for AR and VDR as these receptors are stabilized upon ligand binding. Despite the significance of the ligand binding domain in receptor proteolysis, the deletion of the amino-terminus of receptors such as ER, RARγ2 and PPARγ renders them insensitive to ligand-dependent degradation suggesting that the activation domain of NHRs located in the amino-terminus and the ligand binding domain are necessary for their turnover by the 26S proteasome.[37] In this sense, NHRs are similar to other transcription factors such as Myc, VP16 and E2F-1 wherein the degradation signal or degron overlaps with their transcription activation domains (TADs).[38]

Post-translational modifications such as acetylation and phosphorylation are known to act as signals for proteasomal degradation of NHRs. Mutation of mitogen-activated protein kinase (MAPK) phosphorylation site on PR and the site of acetylation on ER has been shown to impair hormone-dependent degradation of these receptors. In the case of ER,

PKC (Protein Kinase C) is known to enhance proteasome-mediated degradation of both unliganded and liganded receptor. On the contrary, phosphoinositide-3 kinase (PI3K) increases the stability of the ER. Additionally, PKA (Protein Kinase A) mediated increase in transcriptional activity of ER is independent of proteasomal degradation and in fact, it protects the receptor against ligand-induced degradation.[37] Furthermore, inhibition of the kinase activity of constitutively photomorphogenic 9 (COP9) signalosome subunit 5 (CSN5/Jab1), a subunit of the COP9 signalosome by curcumin prevented estrogen-induced downregulation of ER, which correlated with loss of phosphorylation, although the specific phosphorylation site was not defined.[37] Even with the GR, it has been demonstrated that multiple phosphorylation sites in the amino-terminal domain are responsible for the regulation of GR protein stability. Other examples of kinase-induced regulation of receptor stability include Akt (also known as protein kinase B) -induced degradation of AR and okadaic acid-induced stabilization of TR reportedly by MAPK. All these studies point to the fact that phosphorylation can be a signal for degradation of NHRs.[37]

Other signals for NHR degradation include coactivator binding. Coactivators bind to NHRs through specific residues in helix 12 within the ligand binding domain, and mutations of these residues in ER and RAR have been shown to prevent hormone-dependent degradation of these receptors, implicating coactivator binding as a possible signal for degradation.[29] Recent reports suggest that cross talk between different members of the NHR family can also modulate receptor proteolysis. For example, GR, AR and VDR stability is regulated by ER, ER stability by TR and TR in turn is regulated by its heterodimeric partner RXR. The mechanism(s) by which one receptor controls the stability of another receptor has not been fully worked out. However, the regulation of GR protein levels by ER is known in some detail. Estrogen bound ER induces the expression of two E3 ligases, namely MDM2 and Siah2 (Seven in absentia homolog 2). Increase in MDM2 protein levels accelerates the turnover of GR. Since MDM2 is also implicated in AR degradation, it is possible that a similar mechanism is operating. On the other hand, the ER induced E3 ligase Siah2 is known to downregulate the expression of corepressor NCoR (Nuclear Corepressor), leading to a decrease in its half-life, which in turn leads to de-repression of VDR-mediated transcription.[37] Along with receptor degradation, the coregulators that act on NHR are degraded by UPP and the relevance of this function will be discussed in the next section.

3.3.5 *Degradation of coregulators by the ubiquitin-proteasome pathway*

Coregulators serve to modulate the transactivation activity of NHRs, and they are an integral part of NHR signaling. So it is reasonable to think that coregulators themselves are targets of proteasomal degradation. It has been demonstrated that all three members of the p160 family of coactivators, SRC-1 (steroid receptor coactivator-1), SRC-2 and SRC-3, as well as E6-AP are targets of the 26S proteasome.[29] Inhibitors that block the UPP could also increase the steady-state levels of all these coactivators that lead to a concomitant increase in their intrinsic transcriptional activation function. In this context, it has been reported that SRC-2 interacts with the proteasome via its activation function domain, which suggests that proteasome activity could be linked to its coactivation function. Additionally, it has also been shown that E6-AP is directly involved in the degradation of SRC-3 via the UPP.[39] SRC-3 is an oncogene amplified in breast and other cancers and its protein levels appears to be regulated by multiple pathways in the cell. Another coactivator, p300 is phosphorylated by the PI3K/Akt pathway, which in turn increases its degradation by the 26S proteasome and impedes its coactivation function. In line with these findings, the NHR corepressors, NCoR and SMRT (silencing mediator for retinoid and thyroid hormone receptors) have also been shown to be ubiquitinated by two different E3 ligases, leading to subsequent degradation.[4] Furthermore, it has been demonstrated that similar to other protein substrates targeted for degradation by the proteasome, both coactivators and corepressors are also targeted by specific combinations of E2s and E3s.[4] Of particular interest is the recent finding which shows that SRC-3 is degraded in a ubiquitin and ATP-independent manner by the REGγ proteasome.[40] The REGγ proteasome, also known as PA28γ or Ki antigen, belongs to the 11S family of proteasome activators, which along with the 20S catalytic core is involved in the degradation of peptide substrates. This report is the first one to show the regulation of a coactivator protein by the proteasome, independent of the 19S regulatory particle.

3.3.6 *Proteolysis independent roles of ubiquitin and 26S proteasome in transcription*

In recent years, it has become evident that tagging of proteins with ubiquitin has many more functions in the cell rather than just targeting them for degradation via the 26S proteasome. However, it was not clear how a cell determines which ubiquitin-tagged proteins should be degraded and which ones should not? It has been proposed that different ubiquitin-chain lengths are associated with different processes. As a general rule, proteasomal degradation requires that the target protein to be linked by a Lys48-mediated polyubiquitin chain of at least four ubiquitin moieties long. In contrast, non-proteasomal functions are signaled when the substrate protein is linked to a single ubiquitin (monoubiqutination), single ubiquitin at several different Lys of the substrate (multi-monoubiqutination) or multi-ubiquitination of the target protein tagged with a polyubiquitin chain linked using lysine residues other than Lys48 like Lys6, Lys11, Lys29 and Lys63 of the ubiquitin polypeptide.[41]

Mono and di-ubiquitination have been implicated in endocytosis, budding of retroviruses and histone regulation. Ubiquitin is also known to participate in protein trafficking to endosomal compartments either from plasma membrane, or from the trans-Golgi network.[42,43] One example of ubiquintin's role in a non-proteolytic process is the mono-ubiquitination of proliferating cell nuclear antigen (PCNA) by Rad6 (RADiation sensitive 6 is an E2) and Rad18 (an E3). PCNA is a DNA polymerase sliding clamp involved in DNA replication and repair. Mono-ubiquitination of PCNA at Lys164 allows a switch of polymerases at stalled replication forks allowing translesion synthesis of DNA. Polyubiquitination can be involved in non-proteolytic processes as well. For example the Lys63 linked polyubiquitination of PCNA by Rad5 (a RING domain E3) is required for error-free repair of damaged DNA. Another example is the polyubiquitination of TRAF6 (tumor necrosis factor receptor-associated factor 6), a RING finger domain E3, through Lys63 of ubiquitin, which leads to activation of a protein kinase cascade. The cascade leads to phosphorylation and activation of IκB kinase β (IKKβ) ultimately allowing NF-κB to enter the nucleus to regulate gene expression.[3]

In addition to acting as a degradation signal for proteins, the 76aa polypeptide ubiquitin has been found to function in a proteasome-independent mechanism in transcription regulation. One of the first ubiquitinated proteins to be identified was histone H2A, but its function still remains obscure. Recently, the mono-ubiquitylation of histone H2B was demonstrated to be essential for the methylation of histone H3 at Lys4. In addition, ubiquitin is also known to participate in transcription activation through ubiquitylation of TADs of transcription factors like VP-16. Investigation of the mechanism of action of the viral protein VP16 TAD in *S. cerevisiae* showed that the yeast strain lacking the components of the SCF complex failed not only to be efficiently ubiquitylated and degraded by the proteasome, but also to function as a transcription activator. Furthermore, the requirement of the SCF complex was circumvented by fusing a single ubiquitin moiety to the VP16 activator, demonstrating that the attachment of ubiquitin was essential for transcription activation.[2] Another piece of evidence comes from monoubiquitination of the AR itself. Monoubiquitylated AR is critical for efficient activation of AR protein and is stabilized by TSG101 (tumor susceptibility gene 101), which prevents polyubiquitination of AR, thereby blocking AR degradation.[34] It has been hypothesized that the signal for ubiquitylation of AR might come from TAF II 250 subunit of the TFIID, a general transcription factor possessing ubiquitin-conjugating activity. Another possibility is ubiquitination via p300, which apart from being a coactivator also has ubiquitin-ligase activity.

There are also evidences to demonstrate that the proteasome itself can be intimately involved in transcription independent of its proteolytic function. Using yeast as a model system, it has been shown that components of the 19S regulatory cap of the proteasome is capable of activating Pol II transcription elongation *in vitro*, by a mechanism independent of proteolysis.[44] Consistent with these findings, it has been demonstrated that the 19S regulatory particle recruits the SAGA histone acetyltransferase complex to target gene promoters implicating a non-proteolytic function of the 19S subunit in transcription.[45] Another report shows that the 20S proteolytic core may facilitate transcription by remodeling the transcriptional complexes or chromatin.[46] Furthermore, it has been shown that both 19S regulatory subunit and 20S subunit are physically associated with target genes of ER, and the LMP2 subunit of the 20S core was present along the entire sequence of ER target genes.[7,27] Therefore, it is tempting to speculate that the proteasome (19S and 20S) might be employing its ATPases as chaperones for unfolding or restructuring proteins which may result in activation of transcription initiation or elongation. More studies are needed to clearly define the role of the proteasome in gene transcription. Apart from ubiquitination, there are certain ubiquitin-like pathways that are involved in NHR signaling. The next section discusses the role of ubiquitin-like pathways in NHR transcription regulation.

3.4 Sumoylation and Neddylation

In addition to ubiquitin, there are a few ubiquitin-like small proteins that can bind covalently to substrate proteins and alter their function. One such protein is the small ubiqutin related modifier (SUMO) that is known to be covalently attached to proteins and play important roles in modulating chromatin structure, gene expression and signal transduction. The sumoylation enzymatic machinery is similar but not identical to the ubiquitination machinery. Although there is 18% homology in the structure of ubiquitin and SUMO, the distribution of charged residues on the surface of SUMO is different from that of ubiquitin. In addition, SUMO also has an amino-terminal extension not found in ubiquitin. In contrast to ubiquitin wherein attachment of a minimum of four ubiquitin moieties is essential for the protein to be recognized by the proteasome, SUMO generally functions as a monomer. Another striking difference between ubiquitin and SUMO is the presence of only a single E2 (Ubc9) specific for SUMO, unlike UPP, which has several E2 enzymes. Multiple E3 ligases have recently been found to act as adaptors between Ubc9 and the substrate proteins. A consensus sumoylation motif, ψKxE/D, has been identified in many target proteins, where ψ represents a large hydrophobic residue, K is the SUMO acceptor site, and x may be any amino acid.[47]

A number of transcriptional factors including certain members of NHR superfamily such as AR, GR, PPARγPR, and MR have been shown to be targets of sumoylation, and in most cases repressed by sumoylation. Furthermore, coregulators of nuclear receptors, like SRC-1, p300, HDAC-1 (histone-deacetylase-1), HDAC-4 and HDAC-6, have also been shown to be regulated by sumoylation.[48,49] Therefore, the process of sumoylation seems to be an important regulatory mechanism for nuclear receptor-mediated transcription.

Like ubiquitination and sumoylation, neddylation is another pathway that is involved in the covalent conjugation of the 81aa polypeptide NEDD8 (neural precursor cell expressed, developmentally down-regulated 8) to proteins. Neddylation is a process by which the carboxyl-terminal glycine of the ubiquitin-like protein NEDD8 is linked through an isopeptide bond to lysine residues of proteins, analogous to protein ubiquitination. However, unlike ubiquitination, neddylation does not generate polymeric chains. Until recently cullin proteins have been the only known substrates for neddylation. NEDD8 has been reported to play a role in degradation through its activator function of the cullin subunit of the SCF ubiquitin-ligases. Once cullin is neddylated, its ability to promote ubiquitination of its substrates is greatly enhanced. It has been shown that the activating enzyme of NEDD8 pathway, Uba3 interacts directly with ligand-bound NHRs such as ER, AR and PR to inhibit their transcriptional activities.[50] Furthermore, both Uba3 and Ubc12 (ubiquitin-conjugating enzyme equivalent of NEDD8 pathway) are required for ubiquitination and degradation of liganded ER.[51] However, in depth studies are needed to decipher the precise role of neddylation in steroid receptor-dependent cell proliferation, and the biological relevance of this pathway in the development and progression of uterine and breast cancers.

3.5 What We Know and What We Need to Know About UPP's Role in NHR Function?

Although our knowledge about the role of UPP in NHR function has tremendously improved in the past few years, there are still lots of pieces in the puzzle that are missing. Important contributions from different laboratories have demonstrated that the UPP components play a significant role in the degradation of both NHRs and their coregulators. In light of all the studies cited above, the simplistic notion that receptors and their coregulators are polyubiquitinated and then degraded by the proteasome does not hold true. Also, the available literature in the field regarding the role of UPP and NHR transactivation functions is controversial which leads to a lot of unanswered questions and lack of precise mechanistic details.

3.5.1 *Nature of the ubiquitination site(s) on different receptors*

Several ubiquitin-conjugating and ubiquitin-ligating enzymes have been reported to act as coactivators of NHRs. For example the E2s, UbcH5 and UbcH7 are known to enhance transactivation of a number of receptors. Similarly, E3 ligases E6-AP, MDM2, RPF-1 and breast cancer 1 (BRCA1) are recruited to the promoters of NHR target genes and influence the ligand-dependent transcriptional activation function of NHRs. The puzzling fact about the E3 ligases, RPF1 and E6-AP is that their ligase activity is dispensable for their coactivation function, suggesting that these two activities are separable. However, the ubiquitin-conjugating activity of UbcH7 was indispensable for its coactivation function, but the E2s themselves do not interact directly with their substrate receptors. Hence, it is not clear if these E2s and E3s are actually involved in ubiquitinating specific lysine residues of receptors. Identification of combinations of specific E2-E3 enzymes and their target lysines on the receptor will give us a better understanding of the involvement of the UPP in NHR function.

3.5.2 *Why do different NHRs behave differently with respect to the role of the UPP on the transcriptional activation of their target genes?*

As discussed above, different receptors are affected differently by the UPP. Similarly, ligand binding also induces differential effects on various receptors. For example, ER, PR and GR undergo increased proteolysis via the 26S proteasome upon ligand binding, whereas AR and VDR protein levels are stabilized when bound to their cognate ligand. In addition, inhibition of proteasome activity can either increase or decrease the transactivation of their target genes. Furthermore, different target genes of the same receptor (e.g. ER) also exhibit differential effects upon inhibition of the proteasome. Recently, genome wide transcript profiling analyses of cells treated with proteasome inhibitors demonstrated that expression of some genes increased, some decreased and some were unchanged.[14] Hence, further investigations on the differential expression of target gene upon proteasome inhibition are warranted to understand the complexity of this system.

3.5.3 *Where does receptor degradation occur: On promoters of target genes or another place such as the nuclear matrix?*

It has been demonstrated that the different enzymes involved in ubiquitination such as E2s (i.e. UbcH7) and E3s (i.e. E6-AP) and components of the 19S and 20S proteasome subunits (i.e. LMP2) are recruited onto the promoters of NHR target genes. LMP2 is only one of the many subunits of the 19S regulatory cap and 20S catalytic core that were found to be associated with target gene promoters. This raises the possibility that receptor degradation might occur at NHR target gene promoters. Other components of the proteasome such as the Rpn12 protein of 19S lid and HC3 protein of the $\alpha 2$ subunit of 20S were not recruited onto ER responsive target promoters, thus leading to speculations that receptor degradation might occur at another location besides the promoter. Recently, it has been demonstrated that the nuclear matrix plays an essential role in ER turnover and both unliganded and liganded ER are associated with the nuclear matrix protein HET/SAF-B with both populations of ER being highly mobile. With these revelations, it is plausible that polybiquitinated ER is translocated from the promoter to distinct specialized locations like the nuclear matrix for degradation. Similar observations were made with the GR that is also rapidly exchanged between its target gene promoter and the nucleoplasm. Furthermore, the nuclear matrix may provide a scaffold role in this movement. In addition, the yeast Rad23 and Dsk2 (proteins which are involved in nucleotide excision repair and spindle-pole body duplication respectively) contain a proteasome-interacting ubiquitin-like domain capable of binding polyubiquitin chains and may function as guides to bring the substrate protein to the 26S proteasome for destruction.[14] Proteins like Rad23 may be operating in mammalian cells also, but we do not have evidence of that yet. Further in depth studies with other NHRs are needed to identify the NHR's degradation site in the cell.

3.5.4 *UPP's role in human disease and as a potential target for drug discovery*

In view of the central role played by the UPP in a number of cellular processes, it is not surprising that deregulation or malfunction of the UPP contributes to the pathogenesis of various human disorders, such as cancer, neurodegenerative, autoimmune, genetic and metabolic diseases. It was thought that developing drugs that modulate the activity of the UPP could have great potential for the treatment of various disorders, specifically in cancer. Proteasome inhibitors thus represent a novel class of agents that may arrest the progression of cancer by interfering with the degradation of cell cycle proteins. Interestingly, many researchers have recently demonstrated that proteasome inhibition is more toxic to transformed cells than to normal cells. Bortezomib, also known as PS-341 or Velcade, is the first proteasome inhibitor to enter clinical trials and has been recently approved by the FDA for the treatment of patients with relapsed multiple myeloma. The mechanism of action of bortezomib appears to be the induction of apoptosis in a dose-dependent and cell-type specific manner.[52]

Knowing the fact that the UPP serves a lot of different functions in the cell, inhibiting the 26S proteasome itself or a major component of the UPP such as the ubiquitin-activating enzyme E1 may affect cellular proteins/processes nonspecifically. On the other hand, the ubiquitin ligases or E3s serve as the specific substrate-recognition element of the UPP, and these enzymes confer the utmost specificity for protein degradation by the UPP. Attacking a single E3 ligase might allow for selective stabilization of a subset of target proteins. This specific targeting could potentially boost the effectiveness of treatment and also eliminate nonspecific side effects. Furthermore, accumulating data strongly suggests that the deregulation of E3 ligases contributes to cancer development and that overexpression of these enzymes is often associated with poor prognosis. Therefore, it is thought that the E3 ligases can serve both as potential cancer targets as well as cancer biomarkers.

The disease states associated with UPP can be classified into two mechanism-based categories: (1) those resulting from mutation in either the target protein or the UPP enzyme, resulting in loss of function; and (2) those resulting from excessive degradation of the substrate protein, leading to gain of function. An example of mutation in the target protein is the epithelial Na+ channel which is unable to bind to its E3 ligase NEDD4, leading to its accumulation, in the rare disorder termed "The Liddle syndrome." Defects in the enzymatic activity of specific E3 ligases causes the Angelman's syndrome and Parkinson's disease, wherein their specific E3s namely E6-AP and Parkin are defective in ligase activity respectively. Although unconfirmed, it is believed that the neurological symptoms observed in these diseases are due to the accumulation of toxic substrates of these defective E3s.[52]

In addition, the UPP is also implicated in cancer. Cancer can result either from the stabilization of oncoproteins or destabilization of tumor suppressors. A typical example of gain of function of the UPP is the accelerated degradation of the cell cycle inhibitor p27. The mechanism underlying the low levels of p27 observed in a number of human cancers is its abnormally increased degradation, resulting from overexpression of Skp2, the E3 ligase specific for recognition and ubiquitination of p27. Other cancer-related proteins that are substrates of the proteasome are tumor suppressors (e.g. p53), cell-cycle regulators (e.g. p21) and cell surface receptors for growth factors (e.g. EGFR, TGF-βR). When the degradation of these proteins listed above are disrupted, the effect is potentially grave, particularly in rapidly dividing cancer cells, which require increased availability of growth-promoting factors to sustain accelerated and uncontrolled cell division. Consistent with this idea, it has been demonstrated that in certain cancers, specific E3 ligases have either altered expression or mutation in their genes. We will restrict our discussion to only two UPP enzymes namely, MDM2 and E6-AP since they act as NHR coregulators and are known to play a critical role in different types of cancers.

3.5.4.1 *MDM2*

MDM2 or HDM2 (the human homolog of MDM2) is a 90-kDa oncogenic RING finger protein whose expression is transcriptionally induced by p53 to generate a negative feedback loop that results in the degradation of p53 protein. p53 is a well known tumor suppressor and is mutated in 50% of human cancers. It has been shown that some of the tumors that retain wild-type p53 frequently disrupt other elements of the p53 network by promoting its degradation. Therefore, p53 is aptly named as the "guardian of the genome" that prevents mutagenesis leading to cancer by promoting cell cycle arrest and apoptosis. Since MDM2 is the major E3 ligase involved in the degradation of p53, it is the most attractive target for drug development against cancer. It has been thought that either inhibiting the MDM2-p53 interaction or blocking the conjugation of ubiquitin to p53 might result in reactivation of the tumor-suppressor activity of p53. In addition, inactivating MDM2 might be beneficial in inhibiting the degradation of other tumor suppressor proteins for which MDM2 might act as an E3 ligase.

Initially, anti-MDM2 antisense oligonucleotides were used as a proof-of-principle strategy to block MDM2 and p53 interaction. However, the major challenge is to develop small bio-available molecules that could block interaction of p53-MDM2 interaction. From a structural point of view, the MDM2-p53 interface looked very inviting since MDM2 has an open-pocket for p53 binding, which could be accessible to small molecules. The first small molecule MDM2 inhibitor is called "Nutlin" which was identified in a chemical library screen and is able to bind to the p53 binding pocket on MDM2, displacing p53.[9] Nutlins are now considered to be promising drugs as their efficacy was proven in both *in vitro* and *in vivo* systems. They were shown to activate p53-dependent cell cycle arrest and apoptosis in cancer cell lines and also arrest the growth of xenografts in nude mice without noticeable toxicity to healthy tissues. Another p53-stabilizing small molecule is RITA (reactivation of p53 and induction of tumor cell apoptosis) which has been demonstrated to possess growth arresting properties. The mode of action of RITA is distinct from Nutlin, it binds to amino-terminus of p53, thereby preventing p53 binding to MDM2. However, the synergistic effect of Nutlin and RITA if any, has not yet been tested. These small molecules appear to be promising drugs for p53 activation in cancers, but their efficacy in the treatment of human cancer remains to be proven.

3.5.4.2 *E6-AP*

E6-AP is the founding member of the *HECT* domain containing E3 ligases that was discovered as a protein associated with the human papillomavirus (HPV) E6 oncoprotein. E6-AP is known to target p53 for degradation by the 26S proteasome in cells infected by HPV and expressing the E6 protein. Till date, more than three dozen *HECT* ligases have been identified in the human genome, but very little is known about this family of ligases and very few of them have been linked to human disease. One genetic disorder in which E6-AP is implicated is the Angelman's syndrome, which results from mutation in the maternal copy of the E6-AP gene. It has been demonstrated that Angelman's syndrome is caused by the accumulation of mutations around the E6-AP active site leading to a defect in the enzymatic activity of E6-AP. Patients suffering from this disorder exhibit a variety of symptoms, which include mental retardation, neurodegeneration and puppet-like movements.

The role of E6-AP in HPV-induced cervical cancer is well studied. Certain strains of HPV like HPV-16 and HPV-18 are highly oncogenic due to degradation of p53, ultimately leading to malignant transformation of cervical epithelial cells. Hence, both E6-AP and HPV are pharmacological targets for cervical cancer. One possible strategy is the use of zinc-ejecting compounds which would inactivate HPV E6 by removing the functionally critical Zn atoms from the E6 protein. This will prevent the interaction between E6 and E6-AP and inhibit degradation of p53. However, the *in vivo* safety and bio-availability of these Zn ejectors has not yet been tested in humans. Another strategy is to take advantage of the *HECT*-dependent ubiquitination of substrates by E6-AP. E6-AP works by having the catalytic cysteine residue in the *HECT* domain accept the ubiquitin from the E2, and then undergo a conformational change to transfer ubiquitin to the substrate protein. Therefore, it is thought that molecules that could block or stabilize the *HECT* domain in a non-functional conformation can be employed to inhibit E6-AP's action.[9]

E6-AP protein plays a significant role in breast and prostate carcinogenesis while the estrogen and androgen receptors are known to play vital roles in breast and prostate cancer respectively. Both ER and AR have been shown to be regulated by the UPP. E6-AP degrades both the receptors, thereby regulating there levels in breast and prostate tissues resulting in a reciprocal relationship in the expression pattern of E6-AP and ER/AR in human breast and prostate tumors.[53] E6-AP has also been shown to regulate PI3K/Akt pathway which is the major pathway involved in cell survival and growth.[54] All of these observations suggest that E6-AP can be a potential target for drug discovery in not only cervical but breast and prostate cancers as well.

3.6 Conclusion

The role of the UPP in the cell has come a long way from an alternative pathway of lysosomal degradation to an intricate part of the cycle of proteins involved in the processes ranging from cell cycle to transcription to differentiation. Its importance to the cell is highlighted by diverse diseases arising from the aberrant regulation of the UPP and its normal protein targets. Normal protein targets of the UPP include the NHRs which depend on the UPP to facilitate the receptors function. The involvement of the UPP in NHR function has led to an interest in drugs that can modulate the UPP action in hope of treating various diseases such as hormone dependent cancers. Continued research is sure to lead to more intricate understanding in the UPPs' general role in the homeostasis of numerous proteins as well as explain how and why the UPP modulates NHRs specifically.

References

1. Hershko A, Ciechanover A, The ubiquitin system, *Annu Rev Biochem* **67**:425–479, 1998.
2. Conaway RC, Brower CS, Conaway JW, Emerging roles of ubiquitin in transcription regulation, *Science* **296**(5571):1254–1258, 2002.
3. Mukhopadhyay D, Riezman H, Proteasome-independent functions of ubiquitin in endocytosis and signaling, *Science* **315**(5809):201–205, 2007.
4. Nawaz Z, O'Malley BW, Urban renewal in the nucleus: Is protein turnover by proteasomes absolutely required for nuclear receptor-regulated transcription? *Mol Endocrinol* **18**(3):493–499, 2004.
5. Kang Z, *et al.*, Involvement of proteasome in the dynamic assembly of the androgen receptor transcription complex, *J Biol Chem* **277**(50):48366–48371, 2002.
6. Nawaz Z, *et al.*, The Angelman syndrome-associated protein, E6-AP, is a coactivator for the nuclear hormone receptor superfamily, *Mol Cell Biol* **19**(2):1182–1189, 1999.
7. Reid G, *et al.*, Cyclic, proteasome-mediated turnover of unliganded and liganded ERalpha on responsive promoters is an integral feature of estrogen signaling, *Mol Cell* **11**(3):695–707, 2003.
8. Verma S, *et al.*, The ubiquitin-conjugating enzyme UBCH7 acts as a coactivator for steroid hormone receptors, *Mol Cell Biol* **24**(19):8716–8726, 2004.
9. Nalepa G, Rolfe M, Harper JW, Drug discovery in the ubiquitin-proteasome system, *Nat Rev Drug Discov* **5**(7):596–613, 2006.
10. Weissman AM, Themes and variations on ubiquitylation, *Nat Rev Mol Cell Biol* **2**(3):169–178, 2001.
11. Wong BR, *et al.*, Drug discovery in the ubiquitin regulatory pathway, *Drug Discov Today* **8**(16):746–754, 2003.
12. Koegl M, *et al.*, A novel ubiquitination factor, E4, is involved in multiubiquitin chain assembly, *Cell* **96**(5):635–644, 1999.
13. Glickman MH, Ciechanover A, The ubiquitin-proteasome proteolytic pathway: Destruction for the sake of construction, *Physiol Rev* **82**(2):373–428, 2002.
14. Dennis AP, O'Malley BW, Rush hour at the promoter: How the ubiquitin-proteasome pathway polices the traffic flow of nuclear receptor-dependent transcription, *J Steroid Biochem Mol Biol* **93**(2–5):139–151, 2005.
15. Pickart CM, VanDemark AP, Opening doors into the proteasome, *Nat Struct Biol* **7**(11):999–1001, 2000.
16. Hill CP, Masters EI, Whitby FG, The 11S regulators of 20S proteasome activity, *Curr Top Microbiol Immunol* **268**:73–89, 2002.
17. Aranda A, Pascual A, Nuclear hormone receptors and gene expression, *Physiol Rev* **81**(3):1269–1304, 2001.

18. McKenna NJ, Lanz RB, O'Malley BW, Nuclear receptor coregulators: Cellular and molecular biology, *Endocr Rev* **20**(3):321–344, 1999.
19. McKenna NJ, *et al.*, Nuclear receptor coactivators: Multiple enzymes, multiple complexes, multiple functions, *J Steroid Biochem Mol Biol* **69**(1–6):3–12, 1999.
20. Meyer ME, *et al.*, Steroid hormone receptors compete for factors that mediate their enhancer function, *Cell* **57**(3):433–442, 1989.
21. Lonard DM, O'Malley BW, The expanding cosmos of nuclear receptor coactivators, *Cell* **125**(3):411–414, 2006.
22. Imhof MO, McDonnell DP, Yeast RSP5 and its human homolog hRPF1 potentiate hormone-dependent activation of transcription by human progesterone and glucocorticoid receptors, *Mol Cell Biol* **16**(6):2594–2605, 1996.
23. Perissi V, *et al.*, A corepressor/coactivator exchange complex required for transcriptional activation by nuclear receptors and other regulated transcription factors, *Cell* **116**(4):511–526, 2004.
24. Ishizuka T, *et al.*, Human immunodeficiency virus type 1 Tat binding protein-1 is a transcriptional coactivator specific for TR, *Mol Endocrinol* **15**(8):1329–1343, 2001.
25. Masuyama H, Hiramatsu Y, Involvement of suppressor for Gal 1 in the ubiquitin/proteasome-mediated degradation of estrogen receptors, *J Biol Chem* **279**(13):12020–12026, 2004.
26. Masuyama H, MacDonald PN, Proteasome-mediated degradation of the vitamin D receptor (VDR) and a putative role for SUG1 interaction with the AF-2 domain of VDR, *J Cell Biochem* **71**(3):429–440, 1998.
27. Zhang H, *et al.*, The catalytic subunit of the proteasome is engaged in the entire process of estrogen receptor-regulated transcription, *EMBO J* **25**(18):4223–4233, 2006.
28. Nawaz Z, *et al.*, Proteasome-dependent degradation of the human estrogen receptor, *Proc Natl Acad Sci USA* **96**(5):1858–1862, 1999.
29. Lonard DM, *et al.*, The 26S proteasome is required for estrogen receptor-alpha and coactivator turnover and for efficient estrogen receptor-alpha transactivation, *Mol Cell* **5**(6):939–948, 2000.
30. Metivier R, *et al.*, Estrogen receptor-alpha directs ordered, cyclical, and combinatorial recruitment of cofactors on a natural target promoter, *Cell* **115**(6):751–763, 2003.
31. Fan M, Nakshatri H, Nephew KP, Inhibiting proteasomal proteolysis sustains estrogen receptor-alpha activation, *Mol Endocrinol*, **18**(11):2603–2615, 2004.
32. Wallace AD, Cidlowski JA, Proteasome-mediated glucocorticoid receptor degradation restricts transcriptional signaling by glucocorticoids, *J Biol Chem* **276**(46):42714–42721, 2001.

33. Ismail A, Nawaz Z, Nuclear hormone receptor degradation and gene transcription: An update, *IUBMB Life* **57**(7):483–490, 2005.
34. Jaworski T, Degradation and beyond: Control of androgen receptor activity by the proteasome system, *Cell Mol Biol Lett* **11**(1):109–131, 2006.
35. Deroo BJ, *et al.*, Proteasomal inhibition enhances glucocorticoid receptor transactivation and alters its subnuclear trafficking, *Mol Cell Biol* **22**(12):4113–4123, 2002.
36. Glickman MH, Getting in and out of the proteasome, *Semin Cell Dev Biol* **11**(3):149–158, 2000.
37. Alarid ET, Lives and times of nuclear receptors, *Mol Endocrinol* **20**(9):1972–1981, 2006.
38. Salghetti SE, *et al.*, Functional overlap of sequences that activate transcription and signal ubiquitin-mediated proteolysis, *Proc Natl Acad Sci USA* **97**(7):3118–3123, 2000.
39. Mani A, *et al.*, E6AP Mediates regulated proteasomal degradation of the nuclear receptor coactivator amplified in breast cancer 1 in Immortalized Cells, *Cancer Res* **66**(17):8680–8686, 2006.
40. Li X, *et al.*, The SRC-3/AIB1 coactivator is degraded in a ubiquitin- and ATP-independent manner by the REGgamma proteasome, *Cell* **124**(2):381–392, 2006.
41. Pickart CM, Fushman D, Polyubiquitin chains: Polymeric protein signals, *Curr Opin Chem Biol* **8**(6):610–616, 2004.
42. Hicke L, Protein regulation by monoubiquitin, *Nat Rev Mol Cell Biol* **2**(3):195–201, 2001.
43. Schnell JD, Hicke L, Non-traditional functions of ubiquitin and ubiquitin-binding proteins, *J Biol Chem* **278**(38):35857–35860, 2003.
44. Ferdous A, *et al.*, The 19S regulatory particle of the proteasome is required for efficient transcription elongation by RNA polymerase II, *Mol Cell* **7**(5):981–991, 2001.
45. Lee D, *et al.*, The proteasome regulatory particle alters the SAGA coactivator to enhance its interactions with transcriptional activators, *Cell* **123**(3):423–436, 2005.
46. Morris MC, *et al.*, Cks1-dependent proteasome recruitment and activation of CDC20 transcription in budding yeast, *Nature* **423**(6943):1009–1013, 2003.
47. Gill G, SUMO and ubiquitin in the nucleus: Different functions, similar mechanisms? *Genes Dev* **18**(17):2046–2059, 2004.
48. Chauchereau A, *et al.*, Sumoylation of the progesterone receptor and of the steroid receptor coactivator SRC-1, *J Biol Chem* **278**(14):12335–12343, 2003.
49. Girdwood D, *et al.*, P300 transcriptional repression is mediated by SUMO modification, *Mol Cell* **11**(4):1043–1054, 2003.

50. Fan M, *et al.*, The activating enzyme of NEDD8 inhibits steroid receptor function, *Mol Endocrinol* **16**(2):315–330, 2002.
51. Fan M, Bigsby RM, Nephew KP, The NEDD8 pathway is required for proteasome-mediated degradation of human estrogen receptor (ER)-alpha and essential for the antiproliferative activity of ICI 182,780 in ERalpha-positive breast cancer cells, *Mol Endocrinol* **17**(3):356–365, 2003.
52. Reinstein E, Ciechanover A, Narrative review: Protein degradation and human diseases: The ubiquitin connection, *Ann Intern Med* **145**(9):676–684, 2006.
53. Gao X, *et al.*, Decreased expression of e6-associated protein in breast and prostate carcinomas, *Endocrinology* **146**(4):1707–1712, 2005.
54. Khan OY, *et al.*, Multifunction steroid receptor coactivator, E6-associated protein, is involved in development of the prostate gland, *Mol Endocrinol* **20**(3):544–559, 2006.

Chapter 4

Coregulators as Oncogenes and Tumor Suppressors

Rakesh Kumar and Anupama E. Gururaj

Cancer is caused by genetic changes that activate oncogenes or inactivate tumor suppressor genes. Transcription factors and their accessory factors, commonly known as coregulators, make up a high proportion of the proteins with dysregulated expression or altered activity in cancer cells. Furthermore, many signaling pathways that are disrupted in cancer converge on coregulators, ultimately leading to altered expression of numerous target genes. In this chapter, we highlight rogue coregulators that circumvent normal controls in cancer cells and discuss their potential as targets for the development of successful therapeutic anticancer strategies.

4.1 Introduction

The eukaryotic nucleus houses a complex information-retrieval system. The timeliness, accuracy, and completeness of this system are inherently important: almost every function within the cell is tightly linked to the interpretation of this retrieved genetic information and its eventual packaging and export in edited form. Eukaryotic genes are packed in the chromatin in the nucleus of a living cell. The basic unit of chromatin is the nucleosome, which consists of a central core histone (two molecules each of H2A, H2B, H3, and H4 to form an octamer) and the DNA helix, which wraps around the core histone twice.[1] In recent years, it has become clear that nucleosomes not only provide structural support for ordered packing of the chromosomal DNA but are also critical sites that control gene activation and repression and probably DNA replication, DNA repair, and chromosomal recombination.[1]

The two classes of cofactors (collectively called coregulators) are the coactivators and the corepressors, which assist transcription factors in turning genes "on" and "off," respectively. Coregulators change the state of the nucleosomes by modifying the chromatin structure either directly or indirectly.[2] One process involves ATP-dependent chromatin-remodeling complexes, in which helicase activity allows sliding of the nucleosomes along the genomic DNA, leaving specific DNA regions accessible to transcription or replication factors. Other coregulators are large enzymatic complexes that regulate histone acetylation, methylation, and phosphorylation and thereby control chromatin remodeling. Some coregulators have intrinsic histone acetylation or methylation capabilities, but most serve as bridges for DNA-bound nuclear receptor (NR) enzymes, modifying enzymes, and the basal transcriptional machinery.[3] A fine balance of these coregulators is crucial for the proper maintenance of gene expression; thus coregulators are often the target of mutations and alterations in human cancer (Fig. 4.1). Here, we briefly review the current understanding of how some of these coregulators modify the chromatin structure and how their functional disruption leads to inappropriate gene expression and cell transformation.

4.2 Chromosomal Abnormalities and Human Cancer

Cancer is a genetic disease characterized by dominant and recessive gene mutations that result in the alteration of normal differentiation and proliferation pathways. Some mutations are inherited and are therefore present in all cells, but many other mutations are acquired by a single somatic cell during the lifetime of the organism and are transmitted in a clonal pattern to the progeny of that cell. Cancer develops from the accumulation of a series of inherited and/or acquired mutations that cause a remarkable change in the behavior of a single cell and its offspring.[4] Cancer cells differ from their normal counterparts in that their normal physiologic processes are abnormally regulated.[4] Cancer occurs through a multistep process during which the genomes of new cancer cells acquire mutant alleles of proto-oncogenes, tumor suppressor genes, and other genes that directly or indirectly control cell proliferation.[5]

Two major classes of genes are involved in the development of cancer, oncogenes and tumor suppressor genes (Fig. 4.1). When mutated or overexpressed, oncogenes release the cell from its normal growth constraints. Often referred to as "gain of function," the activation of oncogenes involves dominant mutations wherein the mutation of a single allele is sufficient for activation.[5] A classic example is the constitutive activation of a signal transducer, *Ras* found in various human tumors.[6] These genes are also "turned on" by other processes such as gene amplification and chromosomal rearrangements. In contrast, the protein products of tumor suppressor genes are often directly involved in the growth inhibition or differentiation of cells. The genetic mutations in

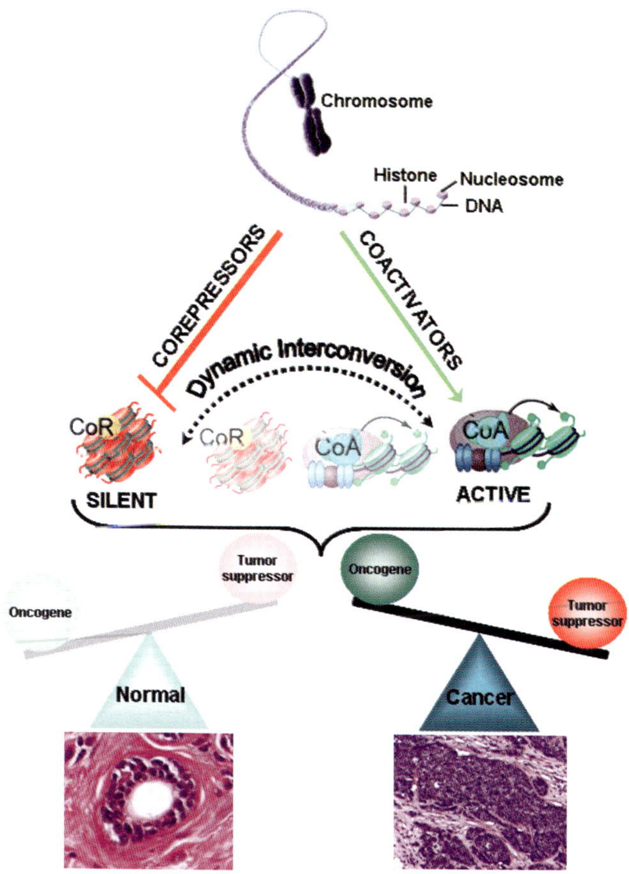

Fig. 4.1. A schematic diagram depicting chromosome organization, and potential oncogenic and tumor suppressor functions of coregulators.

such cases are mainly recessive, with both alleles affected, and often lead to the inactivation of gene function.[5] These genes are also inactivated by the deletions of large portions of DNA or by methylation of promoter regions.[5] *p53* and *pRB* are two widely recognized tumor suppressor genes.[7]

It has become evident that epigenetic processes play an important role in cell development, growth, and differentiation.[8] The term "epigenetic" generally refers to mechanisms that result in heritable alteration of gene expression profiles without an accompanying change in the primary DNA sequence.[9] The core of epigenetic control lies in the chemical modification of DNA, histones, and nonhistone proteins.[10] Most recurring genetic abnormalities in human cancers are point mutations and rearranged genes that encode nuclear factors that interact directly or indirectly with enzymatic complexes involved in histone modification or chromatin remodeling. We will now briefly discuss the normal cellular role of these coregulator complexes and describe how their alteration may lead to cellular transformation.

4.3 Coregulators Functioning as Oncogenes

At the heart of transcription regulation is the assembly of multimeric complexes between sequence-specific DNA-binding transcription factors and families of transcription coactivators or corepressors.[1] Our understanding of how these regulatory complexes reversibly modify the chromatin structure and control gene expression follows the remarkable progress made in defining the nuclear hormone receptors as coregulators of gene transcription.[2,3] We will first discuss NR coregulators that are known to be dysregulated in cancer cells and act as oncogenes.

4.3.1 *NR coregulators*

The NR gene family is a class of rapidly growing family of transcription factors that modulate gene expression in response to lipophilic ligands.[3] NRs are broadly implicated in normal physiologic development and metabolism and represent therapeutic targets for a wide range of human diseases, including cancer, endocrine and metabolic disorders, and heart disease.[11,12] NRs repress or activate transcription by participating in the formation of large multimeric protein complexes on the promoter of their target genes.[11] The functional state of NRs is changed with the binding of ligand so that corepressors are released and coactivators are recruited. Corepressors such as the NR corepressor (NCoR) interface between apo-NRs, Sin3, and histone deacetylase (HDAC) complexes, which harbor HDAC activity and thus lead to condensation of the nucleosome. On the other hand, ligand-bound NRs recruit coactivators, which mediate chromatin remodeling and recruit the basal transcription apparatus.[11]

The best-studied NR coactivator (NCoA) family is the p160 family (SRC1 [NCoA-1/ERAP], SRC2 [TIF2/NCoA-2/GRIP1], SRC3 [AIB1/NCoA-3/RAC-3/ACTR/TRAM1]), and PELP1 [MNAR]. SRC1, SRC2, and SRC3 share a 40% sequence homology, primarily in the N-terminus, which contains the bHLH/PAS domain.[13] All three SRCs are involved in the progression of prostate cancer, but only SRC3 has been directly shown to possess oncogenic properties.[14]

SRC3 was originally identified and referred to as amplified in breast cancer-1 (AIB1).[15] The *SRC3* gene was initially mapped to the 20q12 segment during a search for amplified and overexpressed genes that mapped to chromosome 20q in breast cancer.[15] Later studies showed that the *SRC3* gene was not only amplified in human breast tumors but also overexpressed.[16] Of interest, overexpressed SRC3 was detected in both hormone-sensitive and hormone-insensitive breast tumors.[16] This finding paved the way for further investigations of the role of SRC3 in hormone-independent tumors. In addition to the presence of SRC3 in breast tumors, amplified or overexpressed SRC3 has also been seen in ovarian tumors, prostrate tumors, and meningiomas (hormone sensitive) and in many types of non-steroid-targeted cancers, such as pancreatic cancer, gastric cancer, colorectal carcinoma, and hepatocellular carcinoma.[16,17] In most of these tumors and cancers, SRC3 expression correlated with tumor grade, disease progression and recurrence, and survival time.[16,17] Of interest is the fact that increased SRC3 expression in breast tumors indicated tamoxifen resistance and poor prognosis.[18] Because SRC3 overexpression was also correlated with high levels of HER2/neu expression[19] and because patients with increased expression of both molecules exhibited the worst response to tamoxifen therapy, it has been postulated that cross-talk between SRC3, HER2/neu, and estrogen receptor α (ERα) signaling pathways is important in breast cancer.[20] Mechanistic explanations for tamoxifen resistance due to SRC3 overexpression have been developed and are the subject of an excellent review.[20]

The functional importance of SRC3 in the genesis and progression of tumors has also been demonstrated in mouse model systems. MMTV-SRC3 transgenic mice had a high incidence of tumors in organs with the transgene, namely the breast, pituitary gland, uterus, and lungs.[21] Furthermore, SRC3 deficiency significantly suppressed the incidence of MMTV-v-Ha-ras oncogene-induced mammary gland ductal hyperplasia, tumorigenesis, and metastasis to the lung as well as mammary tumorigenesis induced by 7,12-dimethylbenz(a)anthracene (DMBA).[22,23] The amplification and overexpression of SRC3 in various cancers underscore its functional importance as a prosurvival gene that is necessary for normal growth and development. Exhaustive studies of the role of SRC3 in tumorigenesis and progression have provided proof beyond doubt that SRC3 is a *bona fide* oncogene.

To date, more than 30 coactivator molecules have been described. A recently described coactivator of ERα, breast carcinoma amplified sequence 3 (BCAS3), resides in another frequently amplified region in breast tumors, chromosome 17q23. The involvement of the *BCAS3* gene in breast cancer progression has been suggested by other findings that this gene is amplified in breast cancer cell lines (mainly ERα-positive cell lines)[24,25] and in approximately 10% of primary breast tumors analyzed in one study.[24] Furthermore, overexpression of BCAS3 in primary breast tumors is associated with tumor grade and proliferation.[25] Similar to SRC3, BCAS3 overexpression in hormone receptor-positive premenopausal breast cancer seems to be associated with impaired responses to tamoxifen treatment.[26] Because copy number gains at 17q23 have also been reported in tumors of the brain, lungs, bladder, testis, and liver; BCAS3 amplification or overexpression in these tumors has been suggested.

As in breast cancer, multiple coregulators have been discovered to be deregulated in other hormone-responsive tumors. It has been proposed that alterations in the expression or function of several cofactors occur in prostate tumors. In this context, the cofactor ARA70, which potentiates androgen receptor (AR) activation,[27] has been studied extensively. Decreased expression of ARA70 in prostate cancer tissues has been reported.[28] Consistent with these results, the proliferation of LNCaP cells and colony formation were inhibited after overexpression of ARA70, suggesting its role as a tumor suppressor.[29] In contrast, data from Hu *et al.*[30] showed an increase in ARA70 protein expression in high-grade prostate cancers and cells cultured in androgen-deficient conditions. More research is needed to clarify the reasons for these contrasting results.

In addition to ARA70, two more coactivators have been implicated in prostate tumor progression. A recent report showed that lysine-specific histone demethylase 1 (LSD1), and four and a half LIM-domain protein 2 (FHL2) may serve as novel biomarkers for aggressive prostate cancer.[31] This observation is interesting because histone demethylases have the potential to epigenetically alter gene expression, as do HDACs, and thus could be at least as important as HDACs in terms of therapy. Numerous investigations about the role of FHL2 in cancer exist.[32] Although FHL2 seems to function as a potential tumor suppressor in prostate cancer and rhabdomyosarcoma,[33] as evidenced by its decreased expression in these cancers, it was overexpressed in ovarian, lung, breast, and colon cancers and in melanoma.[32] FHL2 is intriguing because it may function as both an oncogene and a tumor suppressor in a tissue-dependent fashion. The finding that FHL2 can function as a repressor or activator of transcriptional activity depending on cell type further emphasizes its dual functionality.[33] In addition to FHL2, other coregulators are being identified that function as coactivators or corepressors in a cell- or promoter-specific context.

The metastasis-associated (MTA) family of coregulators includes three key genes in vertebrates that code for five proteins.[34] *MTA1*, the founder of this family, was discovered approximately a decade ago in a differential display screen comparing mRNA from rat breast cancer cell lines with different growth properties.[35] Subsequent studies of human *MTA1* have revealed a correlation between expression level and invasive growth properties in several tumor types.[36] Overexpression of the *MTA1* gene was closely correlated with tumor invasion and lymph node metastasis in colorectal, gastric, esophageal, ovarian, non-small cell lung, and hepatocellular carcinomas, pancreatic cancer, thymoma, and prostate tumor,[37] suggesting that the overexpression of this gene may be a useful indicator of the malignant potential of these tumors. Because MTA family members were originally identified as part of the nucleosome remodeling complex,[38] these molecules were hypothesized to be corepressors.[39] Recent reports, however, have indicated that MTA1 can also behave as a coactivator, depending on the acetylation status of MTA1 and promoter context of the target gene, thus reflecting its dual nature as a coregulator.[25,40]

MTA1 also plays a role in the development of mammary glands and in the process of tumorigenesis. Mammary glands of virgin transgenic mice expressing MTA1 exhibit extensive side branching and precocious differentiation, whereas mice with the MTA1 transgene develop hyperplastic nodules and mammary tumors within 18 weeks.[41] Likewise, the forced expression of MTA1 in noninvasive breast and pancreatic cancer cells and in keratinocytes results in their acquiring an invasive phenotype, as assessed by various biologic end points.[42] Furthermore, induced levels of MTA1 are sufficient to transform Rat1 fibroblasts, and this transforming potential depends on the acetylation of MTA1 at lysine 626 (Suresh Rayala and Rakesh Kumar, personal communication). In addition, studies using MTA1 antisense oligonucleotides have demonstrated that MTA1 protein plays a role in cellular signaling processes that are important for the progression and growth of cancer cells.[36,42] Of interest, a naturally occurring variant of MTA1, MTA1s, functions as an ERα corepressor by sequestering ERα in the cytoplasm. The expression of MTA1s has been shown to increase in ERα-negative human breast tumors and to stimulate malignant phenotypes.[43] Furthermore, both MTA1 and MTA1s appear to regulate mammary tumorigenesis by the hyperactivation of the Wnt signaling pathway. (Seethraman Balasenthil and Rakesh Kumar, personal communication). These results suggest that overexpression of MTA1 promotes invasive and metastatic phenotypes in different cancer cell lines.

There is no evidence of a direct role of MTA2, another member of the MTA family, in malignancy. However, a role for MTA2 in regulating p53-mediated growth arrest and apoptosis was reported by Luo *et al.*;[44] specifically, MTA2 was shown to antagonize p53 functions by deacetylating p53 via HDACs.[45] Likewise, not much is known about MTA3, a recently discovered member of the MTA family. Studies have shown that MTA3 acts as a corepressor of Snail, a master regulator of epithelial-to-mesenchymal transition that itself is a corepressor of E-cadherin.[46] Thus, decreased amounts of MTA3 lead to decreased E-cadherin expression, eventually resulting in a more motile phenotype. In this context, MTA3 seems to behave more as a tumor suppressor. This idea is supported by results from an elegant study in which forced overexpression of MTA3 in the mammary epithelium resulted in the inhibition of ductal side branching in virgin glands due to differentiation, a reduced rate of cell proliferation, and the repression of Wnt4 expression.[47] These effects of MTA3 are thus the opposite of those of

MTA1 and raise the issue of distinct and opposite physiologic functions of structurally similar members of the same family.

Similar to MTA1, proline-, glutamic acid-, and leucine-rich protein 1 (PELP1) is a bifunctional coregulator that was first identified as a coactivator for ER.[48] Upon E2 binding, PELP1 interacts with both ERα and the coactivators CBP/p300.[48] Recent data indicate that PELP1 participates in chromatin remodeling activity via the displacement of histone H1 in cancer cells.[49] Besides its role as an ER coactivator, PELP1 has been shown to act as a corepressor for other nuclear hormone receptors (e.g. glucocorticoid receptor) or other transcription factors (e.g. activating protein 1 and nuclear factor κB [NF-κB]). The repressive activity of PELP1 involves both the N-terminal and C-terminal regions of protein binding to HDAC2 and to histones H3 and H4, respectively. These interactions block the access of histone acetyltransferases to the histones, thus inhibiting gene transcription.[50] PELP1 is overexpressed in breast tumors, and evidence indicates that it plays a permissive role in E2-mediated cell-cycle progression via its regulatory interaction with the retinoblastoma protein pathway.[51] In addition to breast cancers, a role for PELP1 in prostate and endometrial cancer has been demonstrated.[52] Studies from prostate tumors suggest that PELP1 dependent extranuclear signaling may play a role in the transition of prostate cancer cells to androgen independence.[52] Furthermore, PELP1 was recently shown to interact with FHL2, raising the possibility that nuclear functions of PELP1 may also be relevant to its role in the progression of prostate tumors.[53]

4.3.2 *Chimeric transcription factors and coregulators: Chromatin condensation and cell transformation*

Recurring chromosomal translocations are perhaps the most frequent genetic lesions in human leukemia and in a subgroup of sarcomas. Many genes involved at the breakpoint of chromosomal translocations have been cloned with the use of molecular tools. Chromosomal rearrangements have been shown to often target proteins that have strong tyrosine kinase activity (either transmembrane receptors or receptor-associated proteins) and nuclear factors that interact with chromatin-remodeling enzymes and result in the expression of a fusion protein that contains functional domains from both parental proteins.

The retinoic acid receptors RXR and RAR form the functionally active RXR-RAR heterodimer that belongs to the superfamily of NRs, together with glucocorticoid and thyroid receptors. Among these, RAR is the target of virtually all chromosomal translocations associated with acute promyelocytic leukemia.[54] Unliganded RXR-RAR interacts with the DNA retinoic acid response elements in promoter sites and is constitutively bound to the highly conserved and closely related corepressors NCoR and the silencing mediator for retinoic acid and thyroid (SMRT) hormone receptors. Five chromosomal translocations involving RARα have been described so far, and five different genes (promyelocytic leukemia [*PML*; 98% of cases], *PLZF*, *NPM*, *NuMA*, and *STAT5b*) are rearranged upstream of RARα and expressed as chimeric RARα receptors.[55] The chimeric receptor N/RARα, in which N is any of the partner genes rearranged with the RARα receptor in promyelocytic leukemia, is also bound to retinoic acid response elements. However, in the absence of the retinoic acid ligand, the chimeric receptor N/RARα has a much higher affinity for NCoR and SMRT than the wild-type RXR-RARα does and is therefore a much stronger repressor that requires a higher concentration of ligand to release the corepressor complex.[56] The overwhelming correlation between the disruption of RAR signaling due to chromosomal translocations and leukemia links very strongly HDAC-mediated transcription repression with the pathogenesis of leukemias.[56]

Chromosomal translocations also rearrange genes encoding transcription factors that recruit corepressors. The nuclear factor AML1 (RUNX-1), located on chromosome band 21q22, is fused to the zinc finger nuclear factor ETO in t(8;21) associated with acute myeloid leukemia (AML) subtype M2[57] and to the v-ets erythroblastosis virus E26 oncogene homolog (ETS)-protein TEL (ETV6) in the t(12;21) associated with childhood leukemia.[58] The chimeric protein AML1/ETO loses the C-terminus of AML1 and the ability to interact with CBP/p300 but acquires the potential to form homodimers and recruit multiple copies of NCoR and SMRT through the ETO domains.[57] This complex is a very strong repressor and is necessary to block hematopoietic differentiation, thereby reinforcing the correlation between an inappropriate multimeric repressor complex and a differentiation block leading to leukemic transformation. The fusion protein TEL/AML1 contains the entire AML1 fused downstream of a truncated form of TEL that lacks the ETS-DNA-binding domain. Thus, like ETO, TEL is characterized by an oligomerization region and by a repression domain capable of recruiting corepressors.[58] There are no conclusive data yet that show that the leukemic property of TEL/AML1 depends on the ability of the fusion protein to multimerize. However, the results with the chimeric RARα receptors and with AML1/ETO suggest that the efficient recruitment of the NCoR-HDAC complex and the inappropriate formation of multimeric complexes are required to activate the oncogenic potential of the chimeric proteins. The nonphysiologic recruitment of corepressors could lead to a chromatin configuration of target promoters' refractory to activating signals from other *cis*-acting elements.

Aberrant regulation of HDACs plays an important role in human carcinogenesis. Results from a study of a comprehensive panel of normal tissues, cancer cell lines, and primary tumors have demonstrated that cancer cells have a loss of monoacetylated and trimethylated forms of histone H4. This loss is associated with the hypomethylation of DNA repetitive sequences, a well known characteristic of cancer cells.[59] Such an aberrant modification of histone H4 appears early and accumulates during tumorigenesis, suggesting that the global loss of monoacetylated and trimethylated of histone H4 is a common hallmark of human tumor cells.[59] The level of histone hypoacetylation is significantly associated with the depth of tumor invasion and the extent of metastasis.[60] Histone hypoacetylation can result from overexpression of HDACs, mutation or silencing of histone acetyltransferase genes, or amplification or translocation of histone acetyltransferase or HDAC genes. Overexpression of HDACs has been reported in various solid tumor samples.[61] HDAC2 overexpression is observed in gastric adenocarcinoma tumors and is associated with tumor aggressiveness.[62] Mice lacking the adenomatosis polyposis coli (*APC*) tumor suppressor gene overexpressed HDAC2, which prevented apoptosis in cultured colon cells. Treatment with HDAC inhibitors reduced the formation of adenomas in these mice.[63] Elevated expression of HDAC2 has also been detected in human colon carcinomas.[63] Therefore, HDACs appear to be appropriate targets for cancer therapy. Indeed, inhibiting HDAC activity has been proven to be very effective in inhibiting cancer cell growth *in vitro* and *in vivo*. Some excellent reviews provide more detailed explanations.[64,65]

4.3.3 *Chimeric coactivators: Chromatin relaxation and cancer*

Transcription coactivators are at the heart of biologic pathways and serve as interpreters of physiologic stimuli to regulate the cell cycle, DNA repair, cell differentiation, and apoptosis. Thus, it is not surprising that their functions are also altered in cancer. Three coactivators (SRC2, CBP, and p300) are directly rearranged by chromosomal translocations in leukemia. This mechanism of leukemogenesis contrasts with that described for corepressors, which are recruited unaltered to core promoters by chimeric transcription factors. The p300 missense mutations have been reported in colorectal and gastric carcinomas.[66]

So far, rearranged coactivators have been identified in four chromosomal translocations associated with leukemia. In the t(8;16) and inv(8) translocations, CBP and SRC2, respectively, are rearranged downstream of the monocytic leukemia zinc (MOZ) finger gene.[67] The fusion products encode a protein consisting of the N-terminus of MOZ fused to the C-terminus of SRC2, which contains the CBP-interacting domain and an activation domain. In the t(8;16) translocation, the N-terminus of MOZ is fused to almost the entire open reading frame of CBP. CBP is also rearranged in the t(11;16) translocation, which is associated with therapy-related leukemia.[68] Two breakpoints have been described for CBP, and in both cases, the resulting proteins contain the region from the bromodomain to the C-terminus. The gene that is fused upstream of CBP is MLL (also known as HRX and ALL) and is involved also in the t(11;22) translocation with p300.[68] From a functional point of view, the breakpoints of CBP and p300 are analogous and include the same functional domains, indicating that the movement of the region from the bromodomain to the C-terminus is necessary for the pathogenesis of leukemia. Indeed, an analysis of murine bone marrow expressing various C-terminal and internal CBP deletion mutants of an MLL/CBP chimeric protein confirmed that the bromodomain and the histone acetyltransferase domain are necessary and sufficient to transform the murine bone marrow and to induce leukemia in recipient mice.[69]

The MLL chromosomal breakpoints in the t(11;16) and t(11;22) translocations are not identical, but they retain the MLL regions with homology to methyltransferase and three A/T hook repeats, originally described in HMGI(Y) proteins and histones. The A/T hook motif is presumed to bind nonspecifically to nucleosomal DNA rather than to a specific DNA sequence. The function of MLL was unknown until recently, when it was identified as a histone methyltransferase. Because of its functional motifs, MLL could be involved in chromatin remodeling. Thus, the chimeric proteins resulting from 11q23 translocations could possibly perturb the chromatin structure of regions targeted by MLL and inappropriately regulate the expression of developmental genes. The members of the MLL family, including MLL3 and MLL4, have also been shown to be subjected to translocation, amplification, and deletion in various tumor types.[70]

4.3.4 *Non-NR coactivators*

Tumors of diverse origin contain genetic rearrangements involving *myc* family genes. In recent studies, a myc coactivator, SNIP1, was shown to be indispensable for c-myc–induced transformation.[71,72] The same studies also showed overexpression of both SNIP1 and c-myc in lung tumor samples and other tumor types. Evidence is increasing for the role of NF-κB/Rel transcription factors in malignant transformation, especially in lymphomas. Bcl3, a coactivator of NF-κB, is a candidate proto-oncogene involved in the recurring t(14;19) translocation found in patients with chronic lymphocytic leukemia. Bcl3 overexpression and amplification have been found in multiple hematopoietic neoplasms, including Hodgkin's lymphoma and T-cell non-Hodgkin's lymphoma.[73] Jun-activation domain-binding protein 1 (JAB1), an evolutionary conserved nuclear protein, was shown to interact with AP-1 family members and to potentiate c-Jun and JunD activity.[74] Jab1 has been shown to be overexpressed in multiple tumors such as neuroblastoma, ovarian tumor, rhabdomyosarcoma, hepatoblastoma, lymphoma, oral cancer, breast cancer, head and neck cancer, and melanoma.[75]

4.4 Coregulators Functioning as Tumor Suppressors

Tumor-suppressor genes are frequently lost or inactivated in human cancers. Considerably less evidence is available concerning coregulators functioning as negative regulatory elements in cancers and behaving as tumor suppressor genes. However, sufficient examples exist to merit a discussion of this subject.

4.4.1 *NR coregulators in growth suppression*

Steroid hormone-associated coregulators that have been described as tumor suppressors have been generally those of ER and AR. SWI/SNF was the first chromatin-remodeling complex identified as being directly involved in the regulation of several NRs, including ER. The SWI/SNF family comprises a number of large multiprotein complexes containing between eight and 10 subunits.[76] These complexes always contain a single catalytic subunit, BRM/SNFα or BRG1/SNFβ, and several other variable BRG1-associated-factors (BAFs) that contribute to the enzymatic activity of the complex and facilitate its recruitment to sequence-specific transcription factors, although the detailed functions of each individual subunit remain to be determined. One of these subunits, BAF57, behaves as an ERα-specific coactivator[77] that interacts directly with the p160 family of coactivators. ERα and BAF57 have been shown to be recruited to estrogen-responsive promoters in a ligand-dependent manner.[77] BAF57 has also been shown to potentiate AR transactivation functions.[78] BAF57 expression is downregulated in a pancreatic cell line.[79] Recently, mutations in BAF57 were identified in breast cancer cell lines, indicating that BAF57 functions could be compromised in these cells.[80] More research is needed to evaluate whether mutations in BAF57 could contribute to oncogenic transformation in primary breast cancers.

Multiple endocrine neoplasia type 1 (MEN1) is a familial disorder characterized by tumors of the parathyroid glands, pancreatic islets, pituitary gland, and adrenal glands as well as by neuroendocrine carcinoid tumors.[81] MEN1 is caused by germline mutations of the *MEN1* tumor-suppressor gene, which is localized on chromosome 11q13.[81] The protein product of the *MEN1* gene is termed "menin." Most *MEN1* gene mutations result in the production of nonfunctional menin. In studies addressing the function of menin, several proteins have been found to interact with it, indicating that the protein is involved in the regulation of gene transcription.[82] Several DNA-binding transcription factors (JunD, NF-κB, mothers against decapentaplegic homolog 3 [Smad3], Pem) and chromatin-modifying proteins (mSin3A, HDACs, and mixed-lineage leukemia proteins [MLL1 and MLL2]) have been described as being able to interact with menin. Of interest, subsequent functional analyses have indicated that menin can have a dual role in the regulation of transcription (as a repressor and an activator of gene expression) and that this duality might be explained by a model in which menin links transcription regulation with chromatin modification.[82,83]

The tumor suppressor gene *BRCA1* causes approximately 50% of heritable breast cancers,[84] which in turn account for 5% to 10% of all breast cancers. BRCA1 functions as an ERα corepressor[85] and acts as a ligand-independent corepressor for ER, AR, and progesterone receptors.[86] Studies using breast cancer cell lines and fibroblasts from BRCA1 knockout mice elegantly showed that BRCA1 is involved in regulating the ligand-dependent and ligand-independent activity of ERα. In contrast to wild-type BRCA1, BRCA1 mutants carrying familial breast cancer-derived missense mutations fail to repress ERα.[87] Some of these mutations map to the exon 11-encoded region that binds Rad50, a domain previously shown to be involved in DNA damage repair. Inherited mutations in *BRCA1* are correlated with an increased risk of ovarian cancer.[86]

4.4.2 *Non-NR coregulators*

The C-terminal binding protein (CtBP) family of proteins consists of conserved transcriptional regulators.[88] The CtBP functions as a corepressor for a wide array of DNA-binding transcriptional factors and has roles in the development cancer.[88] CtBP mediates its corepressor function via multiple mechanisms using HDACs, mSin3, and polycomb complexes. CtBP expression is lost in malignant melanoma.[88] The CtBP-interacting protein (CtIP) was originally identified as a binding partner for CtBP.[89] CtIP acts as a molecular bridge to recruit CtBPs to transcriptional repressors such as the G_1 checkpoint regulator, Rb, and the hemolymphoid factor Ikaros.[89] In both cases, the CtBP-CtIP complex is thought to cooperate with these factors to repress target gene transcription. Mice with heterozygous inactivation of the *Ctip* gene show an increased incidence of multiple tumor types, particularly large lymphomas. This observation is consistent with a tumor suppressor role for the protein, although the wild-type *Ctip* allele has so far not been shown to be lost in cancers.[89] Some of these effects of CtIP loss might be accounted for by its interaction with the BRCT domains of the BRCA1 protein.[89]

The *PML* gene is a tumor suppressor originally identified in patients with acute promyelocytic leukemia with a reciprocal t(15;17) chromosomal translocation in which *PML* is fused to the *RARα* gene (discussed earlier). Several studies have suggested that PML is an upstream regulator of p53. For example, $pml^{-/-}$ thymocytes are severely impaired in p53 transcriptional activity and p53-dependent apoptosis.[90] In addition, cells derived from $pml^{-/-}$ mice are resistant to ionizing radiation-induced apoptosis.[90] Conversely, p53-deficient cells are resistant to PML-induced arrest and apoptosis,[91] and after DNA damage, PML recruits p53 to PML nuclear bodies, which coincides with the increased transcription of p53 target genes.[91] This is thought to be partly due to the ability of PML to recruit CBP to PML nuclear bodies, which then prompts p53 acetylation and stability.[92]

Tetradecanoylphorbol acetate (TPA)-inducible sequence 21 (*TIS21*; also called *BTG2* or *PC3*) is one of the early growth response genes, and it has been isolated from 3T3 fibroblasts treated with tissue plasminogen activator. Together with *BTG1*, *BTG3*/*ANA*/*Rbtg3*, and *Tob* genes, *TIS21* is an antiproliferative gene.[93] The fact that mRNA of TIS21 was lost in thymic carcinoma tissues but the expression was very high and constitutive in mouse thymus suggested a potential role of TIS21 during carcinogenesis in the thymus.[93] Of interest, the constitutive expression of TIS21 mRNA in renal proximal tubules and prostate acini was lost in renal cell carcinoma and the early stage of carcinogenesis in the prostate, respectively.[94] Moreover, TIS21 expression is very low or undetectable in human breast cancer cell lines compared with its expression in nontumorigenic mammary epithelial cell lines. These findings clearly imply that the loss of TIS21 expression plays an important role in the initiation of the carcinogenic process.[94]

The inhibitor of growth (ING) family of proteins are candidate tumor-suppressor proteins. At the molecular level, ING proteins are thought to function as chromatin-regulatory molecules by acting as cofactors for distinct histone acetyltransferases and HDAC enzyme complexes.[95] ING1 mutations occur infrequently, and most are found in the nuclear localization signal (NLS) or the PHD domains. Growing evidence shows that naturally occurring mutations in the PHD domains of different proteins lead to cancer.[96] Apart from mutations, ING1 downregulation is observed very frequently in several tumor types, including breast, gastric, esophageal, blood, lung, and brain tumors.[97] Loss and transcriptional repression are also seen for the ING3 gene and protein.[96]

Moreover, levels of ING4 messenger RNA are two to three times lower in low-grade gliomas and six times lower in glioblastomas than in normal brain tissue,[97] suggesting that the expression of ING4 is progressively suppressed in brain tumors with increasing tumor grade and degree of malignancy. In addition, mislocalization of ING1 proteins has been observed in brain tumors. Consistent with this observation, in melanoma, papillary thyroid carcinoma, and ductal breast carcinoma, increased levels of cytoplasmic $p33^{ING1b}$ were noted to be concomitant with the loss of nuclear localization.[96]

The NGFI-A-binding protein family of corepressor molecules (NAB1 and NAB2) physically interact with the early growth response genes and influence their transcriptional activity in prostrate cells.[98] The *NAB2* gene has been localized to human chromosome 12q13, within a region thought to harbor a tumor suppressor gene for prostate cancer. In fact, the transfer of this region into DU145 human prostate carcinoma cells suppresses their tumorigenicity in mice.[99] The expression of NAB2 in primary human prostate carcinomas was lost.[99]

4.5 Conclusions and Future Perspective

The identification of proteins involved in chromatin structure as targets of genetic lesions associated with human cancers suggests that the biochemical alteration of these factors is a dominant event in the pathogenesis of the disease. With our increasing understanding of the chromatin architecture, it appears that the lack of physiologic balance between coregulator and chromatin remodeling complexes brought about by mutated factors leads to the inappropriate alteration of the chromatin structure and to abnormal gene regulation. This view of illegitimate chromatin restructuring as a cause of human cancer provides us with new and exciting means of attacking and controlling human disease.

The results of cell lines and murine studies proposing the use of HDAC inhibitors to control rearranged transcription factors have been very encouraging, and the first clinical trials of leukemia patients with a t(15;17) or t(8;21) translocation are underway. At this time, however, even in preliminary *in vitro* systems, no drugs are effective enough to reverse the effects of activators and components of the chromatin remodeling complexes that have been altered by genetic lesions. It is clear, however, that a novel road has been mapped, and eventually new "designer" drugs will be available to repair the genetic lesions that contribute to human cancers.

It has become increasingly clear that numerous transcriptional cofactors positively and negatively affect the activity of a variety of transcription factors. Although a more complete understanding of the factors themselves and the signals that control their activity is required to gain a deeper understanding of the multiple regulatory levels that transcription factors are subjected to, we can still conclude that coregulators are potentially attractive targets for antitumor therapy. An important area of research that needs to be more fully explored is the role of tissue-specific differences in the levels of cofactors in the development and maintenance of tumors. Differences in the tissue-specific distribution of these regulators may potentially contribute to the differing responses to chemotherapeutic agents. A more complete understanding of the regulation of the factors will be important for developing a better understanding of therapeutic strategies to treat cancer. Finally, it will be important to define the nature of post-translational modifications of coregulators, and the roles these modifications play in protein–protein interactions and in conferring coactivator versus corepressor function to a given coregulator. Furthermore, the interplay between chromatin modifiers is likely to shed light on the transregulation and feedback control of the NR coregulator network. We await with interest new information on the pathways and physiologic cues that control the expanding group of coregulators implicated in growth controls.

Acknowledgments

The work in Kumar laboratory is supported by grant number CA98823 from the National Institutes of Health to R.K., and The Norman Brinkler Award for Research Excellence.

References

1. Li B, Carey M, Workman JL, The role of chromatin during transcription, *Cell* **128**:707–719, 2007.
2. Rosenfeld MG, Lunyak VV, Glass CK, Sensors and signals: A coactivator/corepressor/epigenetic code for integrating signal-dependent programs of transcriptional response, *Genes Dev* **20**:1405–1428, 2006.
3. Nettles KW, Greene GL, Ligand control of coregulator recruitment to nuclear receptors, *Annu Rev Physiol* **67**:309–333, 2005.

4. Hanahan D, Weinberg RA, The hallmarks of cancer, *Cell* **100**:57–70, 2000.
5. Wang H, Han H, Mousses S, *et al.*, Targeting loss-of-function mutations in tumor-suppressor genes as a strategy for development of cancer therapeutic agents, *Semin Oncol* **33**:513–520, 2006.
6. Downward J, Targeting RAS signalling pathways in cancer therapy, *Nat Rev Cancer* **3**:11–22, 2003.
7. Sherr CJ, Principles of tumor suppression, *Cell* **116**:235–246, 2004.
8. Jones PA, Baylin SB, The epigenomics of cancer, *Cell* **128**:683–692, 2007.
9. Goldberg AD, Allis CD, Bernstein E, Epigenetics: A landscape takes shape, *Cell* **128**:635–638, 2007.
10. Kouzarides T, Chromatin modifications and their function, *Cell* **128**:693–705, 2007.
11. Belandia B, Parker MG, Nuclear receptors: A rendezvous for chromatin remodeling factors, *Cell* **114**:277–280, 2003.
12. Lonard DM, O'Malley BW, The expanding cosmos of nuclear receptor coactivators, *Cell* **125**:411–414, 2006.
13. Lonard DM, O'Malley BW, Expanding functional diversity of the coactivators, *Trends Biochem Sci* **30**:126–132, 2005.
14. Liao L, Kuang SQ, Yuan Y, *et al.*, Molecular structure and biological function of the cancer-amplified nuclear receptor coactivator SRC3/AIB1, *J Steroid Biochem Mol Biol* **83**:3–14, 2002.
15. Xu J, Li Q, Review of the *in vivo* functions of the p160 steroid receptor coactivator family, *Mol Endocrinol* **17**:1681–1692, 2003.
16. Yan J, Tsai SY, Tsai MJ, SRC3/AIB1: Transcriptional coactivator in oncogenesis, *Acta Pharmacol Sin* **27**:387–394, 2006.
17. Zhou HJ, Yan J, Luo W, *et al.*, SRC3 is required for prostate cancer cell proliferation and survival, *Cancer Res* **65**:7976–7983, 2005.
18. Smith CL, Nawaz Z, O'Malley BW, Coactivator and corepressor regulation of the agonist/antagonist activity of the mixed antiestrogen, 4-hydroxytamoxifen, *Mol Endocrinol* **11**:657–666, 1997.
19. Osborne CK, Bardou V, Hopp TA, *et al.*, Role of the estrogen receptor coactivator AIB1 (SRC3) and HER-2/neu in tamoxifen resistance in breast cancer, *J Natl Cancer Inst* **95**:353–361, 2003.
20. Schiff R, Massarweh S, Shou J, *et al.*, Breast cancer endocrine resistance: How growth factor signaling and estrogen receptor coregulators modulate response, *Clin Cancer Res* **9**:447S–454S, 2003.
21. Torres-Arzayus MI, Font dM, Yuan J, *et al.*, High tumor incidence and activation of the PI3K/AKT pathway in transgenic mice define AIB1 as an oncogene, *Cancer Cell* **6**:263–274, 2004.
22. Kuang SQ, Liao L, Zhang H, *et al.*, AIB1/SRC3 deficiency affects insulin-like growth factor I signaling pathway and suppresses v-Ha-ras-induced breast cancer initiation and progression in mice, *Cancer Res* **64**:1875–1885, 2004.

23. Kuang SQ, Liao L, Wang S, *et al.*, Mice lacking the amplified in breast cancer 1/steroid receptor coactivator-3 are resistant to chemical carcinogen-induced mammary tumorigenesis, *Cancer Res* **65**:7993–8002, 2005.
24. Barlund M, Monni O, Weaver JD, *et al.*, Cloning of BCAS3 (17q23) and BCAS4 (20q13) genes that undergo amplification, overexpression, and fusion in breast cancer, *Genes Chromosomes Cancer* **35**:311–317, 2002.
25. Gururaj AE, Singh RR, Rayala SK, *et al.*, MTA1, a transcriptional activator of breast cancer amplified sequence 3, *Proc Natl Acad Sci USA* **103**:6670–6675, 2006.
26. Gururaj AE, Holm C, Landberg G, *et al.*, Breast cancer-amplified sequence 3, a target of metastasis-associated protein 1, contributes to tamoxifen resistance in premenopausal patients with breast cancer, *Cell Cycle* **5**:1407–1410, 2006.
27. Culig Z, Comuzzi B, Steiner H, *et al.*, Expression and function of androgen receptor coactivators in prostate cancer, *J Steroid Biochem Mol Biol* **92**:265–271, 2004.
28. Nessler-Menardi C, Jotova I, Culig Z, *et al.*, Expression of androgen receptor coregulatory proteins in prostate cancer and stromal-cell culture models, *Prostate* **45**:124–131, 2000.
29. Rahman MM, Miyamoto H, Takatera H, *et al.*, Reducing the agonist activity of antiandrogens by a dominant-negative androgen receptor coregulator ARA70 in prostate cancer cells, *J Biol Chem* **278**:19619–19626, 2003.
30. Hu YC, Yeh S, Yeh SD, *et al.*, Functional domain and motif analyses of androgen receptor coregulator ARA70 and its differential expression in prostate cancer, *J Biol Chem* **279**:33438–33446, 2004.
31. Kahl P, Gullotti L, Heukamp LC, *et al.*, Androgen receptor coactivators lysine-specific histone demethylase 1 and four and a half LIM domain protein 2 predict risk of prostate cancer recurrence, *Cancer Res* **66**:11341–11347, 2006.
32. Kleiber K, Strebhardt K, Martin BT, The biological relevance of FHL2 in tumour cells and its role as a putative cancer target, *Anticancer Res* **27**:55–61, 2007.
33. Johannessen M, Moller S, Hansen T, *et al.*, The multifunctional roles of the four-and-a-half-LIM only protein FHL2, *Cell Mol Life Sci* **63**:268–284, 2006.
34. Manavathi B, Kumar R, Metastasis tumor antigens, an emerging family of multifaceted master coregulators, *J Biol Chem* **282**:1529–1533, 2007.
35. Toh Y, Pencil SD, Nicolson GL, A novel candidate metastasis-associated gene, mta1, differentially expressed in highly metastatic mammary adenocarcinoma cell lines. cDNA cloning, expression, and protein analyses, *J Biol Chem* **269**:22958–22963, 1994.

36. Singh RR, Kumar R, Steroid hormone receptor signaling in tumorigenesis, *J Cell Biochem* **96**:490–505, 2005.
37. Kumar R, Wang RA, Bagheri-Yarmand R, Emerging roles of MTA family members in human cancers, *Semin Oncol* **30**:30–37, 2003.
38. Xue Y, Wong J, Moreno GT, *et al.*, NURD, a novel complex with both ATP-dependent chromatin-remodeling and histone deacetylase activities, *Mol Cell* **2**:851–861, 1998.
39. Yao YL, Yang WM, The metastasis-associated proteins 1 and 2 form distinct protein complexes with histone deacetylase activity, *J Biol Chem* **278**:42560–42568, 2003.
40. Balasenthil S, Gururaj AE, Talukder AH, *et al.*, Identification of Pax5, as a target of MTA1 in B-cell lymphomas, *Cancer Res* **67**:7132–7138, 2007.
41. Bagheri-Yarmand R, Talukder AH, Wang RA, *et al.*, Metastasis-associated protein 1 deregulation causes inappropriate mammary gland development and tumorigenesis, *Development* **131**:3469–3479, 2004.
42. Barnes CJ, Vadlamudi RK, Kumar R, Novel estrogen receptor coregulators and signaling molecules in human diseases, *Cell Mol Life Sci* **61**:281–291, 2004.
43. Kumar R, Wang RA, Mazumdar A, *et al.*, A naturally occurring MTA1 variant sequesters oestrogen receptor-alpha in the cytoplasm, *Nature* **418**:654–657, 2002.
44. Luo J, Su F, Chen D, *et al.*, Deacetylation of p53 modulates its effect on cell growth and apoptosis, *Nature* **408**:377–381, 2000.
45. Gu W, Luo J, Brooks CL, *et al.*, Dynamics of the p53 acetylation pathway, *Novartis Found Symp* **259**:197–205, 2004.
46. Fujita N, Kajita M, Taysavang P, *et al.*, Hormonal regulation of metastasis-associated protein 3 transcription in breast cancer cells, *Mol Endocrinol* **18**:2937–2949, 2004.
47. Zhang H, Singh RR, Talukder AH, *et al.*, Metastatic tumor antigen 3 is a direct corepressor of the Wnt4 pathway, *Genes Dev* **20**:2943–2948, 2006.
48. Vadlamudi RK, Wang RA, Mazumdar A, *et al.*, Molecular cloning and characterization of PELP1, a novel human coregulator of estrogen receptor alpha, *J Biol Chem* **276**:38272–38279, 2001.
49. Nair SS, Mishra SK, Yang Z, *et al.*, Potential role of a novel transcriptional coactivator PELP1 in histone H1 displacement in cancer cells, *Cancer Res* **64**:6416–6423, 2004.
50. Choi YB, Ko JK, Shin J, The transcriptional corepressor, PELP1, recruits HDAC2 and masks histones using two separate domains, *J Biol Chem* **279**:50930–50941, 2004.

51. Balasenthil S, Vadlamudi RK, Functional interactions between the estrogen receptor coactivator PELP1/MNAR and retinoblastoma protein, *J Biol Chem* **278**:22119–22127, 2003.
52. Vadlamudi RK, Kumar R, Functional and biological properties of the nuclear receptor coregulator PELP1/MNAR, *Nucl Recept Signal* **5**:e004, 2007.
53. Nair SS, Guo Z, Mueller JM, *et al.*, Proline-, glutamic acid-, and leucine-rich protein-1/modulator of nongenomic activity of estrogen receptor enhances androgen receptor functions through LIM-only coactivator, four-and-a-half LIM-only protein 2, *Mol Endocrinol* **21**:613–624, 2007.
54. Redner RL, Variations on a theme: The alternate translocations in APL, *Leukemia* **16**:1927–1932, 2002.
55. Marill J, Idres N, Capron CC, *et al.*, Retinoic acid metabolism and mechanism of action: A review, *Curr Drug Metab* **4**:1–10, 2003.
56. Parmar S, Tallman MS, Acute promyelocytic leukemia: A review, *Expert Opin Pharmacother* **4**:1379–1392, 2003.
57. Yamagata T, Maki K, Mitani K, Runx1/AML1 in normal and abnormal hematopoiesis, *Int J Hematol* **82**:1–8, 2005.
58. Bohlander SK, ETV6: A versatile player in leukemogenesis, *Semin Cancer Biol* **15**:162–174, 2005.
59. Fraga MF, Ballestar E, Villar-Garea A, *et al.*, Loss of acetylation at Lys16 and trimethylation at Lys20 of histone H4 is a common hallmark of human cancer, *Nat Genet* **37**:391–400, 2005.
60. Yasui W, Oue N, Ono S, *et al.*, Histone acetylation and gastrointestinal carcinogenesis, *Ann N Y Acad Sci* **983**:220–231, 2003.
61. Mehnert JM, Kelly WK, Histone deacetylase inhibitors: Biology and mechanism of action, *Cancer J* **13**:23–29, 2007.
62. Song J, Noh JH, Lee JH, *et al.*, Increased expression of histone deacetylase 2 is found in human gastric cancer, *APMIS* **113**:264–268, 2005.
63. Zhu P, Martin E, Mengwasser J, *et al.*, Induction of HDAC2 expression upon loss of APC in colorectal tumorigenesis, *Cancer Cell* **5**:455–463, 2004.
64. Marks P, Rifkind RA, Richon VM, *et al.*, Histone deacetylases and cancer: Causes and therapies, *Nat Rev Cancer* **1**:194–202, 2001.
65. Minucci S, Pelicci PG, Histone deacetylase inhibitors and the promise of epigenetic (and more) treatments for cancer, *Nat Rev Cancer* **6**:38–51, 2006.
66. Muraoka M, Konishi M, Kikuchi-Yanoshita R, *et al.*, p300 gene alterations in colorectal and gastric carcinomas, *Oncogene* **12**:1565–1569, 1996.
67. Troke PJ, Kindle KB, Collins HM, *et al.*, MOZ fusion proteins in acute myeloid leukaemia, *Biochem Soc Symp* **73**:23–39, 2006.

68. Iyer NG, Ozdag H, Caldas C, p300/CBP and cancer, *Oncogene* **23**:4225–4231, 2004.
69. Lavau C, Du C, Thirman M, *et al.*, Chromatin-related properties of CBP fused to MLL generate a myelodysplastic-like syndrome that evolves into myeloid leukemia, *EMBO J* **19**:4655–4664, 2000.
70. Popovic R, Zeleznik L, MLL: How complex does it get? *J Cell Biochem* **95**:234–242, 2005.
71. Fujii M, Lyakh LA, Bracken CP, *et al.*, SNIP1 is a candidate modifier of the transcriptional activity of c-Myc on E box-dependent target genes, *Mol Cell* **24**:771–783, 2006.
72. Larsson LG, SNIP1: Myc's new helper in transcriptional activation, *Mol Cell* **24**:811–812, 2006.
73. Ohno H, Nishikori M, Maesako Y, *et al.*, Reappraisal of BCL3 as a molecular marker of anaplastic large cell lymphoma, *Int J Hematol* **82**:397–405, 2005.
74. Claret FX, Hibi M, Dhut S, *et al.*, A new group of conserved coactivators that increase the specificity of AP-1 transcription factors, *Nature* **383**:453–457, 1996.
75. Larsen M, Hog A, Lund EL, *et al.*, Interactions between HIF-1 and Jab1: Balancing apoptosis and adaptation. Outline of a working hypothesis, *Adv Exp Med Biol* **566**:203–211, 2005.
76. Roberts CW, Orkin SH, The SWI/SNF complex — chromatin and cancer, *Nat Rev Cancer* **4**:133–142, 2004.
77. Garcia-Pedrero JM, Kiskinis E, Parker MG, *et al.*, The SWI/SNF chromatin remodeling subunit BAF57 is a critical regulator of estrogen receptor function in breast cancer cells, *J Biol Chem* **281**:22656–22664, 2006.
78. Link KA, Burd CJ, Williams E, *et al.*, BAF57 governs androgen receptor action and androgen-dependent proliferation through SWI/SNF, *Mol Cell Biol* **25**:2200–2215, 2005.
79. Decristofaro MF, Betz BL, Rorie CJ, *et al.*, Characterization of SWI/SNF protein expression in human breast cancer cell lines and other malignancies, *J Cell Physiol* **186**:136–145, 2001.
80. Kiskinis E, Garcia-Pedrero JM, Villaronga MA, *et al.*, Identification of BAF57 mutations in human breast cancer cell lines, *Breast Cancer Res Treat* **98**:191–198, 2006.
81. Lakhani VT, You YN, Wells SA, The multiple endocrine neoplasia syndromes, *Annu Rev Med* **58**:253–265, 2007.
82. Dreijerink KM, Hoppener JW, Timmers HM, *et al.*, Mechanisms of disease: Multiple endocrine neoplasia type 1-relation to chromatin modifications and transcription regulation, *Nat Clin Pract Endocrinol Metab* **2**:562–570, 2006.

83. Balogh K, Racz K, Patocs A, *et al.*, Menin and its interacting proteins: Elucidation of menin function, *Trends Endocrinol Metab* **17**:357–364, 2006.
84. Levy-Lahad E, Friedman E, Cancer risks among BRCA1 and BRCA2 mutation carriers, *Br J Cancer* **96**:11–15, 2007.
85. Fan S, Wang J, Yuan R, *et al.*, BRCA1 inhibition of estrogen receptor signaling in transfected cells, *Science* **284**:1354–1356, 1999.
86. Rosen EM, Fan S, Isaacs C, BRCA1 in hormonal carcinogenesis: Basic and clinical research, *Endocr Relat Cancer* **12**:533–548, 2005.
87. Zheng L, Annab LA, Afshari CA, *et al.*, BRCA1 mediates ligand-independent transcriptional repression of the estrogen receptor, *Proc Natl Acad Sci USA* **98**:9587–9592, 2001.
88. Chinnadurai G, CtBP, an unconventional transcriptional corepressor in development and oncogenesis, *Mol Cell* **9**:213–224, 2002.
89. Chinnadurai G, CtIP, a candidate tumor susceptibility gene is a team player with luminaries, *Biochim Biophys Acta* **1765**:67–73, 2006.
90. Guo A, Salomoni P, Luo J, *et al.*, The function of PML in p53-dependent apoptosis, *Nat Cell Biol* **2**:730–736, 2000.
91. de Stanchina E, Querido E, Narita M, *et al.*, PML is a direct p53 target that modulates p53 effector functions, *Mol Cell* **13**:523–535, 2004.
92. Salomoni P, Pandolfi PP, The role of PML in tumor suppression, *Cell* **108**:165–170, 2002.
93. Matsuda S, Rouault J, Magaud J, *et al.*, In search of a function for the TIS21/PC3/BTG1/TOB family, *FEBS Lett* **497**:67–72, 2001.
94. Tirone F, The gene PC3(TIS21/BTG2), prototype member of the PC3/BTG/TOB family: Regulator in control of cell growth, differentiation, and DNA repair? *J Cell Physiol* **187**:155–165, 2001.
95. Shi X, Gozani O, The fellowships of the INGs, *J Cell Biochem* **96**:1127–1136, 2005.
96. Gong W, Suzuki K, Russell M, *et al.*, Function of the ING family of PHD proteins in cancer, *Int J Biochem Cell Biol* **37**:1054–1065, 2005.
97. Russell M, Berardi P, Gong W, *et al.*, Grow-ING, Age-ING and Die-ING: ING proteins link cancer, senescence and apoptosis, *Exp Cell Res* **312**:951–961, 2006.
98. Svaren J, Sevetson BR, Apel ED, *et al.*, NAB2, a corepressor of NGFI-A (Egr-1) and Krox20, is induced by proliferative and differentiative stimuli, *Mol Cell Biol* **16**:3545–3553, 1996.
99. Abdulkadir SA, Carbone JM, Naughton CK, *et al.*, Frequent and early loss of the EGR1 corepressor NAB2 in human prostate carcinoma, *Hum Pathol* **32**:935–939, 2001.

Chapter 5

A Central Role of SRC-3/AIB1 in Tumorigenesis

Jun Yan, Sophia Y. Tsai and Ming-Jer Tsai

The steroid receptor coactivator-3 (SRC-3), initially identified as coregulator of steroid receptors and later also cloned independently as amplified in breast cancer 1 (AIB1), belongs to the p160 nuclear receptor coactivator family. SRC-3/AIB1 regulates a variety of cellular processes including cell growth and proliferation, cell cycle progression, differentiation, and apoptosis. Deregulated expression of SRC-3/AIB1 occurs in a broad range of human cancers, not limited to steroid targeted cancer, and is often associated with poor prognosis, indicating a key role for this oncogene in tumor progression. Recent *in vitro* and *in vivo* studies have provided new insights into the mechanisms through which this factor acts at various levels of gene regulation and modulation of signal pathways. In addition to its roles in growth promotion, unexpectedly, SRC-3/AIB1 deficiency induces B-lymphoma in mice suggesting that SRC-3/AIB1 has a tumor suppressor role. Thus, how SRC-3/AIB1 works in tumorigenesis could be cell type or context dependent.

5.1 Introduction

Nuclear receptors (NRs) comprise a large family of transcription factors that play central roles in development, reproduction and homeostasis.[1,2] Many NRs, such as estrogen receptor (ER) and androgen receptor (AR), also play essential roles in carcinogenesis.[3,4] The activities of many NRs are regulated by specific steroid ligands. The ligand-loaded NRs dissociate from corepressors and recruit coactivators, eventually causing the stepwise remodeling of chromatin structure at target promoters, recruitment of basal transcription machinery and initiation of RNA polymerase activity.[5]

The most widely studied coactivators are members of the steroid receptor coactivator (SRC) p160 family.[6] This family consists of three 160 kDa proteins: SRC-1, SRC-2/TIF2/GRIP1, and SRC-3/AIB1/pCIP/ACTR/RAC-3/TRAM-1. SRC-3/AIB1 was also identified as a gene amplified in breast and ovarian cancer.[7] Cumulative evidences have indicated that amplification and/or overexpression of SRC-3/AIB1 in various cancers, not limited to steroid-targeted cancers.[8] Recent studies also demonstrated that SRC-3/AIB1 can interact not only with NRs, such as ER, AR and progesterone receptor (PR), but also with other transcription factors, including activator protein-1 (AP-1), E2F1, nuclear factor-κB (NFκB), signal transducer and activator of transcription (STAT), TEF-4, and ER81.[1, 8–14] In addition, it was observed that phosphorylation plays a critical role in regulating SRC-3/AIB1 activity for steroid and growth factor signaling and cell transformation.[15] Therefore, deregulation and/or activation of SRC-3/AIB1 may greatly affect the expression levels of numerous genes and, as a consequence, influence a variety of cellular process.

Overall, SRC-3/AIB1 can serve as adaptor proteins to recruit additional coactivators and basal transcriptional machinery onto the promoter. Such recruitment may be critical for NR-directed or other transcription factors-directed local chromatin remodeling and assembly of the transcriptional machinery around the promoter.

5.2 Structure of SRC-3/AIB1

The human SRC-3/AIB1 gene encodes a protein that contains several structurally and functionally distinct domains (Fig. 5.1). At the N-terminus, there is a basic helix-loop-helix-Per/ARNT/Sim (bHLH-PAS) domain, which is most conserved region among SRC family members. The bHLH-PAS domain functions as DNA-binding, protein-protein interaction surfaces for various bHLH-PAS-containing factors.[16] Recent studies revealed that the bHLH domain contains a bipartite nuclear localization signal[17,18] and two residues (K17 and R18) within bHLH domain are essential for SRC-3/AIB1 proteasome-dependent stability.[18] The central region of the SRC-3/AIB1 is the receptor interaction domain

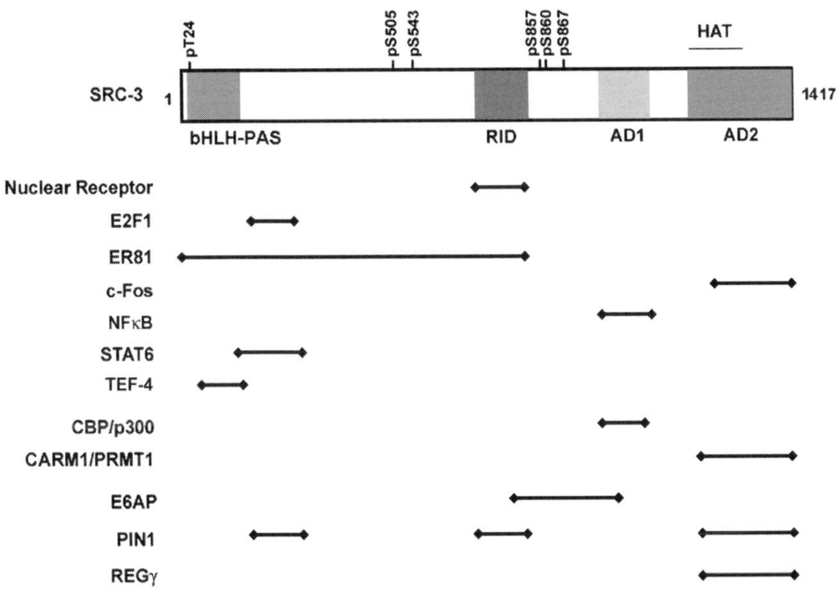

Fig. 5.1. Interactions between SRC-3/AIB1 and other proteins. The functional domains of SRC-3/AIB1 are shown. Double arrowed lines indicate the amino acids of SRC-3/AIB1 that have been reported to interact with the indicated protein. The specific amino acids of phosphorylation sites are indicated.

(RID), containing multiple LxxLL motifs (where L = leucine and x is any amino acid), which is critical for its ability to bind to ligand-bound steroid receptors via their coactivator-binding groove within the ligand binding domain (LBD).[19] At the C-terminal, there exist two intrinsic transcriptional activation domains (AD1 and AD2). The AD1 domain contains multiple LxxLL motifs that are responsible for interactions with the histone acetyltransferase (HAT) CBP and p300.[20] The AD2 domain can interact with protein arginine methyltransferases (PRMT), such as CARM1 and PRMT1.[21,22] It also possesses HAT activity, although it is weaker than those of CBP, p300 and p/CAF.[20] In addition, SRC-3/AIB1 can interact with many transcription factors, such as E2F1 and NFκB, through different domains (Fig. 5.1).

Moreover, SRC-3/AIB1 protein stability is finely regulated through direct interaction with proteins involving in protein degradation. E6AP, an ubiquitin ligase, interacts with the a.a. 723–1034 region of SRC-3/AIB1, partially overlapping the RID domain.[23] REGγ interacts with the C-terminus of SRC-3/AIB1 and promotes the degradation of SRC-3/AIB1 in an ubiquitin- and ATP-independent manner.[24] Peptidyl-prolyl isomerase 1 (Pin1), which induces conformational changes of its target proteins, can interact with three different parts of SRC-3/AIB1, including the RID and AD2 domain.[25] Pin1 functions as a transcriptional coactivator of nuclear receptors by modulating SRC-3/AIB1 coactivator protein-protein complex formation and ultimately promoting the turnover of the activated SRC-3/AIB1.

5.3 Oncogenic Functions of SRC-3/AIB1

Since the first report in 1997, SRC-3/AIB1 is found to be the gene amplified in breast cancer on human chromosome 20q21 in 1999.[7] It has been detected not only in tumors derived from steroid target tissues but also from non-steroid dependent tissues. More studies revealed that SRC-3/AIB1 is a prognostic marker for some cancers and could be a potential drug targets (Table 5.1).

5.3.1 *Breast cancer*

SRC-3/AIB1 was identified as amplified in breast cancer 1 (AIB1). As its name implies, the gene for SRC-3/AIB1 was first reported to be amplified in 9.5% (10 out of 105), whereas the overexpression of SRC-3/AIB1 at mRNA level in 64% (48 out of 75) of human breast cancer specimens, in comparison with normal mammary epithelium.[7] Another study has found that the amplification of SRC-3/AIB1 in 4.8% (56 out of 1157) of breast cancer and its expression is correlated with ER and PR expressions, as well as with tumor size.[26] Similarly, *in situ* hybridization study on 93 unselected human breast carcinomas revealed that the upregulation of SRC-3/AIB1 transcripts in 31% (26 out of 83) of specimens. The elevated expression level of SRC-3/AIB1 is associated with high tumor grade. Of note, the overexpression of SRC-3/AIB1 is not correlated with ER and PR in this study.[27] This raises the possibility that SRC-3/AIB1 may contribute to breast cancer progression through other oncogenic events beyond nuclear receptor signal pathway.

Interestingly, a splice variant of SRC-3/AIB1 (SRC-3-Δ3) was shown to be over-expressed in a series of eight human breast cancer specimens, whereas full-length SRC-3/AIB1 expression levels were not significantly

Table 5.1. Overexpression and Amplification of SRC-3/AIB1 in Human Tumors

Type of Cancer		Frequency (*n*)	Correlation with Clinical Parameters and Biomarkers	Refs
Breast Cancer	A[1]	9.5% (*105*)	N.D.	7
	O[2]		64% (*75*)	
		Tumor size		
	A	4.8% (*1157*)	ER and PR expressions	26
		MDM2 and FGFR1		
	O	31% (*83*)	Not with ER and PR	
		High tumor grade	27	
		p53 and HER2		
	A	0% (*127*)	N.D.	29
	O	13% (*23*)		
	A	10% (*85*)	Later clinical stages	
Colorectal Cancer	O	35% (*85*)	Aneuploid DNA content	36
	A	31.82% (22)	p53 overexpression	
	A	0% (*30*)	CIN-type sporadic CRC	37
		N.D.	29	
	O	17% (*30*)		
Endometrial Cancer	O	Yes (*82*)	Age, peri- or postmenopausal status	
			Tumor grade and poor prognosis	30
			ER, not PR expression	

(Continued)

Table 5.1. (*Continued*)

Type of Cancer		Frequency (*n*)	Correlation with Clinical Parameters and Biomarkers	Refs
Esophageal Squamous Cell Carcinoma	A	4.9% (*41*)	20q gain with lung and liver metastases, pleural effusion,	42
		and poorer prognosis		
	A	7% (*72*)	Lymph node metastases	43
Gastric Cancer	O	40% (*72*)	Liver metastases	
		Poor prognosis		
Hepatocellular		23% (*35 primary HCC*)		
Carcinoma (HCC)	A	41% (*33 metastatic HCC*)	Poor prognosis	41
Meningioma	O	76% (*21*)	N.D.	35
Oral Squamous Cell	A	Yes (*13*)	Lymph node metastasis	45
Carcinoma		7.4% (*122*)	N.D.	26
Ovarian Cancer	A	25% (*24*)	Poor survival	31
			ER, but not PR expression	
	O	64% (*83 high-grade*)	Tumor grade	32
Pancreatic Cancer	A	37% (*46*)	N.D.	40
	O	>65%		
Prostate Cancer	O	Yes (*480*)	PSA recurrence	34
			Ki-67 and pAKT	
	O	Yes (*37*)	Prostate cancer grade and stage	33

[1] A, Amplification. [2] O, Overexpression. [3] N.D., not determined.

higher than normal breast specimens.[28] This splice variant lacks the exon three sequence, resulting in a 130-kDa protein which lacks the N-terminal bHLH and a portion of the PAS dimerization domain, but still contains nuclear receptor interacting domain. *In vitro* data revealed that SRC-3-Δ3 can promote transcription mediated by ER, PR or induced by EGF, suggesting, similar to full-length SRC-3/AIB1, it may sensitize breast cancer cells to hormone or growth factor stimulation. Interestingly, since SRC-3-Δ3 lacks bHLH domain which is shown to be important for nuclear localization and for proteasome-dependent stability,[20,21] the SRC-3-Δ3 dependent coactivation of nuclear receptor might derive from a non-canonical pathway.

5.3.2 *Endometrial cancer*

SRC-3/AIB1 was overexpressed in 17% (5 out of 30) of endometrial carcinomas, but no SRC-3/AIB1 gene amplification was detected.[29] Another study also found a significantly higher expression of SRC-3/AIB1 in endometrial carcinoma, compared to cancer-associated normal endometrium.[30] Its expression level correlates with age, peri- or post-menopausal status and higher grade of carcinoma. Although SRC-3/AIB1 expression does not correlated with the expression of HER2/neu or total PR and PRB isoform, its expression correlates with ER expression suggesting that overexpression of SRC-3/AIB1 could cause increased sensitivity to estrogen exposure leading to endometrial hyperplasia and cancer. It is interesting to note that SRC-3/AIB1 expression level is higher in Type II than Type I endometrial carcinoma, which is more estrogen dependent and less aggressive. It raises the possibility that higher SRC-3/AIB1 can activate ER signaling even in lower estrogen condition or it can activate other signals for cancer progression.

5.3.3 *Ovarian cancer*

The SRC-3/AIB1 gene was amplified in 25% (6 out of 24) of sporadic ovarian tumors. Amplification was significantly associated with ER status of the tumor, but not PR.[31] Ovarian carcinomas arise from the ovarian surface epithelium, which expresses ER protein and undergoes rapid proliferation after ovulation. This suggests that SRC-3/AIB1 amplification may activate an estrogen-dependent, growth-promoting pathway in transformed epithelial cells to form the malignant tumors.

SRC-3/AIB1 amplification also showed statistically significant correlation with poor patient survival, underscoring the importance of SRC-3/AIB1 in the ovarian carcinogenesis. Another study found that SRC-3/AIB1 is preferentially overexpressed in 64% (53 out of 83) high-grade ovarian carcinoma, whereas no of modestly increased expression of SRC-3/AIB1 is found in normal ovarian surface epithelium and benign cystadenomas, respectively.[32]

5.3.4 *Prostate cancer*

Androgen receptor (AR) signaling plays an important role in the progression of prostate cancer. The overexpression of SRC-3/AIB1, as a coactivator for AR, is correlated significantly with tumor grade, stage of disease, poorer disease-specific survival in 37 prostate cancer specimens.[33] Recent study using a cohort of 480 clinically localized prostate cancers revealed that SRC-3/AIB1 is overexpressed in prostate cancer patients, which correlates with prostate-specific antigen (PSA) recurrence, prostate cancer cell proliferation and survival.[34] Importantly, strong SRC-3/AIB1 expression significantly associates with p-AKT level, suggesting that SRC-3/AIB1 may stimulate AKT signal pathway to facilitate prostate cancer progression.

5.3.5 *Meningioma*

Meningiomas account for 18% of all primary intracranial tumors and 25% of all intrspinal tumors. The incidence of meningioma is higher in women than in men is related to hormonal regulation. There is also an association between meningiomas and breast cancer. One study revealed that SRC-3/AIB1 is expressed in 76% (16 out of 21) of meningiomas (the majority is PR positive), while all normal brain specimens (n = 7) were negative for SRC-3/AIB1.[35] The correlation between SRC-3/AIB1 and PR is not significant. Although SRC-1 and TIF2 were also found to be overexpressed in the majority of meningiomas, their expression is significantly related to PR, but not ER expression.

5.3.6 *Colorectal cancer*

The amplification and overexpression of SRC-3/AIB1 were detected in 10% to 35% of human colorectal cancers (CRCs, n = 85).[36] In addition, the overexpression of AIB1 is significantly associated with later clinical stages, p53 overexpression as well as with aneuploid DNA content. Recently, genome-wide array comparative genomic hybridization (aCGH) study revealed that chromosomal region 20q, where SRC-3/AIB1 is located, is preferentially amplified in chromosomal-unstable (CIN), but not in microsatellite-unstable (MIN) sporadic colorectal cancers (sCRC).[37] Both findings suggest that amplified and/or overexpression of SRC-3/AIB1 might provide a selective advantage for the developmental growth and/or progression of a subset of CRCs.

5.3.7 *Pancreatic cancer*

Pancreatic cancer is the fourth most common cause of cancer-related mortality in the United States, accounting for >30 000 deaths each year.[38] The etiology of pancreatic cancer remains largely elusive. SRC-3/AIB1 was detected to be amplified in six out of nine pancreatic cancer cell lines.[39] Furthermore, SRC-3/AIB1 is progressively overexpressed during progression of human pancreatic adenocarcinoma, from pancreatitis, pancreatic intraepithelial neoplasia to invasive ductal adenocarcinomas.[40] Remarkable concordance of SRC-3/AIB1 mRNA and protein expression was found in the samples. Moreover, an increased copy number of the SRC-3/AIB1 gene was observed in 37% (17 out of 46) of pancreatic cancers. Of note, since <15% of pancreatitis samples show low positive SRC-3/AIB1 protein staining, Henke *et al.* suggested that high levels of SRC-3/AIB1 protein (especially with pronounced nuclear localization) might serve as an important diagnostic indicator for pancreatic cancers.

5.3.8 *Other cancers*

In addition to those cancer types mentioned above, elevated expression and/or amplification of SRC-3/AIB1 were also reported in other non-steroid targeted cancers, such as hepatocellular carcinoma (HCC), esophageal squamous cell carcinoma (ESCC), gastric cancer and oral squamous cell carcinoma (OSCC).[41–45] Notably, amplification of SRC-3/AIB1 is associated with metastatic HCC (41%),[41] and with lymph node metastasis in OSCC.[42] Moreover, overexpression of SRC-3/AIB1 is more frequently observed in primary ESCCs in late stages (T3/T4) than those in earlier T1/T2 stages, suggesting that it might provide a selective advantage for the development and local invasion of certain subsets of ESCC.[44] In HCC, the amplification of SRC-3/AIB1 is also correlated with a poor prognosis.[41]

5.4 Transgenic Mouse Model

Since the amplification and/or overexpression of SRC-3/AIB1 are highly associated with tumorigenesis in human patients, Brown and colleagues generated mouse mammary tumor virus (MMTV) LTR driven SRC-3/AIB1 transgenic (SRC-3-tg) mouse in FVB/N background to investigate the oncogenic role of SRC-3/AIB1 in mouse model.[46] SRC-3-tg mice showed an increase in size of each individual mammary gland cell at four weeks of age, hyperplasia of mammary gland during pregnancy and delayed involution, which do not result from differences in the levels of circulating prolactin, luteinizing hormone or growth hormone. Tumors developed in mice from all seven founder lines. High tumor incidence was found in mammary glands (48/145 detected tumors), pituitary (42/145), uterus (18/145) and lung (18/145), due to the high expression level of SRC-3/AIB1 in these organs. The hyperplasia of mammary gland is due to the high proliferative and low apoptotic rate. Most of the mammary tumors were invasive, and several adenocarcinomas were metastatic. Further analysis revealed that serum IGF-1 level was increased and the IGFR/PI3K/AKT/mTOR pathway and GSK3 were activated in mammary glands and tumors from SRC-3-tg mice.

Notably, the mammary tumors formation is not dependent on pituitary adenomas, or the reproductive history of the mice. Furthermore, although approximately 85% of the tumors were ER-positive, there are still 15% of the mammary gland tumors which were ER-negative. This phenotype mimics the clinical observation that the overexpression of SRC-3/AIB1 was also found in ER-negative human breast cancers. Overall, these data suggests that SRC-3/AIB1 may activate other receptors and/or signal pathways, especially in non-mammary gland tumors and ER negative mammary tumors.

Recently, another SRC-3-tg mouse model was established by Font de Mora J and colleagues, using the same method as described above, but kept in a C57/B16 genetic background.[47] These transgenic mice showed moderate increase (2.5± 0.3 fold) in recombinant SRC-3/AIB1 expression, compared to 7.6-fold increases of SRC-3/AIB1 in the mammary gland of mouse with FVB/N background. The SRC-3-tg mice with C57/B16 background display mammary hyperplasia at the onset of puberty, but they do not develop breast tumors. Nuclear localization of SRC-3/AIB1 is associated with BrdU incorporation, and higher expression level of cyclin D1, indicating its role for cell proliferation.

Since SRC-3/AIB1 splice variant, SRC-3-Δ3, was shown to be overexpressed in the breast cancer specimens, Riegel and colleagues generated CMV promoter driven SRC-3-Δ3-tg mice, from FVB inbred mouse embryo.[48] The expression level of SRC-3-Δ3 was not more than two-fold increase over the endogenous level in the transgenic mice. Because *in vitro* study revealed that SRC-3-Δ3 is a more potent coactivator, even in relatively low amounts, the SRC-3-Δ3-tg mice showed significantly increased mammary epithelial cell proliferation and mammary gland mass with increased cyclin D1 and IGF1R expression as well as altered expression of CCAT/enhancer binding protein isoforms. However, no tumor was observed in CMV- SRC-3-Δ3-tg mice, probably because of the lower expression level of SRC-3-Δ3 and lack of elevated systemic IGF-1 level.

Collectively, a high-level overexpression SRC-3-tg mouse model provide evidence that SRC-3/AIB1 is a *bone fade* oncogene and its high overexpression is sufficient to initiate tumorigenesis *in vivo*. Given that SRC-3/AIB1 overexpression is frequent in the precursor lesions of pancreatic adenocarcinoma, low level overexpression of SRC-3/AIB1 or SRC-3-Δ3 may contribute to the early stages of breast cancer. Since no tumors were detected in low-level overexpression mouse model, the additional genetic-epigenetic alterations are required to progress from preneoplastic changes in mammary epithelium to pathological stages of breast cancer.

5.5 Knockout Mouse Model

SRC-3/AIB1 knockout mice displayed many developmental defects, including growth retardation and reduction in body size, suggesting that SRC-3/AIB1 might play a proliferative role *in vivo*.[49,50] In addition, SRC-3/AIB1 knockout mice showed the reduction of mammary gland alveolar development during pregnancy, and resistance to growth hormone and estrogen.[49,50]

Most importantly, SRC-3/AIB1 deficiency significantly reduces the tumor incidence in mice induced by MMTV-v-Ha-ras oncogene or chemical carcinogen 7,12-dimethylbenz[a]-anthracene (DMBA).[51,52] In addition, loss of SRC-3/AIB1 also impairs the metastasis to the lung induced by Ha-ras. Similar to the findings in MMTV-SRC-3-tg mice model, SRC-3/AIB1 expression is important for the expression of IRS-1 and IRS-2, the two major components in IGF/AKT signaling, and consequently, the activation of AKT signaling pathway. Of note, various hormonal conditions (such as normal, elevated, or depleted ovarian hormones and pituitary isografts) do not significantly affect the tumor incidences in SRC-3/AIB1 knockout mice under Ha-ras genetic background. Overall, the oncogenic role of SRC-3/AIB1 in the mammary gland is not limited to its coactivator function for ovarian steroid-activated ER and PR. Moreover, these *in vivo* results also strongly suggest that SRC-3/AIB1 could be a good drug target for breast cancers.

5.6 Molecular Mechanisms of SRC-3/AIB1 Function

5.6.1 *Nuclear receptor (NR) signal pathway*

Hormone modulation plays an essential role in tumorigenesis of steroid targeted tissues, such as mammary gland and prostate gland. Since SRC-3/AIB1 is one of the p160 steroid receptor coactivators, overexpression of SRC-3/AIB1 may play an important role in hormone-stimulated proliferation. In breast cancer cells, overexpressed SRC-3/AIB1 can enhance estrogen-dependent transcription of cyclin D1 through interaction with ERα.[53] Furthermore, inhibition of SRC-3/AIB1 expression dramatically affects ERα target gene than other p160 proteins.[54] Recent study identified the direct target genes of SRC-3/AIB1 in estradiol (E2)-treated MCF-7 breast cancer cells, by using chromatin immunoprecipitation (ChIP)-based approaches with mapping the genomic locations of immunoprecipitated DNA.[55] Interestingly, two novel SRC-3/AIB1 direct genes through coactivation of E2-bound ERα are IER3/IEX3 and PARD6B /Par6, which are involved in cell apoptosis and cell polarity.

Overexpression of SRC-3/AIB1 is one of the alternatives for the adaptation of NR pathway in a low hormone microenvironment, e.g. induced by the androgen ablation therapy for advanced prostate cancer patients. In line with this notion, increased SRC-3/AIB1 expression is associated with prostate tumor grade and stage. Moreover, overexpression of SRC-3/AIB1 can act together with AR to stimulate cyclin A2 expression in an androgen-independent manner.[56] Taken together, deregulation of SRC-3/AIB1 plays an essential role for maximal NR activation under both high and low hormone condition.

5.6.2 *NR-independent signal pathways*

5.6.2.1 *IGF/AKT*

IGF/AKT signaling controls many important cellular processes, such as cell growth, proliferation and survival. IGF-1 and IGF-2 can interact with and activate its tyrosine kinase receptor, IGF1R, and stimulate a cascade of phosphorylation from IRS-1, IRS-2 through PI3K to AKT. Deregulation of this signaling has been found in many types of cancer.[57] SRC-3/AIB1 can activate the IGF/AKT signal pathway in different cell and mouse models. In SRC-3$^{-/-}$ mice, serum IGF-1 and IGF mRNA in liver tissues were reduced.[49] SRC-3$^{-/-}$ mouse embryonic fibroblasts are resistant to IGF-1, suggesting that other downstream components in this signal pathway may also be affected. In line with this notion, SRC-3/AIB1 deficiency reduces IRS-1 and IRS-2 expression in mammary glands and breast tumors induced by Ha-ras or DMBA.[51,52] Furthermore, in MMTV-SRC-3-tg mouse model, serum IGF-1 was increased and the whole IGF/AKT signal pathway was activated in mammary tumors.[46] Consistently, SRC-3/AIB1 expression level can affect the IGF/AKT signal pathway in many prostate and breast cancer cell lines.[58,59] Finally, and the most importantly, the expression of SRC-3/AIB1 in patient samples correlates well with the p-AKT level.[34] Therefore, SRC-3/AIB1 expression intimately correlates with the IGF/AKT signaling. Recent studies revealed that SRC-3/AIB1 can coordinately and directly regulate the transcription of many components of the IGF/AKT signal pathway.[60] The direct recruitments of SRC-3/AIB1 onto the promoters of IGF1 and IRS-2 require the transcription factor, AP-1. The activated IGF/AKT signal by SRC-3/AIB1 triggers the phosphorylation or upregulation of numerous downstream targets, such as mTOR, GSK3β and cyclin D1.[46] Overall, the deregulation of SRC-3/AIB1 expression is closely correlated with the expression level and activation of IGF/AKT signal pathway, and eventually contributes to carcinogenesis.

5.6.2.2 *NFκB*

NFκB signaling is another major signaling which is frequently activated in cancers. NFκB consists of dimers of proteins containing the Rel dimerizaion domain, such as NFκB1 (p50/p105), NFκB2 (p52/p100), RelA (p65), RelB and c-Rel proteins. Inactive NFκB is trapped in the cytoplasm by association with IκB inhibitor proteins. Phosphorylation of IκB by specific kinases activated by extracellular signals targets IκB for degradation, thereby allowing the translocation and activation of the NFκB complex in the nucleus.[61]

SRC-3/AIB1 can coactivate RelA (p65) *in vitro*.[11] Moreover, SRC-3/AIB1 is associated with IKK. In response to tumor necrosis factor-a (TNFα), SRC-3/AIB1 is phosphorylated by IKK complex and, subsequently, SRC-3/AIB1 is translocated with NFκB into the nucleus[62] and drive the NFκB downstream targets, such as IL-6, which is important for tumor metastasis. Most importantly, ectopic expression of wild-type SRC-3/AIB1 in SRC-3$^{-/-}$ MEFs can restore the TNFα induced IL-6, whereas phosphorylation mutants cannot.[15] Taken together, SRC-3/AIB1 may affect NFκB signal pathway through transcriptional activity and its phosphorylation status is important for its function.[63]

5.6.2.3 *E2F1*

In normal mammalian cells, cell cycle is tightly regulated. E2F1 transcription factor is one of the most important proteins control cell cycle, controlling the expression of proteins required for the G1/S transition and DNA synthesis. In breast cancer, deregulation of SRC-3/AIB1 is correlated with estrogen-dependent and independent cell proliferation. The finding that SRC-3/AIB1 can interact with and coactivate E2F1, expands the spectrum of coactivation partner for SRC-3/AIB1 and accounts for the NR independent cell proliferation.[10] Overexpression of SRC-3/AIB1 alone can transform human mammary epithelial cells, MCF10A, which is ER negative. The transformed activity requires the association of SRC-3/AIB1 with E2F1.[64] Consistently, SRC-3/AIB1 is recruited to the promoters of many E2F1 target genes involved in cell cycle, such as cyclin E, Cdk2, cdc25A and E2F1 in breast cancer cells. The data were also confirmed in prostate cancer cells, suggesting the coactivation of E2F1 by SRC-3/AIB1 is not cell-type dependent.[56] Most interestingly, SRC-3/AIB1 expression level is regulated during cell cycle, induced at late G1 and attenuated during S phase, which involves E2F1. Sp1 binding site located at SRC-3/AIB1 promoter (+150/+160), is necessary for the recruitment of Sp1, E2F1 and SRC-3/AIB1 to the SRC-3/AIB1 promoter.[65] These data suggested that in cancer cells, hyperactive E2F1 and overexpressed SRC-3/AIB1 can form a positive feedback regulatory loop to maintain high levels of SRC-3/AIB1 and E2F1 activity.

5.6.2.4 *MAPK*

Although SRC-3/AIB1 expression level does not affect MAPK activity *in vitro* and *in vivo* (41,58), SRC-3/AIB1 activity can be regulated by MAPK.[66] MAPK phosphorylates SRC-3/AIB1 and augments ER activity through increasing recruitment of p300 and associated histone acetyltransfease activity. Clinical investigations also revealed that HER2/neu and SRC-3/AIB1 expression level are closely associated with the development of tamoxifen resistance.[67] In HER2 overexpressed MCF-7 cells (with amplified SRC-3/AIB1), tamoxifen behaves as an estrogen agonist, recruiting coactivator complex (ER, SRC-3/AIB1, CBP, p300) to the ER-regulated pS2 gene promoter; while tamoxifen recruits corepressor complexes (NCoR, HDAC3) in control MCF-7 cells. Most importantly, the EGFR inhibitor, gefitinib, presumably blocks HER-2 to ER cross talk and restores the ER antagonistic properties of tamoxifen.[67] These data suggest that SRC-3/AIB1 may serve as a conduit from MAPK signal pathway to ER.

5.7 Potential Therapeutic Applications

Mutant or overexpressed oncogene products have been chosen as potential anticancer drug targets. In cases where the oncogenes are not good for drug targets, factors downstream of the activated oncogene could be more amendable for drug targeting. As for SRC-3/AIB1 case, the second approach will be a more feasible method to treat cancers dependent on SRC-3/AIB1. As described above, the amplified and/or overexpression of SRC-3/AIB1 can activate the IGF/PI3K/AKT/mTOR pathway *in vitro* and *in vivo*.[34,46,51,52,58] Since IGF signaling are responsible for cell growth, proliferation and survival, interfering this signal pathway would be a good method. Brown and colleagues utilize the RAD001 (everolimus) which is a rapamycin derivative currently under evaluation in phase I and II clinical trails.[68] RAD001 alone can reverse the premalignant hyperplastic phenotype of the mammary gland in SRC-3-tg mice. Furthermore, RAD001 can inhibit the tumorigenecity of ER^+ breast cancer cells derived from SRC-3-tg mice. Moreover, mouse model shows that RAD001 inhibits tumor growth more efficiently than 4-Hydroxytamoxifen (OHT) alone and augments anti-tumor effects of OHT in combination. The combination treatment might also prevent the formation of tamoxifen resistance in these ER^+ breast cancer cells. In addition, the clinical study discovered that patients whose breast cancers express high levels of both AIB1 and HER-2 had worse outcomes with tamoxifen therapy than all other patients combined.[69] Gefitinib can restore the anti-tumor effect of tamoxifen in MCF-7/HER2 tumor. The molecular mechanism of these tumors which overexpress HER2 and with amplified SRC-3/AIB1, may be through HER-2 to ER cross talk by the phosphorylation of SRC-3/AIB1.[67] This justifies the clinical trials of the combination of tamoxifen with gefitinib on ER-positive, HER-2–positive breast cancers.

5.8 Two Faces of the Same Coin

In contrast to the proliferation role of SRC-3/AIB1 in carcinomas, recent study showed that SRC-$3^{-/-}$ mice have a hematological deficit, characterized by a decrease in platelets and a striking increase in lymphocyte numbers.[70] Both T and B cells were expanded in SRC-3/AIB1 null mice, which is not due to an autoimmune disease. Consequently, the lymphoid expansion progresses with age into a B-cell lymphoma. This phenotype is specific for SRC-$3^{-/-}$ mice, since no similar abnormalities were found in SRC-$1^{-/-}$ and SRC-$2^{-/-}$ mice. These data reinforce the notion that SRC-3/AIB1 plays a non-redundant role in carcinogenesis, compared to its p160 family members, SRC-1 and SRC-2.

Unexpectedly, the induction of proliferation and survival in lymphoid lineage is due to constitutive NFκB activation. The phosphorylation of IκB by IKK was increased and more nuclear levels of p65, a subunit of NFκB, were induced in the thymus, bone marrow and spleen of SRC-3/AIB1 null mice. Consequently, its target genes, involved in cell proliferation (c-myb, c-myc) and survival (Bcl-2, Bcl-XL and c-FLIP) were upregulated. Consistently, the restoration of SRC-3/AIB1 in T and B cells of SRC-$3^{-/-}$ mice reduced the thymidine incorporation and the expression of cell survival genes (Bcl-2, Bcl-XL and c-FLIP). However, the nuclear levels of p65 were reduced in breast and prostate glands of SRC-3/AIB1 null mice, as well as the reduction of phosphor-IκB in prostate. In skin or liver, depletion of SRC-3/AIB1 also did not affect expression of c-myb, c-myc, Bcl-2, Bcl-XL and c-FLIP. Taken together, these data suggest that SRC-3/AIB1 has diverging effects on cell proliferation and survival depending on the cellular and molecular contexts, and indicated that SRC-3/AIB1 can act as a tumor suppressor specifically in the hematopoietic systems.

5.9 Conclusion and Perspective

Abnormal SRC-3/AIB1 expression results in oncogenic activation and contributes to the progression of a variety of carcinomas. These evidences strongly indicated that SRC-3/AIB1 is a *bona fide* oncogene in carcinoma. Overexpression of SRC-3/AIB1 can coordinately regulate and activate many components of IGF/PI3K/AKT signal pathway both *in vitro* and *in vivo*, indicating that IGF/PI3K/AKT signal pathway is one of the major signal pathway controlled by SRC-3/AIB1 to promote growth.[34,46,51,52,58] Moreover, SRC-3/AIB1 can coactivate not only nuclear receptors, but also other transcription factors, such as AP-1, E2F1 and NFκB, suggesting that deregulated SRC-3/AIB1 can activate multiple signal pathways to contribute to carcinogenesis.[56,60,61]

In carcinoma SRC-3/AIB1 itself or its downstream targets or upstream regulating signals can be very attractive chemotherapeutic targets. The findings that RAD001 suppressed SRC-3/AIB1 induced tumorigenesis and that gefitinib blocks the cross talk between HER2 and ER, presumably through SRC-3/AIB1, provides good paradigms to treat tamoxifen resistant breast cancer.[67,68] Of note, the unexpected role of SRC-3/AIB1 as tumor suppressor in the hematopoietic systems suggests that the function of SRC-3/AIB1 might be cell context dependent, at least for NFκB signaling.[70] Further study on tumor suppressor role of SRC-3/AIB1 is needed. There are still many important questions that remains to be addressed. Overexpression of SRC-3/AIB1 in cancer is partially due to the amplification. There could be other mechanisms for the overexpression, at the transcriptional level and/or stability. SRC-3/AIB1 is a phosphoprotein.[15] The phosphorylation status of SRC-3/AIB1 in cancer could result in its stability in cells and thus the different behavior in cancer. Most recent finding indicated that coordinated phosphorylation-dependent ubiquitination regulates SRC-3/AIB1 coactivator activation and transcriptional specificity by GSK3 and $SCF^{Fbw7\alpha}$ (Wu RC, *et al.* Cell In press). The combination of new proteomic techniques and available genomic data should allow the identification of numerous additional SRC-3/AIB1-binding regulated proteins with potential relevance to cancer biology.

Ackowledgments

This work is supported by NIH grants to MJT and SYT as well as Prostate SPORE grant (P50 CA058204). We apologize to our colleagues whose work could not be cited due to page limitation.

References

1. Tsai MJ, O'Malley BW, Molecular mechanisms of action of steroid/thyroid receptor superfamily members, *Annu Rev Biochem* **63**:451–486, 1994.
2. Glass CK, Rosenfeld MG, The coregulator exchange in transcriptional functions of nuclear receptors, *Genes Dev* **14**:121–141, 2000.
3. Powles TJ, Anti-oestrogenic prevention of breast cancer — the make or break point, *Nat Rev Cancer* **2**:787–774, 2002.
4. Feldman BJ, Feldman D, The development of androgen-independent prostate cancer, *Nat Rev Cancer* **1**:34–45, 2001.
5. Smith CL, O'Malley BW, Coregulator function: A key to understanding tissue specificity of selective receptor modulators, *Endocrine Rev*, **25**: 45–71, 2004.
6. McKenna, NJ, O'Malley BW, Combinatorial control of gene expression by nuclear receptors and coregulators, *Cell* **108**:465–474, 2002.
7. Anzick SL, Kononen J, Walker RL, *et al.*, AIB1, a steroid receptor coactivator amplified in breast and ovarian cancer, *Science* **277**:965–968, 1997.
8. Yan J, Tsai SY, Tsai MJ, SRC-3/AIB1: Transcriptional coactivator in oncogenesis, *Acta Pharmacol Sin* **27**:387–394, 2006.
9. Lee SK, Kim HJ, Na SY, *et al.*, Steroid receptor coactivator-1 coactivates activating protein-1-mediated transactivations through interaction with the c-Jun and c-Fos subunits, *J Biol Chem* **273**:16651–16654, 1998.
10. Louie MC, Zou JX, Rabinovich A, *et al.* ACTR/AIB1 functions as an E2F1 coactivator to promote breast cancer cell proliferation and antiestrogen resistance, *Mol Cell Biol* **24**:5157–5171, 2004.

11. Werbajh S, Nojek I, Lanz R, *et al.*, RAC-3 is a NF-kappa B coactivator, *FEBS Lett* **485**:195–199, 2000.
12. Arimura A, van Peer M, Schroder AJ, *et al.*, The transcriptional co-activator p/CIP (NCoA-3) is up-regulated by STAT6 and serves as a positive regulator of transcriptional activation by STAT6, *J Biol Chem* **279**:31105–31112, 2004.
13. Belandi B, Parker MG, Functional interaction between the p160 coactivator proteins and the transcriptional enhancer factor family of transcription factors, *J Biol Chem* **275**:30801–30805, 2000.
14. Goel A, Janknecht R, Concerted activation of ETS protein ER81 by p160 coactivators, the acetyltransferase p300 and the receptor tyrosine kinase HER2/Neu, *J Biol Chem* **279**:14909–14916, 2004.
15. Wu RC, Qin J, Yi P, *et al.*, Selective phosphorylations of the SRC-3/AIB1 coactivator integrate genomic responses to multiple cellular signaling pathways, *Mol Cell* **15**:937–949, 2004.
16. Huang ZJ, Edery I, Rosbash M, PAS is a dimerization domain common to Drosophila period and several transcription factors, *Nature* **364**:259–262, 1993.
17. Qutob MS, Bhattacharjee RN, Pollari E, *et al.*, Microtubule-dependent subcellular redistribution of the transcriptional coactivator p/CIP, *Mol Cell Biol* **22**:6611–6626, 2002.
18. Li C, Wu RC, Amazit L, *et al.*, Specific amino acid residues in the basic helix-loop-helix domain of SRC-3 are essential for its nuclear localization and proteasome-dependent turnover, *Mol Cell Biol* **27**:1296–1308, 2007.
19. Shiau AK, Barstad D, Loria PM, *et al.*, The structural basis of estrogen receptor/coactivator recognition and the antagonism of this interaction by tamoxifen, *Cell* **95**:927–937, 1998.
20. Spencer TE, Jenster G, Burcin MM, *et al.*, Steroid receptor coactivator-1 is a histone acetyltransferase, *Nature* **389**:194–198, 1997.
21. Chen D, Ma H, Hong H, *et al.*, Regulation of transcription by a protein methyltransferase, *Science* **284**:2174–2177, 1999.
22. Koh SS, Chen D, Lee YH, *et al.*, Synergistic enhancement of nuclear receptor function by p160 coactivators and two coactivators with protein methyltransferase activities, *J Biol Chem* **276**:1089–1098, 2001.
23. Mani A, Oh AS, Bowden ET, *et al.*, E6AP mediates regulated proteasomal degradation of the nuclear receptor coactivator amplified in breast cancer 1 in Immortalized Cells, *Cancer Res* **66**:8680–8686, 2006.
24. Li X, Lonard DM, Jung SY, *et al.*, The SRC-3/AIB1 coactivator is degraded in a ubiquitin- and ATP-independent manner by the REGgamma proteasome, *Cell* **124**:381–392, 2006.

25. Yi P, Wu RC, Sandquist J, *et al.*, Peptidyl-prolyl isomerase 1 (Pin1) serves as a coactivator of steroid receptor by regulating the activity of phosphorylated steroid receptor coactivator 3 (SRC-3/AIB1), *Mol Cell Biol* **25**:9687–9699, 2005.
26. Bautista S, Valles H, Walker RL, *et al.*, In breast cancer, amplification of the steroid receptor coactivator gene AIB1 is correlated with estrogen and progesterone receptor positivity, *Clin Cancer Res* **4**:2925–2929, 1998.
27. Bouras T, Southey MC, Venter DJ, Overexpression of the steroid receptor coactivator AIB1 in breast cancer correlates with the absence of estrogen and progesterone receptors and positivity for p53 and HER2/neu, *Cancer Res* **61**:903–907, 2001.
28. Reiter R, Wellstein A, Riegel AT, An isoform of the coactivator AIB1 that increases hormone and growth factor sensitivity is overexpressed in breast cancer, *J Biol Chem* **276**:39736–39741, 2001.
29. Glaeser M, Floetotto T, Hanstein B, *et al.*, Gene amplification and expression of the steroid receptor coactivator SRC3 (AIB1) in sporadic breast and endometrial carcinomas, *Horm Metab Res* **33**:121–126, 2001.
30. Balmer NN, Richer JK, Spoelstra NS, *et al.*, Steroid receptor coactivator AIB1 in endometrial carcinoma, hyperplasia and normal endometrium: Correlation with clinicopathologic parameters and biomarkers, *Mod Pathol* **19**:1593–1605, 2006.
31. Tanner MM, Grenman S, Koul A, *et al.*, Frequent amplification of chromosomal region 20q12-q13 in ovarian cancer, *Clin Cancer Res* **6**:1833–1839, 2000.
32. Yoshida H, Liu J, Samuel S, *et al.*, Steroid receptor coactivator-3, a homolog of Taiman that controls cell migration in the Drosophila ovary, regulates migration of human ovarian cancer cells, *Mol Cell Endocrinol* **245**:77–85, 2005.
33. Gnanapragasam VJ, Leung HY, Pulimood AS, *et al.*, Expression of RAC 3, a steroid hormone receptor co-activator in prostate cancer, *Br J Cancer* **85**:1928–1936, 2001.
34. Zhou HJ, Yan J, Luo W, *et al.*, SRC-3 is required for prostate cancer cell proliferation and survival, *Cancer Res* **65**:7976–7983, 2005.
35. Carroll RS, Brown M, Zhang J, *et al.*, Expression of a subset of steroid receptor cofactors is associated with progesterone receptor expression in meningiomas, **6**:3570–3575, 2000.
36. Xie D, Sham JS, Zeng WF, *et al.*, Correlation of AIB1 overexpression with advanced clinical stage of human colorectal carcinoma, *Hum Pathol* **36**:777–783, 2005.
37. Lassmann S, Weis R, Makowiec F, *et al.*, Array CGH identifies distinct DNA copy number profiles of oncogenes and tumor suppressor genes in

chromosomal- and microsatellite-unstable sporadic colorectal carcinomas, *J Mol Med* **85**:289–300, 2007.

38. Greenlee RT, Hill-Harmon MB, Murray T, *et al.*, Cancer statistics, 2001, CA, *Cancer J Clin* **51**:15–36, 2001.
39. Ghadimi MB Schröck E, Walker RL, *et al.*, Specific chromosomal aberrations and amplification of the AIB1 nuclear receptor coactivator gene in pancreatic carcinomas, *Am J Pathol* **154**:525–536, 1999.
40. Henke RT, Haddad BR, Kim SE, *et al.*, Overexpression of the nuclear receptor coactivator AIB1 (SRC-3) during progression of pancreatic adenocarcinoma, *Clin Cancer Res* **10**:6134–6142, 2004.
41. Wang Y, Wu MC, Sham JS, *et al.*, Prognostic significance of c-myc and AIB1 amplification in hepatocellular carcinoma. A broad survey using high-throughput tissue microarray, *Cancer* **95**:2346–2352, 2002.
42. Fujita Y , Sakakura C, Shimomura K, *et al.*, Chromosome arm 20q gains and other genomic alterations in esophageal squamous cell carcinoma, as analyzed by comparative genomic hybridization and fluorescence *in situ* hybridization, *Hepatogastroenterology* **50**:1857–1863, 2003.
43. Sakakura C, Hagiwara A, Yasuoka R, *et al.*, Amplification and over-expression of the AIB1 nuclear receptor co-activator gene in primary gastric cancers, *Int J Cancer* **89**:217–223, 2000.
44. Xu FP, Xie D, Wen JM, *et al.*, SRC-3/AIB1 protein and gene amplification levels in human esophageal squamous cell carcinomas, *Cancer Lett* **245**:69–74, 2007.
45. Chen YJ, Lin SC, Kao T, *et al.*, Genome-wide profiling of oral squamous cell carcinoma, *J Pathol* **204**:326–332, 2004.
46. Torres-Arzayus MI, De Mora JF, Yuan J, *et al.*, High tumor incidence and activation of the PI3K/AKT pathway in transgenic mice define AIB1 as an oncogene, *Cancer Cell* **6**:263–274, 2004.
47. Avivar A, Garcia-Macias MC, Ascaso E, *et al.*, Moderate overexpression of AIB1 triggers pre-neoplastic changes in mammary epithelium, *FEBS Lett* **580**:5222–5226, 2006.
48. Tilli MT, Reiter R, Oh AS, *et al.*, Overexpression of an N-terminally truncated isoform of the nuclear receptor coactivator amplified in breast cancer 1 leads to altered proliferation of mammary epithelial cells in transgenic mice, *Mol Endocrinol* **19**:644–656, 2005.
49. Wang Z, Rose DW, Hermanson O, *et al.*, Regulation of somatic growth by the p160 coactivator p/CIP, *Proc Natl Acad Sci USA* **97**:13549–13554, 2000.
50. Xu J, Liao L, Ning G, Yoshida-Komiya H, *et al.*, The steroid receptor coactivator SRC-3 (p/CIP/RAC3/AIB1/ACTR/TRAM-1) is required for normal growth, puberty, female reproductive function, and mammary gland development, *Proc Natl Acad Sci USA* **97**:6379–6384, 2000.

51. Kuang SQ, Liao L, Zhang H, *et al.*, AIB1/ SRC-3 deficiency affects insulin-like growth factor I signaling pathway and suppresses v-Ha-ras-induced breast cancer initiation and progression in mice, *Cancer Res* **64**:1875–1885, 2004.
52. Kuang SQ, Liao L, Wang S, *et al.*, Mice lacking the amplified in breast cancer 1/steroid receptor coactivator-3 are resistant to chemical carcinogen-induced mammary tumorigenesis, *Cancer Res* **65**:7993–8002, 2005.
53. Planas-Silva MD, Shang Y, Donaher JL, *et al.*, AIB1 enhances estrogen-dependent induction of cylcin D1 expression, *Cancer Res* **61**:3858–3862, 2001.
54. Shao W, Keeton EK, McDonnell DP, *et al.*, Coactivator AIB1 links estrogen receptor tanscriptional activity and stability, *Proc Natl Acad Sci USA* **101**:11599–11604, 2004.
55. Labhart P, Karmakar S, Salicru EM, *et al.*, Identification of target genes in breast cancer cells directly regulated by the SRC-3/AIB1 coactivator, *Proc Natl Acad Sci USA* **102**:1339–1344, 2005.
56. Zou JX, Zhong Z, Shi XB, *et al.*, ACTR/AIB1/SRC-3 and androgen receptor control prostate cancer cell proliferation and tumor growth through direct control of cell cycle genes, *Prostate* **66**:1474–1486, 2006.
57. Grimberg A, Mechanisms by which IGF-I may promote cancer, *Cancer Biol Ther* **2**:630–635, 2003.
58. Zhou G, Hashimoto Y, Kwak I, *et al.*, Role of the steroid receptor coactivator SRC-3 in cell growth, *Mol Cell Biol* **23**:7742–7755, 2003.
59. Oh A, List HJ, Reiter R, *et al.*, The nuclear receptor coactivator AIB1 mediates insulin-like growth factor I-induced phenotypic changes in human breast cancer cells, *Cancer Res* **64**:8299–8308, 2004.
60. Yan J, Yu CT, Ozen M, *et al.*, Steroid receptor coactivator-3 and activator protein-1 coordinately regulate the transcription of components of the insulin-like growth factor/AKT signaling pathway, *Cancer Res* **66**:11039–11046, 2006.
61. Lin A, Karin M, NF-kB in cancer: A marked target, *Semin Cancer Biol* **13**:107–114, 2003.
62. Wu RC, Qin J, Hashimoto Y, *et al.*, Regulation of SRC-3 (pCIP/ACTR/AIB-1/RAC-3/TRAM-1) coactivator activity by I kappa B kinase, *Mol Cell Biol* **22**:3549–3561, 2002.
63. Wu RC, Smith CL, O'Malley BW, Transcriptional regulation by steroid receptor coactivator phosphorylation, *Endocrine Rev* **26**:393–399, 2005.
64. Louie MC, Revenko AS, Zou JX, *et al.*, Direct control of cell cycle gene expression by proto-oncogene product ACTR, and its autoregulation underlies its transforming activity, *Mol Cell Biol* **26**:3810–3823, 2006.
65. Mussi P, Yu C, O'Malley BW, *et al.*, Stimulation of steroid receptor coactivator-3 (SRC-3) gene overexpression by a positive regulatory loop of E2F1 and SRC-3, *Mol Endocrinol* **20**:3105–3119, 2006.

66. Font de Mora, J, Brown M, AIB1 is a conduit for kinase-mediated growth factor signaling to the estrogen receptor, *Mol Cell Biol* **20**:5041–5047, 2000.
67. Shou J, Massarweh S, Osborne CK, *et al.*, Mechanisms of tamoxifen resistance: Increased estrogen receptor-HER2/neu cross-talk in ER/HER2-positive breast cancer, *J Natl Cancer Inst* **96**:926–935, 2004.
68. Torres-Arzayus MI, Yuan J, DellaGatta JL, *et al.*, Targeting the AIB1 oncogene through mammalian target of rapamycin inhibition in the mammary gland, *Cancer Res* **66**:11381–11388, 2006.
69. Osborne CK, Bardou V, Hopp TA, *et al.*, Role of the estrogen receptor coactivator AIB1 (SRC-3) and HER-2/neu in tamoxifen resistance in breast cancer, *J Natl Cancer Inst* **95**:353–361, 2003.
70. Coste A, Anal MC, Chan S, *et al.*, Absence of the steroid receptor coactivator-3 induces B-cell lymphoma, *EMBO J* **25**:2453–2464, 2006.

Chapter 6

Thyroid Hormone Receptors, Coregulators, and Disease

Martin L. Privalsky

Thyroid hormone receptors are hormone-regulated transcription factors that play key roles in normal vertebrate physiology and development. Thyroid hormone receptors can either repress or activate target gene expression, reflecting their ability to alternatively recruit corepressor and coactivator proteins; these coregulators, in turn, confer the specific molecular events responsible for the modulation of target gene expression. Notably, defects in this coregulator exchange manifest as human endocrine and neoplastic disease. This chapter describes the machinery that underlies transcriptional regulation by the normal thyroid hormone receptors, and the aberrations in this machinery that result in resistance to thyroid hormone, erythroleukemia, hepatocellular carcinoma, renal clear cell carcinoma, and thyroid neoplasia. Intriguingly, these diseases appear to share a common molecular flaw: an improper retention of corepressor by mutant thyroid hormone receptors under circumstances in which the wild-type receptors release corepressor and acquire coactivators. The implications of these observations are discussed.

6.1 Introduction

Thyroid hormone receptors (TRs) regulate many key processes in vertebrate homeostasis and development. Not unexpectedly, defects in the functions of these receptors result in disease. Many of these diseases are, in fact, due to the abnormalities in the interactions of TRs with their coregulators. This chapter describes how a wide variety of endocrine and oncogenic disorders arise from a common defect: an impaired ability of mutant TRs to exchange corepressor for coactivator in response to hormone agonist. These mutant TRs can function as dominant-negatives and, when coexpressed, can interfere with the actions of the corresponding wild type receptors. The default consequence is an endocrine disorder, resistance to thyroid hormone (RTH-) syndrome. Further elaboration of this basic dominant-negative phenotype confers oncogenic capabilities on certain TR mutants, and has been implicated in the initiation and progression of a diverse series of neoplasia, including erythroleukemia, hepatocellular carcinoma, renal clear cell carcinoma, and thyroid malignancies. Dominant negative mutations in related nuclear receptors, such as the retinoic acid receptors and peroxisome proliferator-activated receptors, are associated with human acute promyelocytic leukemia and with an inherited form of type II diabetes/lipodystrophy. In summary, TRs serve as an excellent model of how a basic defect in the exchange of corepressors and coactivators can generate a diverse spectrum of clinically relevant disorders.

6.2 Primer on Thyriod Hormone Receptors and their Coregulators

The thyroid hormone T3 (3,5,3′ triiodothyronine) and its prohormone T4 (3,5,3′,5′ tetraiodothyronine) contribute to the regulation of many diverse aspects of vertebrate physiology and development, including thermogenesis, glucose utilization, lipid metabolism, cardiac output, and development of the central nervous system, inner ear, and retina.[1] The actions of T3 are mediated primarily through thyroid hormone receptors (TRs), which are founding members of the nuclear receptor family and obey many of the same rules and utilize many of the same coregulators detailed elsewhere in this text. However, a few specific TR features and idiosyncrasies are important to note at the onset of this narrative and are discussed below.

TRs are expressed from two genetic loci, denoted α and β, and are further modified by alternative mRNA splicing, promoter utilization, and translational start sites to generate a series of interrelated isotypes or isoforms.[2–4] Three mammalian TR isoforms have been characterized in greatest detail and are thought to play primary roles in responding to circulating thyroid hormone: TRα1, TRβ1, and TRβ2 [Fig. 6.1(A)]. TRα1 is expressed very early in development and is widely expressed in adult tissues. TRα1 is known to contribute to regulation

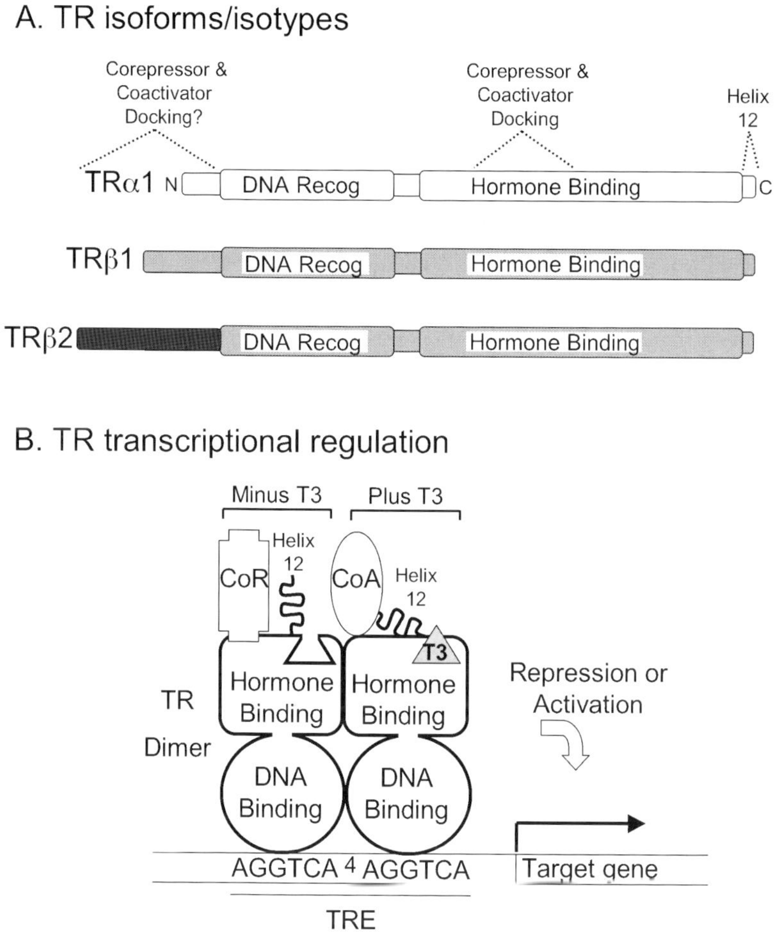

Fig. 6.1. Schematic of TR structure and function. (A) Schematic representations of the three major mammalian TR isoforms. The proteins are depicted from N- to C-terminus, and the domains responsible for DNA recognition, hormone binding, and coregulator interactions are shown, as is the location of helix 12. (B) Schematic representation of transcriptional regulation by TRs. Two TR molecules are shown bound as a receptor dimer to a TRE upstream of a target gene. The left TR molecule is shown in the absence of hormone; under these conditions helix 12 assumes a conformation that permits docking of corepressor ("CoR") to the receptor. The right TR molecule is shown bound to T3 agonist; under these conditions helix 12 reorients to occlude the corepressor binding site, generating a new docking surface for coactivators ("CoA"). In reality, both receptors in the dimer likely would be in one or the other conformation, or the TRE would be bound by an RXR/TR heteromer.

of cardiac output, bone growth, intestinal maturation, erythropoiesis, and adaptive thermogenesis.[2–4] TRβ1 expression begins later in development, often paralleling the onset of T3/T4 synthesis, and although detectable in many different tissues, it is found at highest levels in brain, lung, kidney, and liver. TRβ1 is an important regulator of lipid metabolism and elements of central nervous system development.[2–4] TRβ2 expression is restricted primarily to the hypothalamus, pituitary, inner ear, and the retina. TRβ2 plays a central role in the negative feedback sensing of circulating thyroid hormone levels, has been implicated in the timing and differentiation of M versus S cone cells in the retina, and contributes to the establishment of proper ion conductance in the inner hair cells of the cochlea.[2–4] In addition to these "big three" TR isoforms, a variety of additional TR variants have been discovered that can further modify or diversify target gene regulation.[2–4] As described below, different TR isoforms display different transcriptional properties, a phenomenon that has important consequences in determining the disease phenotype when these receptors fail to function properly.

6.2.1 *DNA, hormone, and coregulator binding by TRs*

In common with most other nuclear receptors, TRα1, β1, and β2 are comprised of relatively conserved DNA and hormone binding domains embedded in more divergent N-terminal and linker sequences [Figs. 6.1(A) and 6.1(B)]. The DNA binding domain in the receptor consists of a series of α-helical domains stabilized by interactions with coordinated zinc ions.[5] The prototypic DNA binding site for all three TR isoforms (a thyroid hormone response element, or TRE) consists of a direct repeat of an AGGTCA half-site with a four base spacer (a DR4 TRE) and can recruit either TR homodimers or TR/retinoid X receptor (RXR) heterodimers [Fig. 6.1(B)].[5] However, a surprising variety of non-consensus half-sites, spacings and orientations can also function as TREs.[6,7] Paralleling this diversity at the DNA level, specific TR isoforms can also bind to DNA as monomers, as dimers with yet other members of the nuclear receptor family (such as retinoic acid receptors, RARs), as trimers, or indirectly through protein-protein contacts with other transcription factors.[6,8–10] TRs are generally believed to bind to their target genes in both the absence and presence of thyroid hormone.

The hormone binding domain of the TRs is composed of a triple sandwich of α-helices, and binds T3 by engulfing the hormone in a clamshell-like embrace.[11] The conformational changes induced by the opening and closing of the clamshell help couple hormone binding to corepressor/coactivator recruitment and release. Receptor "helix 12" serves as a key toggle switch in this T3-driven corepressor/coactivator exchange.[12,13] In the absence of T3, helix 12 is positioned to allow corepressor access to an adjacent docking surface on the receptor; conversely, binding of T3 induces a repositioning of helix 12, causing it to occlude the corepressor docking surface and to form a novel docking site for coactivators [Figs. 6.1(B) and 6.2]. Additional, hormone-independent corepressor and/or coactivator interaction

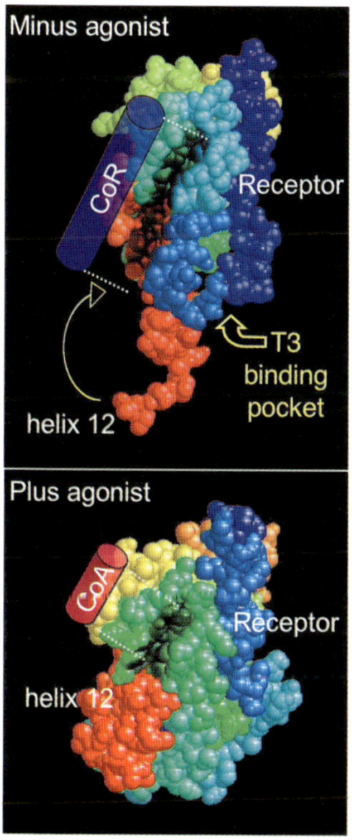

Fig. 6.2. X-ray diffraction structure of a prototypic nuclear receptor hormone binding domain in the absence or presence of hormone agonist. The agonist-induced alteration in the position of helix 12 is indicated, as are the locations of the binding site for agonist (T3 for the TRs), the corepressor (CoR) docking surface, and the coactivator (CoA) docking surface. The structure of the unliganded TR receptor hormone binding domain has not been reported; the corresponding RXR structure is provided instead as a prototype; the precise position of helix 12 in the unliganded TR may differ from that shown.

surfaces are located in the N-terminal domains of many of the TR isoforms.

The above description accounts for TR function on "positive-acting" TREs, such as the DR4 prototype, which are repressed in the absence of hormone and activated in the presence of T3. However, many T3 responsive genes are "negative-acting," i.e. activated by TRs in the absence of hormone but repressed by TRs in the presence of T3.[14] The precise structure of negative-acting TREs, and the contributions of corepressors and coactivators to gene regulation through these negative elements, are incompletely understood. It is possible that the corepressor/coactivator exchange is reversed on negative elements, allowing corepressor complexes to be recruited in response to hormone agonist.[15] Alternatively, it may be the function of the coregulators themselves that is reversed at negative TREs, allowing, for example, coactivators to manifest otherwise cryptic corepressor functions.[16,17] Many negative-acting elements are under complex combinatorial regulation; TRs may operate on these elements by suppressing the actions of other, positive-acting transcription factors arrayed on the same promoter.[18]

6.2.2 *Coregulators: The cast of characters*

The best characterized corepressors for TRs are SMRT and its ortholog N-CoR.[13] These two proteins share approximately 40% amino acid sequence identity, assemble into similar corepressor complexes, and display overlapping, but non-identical biological and biochemical properties.[19–21] Although N-CoR appears to be the preferred corepressor partner for TRs in several experimental scenarios, alternative mRNA splicing produces a multiplicity of SMRT and N-CoR variants that possess differing affinities for TRs.[22] As a result, different splice versions of both SMRT and N-CoR are likely to function as TR corepressors in different contexts. The most extensively studied coactivators for TRs are the p160 and the CBP/p300 families of histone acetyltransferases, the PRMT1 and CARM1 histone methyltransferases, and the DRIP/TRAP/mediator complexes implicated in assembly of the preinitiation complex.[23–27] Interestingly, ATP-dependent chromatin remodeling complexes, such as SWI/SNF and NURD, have been implicated in both repression and activation by TRs, and yet other proteins, such as TBL1 and TBLR1, appear to assist in the corepressor/coactivator exchange itself.[28–30] In addition to these major players, there is a diverse assemblage of additional proteins (and one RNA, denoted SRA) that have been identified as possessing either corepressor or coactivator function; unfortunately, space constraints prohibit detailed discussion of these supporting cast members.

6.3 Resistance to Thyroid Hormone (RTH)-Syndrome: A Corepressor Disease

6.3.1 *Clinical and genetic features of RTH-Syndrome*

The majority of thyroid endocrine abnormalities observed in humans reflect an improper production of T3/T4 hormone by the thyroid gland, leading to the classical clinical symptoms of hypothyroidism or hyperthyroidism. More rarely, the abnormality is not in the thyroid *per se*, but instead reflects an inherited inability to respond properly to the T3/T4 hormone itself, a disorder denoted resistance to thyroid hormone (RTH) syndrome.[31–33] Over 400 families with this disease have been identified worldwide, and the vast majority of the associated genetic lesions map to autosomal dominant mutations in the TRβ locus. Different RTH-Syndrome kindreds can display somewhat different biochemical and clinical features. Nonetheless, a key commonality has emerged: virtually all known RTH-Syndrome mutations are impaired in their ability to release corepressor and to bind coactivator in response to physiological levels of T3.[34–36] As a consequence, the RTH-mutant receptors display dominant negative properties and can interfere with the actions of wild-type TRs when coexpressed in the same cells.

The physiological consequences of this defect are several fold. In the normal individual, T3/T4 levels are regulated through a negative feedback loop involving the hypothalamus and pituitary (Fig. 6.3). Surges in circulating T3/T4 are sensed primarily by TRβ2 in these organs, which negatively regulates expression of the genes for thyroid releasing hormone (TRH) and thyroid stimulating hormone (TSH), leading in turn to the reductions in the synthesis of T3 and T4 by the thyroid gland.[1] The mutant TRs expressed in RTH-Syndrome disrupt this negative feedback loop, resulting in significantly increased levels of circulating T3/T4 and manifestation of a spectrum of associated physiological consequences.[31–33] At one end of the spectrum, denoted generalized RTH-Syndrome, the peripheral tissues share the insensitivity to T3 found in the hypothalamus/pituitary axis. This results in a complex endocrine phenotype resembling certain aspects of hypothyroidism, including growth delay, goiter, hearing defects, and bone abnormalities. At the other

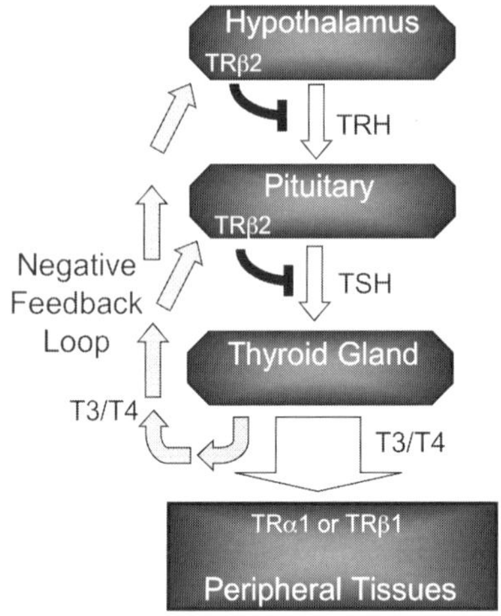

Fig. 6.3. T3/T4 homeostasis as regulated by the hypothalamus-pituitary-thyroid gland axis. As shown schematically, the hypothalamus secretes TRH, which stimulates TSH production by the pituitary, which is turn stimulates T3/T4 production by the thyroid gland. T3/T4 have a wide range of effects on other tissues in the body, mediated primarily by TRα1 and/or TRβ1 ("peripheral tissues"). T3/T4 also participate in a negative feedback loop, mediated primarily by TRβ2 in the hypothalamus and pituitary, which results in suppression of TRH and TSH production in response to rising T3/T4 levels, as shown.

extreme, denoted pituitary (or central) RTH-Syndrome, the peripheral tissues remain relatively responsive to T3 despite the loss of TRβ2-mediated negative feedback in the hypothalamus/pituitary; the resulting combination of high circulating T3/T4 levels and a relatively responsive periphery produces symptoms of peripheral hyperthyroidism/thyrotoxicity. Nonetheless, RTH-Syndrome is a complex disorder that can differ in presentation and severity even among individuals bearing the same mutation. The symptoms can also vary in any one individual at different times or in different tissues.[31–33] Additional genetic and physiological factors undoubtedly influence how a given TRβ mutation manifests the resistance syndrome.

6.3.2 *Molecular biology of the disease: A flawed toggle switch and defective coregulator exchange*

Approximately 85% of the known RTH-Syndrome genetic lesions map to TRβ, with most of these mutations falling within a defined series of genetic "hot spots" within the receptor coding region.[31–33] One subset of these RTH-mutations maps to the hormone binding cavity of the receptor and impair the ability of the receptor to bind to T3. As a result these mutant receptors require higher than normal levels of T3 to release corepressors and to bind coactivators [Fig. 6.4(A)]. A second subset of RTH mutations maps to the sequences that operate the helix 12 toggle switch itself; the resulting mutant receptors are generally permissive for hormone binding, but are defective in the subsequent conformational changes necessary for corepressor release and coactivator recruitment [Fig. 6.4(A)].[34,36–38] Therefore RTH-Syndrome at the molecular level represents primarily a failure to correctly couple T3 binding to the

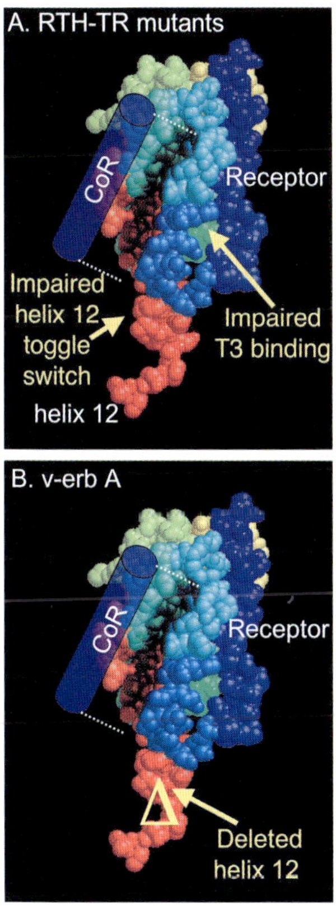

Fig. 6.4. Location of genetic lesions in mutant TRs associated with RTH-Syndrome and avian erythroblastosis virus. The location of the mutated amino acids in the TRs associated with these diseases/disease agents is indicated relative to the prototypic nuclear receptor structure depicted in Fig. 6.2.

operation of the receptor helix 12 toggle, resulting in the retention of corepressor by the mutant receptor under T3 concentrations that would, in the wild type receptor, result in coactivator recruitment.

Although this model explains the failure of most RTH-mutant receptors to correctly switch from repressors to activators at physiological T3 concentrations, RTH-Syndrome is classically a disease of heterozygotes. One must, therefore, also account for how these mutant TRs interfere *in trans* with the functions of the wild type TRs expressed from the unaffected chromosome. Several models are possible. A leading contender proposes that the RTH-TR mutant competes with the wild type TRs for binding to TREs, either as homodimers of mutant receptors or as heterodimers of the mutant TR with RXR.[39] Alternatively, RTH-mutant receptors may dimerize with the wild type TRs and cripple the transcriptional activation properties of their wild-type TR partners at "point-blank" range.[40] Yet other mechanisms may also contribute to the dominant-negative actions of RTH-TRs, such as a competition for limiting amounts of RXRs or the sequestration of other critical cofactors, and these additional mechanisms may predominate in certain tissue or promoter contexts. No matter what the specific mechanism is, the dominant-negative activity of a given RTH mutant TR is likely to reflect both the repressive properties of the RTH-mutant itself, and its abundance relative to wild type receptors. Notably, RTH mutant receptors may be expressed in excess compared to the wild type allele, enhancing the potency of the mutant allele.

6.3.3 *RTH-Syndrome is primarily a disease mediated by inappropriate corepressor retention*

Is it the failure to release corepressors, or the inability to bind coactivators, that is most responsible for the dominant-negative activity of the RTH-TRβ mutants? To answer this question, secondary mutations have been introduced into RTH-TRβ receptors to disrupt corepressor binding; alternatively, dysfunctional corepressor fragments have been introduced into cells to impair endogenous corepressor function.[34–36,41,42] Preventing corepressor binding by RTH-TRs by either means significantly reduces their ability to interfere with wild type TR function. Further highlighting the role of corepressors in RTH disease, several

RTH-mutants not only fail to release corepressor in response to T3, but also bind corepressor with an inherently higher affinity than do wild-type receptors. These "super-repressor" TR mutants typically contain single codon deletions that shorten and rotate portions of helix 11; these alterations reposition helix 12 and probably increase affinity for corepressor by increasing access to the corepressor docking surface on the mutant receptor.[43] These single codon RTH-TRβ deletions have been found in kindreds, displaying relatively severe clinical phenotypes, but it has not been established if there is a link between the clinical presentation and the greater affinity of these mutant receptors for corepressor. A further interesting twist to this story comes from yet another RTH-mutant that bears a frame shift in the same helix 11 region. This frame-shift mutation retains the ability to bind to SMRT, but exhibits a much reduced affinity for N-CoR, compared to the wild type TRβ1.[44] This RTH mutant, therefore has the potential to mediate repression in cell types in which SMRT, rather than N-CoR, predominates. Clinically, this unusual mutation was associated with a severe, predominantly pituitary resistance, but the relationship between this mutant's change in corepressor specificity and its biological effects remains to be determined. Notably, a related alteration in corepressor specificity has been observed for the v-erb A oncoprotein variant of TRα1, a point that will be revisited below.

Studies in genetically manipulated mice provide further enlightenment on these issues. Mice lacking all TR isoforms display a less severe phenotype than mice expressing wild type TRs, but lacking thyroid hormone. Similarly, strong dominant-negative alleles of either TRα or TRβ have more severe effects when introduced into mice than do the corresponding homozygous null alleles.[4,45] These results indicate that inappropriate retention of corepressor by unliganded TRs is more disruptive to development and homeostasis than is a total lack of TR function. However, it should be noted that even TR mutants which are unable to recruit corepressor do have some dominant-negative potential, and a least one identified RTH-mutant has been proposed to exert the majority of its dominant negative properties as a consequence of its loss of coactivator binding.[46]

RTH-TR mutants are not only defective in T3-mediated activation of positive response genes, but also are defective in the T3-mediated repression of negative response genes. This defect, manifested as a failure of the mutant TRs to inhibit TSH and TRH expression in response to T3, is the cause of the elevated thyroid hormone levels seen in RTH-Syndrome.[31–33] Unfortunately, our incomplete understanding of the normal molecular mechanisms operating in negative gene regulation by the wild type TRs makes it difficult to interpret the basis behind these RTH-defects. One suggestion, noted previously, is that a fundamental functional dualism may exist for coregulators such that corepressors may function as coactivators under certain circumstances. Indeed, SMRT and N-CoR have been implicated in the transcriptional activation of negative response elements in the absence of T3. In this mirror world, a genetic inability to release corepressor would therefore resulted in the retention of gene expression in the presence of T3, as is, in fact, observed for TSH and TRH in RTH-Syndrome.[16,17] Alternatively, other studies have suggested that coactivators and coregulators remain true to their names even on negative response elements, but somehow their recruitment is reversed relative to that seen on positive elements.[15] By this model, it would be the T3-driven recruitment of corepressor to negative elements that is disrupted by the RTH-mutations. Ultimately, our simple binary notion of positive- versus negative-acting response elements may itself prove too limited. It is possible that many response elements are capable of hosting a multitude of T3-mediated transcriptional responses, both up and down, in different cells and in response to different signals.

6.3.4 *Why is RTH-Syndrome exclusively a disease of the TRβ locus?*

The TRβ mutants found in RTH-Syndrome are dominant-negative inhibitors of wild-type TRβ1 and TRβ2 function; do these mutant TRβ receptors also interfere with wild-type TRα1 function? The short answer is yes, when assayed in transfected cells using prototypic reporter genes.[3,4,47] The situation in the intact animal is a bit more complex. At the organismal level, RTH-TRβ mutations mediate wide-reaching dominant-negative effects that, in many tissues, include the ability to interfere with TRα1 function. However, these inhibitory effects can, of course, only be manifested in cell types where both isoforms are expressed.[48] Mutant TRβ expression generally parallels that of wild-type TRβ; therefore tissues that express primarily the TRα locus retain much of their normal T3 responsiveness in RTH-Syndrome, whereas tissues that express both the TRα and TRβ loci appear to have the T3 response of both isoforms attenuated by the RTH TRβ mutant. This concept does not exclude the possibility that certain TRα target genes are more susceptible to inhibition by RTH-TRβ mutants than others, although relatively few such isoform-specific genes have been reported to date.[45]

Why then is the TRβ locus found mutated in RTH-Syndrome and not the TRα locus? Given the early and widespread expression of TRα1 in normal development, it is possible that dominant-negative mutations of TRα produce a highly detrimental phenotype that would be rapidly selected against and lost from the human population. Alternatively, an RTH-like TRα mutation might have no observable phenotype, or might present in an unanticipated manner that would not be diagnosed as T3 resistance. This issue has been experimentally tested by artificially introducing a strong human RTH-TRβ mutation (fs394) into the mouse TRα and examining the effects in a gene knock-in mouse model. Interestingly, the artificial RTH-TRα mutation was embryonically lethal as a homozygote, and produced severe bone ossification defects, dwarfism, and impaired adipogenesis as a heterozygote.[49,50] This mouse model suggests that the predominance of TRβ lesions in human RTH-Syndrome may reflect at least two phenomena: a fitness-based selection against the more severe developmental defects produced by TRα mutations compared to TRβ mutations, combined with the likelihood that any mutant TRα alleles that do survive in the human population may not be recognized under the diagnostic rubric of RTH-Syndrome.

6.3.5 *Pituitary versus general resistance to thyroid hormone: A contribution of coactivators?*

As previously noted, RTH-Syndrome has a variable clinical presentation. Much of this diversity probably arises from differences in the genetic background or physiology of the affected individuals. Nonetheless, certain RTH-Syndrome mutations appear to be more often associated with the pituitary form of the disease (PRTH), whereas others associate more often with the generalized syndrome (GRTH). Why? As noted previously, TRβ1 is widely expressed in the adult, whereas TRβ2 is more restricted in its expression pattern and plays a key role in the negative feedback regulation of T3/T4 synthesis by the hypothalamus and pituitary.[4] The TRβ1 and TRβ2 isoforms share identical DNA and hormone binding domains and differ only in their N-terminal sequences [Fig. 6.1(A)]. As a result, the mutations found in RTH-Syndrome are expressed in the TRβ2 context in the pituitary and hypothalamus, but in the TRβ1 context in peripheral tissues.[51] It has been proposed that the nature of the receptor mutation, acting in conjunction with this tissue-specific alternative mRNA splicing, may work to evoke a more PRTH or a more GRTH like phenotype, as described below.

The first point to note is that wild type TRβ2 possesses an enhanced ability to activate positive-response genes (and to repress negative-response genes) compared to wild type TRβ1 or to wild-type TRα1. This reflects, in part, an elevated ability of the wild-type TRβ2 receptor to bind the p160 and CBP/p300 coactivators under limiting T3 concentrations, a consequence of contacts the TRβ2 N-terminal domain makes with these coactivators that are not made by the TRβ1 N-terminal domain.[52–54] As a result, TRβ2 requires significantly less T3 to achieve a specific level of target gene expression than does either TRα1 or TRβ1. When analyzed in the same assays, the RTH mutations fall into two classes. One class impairs coactivator binding proportionally whether expressed as the TRβ1 or as the TRβ2 isoform; the T3 response is therefore reduced for both isoforms, but the TRβ2 mutant still retains its elevated (or even an exaggerated) responsiveness to T3 compared to the corresponding TRβ1 mutant.[55] The second class of RTH mutations, in contrast, impairs coactivator recruitment more severely when expressed as TRβ2 than when expressed as TRβ1, thereby reducing or eliminating the enhanced T3 response observed for the wild type TRβ2 isoform.[55] Significantly, the first class of mutations have typically been isolated from GRTH patients, whereas the second class of mutations have typically been isolated from PRTH patients.

This leads to an admittedly simplistic, but possibly useful model (Fig. 6.5). Under normal physiological conditions, surges in circulating T3/T4 are sensed first by the more-responsive TRβ2 isoform in the pituitary and hypothalamus. This leads to down regulation of TSH and TRH and suppression of T3/T4 synthesis before the more widely

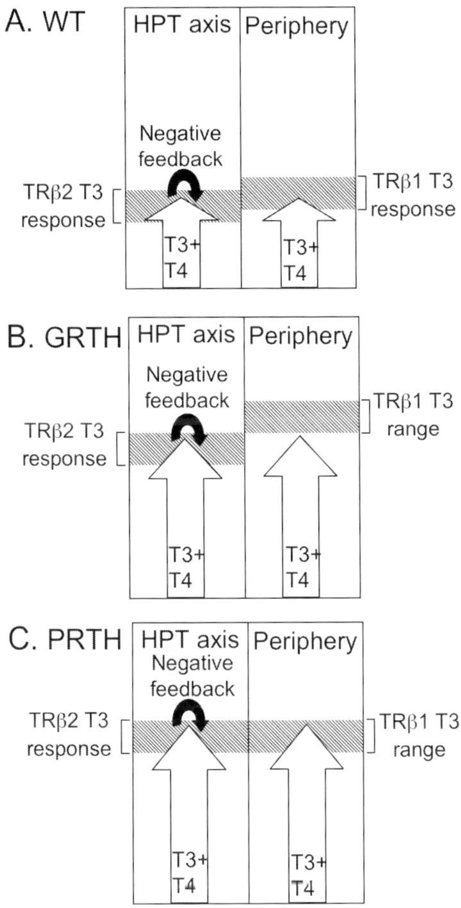

Fig. 6.5. Possible model of defective T3/T4 homeostasis in GRTH and PRTH. (A) Wild-type receptors. A model is proposed by which wild-type TRβ2 in the hypothalamus and pituitary (left panels) senses and suppressed surges of T3/T4 ("negative feedback") before the less responsive wild-type TRβ1 in peripheral tissues responds (right panels). The height of each arrow depicts the circulating T3/T4 concentration, and the range of T3 response for each receptor isoform is shown as a cross-hatched region. (B) GRTH. The ability of both the TRβ1 and TRβ2 isoforms to respond to T3/T4 is impaired, but the difference between the two isoforms is either maintained or exaggerated. As a result, circulating T3/T4 levels increase, but the periphery remains relatively non-responsive. (C) PRTH. The ability of both the TRβ1 and TRβ2 isoforms to respond to T3/T4 is impaired, but the difference between the two isoforms is reduced or eliminated. As a result, circulating T3/T4 levels increase and the periphery remains sufficiently responsive to produce symptoms of thyrotoxicity.

expressed (but less responsive) TRβ1 isoform is fully activated. In this fashion, T3 surges can be dampened before they cause a peripheral thyrotoxicity. For the GRTH mutations, both the TRβ2 and TRβ1 isoforms are proportionally impaired. As a result, circulating T3/T4 levels rise (due to the loss of negative regulation by the mutant TRβ2 in the hypothalamus and pituitary), but the T3 response in the peripheral tissues is also disrupted (reflecting the impact of the same mutation when expressed in TRβ1); depending on the tissue, this leads to symptoms of a mixed peripheral euthyroidism/hypothyroidism. For the PRTH mutations, the TRβ2 isoform is disproportionally impaired compared to TRβ1. As a result, circulating T3/T4 levels rise (due to the loss of negative regulation by the mutant TRβ2 in the hypothalamus and pituitary) but the peripheral tissues remain relatively responsive (reflecting the less severe impairment mediated by the same mutation when expressed as TRβ1), generating a peripheral hyperthyroidism/thyrotoxicity.

Notably, the T3 response curves of several RTH lesions exhibit a more PRTH-like pattern at one T3 concentration, but a more GRTH pattern at another T3 concentration, potentially explaining how a single mutation can produce different clinical symptoms in different patients, or at different times. Clearly, the above model is only a first order approximation, and cannot explain all the variation seen with RTH-Syndrome. Many additional genetic, biochemical, and physiological factors must be involved that influence how a given mutation manifests as a given resistance phenotype. However, this proposal may provide a useful mechanistic foundation on which more sophisticated analyses can be built.

6.3.6 *Atypical RTH-Syndromes*

Approximately 85% of the RTH-Syndromes characterized in humans are associated with dominant-negative TRβ mutations, and the vast majority of these cases are found in individuals heterozygous for the TRβ mutation.[32] More rarely, RTH-Syndrome occurs in a homozygous TRβ mutant background.[56] Not surprisingly, the affected individual in this case displayed an unusually severe physiological and developmental phenotype. What of the remaining 15% of RTH-Syndromes are not due to TRβ mutations? In one example, the mutation has been shown to be in MCT8, a transporter that facilitates the entry of thyroid hormone into cells.[45] In another case, the resistance lesion mapped to a defect in the incorporation of an essential selenocysteine into deiodinase, resulting in impaired conversion of the pro-hormone T4 into active T3.[45] Notably, the RTH-disease caused by these mutations is again due to an impaired corepressor release, although as a result of reductions in local T3 concentration rather than as a result of lesions in the TRs themselves. Additional metabolic derivatives of T3 and T4 have been identified that signal not through the classical nuclear TRs, but through a membrane-associated receptors.[2,45] As a result, there may exist a parallel universe of yet-to-be discovered RTH-Syndromes that represent defects in these extrachromosomal signaling pathways.

6.4 TRs, Corepressors, and Retroviral-Induced Cancer

6.4.1 *Oncogenic retroviruses and identification of TRs as cryptic oncoproteins*

Surprisingly, the first molecular clone of a TR allele came not from endocrinology research, but instead from the studies of avian erythroblastosis virus (AEV) strain ES4. AEV is a retrovirus that induces erythroleukemias and fibrosarcomas in infected birds, and encodes two viral oncogenes, v-erb A and v-erb B.[57] Both of these AEV oncogenes represent captured and mutated versions of normal host cell genes. V-erb B is derived from a host cell gene for epidermal growth factor (EGF) receptor, a tyrosine-kinase involved in control of normal cell proliferation. V-erb B encodes a constitutively-active mutant form of EGF-receptor and induces inappropriate replication in both fibroblasts and erythroid cells. In contrast, v-erb A is derived from the host cell gene for TRα1. V-erb A encodes a dominant-negative mutant version of TRα1 that blocks erythroid cell differentiation and confers long term proliferation on fibroblasts.[57] The dominant-negative properties of the v-erb A protein map primarily to a deletion of helix 12 that prevents the exchange of coactivator for corepressor in response to T3 ligand [Fig. 6.4(B)].[58,59] Therefore, both the v-erb A protein and the mutant TRs found in human RTH-Syndrome share a common molecular defect, which is a failure to release corepressor. Both v-erb and the RTH-TRβ mutants can function as antimorphs of their normal cellular progenitors. Why then do the RTH-Syndrome TR mutants give rise to endocrine, but not neoplastic disease, whereas the v-erb A TR mutant manifests the opposite disease proclivities?

Several features may contribute to the oncogenic status of the v-erb A allele: (A) V-erb A is coexpressed together with v-erb B in AEV-induced neoplasia and it is clear that v-erb B contributes to the oncogenic phenotype in this context.[57,58] Nonetheless, this cooperativity with v-erb B cannot be the sole reason for the neoplastic properties of v-erb A. Notably, v-erb A displays a variety of transforming/oncogenic properties even in the absence of v-erb B.[60,61] Conversely, activated EGF-receptor mutants resembling v-erb B are found in a variety of human tumors,[62] yet these EGF-receptor mutants are not known to cooperate with RTH-TRβ mutants to elevate tumor incidence in RTH-Syndrome. (B) V-erb A is derived from the TRα locus, whereas RTH-Syndrome mutants are derived from the TRβ locus. Although the isoform context may contribute in some fashion to the oncogenic properties of v-erb A, it is unlikely to be the primary determinant. For example, a swap of a very strong RTH mutation from TRβ into the TRα background creates growth and skeletal abnormalities, but not neoplasia when expressed in heterozygous mice.[63] (C) V-erb A is avian derived, whereas RTH-TRs are human in origin. Although a formal possibility, the species of origin seems very unlikely to account for the neoplastic aspects of v-erb A. The sequences of human TRα1 and *Gallus* TRα1 (the v-erb A progenitor) are very similar; in fact, there are fewer differences between these two isoforms from one species to the other than there are between the TRα1 and TRβ1 isoforms within any one species. (D) V-erb A is expressed from a retroviral promoter, whereas the RTH mutants are expressed from the normal cellular TRβ promoter. Might v-erb A cause cancer because it is expressed in cell types that do not express the RTH-TRβ mutants? This mechanism may also play some role, but again falls short of an complete explanation for the oncogenic properties of v-erb A. The v-erb A retroviral and the RTH-mutant TR promoters are expressed in a fairly broad set of overlapping tissues, and as described below, ectopic expression of v-erb A causes hepatocellular carcinoma in transgenic mice, whereas the expression of RTH TRβ mutants in the same tissues does not.[60]

6.4.2 *Oncogenesis by v-erb A: Failed corepressor release and a bit more?*

In accounting for the different diseases induced by v-erb A versus the RTH-TR mutants, one additional distinction between these alleles appears particularly significant; although the specific mutation may vary from kindred to kindred, each RTH-TRβ mutant typically has sustained a single genetic lesion, whereas v-erb A is heavily riddled with multiple mutations [Fig. 6.6(A)]. In addition to the C-terminal helix 12 deletion noted above, v-erb A has sustained an N-terminal deletion, a fusion with retroviral-derived GAG sequences, and 13 internal amino acid substitutions.[58] Many of these mutations operate together with the helix 12 deletion to further enhance the repressive properties of the v-erb A protein compared to the TRα1 progenitor.[59] For example, the K231N substitution in v-erb A reduces T3 binding affinity from that of TRα1, and the P363S and T370A substitutions favor the formation of v-Erb A homodimers (which preferentially recruit corepressors) relative to v-Erb A/RXR heterodimers (which bind to corepressors relatively poorly).[64,65] Interestingly, the helix 12 deletion in v-erb A not only inhibits corepressor release, but also permits v-erb A to bind to SMRT and N-CoR with equal efficiency.[64] Most likely these mutations are the result of a selective pressure for a strong dominant-negative phenotype during the evolution of v-erb A into an oncogene.

In addition to mutations that enhance transcriptional repression, four of the 13 amino acid substitutions found in v-erb A (R12H, Y32C, G61S, K78T) occur in its DNA recognition domain and alter its DNA binding specificity compared to that of TRα1 [Fig. 6.6(A)].[66] Converting these v-erb A sequences to more closely resemble the TRα1 sequence strongly impairs erythroid oncogenic transformation by v-erb A.[67] These results suggest that v-erb A is not simply an antimorph of its TRα1 progenitor, but may also participate in oncogenesis by repressing target genes not normally regulated by wild-type TRs. What genes might these be? Studies have identified a panel of genes that are induced in normal erythroid differentiation, but are repressed in AEV-infected erythroid progenitors cells, including α-globin, ala S, band 3, and carbonic anhydrase.[57,68] At least several of these genes contain binding sites for v-erb A and probably represent direct targets for v-erb A repression. A number of these genes are

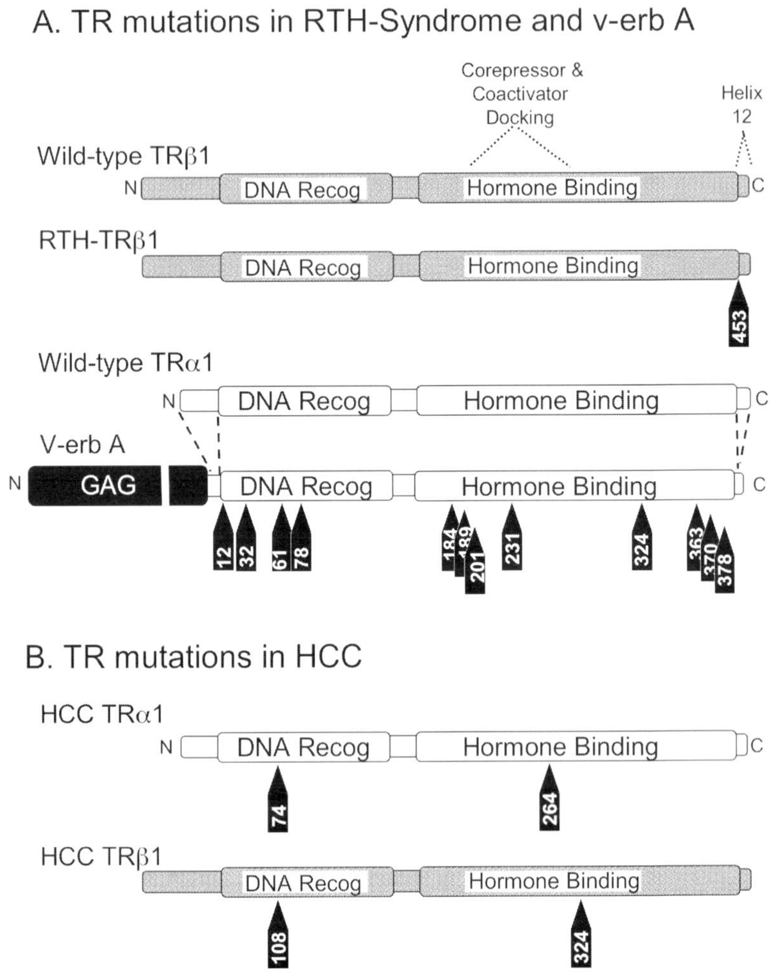

Fig. 6.6. Comparison of TR mutations in different diseases. (A) Comparison of RTH-Syndrome TR mutations to those in v-erb A The wild-type and mutant TRα and TRβ proteins are depicted as linear sequences from N- to C-termini, as in Fig. 6.1. Amino acid substitutions are shown as black flags; numbers refer to the amino acid position, beginning with the N-terminus of the wild-type TR sequences or with the first v-erb A-specific amino acid in the viral oncoprotein. N- and C-terminal deletions and a fusion of retroviral GAG sequences are also indicated for the v-Erb A sequence. Only one of many possible RTH-TRβ mutants is shown. (B) Comparison of HCC TR mutants. Two representative TR mutants isolated from HCC-derived tissues are depicted.

induced by T3 in normal erythroid cells, supporting the notion that part of v-erb A's function in blocking erythroid differentiation is mediated by simple interference with TR function. However, several of these genes are also regulated by all-*trans* retinoic acid, and it has been proposed that the change in DNA binding specificity by v-erb A has enhanced its ability to interfere with an additional panel of genes normally regulated by a second class of nuclear receptors, the retinoic acid receptors (RARs).[69,70] Although somewhat controversial, this proposal is consistent with observations that RARs are powerful modulators of normal hematopoietic differentiation. It is likely that both TR and RAR target genes are involved in the control of normal erythroid differentiation, and v-erb A, through changes in its DNA recognition, can suppress both classes of target genes and thereby help maintain the AEV-erythroid cells in an immature, proliferative state. It is probable that additional v-erb A target genes remain to be discovered; for example, there is evidence that certain actions of v-erb A in erythroid oncogenesis may be mediated through its ability to mimic glucocorticoid or estrogen receptor function.[57]

In addition to its role in erythroleukemia, v-erb A also extends the replicative capacity of fibroblasts in culture, a feature that may contribute to fibrosarcoma formation by AEV *in vivo*.[71] Intriguingly, the effects of v-erb A on fibroblasts also depends on the dominant-negative inhibition of both TRs and RARs, but by means of a distinct pathway from that in the erythroid cell. Normal fibroblast proliferation is positively regulated by the AP-1 protein c-Jun. TRs and RARs stimulate c-Jun transcriptional activation in the absence of hormone, but suppress it in the presence of the corresponding T3 or all-*trans* retinoic acid antagonists (i.e. AP-1 binding sites can act as negative-response elements). As a result, T3 or all-*trans* retinoic acid mediate anti-proliferative effects. When coexpressed with TR or RAR, v-erb A functions as a unliganded TR or RAR would, enhancing c-Jun function, but in both the absence and presence of hormone ligand.[71,72] It will be interesting to determine if the specific mutations in v-erb A that contribute to this ligand-independent, combinatorial stimulation of fibroblast cell proliferation are the same or different from the v-erb A mutations identified as important for erythroid oncogenesis, and if RTH-TRβ mutants can substitute for v-erb A in the fibroblast context.

In summary, we propose that simple dominant-negative mutations in TRs, such as those identified in RTH-Syndrome, result primarily in endocrine disorders. The accumulation of additional mutations that further alter the DNA specificity and transcriptional repression properties of these dominant-negative receptors, as are found in v-erb A, can confer oncogenic capabilities, at least in part by targeting genes not normally regulated by wild-type or RTH-TRs. In erythroleukemogenesis and in fibrosarcomas, these additional v-erb A targets include, but are not limited to genes normally regulated by RARs. Once unleashed in this fashion, the intrinsic oncogenic properties of v-erb A can be still further enhanced by additional factors, such as the presence of a second active oncogene or overexpression from a powerful viral promoter.

6.5 TRs, Corepressors, and Human Cancers

6.5.1 *Dominant-negative TRs and human neoplasia*

If the mutant TRs inherited in RTH-Syndrome are not causative agents of human neoplasia, might other TR mutants exist that are? Indeed, recent evidence has strongly implicated a subset of TR mutations in the initiation and/or progression of several human cancers. Unlike inherited RTH-Syndrome, these mutations are spontaneous and arise in the neoplasia itself. This phenomenon has been best documented for hepatocellular carcinomas (HCC), for renal clear cell carcinomas (RCCC), and for certain thyroid neoplasia, as described below.

6.5.2 *Mutated TRs and aberrant corepressor release in hepatocellular carcinoma (HCC)*

HCC is an significant cause of mortality worldwide and is increasing in incidence in the United States. Known risk factors include hepatitis B or C infection, aflatoxin exposure, and alcohol-related cirrhosis. Despite substantial recent progress in understanding HCC initiation and progression, many of the molecular mechanisms underlying this disease remain incompletely understood. It is therefore both intriguing and remarkable that approximately 70% of HCC tumors and HCC-derived cell lines have sustained mutations in the TRα locus, the TRβ locus, or both.[73,74] Furthermore, the bulk of these HCC-mutant TRs interfere with wild-type TR function in a dominant-negative fashion, and therefore share the same central molecular tenet as found in RTH-syndrome and in v-erb A mediated disease.[73,75]

In fact, the molecular features of the HCC-TR mutants are an interesting mix of the properties first described for the RTH-mutants and those first described for the v-erb A protein. Similar to RTH-TRβ mutants, many of the HCC-TRβ1 mutants analyzed are impaired in their ability to bind T3. As a result, these HCC-TRβ1 mutants retain corepressor and operate as dominant-negative inhibitors at moderate T3 levels, but release from corepressor, bind coactivators, and mediate transcriptional activity at high T3 concentrations.[73,75,76] In contrast with the HCC-TRβ1 mutants, but similar to v-erb A, many HCC-TRα1 mutants are relatively insensitive to hormone, and operate as dominant-negatives over a wide range of T3 concentrations.[75] Curiously, the HCC-TRα mutants can release from corepressor and recruit coactivator *in vitro* in response to sufficiently high T3 levels, so the molecular basis behind their constitutive dominant negative properties in cells remains incompletely defined.[75]

The HCC TR mutants are expressed in a more physiological context than is the v-erb A protein, and therefore provide further insight into the minimal requirements by which a TR might be converted into an oncoprotein. Unlike v-erb A, many of the HCC mutations are in TRβ, the same locus mutated in the non-oncogenic RTH-Syndrome, and are expressed from the same promoters and in the same tissues as are the RTH-Syndrome mutants. Unlike v-erb A, the HCC-TRβ mutations do not come bundled together with a second oncogene; any additional proto-oncogenes that may contribute to carcinogenesis are equally available to the RTH-TRβ mutants and to the HCC-TRβ lesions. So, quite clearly, there must be something intrinsic about the HCC-TR mutants that distinguishes them from the non-oncogenic RTH-TR mutants. The leading candidate for this intrinsic difference? Most of the HCC-mutants are the product of multiple mutations that map to both the hormone binding and the DNA recognition domains of the encoded receptor [Fig. 6.6(B)]. At least several of the HCC mutations detectably alter the DNA recognition specificity of the encoded receptor *in vitro*.[75] We therefore suggest that at least some of the HCC-TRs acquired oncogenic potential in much the same way as did v-erb A: by a modification of their target gene repertoire. Indeed, mice engineered to express v-erb A develop hepatocellular carcinoma, providing a strong biological as well as biochemical link between the v-erb A and the HCC-TR mutants.[60]

Despite the attractiveness of this simple unifying model by which TRs acquire oncogenic potential through changes in their DNA specificity, it must be noted that approximately 20% of HCC-TR mutants have either sustained only a single genetic lesion (typically within their hormone binding domain), or have sustained multiple mutations that nonetheless fail to encompass domains known to influence DNA recognition.[74] This minority class of HCC-TR mutants may acquire an altered target gene specificity by a means not yet fully understood, or may contribute to tumor growth in a fashion different from the receptor mutants that do acquire an altered DNA specificity. The presence of both TRα and TRβ mutations in many HCC tumors, and the divergent properties of these mutant isoforms in cell culture, suggest that the different mutated isoforms may make different contributions to the oncogenic phenotype. It is possible that an alteration in the DNA binding specificity of one isoform alleviates the requirement for an analogous alteration in the other. Although rare, other HCC-TR mutants lack strong dominant-negative properties when tested in cell culture or *in vitro*.[75] These mutants may exert other functions *in vivo*, or may represent the remnants of a hit-and-run mechanism by which a mutant receptor contributed to a specific stage of tumor initiation or progression, only to be later rendered inert by the accumulation of additional genetic lesions.

What are the likely targets of mutant TR function in HCC? Wild type TRs have been reported to exert growth suppressive effects in a number of cell types, and these effects have been attributed to a T3-dependent inhibition of the expression of cyclin D1 or (as noted above) of AP1 target genes.[71,77,78] The HCC-TR mutants may counteract these anti-proliferative effects of T3. At least, several of the HCC-TR mutants also share the ability of v-erb A to interfere with RAR function on model response elements, although the relevance to the HCC phenotype has not yet been investigated. Clearly, much remains to be determined, and a complete understanding of these issues, and the contributions of specific HCC-mutations to the acquisition and progression of the neoplastic state, will have to await the results of microarray, ChIP-on-chip, and transgenic animal studies.

6.5.3 *TR mutants and renal clear cell carcinomas*

Spontaneous TR mutations have also been identified in over 40% of the human renal clear cell carcinomas (RCCC) analyzed.[79] In many ways, the RCCC-TR mutants resemble the HCC-TR mutants: they occur in both TRα and TRβ alleles, most (but not all) RCCC-TR mutants possess two or more genetic lesions, and the majority of RCCC-TR mutants studied display dominant-negative properties in cell culture.[79] Despite these close phenotypic similarities, the specific TR amino acid mutations found in RCCC differ from those found in HCC.[79] This apparent lack of genetic congruence between the two diseases may simply reflect the limited number of mutations cataloged and available for comparison. Alternatively, it is possible that different TR mutations (and therefore different TR functions) are required for initiation or progression of RCCC versus HCC. In partial support of the latter hypothesis, v-erb A induces HCC, but not RCCC, in mice when linked to a ubiquitously expressed β-actin promoter.[60] Nonetheless, it must be noted that these same β-actin-v-erb A transgenic mice do not develop erythroleukemia, despite the prevalence of this malignancy in AEV-infected birds. Factors beyond the nature of the TR mutant itself must also influence the nature of the malignancy. It should also be remembered that there are unavoidable experimental biases in experiments of this type; rapidly lethal forms of neoplasias, for example, can obscure the occurrence of slower onset cancers.

6.5.4 *TR mutants and thyroid neoplasia*

In the thyroid, hormone status and cell proliferation are intertwined. For example, thyroid hyperplasia is a physiologically appropriate and non-malignant increase in thyroid mass that occurs in response to chronically low circulating thyroid hormone; in excess, this hypertrophy presents as an externally visible goiter.[1] Not surprisingly, the disruption of negative feedback T3/T4 sensing in RTH-Syndrome can also lead to chronic stimulation of the thyroid, again giving rise to hyperplasia and goiter.[31–33] Neither of these syndromes are considered neoplasia. However, other TR mutations are associated with overt thyroid malignancies; notably, these TR mutations are spontaneous and restricted to the tumor, rather than inherited as in RTH-Syndrome.[80] Over 90% of human thyroid papillary cancers analyzed have sustained mutations in the TRα locus and over 60% have sustained mutations in the TRβ locus. Suggestive of a role for these TR mutations in tumor progression, only 10% and 22% of thyroid adenomas, a more benign form of the disease, contained TRα or TRβ mutations, and no TR mutations were found in normal thyroids. The majority of the TR mutants characterized from overt thyroid malignancies are impaired for the T3-driven exchange of coactivator for corepressor. They display a dominant-negative phenotype and have sustained two or more mutations per allele.[77,81]

Therefore, the majority of TR mutations in thyroid papillary neoplasia conform to the two (or more) hit mutagenesis model proposed for erythroleukemia, HCC, and RCCC, and differ from the single-hit TR mutations that dominate in RTH-Syndrome. However, the division separating endocrinology and oncology may be more blurry in the thyroid than in the bone marrow, liver, or kidney. Mice engineered to express an RTH-Syndrome TRβ(fs394) allele (representing a frameshift mutation in helix 11) give rise to thyroid follicular carcinoma at a high incidence.[82] Therefore, even an RTH-TRβ allele bearing a single genetic lesion can be oncogenic in the thyroid. It may be that multiple changes in receptor function, rather than multiple mutations *per se*, are involved in unleashing the TR oncoprotein in this example. By altering the downstream reading frame, the fs394 mutation exerts broader effects on receptor structure and potentially more diverse effects on receptor function than would a single base substitution. Consistent with this "one big hit may mimic multiple small hits" conjecture, the fs394 mutation not only confers a strong dominant-negative activity on TRβ but also alters its specificity for SMRT versus N-CoR.[44] Also of interest, the ability of the TRβ(fs394) mutant to induce murine thyroid malignancies has been linked to its dominant negative inhibition of PPARγ, indicating that in the thyroid, as in erythroid cells, oncogenesis by mutant TRs involves interference with non-TR members of the nuclear receptor family.[77] Finally, it should be noted that thyroid malignancy was observed in mice homozygous or hemizygous for the TRβ(fs394) mutation;[82] heterozygous mice displayed T3 resistance, not neoplasia. The human index case for the fs394 mutation, also a heterozygote, similarly presented with RTH-Syndrome and goiter; no malignancy was reported.

Interestingly, the fs394 TRβ homozygous mice also present with TSH-omas.[77] These pituitary tumors are likely to reflect both direct effects of the mutant receptors on pituitary gene expression,[77] and the effects of chronic hormonal stimulation of the pituitary thyrotropes due to the severe disruption of the hypothalamus-pituitary-thyroid gland axis in these animals.

6.6 Other Human Diseases Due to Disruptions in Coregulator Release and Acquistion

6.6.1 *Human acute promyelocytic leukemia (APL), x-RARα fusion proteins, and corepressors*

Human APL is caused by an aberrant proliferation of immature promyelocytes, and is closely linked to chromosomal translocations that fuse ectopic protein coding sequences to the DNA and hormone binding domains of RARα (Fig. 6.7).[83,84] Different chromosomal translocations produce different "x-RARα" fusion proteins; PML-RARα, PLZF-RARα, NPM-RARα, NUMA-RARα, and STAT5b-RARα fusions have been identified.[83,84] Despite this structural diversity, all x-RARα fusions exhibit an impaired ability to release SMRT corepressor in response to physiological levels of all-*trans* retinoic acid (ATRA), and can interfere with wild type RARα function in a dominant-negative manner. The molecular basis behind this defective corepressor release remains somewhat mysterious, but reflects in part the ability of these "x" sequences to confer

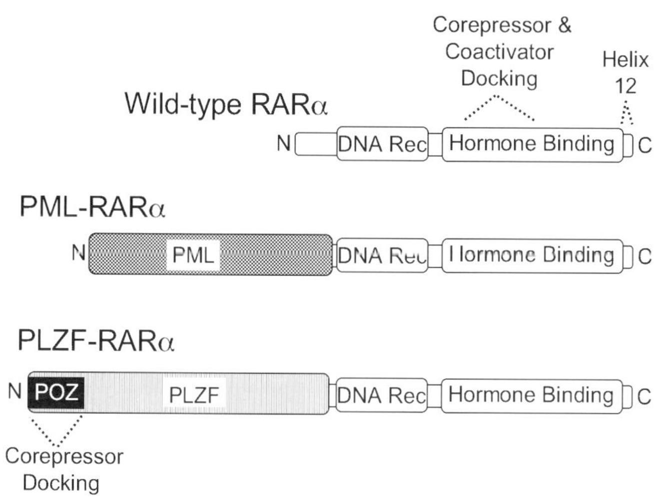

Fig. 6.7. RARα fusion proteins associated with APL. Wild-type RARα is depicted from N- to C-terminus; functional domains are indicated as in Fig. 6.1(A). Two of the x-RARα fusions characteristic of APL are shown below, representing PML- or PLZF fusions. The N-terminus of PLZF contains an additional corepressor docking site (denoted the "POZ" domain), that binds SMRT and N-CoR in a hormone-independent manner.

novel dimerization and oligomerization properties on the RARα moiety of the fusion.[83,84] By an unknown mechanism, the resulting homodimers and oligomers alter the ability of ATRA to operate the helix 12 toggle switch so as to displace corepressors.[85,86] PML-RARα, NPM-RARα, and NUMA-RARα eventually release SMRT and induce target gene expression at sufficiently high ATRA concentrations.[84] PLZF-RARα, in contrast, retains SMRT and acts as a dominant-negative inhibitor even at the highest ATRA levels; this reflects the presence of an ATRA-insensitive SMRT docking surface in PLZF that locks SMRT to the PLZF-RARα fusion even after SMRT is released from the RARα half of the fusion protein (Fig. 6.7).[87–91] Both PML and PLZF, and perhaps STAT5b, also recruit a cadre of additional histone deacetylases and other transcriptional coregulatory molecules that may further contribute to the dominant-negative properties of their corresponding x-RARα fusions.[84]

The SMRT recruitment properties of the various x-RARα fusions correlate remarkably well with their oncogenic properties. For example, PML-RARα fusions recruit corepressor and block APL cell differentiation under normal physiological conditions. Treatment of these cells with supraphysiological levels of ATRA induces the release of corepressor from the PML-RARα protein, expression of previously repressed target genes, differentiation of the immature myeloid cell into macrophages, and the clinical remission of leukemia.[83,84] PLZF-RARα fusions, which bind corepressor constitutively, produce hormone-refractory leukemias; so do PML-RARα fusions bearing secondary mutations that prevent corepressor release regardless of the ATRA concentration. Significantly, even these ATRA-resistant leukemias can be successfully treated with reagents that release corepressor by alternative means. Arsenic trioxide, for example, is used clinically as an adjunct to ATRA therapy, and induces the phosphorylation of SMRT, resulting in an ATRA-independent release of this corepressor from PML-RARα and an abortive differentiation of the leukemic cell.[92] Arsenic trioxide treatment also mediates an accelerated degradation of the PML-RARα protein itself, as well as additional, pro-apoptotic effects through the product of the untranslocated PML locus.[93,94]

What are the targets of x-RARα repression in APL cells? Various panels of genes that have been identified that are repressed by PML-RARα but induced in response to pharmacological levels of ATRA.[83] Some of these PML-RARα target genes play known roles in the differentiation or proliferation of myeloid cells and represent logical candidates for mediators of oncogenic transformation.[83] Others are less obvious in terms of their potential functions. Yet other key targets probably remain to be discovered. Not all of these target genes are normally regulated by RARα, and different x-RARα fusions appear to regulate different, if overlapping, sets of genes. Differences in the dimerization/oligomerization properties of the various x-RARα fusions probably contribute to their divergent gene recognition properties.[95,96] Therefore, as with the neoplastic TR mutants, the acquisition of oncogenic potential by x-RARα fusions appears to extend beyond simple dominant-negative inhibition of its immediate progenitor (in this case, RARα). It should also be noted that in common with TRs, wild type RARα can mediate both positive and negative gene regulation in response to ATRA; at least some of the dominant-negative actions of the x-RARα fusions in leukemia may be by preventing ATRA-mediated inhibition of negative response genes.

6.6.2 *Corepressors, Peroxisome-Proliferator Activator-Receptor (PPAR-γ), and inherited type II diabetes*

PPARγ is an adopted member of the orphan nuclear receptor subfamily and is an important regulator of energy storage in vertebrates.[97] PPARγ agonists promote adipocyte differentiation, cooperate with insulin to enhance glucose uptake, and are employed medically in the treatment of type II diabetes.[97] Autosomal dominant mutations in PPARγ have been found in inherited forms of human type II diabetes; affected individuals can also display an elevated risk for obesity and lipid dystrophy.[31,97] In parallel to the properties of the TRβ mutants in RTH-Syndrome, many of the PPARγ mutants in type II diabetes contain only a single genetic lesion per allele, are defective in the agonist-driven release of corepressor, and can function as dominant-negative inhibitors of wild type PPARγ function.[31] At the molecular level, the resemblance between these PPARγ mutants and the TRβ mutants in RTH-Syndrome can be remarkably close; for example, the same proline, a component of the hinge that allows helix 12 to reorient in response to agonists, is found mutated in a type II diabetes PPARγ allele and in several RTH-Syndrome TRβ alleles. However, yet other PPARγ mutants diverge from the RTH-TRβ model by exerting strong dominant-negative effects in the absence of DNA binding.[98] For a more detailed and eloquent description of the roles of corepressors and of PPARγ in metabolic disease, the reader is referred to several companion chapters in this book.

6.7 Summary and Future Prospectives

TRs play key roles in normal development and physiology, and it is not surprising that mutations in these receptors result in an assortment of human diseases. More remarkable, however, is the reoccurring and vital role of the SMRT and N-CoR corepressors in mediating these disorders. In virtually all cases known, ranging from RTH-Syndrome to human HCC, the predominant TR mutants function as dominant-negative inhibitors, and in most all of these examples, the dominant-negative activity reflects a failure of the mutant TR to release corepressor appropriately in response to T3. In the absence of any additional changes in TR function, this inappropriate corepressor retention produces endocrine disease. Accumulation of yet additional genetic changes can further expand the disease capabilities of these dominant-negative TR mutants to include neoplastic initiation and/or progression. In essence, these sequential changes in TRs in cancer cells are a microcosm of the larger phenomenon of tumor progression: an initiating event (acquisition of dominant-negative ability by TR mutant) is followed by additional mutations in the receptor that further enhance the oncogenic phenotype as the tumor develops. In this chapter, I have highlighted the contributions of the changes in DNA recognition to this progressive acquisition of oncogenic activity by dominant-negative TR mutants, but undoubtedly other processes also participate in unmasking the normally cryptic neoplastic properties of the wild type TRs. Much remains to be learned as to the specific roles these TR mutants play in establishing and maintaining the neoplastic phenotype, and if common or divergent mechanisms operate in erythroleukemia, HCC, RCCC, and in the thyroid malignancies.

Despite the corepressor-centric bias of the previous paragraph, I emphasize that an inherent symmetry applies to the TR mutants discussed here; the same molecular defects that impair corepressor release typically (perhaps inevitably) also impair coactivator recruitment. These defects in coactivator acquisition are likely to contribute to, or in some contexts may dominate, the biological effects of these TR mutants. One example addressed in this chapter is how the clinical bifurcation of RTH-Syndrome into generalized versus pituitary-specific extremes may reflect the ability of the corresponding mutant TRβ1 and TRβ2 receptors to bind coactivators under limiting T3 concentrations.

Finally, I briefly reviewed how defects in coactivator/corepressor exchange are not limited to TRs, but have also been discovered in other nuclear receptors and collectively account for a diverse assortment of human diseases. For example, dominant-negative mutations of PPARγ that are found in inherited forms of type II diabetes prevent proper release corepressor in response to agonist, and can even map to the same amino acid positions in PPARγ as do the TRβ lesions found in RTH-Syndrome. Similarly, dominant-negative x-RARα fusions play causal roles in human APL, and pharmacological strategies that force corepressor release by these mutant RARα proteins produced clinically favorable outcomes. Do yet other nuclear receptors acquire disease potential by losing the ability to release corepressor? Can methodologies that pry corepressors from the grip of these nuclear receptors be used to treat these diseases? Future research will be required to reveal just how broadly these paradigms apply.

Acknowledgments

The author sincerely thanks Brenda Mengeling, Teri Phan, and Meghan Dukerich for many helpful suggestions and comments.

References

1. Larsen PR, Davies TF, Schlumberger MJ, *et al.*, Thyroid physiology and diagnostic evaluation of patients with thyroid disorders, in Larsen PR, Kronenberg HM, Melmed S, Polonsky KS (eds.), *Williams Textbook of Endocrinology*, Saunders, Philadelphia, pp. 331–373, 2003.
2. Bassett JH, Harvey CB, Williams GR, Mechanisms of thyroid hormone receptor-specific nuclear and extra nuclear actions, *Mol Cell Endocrinol* **213**:1–11, 2003.
3. O'Shea PJ, Williams GR, Insight into the physiological actions of thyroid hormone receptors from genetically modified mice, *J Endocrinol* **175**:553–570, 2002.
4. Wondisford FE, Thyroid hormone action: Insight from transgenic mouse models, *J Investig Med* **51**:215–220, 2003.

5. Rastinejad F, Retinoid X receptor and its partners in the nuclear receptor family, *Curr Opin Struct Biol* **11**:33–38, 2001
6. Forman BM, Casanova J, Raaka BM, *et al.*, Half-site spacing and orientation determines whether thyroid hormone and retinoic acid receptors and related factors bind to DNA response elements as monomers, homodimers, or heterodimers, *Mol Endocrinol* **6**:429–442, 1992.
7. Lazar MA, Berrodin TJ, Harding HP, Differential DNA binding by monomeric, homodimeric, and potentially heteromeric forms of the thyroid hormone receptor, *Mol Cell Biol* **11**:5005–5015, 1991.
8. Lee S, Privalsky ML, Heterodimers of retinoic acid receptors and thyroid hormone receptors display unique combinatorial regulatory properties, *Mol Endocrinol* **19**:863–878, 2005.
9. Mengeling BJ, Pan F, Privalsky ML, Novel mode of deoxyribonucleic acid recognition by thyroid hormone receptors: Thyroid hormone receptor β-isoforms can bind as trimers to natural response elements comprised of reiterated half-sites, *Mol Endocrinol* **19**:35–51, 2005.
10. Tini M, Tsui LC, Giguere V, Heterodimeric interaction of the retinoic acid and thyroid hormone receptors in transcriptional regulation on the γF-crystallin everted retinoic acid response element, *Mol Endocrinol* **8**:1494–1506, 1994.
11. Wagner RL, Apriletti JW, McGrath ME, *et al.*, A structural role for hormone in the thyroid hormone receptor, *Nature (London)* **378**:690–697, 1995.
12. Glass CK, Rosenfeld MG, The coregulator exchange in transcriptional functions of nuclear receptors, *Genes Dev* **14**:121–141, 2000.
13. Privalsky ML, The role of corepressors in transcriptional regulation by nuclear hormone receptors, *Annu Rev Physiol* **66**:315–360, 2004.
14. Feng X, Jiang Y, Meltzer P, *et al.*, Thyroid hormone regulation of hepatic genes *in vivo* detected by complementary DNA microarray, *Mol Endocrinol* **14**:947–955, 2000.
15. White JH, Fernandes I, Mader S, *et al.*, Corepressor recruitment by agonist-bound nuclear receptors, *Vitam Horm* **68**:123–143, 2004.
16. Ortiga-Carvalho TM, Shibusawa N, Nikrodhanond A, *et al.*, Negative regulation by thyroid hormone receptor requires an intact coactivator-binding surface, *J Clin Invest* **115**:2517–2523, 2005.
17. Tagami T, Park Y, Jameson JL, Mechanisms that mediate negative regulation of the thyroid-stimulating hormone α gene by the thyroid hormone receptor, *J Biol Chem* **274**:22345–22353, 1999.
18. Xu W, Chen H, Du K, *et al.*, A transcriptional switch mediated by cofactor methylation, *Science* **294**:2507–2511, 2001.
19. Jepsen K, Rosenfeld MG, Biological roles and mechanistic actions of co-repressor complexes, *J Cell Sci* **115**:689–698, 2002.

20. Jones PL, Shi YB, N-CoR-HDAC corepressor complexes: Roles in transcriptional regulation by nuclear hormone receptors, *Curr Top Microbiol Immunol* **274**:237–268, 2003.
21. Ordentlich P, Downes M, Evans RM, Corepressors and nuclear hormone receptor function, *Curr Top Microbiol Immunol* **254**:101–116, 2001.
22. Goodson M, Jonas BA, Privalsky MA, Corepressors: Custom tailoring and alterations while you wait, *Nucl Recept Signal* **3**:e003, 2005.
23. Dilworth FJ, Chambon P, Nuclear receptors coordinate the activities of chromatin remodeling complexes and coactivators to facilitate initiation of transcription, *Oncogene* **20**:3047–3054, 2001.
24. Lee JW, Lee YC, Na SY, *et al.*, Transcriptional coregulators of the nuclear receptor superfamily: Coactivators and corepressors, *Cell Mol Life Sci* **58**:289–297, 2001.
25. Tsai CC, Fondell JD, Nuclear receptor recruitment of histone-modifying enzymes to target gene promoters, *Vitam Horm* **68**:93–122, 2004.
26. McKenna NJ, Xu J, Nawaz Z, *et al.*, Nuclear receptor coactivators: Multiple enzymes, multiple complexes, multiple functions, *J Steroid Biochem Mol Biol* **69**:3–12, 1999.
27. Weiss RE, Ramos HE, Thyroid hormone receptor subtypes and their interaction with steroid receptor coactivators, *Vitam Horm* **68**:185–207, 2004.
28. Alenghat T, Yu J, Lazar MA, The N-CoR complex enables chromatin remodeler SNF2H to enhance repression by thyroid hormone receptor, *EMBO J* **25**:3966–3974, 2006.
29. Kraus WL, Wong J, Nuclear receptor-dependent transcription with chromatin. Is it all about enzymes? *Eur J Biochem* **269**:2275–2283, 2002.
30. Perissi V, Aggarwal A, Glass CK, *et al.*, A corepressor/coactivator exchange complex required for transcriptional activation by nuclear receptors and other regulated transcription factors, *Cell* **116**:511–526, 2004.
31. Chatterjee VK, Resistance to thyroid hormone, and peroxisome-proliferator-activated receptor gamma resistance, *Biochem Soc Trans* **29**:227–231, 2001.
32. Refetoff S, Resistance to thyroid hormone with and without receptor gene mutations, *Ann Endocrinol (Paris)* **64**:23–25, 2003.
33. Yen PM, Molecular basis of resistance to thyroid hormone, *Trends Endocrinol Metab* **14**:327–333, 2003.
34. Safer JD, Cohen RN, Hollenberg AN, *et al.*, Defective release of corepressor by hinge mutants of the thyroid hormone receptor found in patients with resistance to thyroid hormone, *J Biol Chem* **273**:30175–30182, 1998.
35. Tagami T, Jameson JL, Nuclear corepressors enhance the dominant negative activity of mutant receptors that cause resistance to thyroid hormone, *Endocrinology* **139**:640–650, 1998.

36. Yoh SM, Chatterjee VK, Privalsky ML, Thyroid hormone resistance syndrome manifests as an aberrant interaction between mutant T3 receptors and transcriptional corepressors, *Mol Endocrinol* **11**:470–480, 1997.
37. Lin BC, Hong SH, Krig S, *et al.*, A conformational switch in nuclear hormone receptors is involved in coupling hormone binding to corepressor release, *Mol Cell Biol* **17**:6131–6138, 1997.
38. Tagami T, Gu WX, Peairs PT, *et al.*, A novel natural mutation in the thyroid hormone receptor defines a dual functional domain that exchanges nuclear receptor corepressors and coactivators, *Mol Endocrinol* **12**:1888–1902, 1998.
39. Collingwood TN, Adams M, Tone Y, *et al.*, Spectrum of transcriptional, dimerization, and dominant negative properties of twenty different mutant thyroid hormone β-receptors in thyroid hormone resistance syndrome, *Mol Endocrinol* **8**:1262–1277, 1994.
40. Forman BM, Yang CR, Au M, *et al.*, A domain containing leucine-zipper-like motifs mediate novel *in vivo* interactions between the thyroid hormone and retinoic acid receptors, *Mol Endocrinol* **3**:1610–1626, 1989.
41. Matsushita A, Misawa H, Andoh S, *et al.*, Very strong correlation between dominant negative activities of mutant thyroid hormone receptors and their binding avidity for corepressor SMRT, *J Endocrinol* **167**:493–503, 2000.
42. Nagaya T, Fujieda M, Seo H, Requirement of corepressor binding of thyroid hormone receptor mutants for dominant negative inhibition, *Biochem Biophys Res Commun* **247**:620–623, 1998.
43. Yoh SM, Privalsky ML, Resistance to thyroid hormone (RTH) syndrome reveals novel determinants regulating interaction of T3 receptor with corepressor, *Mol Cell Endocrinol* **159**:109–124, 2000.
44. Wu SY, Cohen RN, Simsek E, *et al.*, A novel thyroid hormone receptor-β mutation that fails to bind nuclear receptor corepressor in a patient as an apparent cause of severe, predominantly pituitary resistance to thyroid hormone, *J Clin Endocrinol Metab* **91**:1887–1895, 2006.
45. Flamant F, Gauthier K, Samarut J, Thyroid hormones signaling is getting more complex: STORMs are coming, *Mol Endocrinol* **21**:321–333, 2007.
46. Liu Y, Takeshita A, Misiti S, *et al.*, Lack of coactivator interaction can be a mechanism for dominant negative activity by mutant thyroid hormone receptors, *Endocrinology* **139**:4197–4204, 1998.
47. Cheng SY, Isoform-dependent actions of thyroid hormone nuclear receptors: Lessons from knockin mutant mice, *Steroids* **70**:450–454, 2005.
48. Zhang XY, Kaneshige M, Kamiya Y, *et al.*, Differential expression of thyroid hormone receptor isoforms dictates the dominant negative activity of mutant β receptor, *Mol Endocrinol* **16**:2077–2092, 2002.

49. O'Shea PJ, Bassett JH, Cheng SY, *et al.*, Characterization of skeletal phenotypes of TRα1 and TRβ mutant mice: Implications for tissue thyroid status and T3 target gene expression, *Nucl Recept Signal* **4**:e011, 2006.
50. Ying H, Araki O, Furuya F, *et al.*, Impaired adipogenesis caused by a mutated thyroid hormone α1 receptor, *Mol Cell Biol* **27**:2359–2371, 2007.
51. Safer JD, Langlois MF, Cohen R, *et al.*, Isoform variable action among thyroid hormone receptor mutants provides insight into pituitary resistance to thyroid hormone, *Mol Endocrinol* **11**:16–26, 1997.
52. Oberste-Berghaus C, Zanger K, Hashimoto K, *et al.*, Thyroid hormone-independent interaction between the thyroid hormone receptor β2 amino terminus and coactivators, *J Biol Chem* **275**:1787–1792, 2000.
53. Tian H, Mahajan MA, Wong CT, *et al.*, The N-Terminal A/B domain of the thyroid hormone receptor-β2 isoform influences ligand-dependent recruitment of coactivators to the ligand-binding domain, *Mol Endocrinol* **20**:2036–2051, 2006.
54. Yang Z, Privalsky ML, Isoform-specific transcriptional regulation by thyroid hormone receptors: Hormone-independent activation operates through a steroid receptor-mode of coactivator interaction, *Mol Endocrinol* **15**:1170–1185, 2001.
55. Wan W, Farboud B, Privalsky ML, Pituitary resistance to thyroid hormone syndrome is associated with T3 receptor mutants that selectively impair β2 isoform function, *Mol Endocrinol* **19**:1529–1542, 2005.
56. Ono S, Schwartz ID, Mueller OT, *et al.*, Homozygosity for a dominant negative thyroid hormone receptor gene responsible for generalized resistance to thyroid hormone, *J Clin Endocrinol Metab* **73**:990–994, 1991.
57. Beug H, Bauer A, Dolznig H, *et al.*, Avian erythropoiesis and erythroleukemia: Towards understanding the role of the biomolecules involved, *Biochim Biophys Acta* **1288**:M35–M47, 1996.
58. Privalsky ML, V-erb A, nuclear hormone receptors, and oncogenesis, *Biochim Biophys Acta* **1114**:51–62, 1992.
59. Zenke M, Munoz A, Sap J, *et al.*, V-erb A oncogene activation entails the loss of hormone-dependent regulator activity of c-erb A, *Cell* **61**:1035–1049, 1990.
60. Barlow C, Meister B, Lardelli M, *et al.*, Thyroid abnormalities and hepatocellular carcinoma in mice transgenic for v-erb A, *EMBO J* **13**:4241–4250, 1994, 1991.
61. Gandrillon O, Jurdic P, Pain B, *et al.*, Expression of the v-erb A product, an altered nuclear hormone receptor, is sufficient to transform erythrocytic cells *in vitro*, *Cell* **58**:115–121, 1989.
62. Mosesson Y, Yarden Y, Oncogenic growth factor receptors: Implications for signal transduction therapy, *Semin Cancer Biol* **14**:262–270, 2004.

63. O'Shea PJ, Bassett JH, Sriskantharajah S, *et al.*, Contrasting skeletal phenotypes in mice with an identical mutation targeted to thyroid hormone receptor α1 or β, *Mol Endocrinol* **19**:3045–3059, 2005.
64. Lee S, Privalsky ML, Multiple mutations contribute to repression by the v-Erb A oncoprotein, *Oncogene* **24**:6737–6752, 2005.
65. Zubkova I, Subauste JS, V-erb A homodimers mediate the potent dominant negative activity of V-erb A on everted repeats, *Mol Biol Rep* **31**:131–137, 2004.
66. Judelson C, Privalsky ML, DNA recognition by normal and oncogenic thyroid hormone receptors. Unexpected diversity in half-site specificity controlled by non-zinc-finger determinants, *J Biol Chem* **271**:10800–10805, 1996.
67. Bonde BG, Sharif M, Privalsky ML, Ontogeny of the V-erb A oncoprotein from the thyroid hormone receptor: An alteration in the DNA binding domain plays a role crucial for V-erb A function, *J Virol* **65**:2037–2046, 1991.
68. Rietveld LE, Caldenhoven E, Stunnenberg HG, Avian erythroleukemia: A model for corepressor function in cancer, *Oncogene* **20**:3100–3109, 2001.
69. Sharif M, Privalsky ML, V-erb A oncogene function in neoplasia correlates with its ability to repress retinoic acid receptor action, *Cell* **66**:885–893, 1991.
70. Schroeder C, Gibson L, Zenke M, *et al.*, Modulation of normal erythroid differentiation by the endogenous thyroid hormone and retinoic acid receptors: A possible target for V-erb A oncogene action, *Oncogene* **7**:217–227, 1992.
71. Desbois C, Pain B, Guilhot C, *et al.*, V-erb A oncogene abrogates growth inhibition of chicken embryo fibroblasts induced by retinoic acid, *Oncogene* **6**:2129–2135, 1991.
72. Sharif M, Privalsky ML, V-erb A and c-erb A proteins enhance transcriptional activation by c-Jun, *Oncogene* **7**:953–960, 1992.
73. Lin KH, Zhu XG, Hsu HC, *et al.*, Dominant negative activity of mutant thyroid hormone α1 receptors from patients with hepatocellular carcinoma, *Endocrinology* **138**:5308–5315, 1997.
74. Lin KH, Shieh HY, Chen SL, *et al.*, Expression of mutant thyroid hormone nuclear receptors in human hepatocellular carcinoma cells, *Mol Carcinog* **26**:53–61, 1999.
75. Chan IH, Privalsky ML, Thyroid hormone receptors mutated in liver cancer function as distorted antimorphs, *Oncogene* **25**:3576–3588, 2006.
76. Lin KH, Wu YH, Chen SL, Impaired interaction of mutant thyroid hormone receptors associated with human hepatocellular carcinoma with transcriptional coregulators, *Endocrinology* **142**:653–662, 2001.
77. Cheng SY, Thyroid hormone receptor mutations and disease: Beyond thyroid hormone resistance, *Trends Endocrinol Metab* **16**:176–182, 2005.

78. Gonzalez-Sancho JM, Garcia V, Bonilla F, *et al.*, Thyroid hormone receptors/THR genes in human cancer, *Cancer Lett* **192**:121–132, 2003.
79. Kamiya Y, Puzianowska-Kuznicka M, McPhie P, *et al.*, Expression of mutant thyroid hormone nuclear receptors is associated with human renal clear cell carcinoma, *Carcinogenesis* **23**:25–33, 2002.
80. Cheng SY, Abnormalities of nuclear receptors in thyroid cancer, *Cancer Treat Res* **122**:165–178, 2004.
81. Puzianowska-Kuznicka M, Krystyniak A, Madej A, *et al.*, Functionally impaired TR mutants are present in thyroid papillary cancer, *J Clin Endocrinol Metab* **87**:1120–1128, 2002.
82. Suzuki H, Willingham MC, Cheng SY, Mice with a mutation in the thyroid hormone receptor β gene spontaneously develop thyroid carcinoma: A mouse model of thyroid carcinogenesis, *Thyroid* **12**:963–969, 2002.
83. Insinga A, Pelicci PG, Inucci S, Leukemia-associated fusion proteins. Multiple mechanisms of action to drive cell transformation, *Cell Cycle* **4**:67–69, 2005.
84. Redner RL, Variations on a theme: The alternate translocations in APL, *Leukemia* **16**:1927–1932, 2002.
85. Lin RJ, Evans RM, Acquisition of oncogenic potential by RAR chimeras in acute promyelocytic leukemia through formation of homodimers, *Mol Cell* **5**:821–830, 2000.
86. Minucci S, Maccarana M, Cioce M, *et al.*, Oligomerization of RAR and AML1 transcription factors as a novel mechanism of oncogenic activation, *Mol Cell* **5**:811–820, 2000.
87. He LZ, Guidez F, Tribioli C, *et al.*, Distinct interactions of PML-RARα and PLZF-RARα with co-repressors determine differential responses to RA in APL, *Nat Genet* **18**:126–135, 1998.
88. Hong SH, David G, Wong CW, *et al.*, SMRT corepressor interacts with PLZF and with the PML-retinoic acid receptor α (RARα) and PLZF-RARα oncoproteins associated with acute promyelocytic leukemia, *Proc Natl Acad Sci USA* **94**:9028–9033, 1997.
89. Grignani F, De Matteis S, Nervi C, *et al.*, Fusion proteins of the retinoic acid receptor-α recruit histone deacetylase in promyelocytic leukaemia, *Nature (London)* **391**:815–818, 1998.
90. Guidez F, Ivins S, Zhu J, *et al.*, Reduced retinoic acid-sensitivities of nuclear receptor corepressor binding to PML- and PLZF-RARα underlie molecular pathogenesis and treatment of acute promyelocytic leukemia, *Blood* **91**:2634–2642, 1998.
91. Lin RJ, Nagy L, Inoue S, *et al.*, Role of the histone deacetylase complex in acute promyelocytic leukaemia, *Nature (London)* **391**:811–814, 1998).
92. Hong SH, Yang Z, Privalsky ML, Arsenic trioxide is a potent inhibitor of the interaction of SMRT corepressor with its transcription factor partners,

including the PML-RARα oncoprotein found in human acute promyelocytic leukemia, *Mol Cell Biol* **21**:7172–7182, 2001
93. Chen Z, Zhao WL, Shen ZX, *et al.*, Arsenic trioxide and acute promyelocytic leukemia: Clinical and biological, *Curr Top Microbiol Immunol* **313**:129–144, 2007.
94. Hayakawa F, Privalsky ML, Phosphorylation of PML by mitogen-activated protein kinases plays a key role in arsenic trioxide-mediated apoptosis, *Cancer Cell* **5**:389–401, 2004.
95. Hauksdottir H, Privalsky ML, DNA recognition by the aberrant retinoic acid receptors implicated in human acute promyelocytic leukemia, *Cell Growth Differ* **12**:85–98, 2001.
96. Zhou J, Peres L, Honore N, *et al.*, Dimerization-induced corepressor binding and relaxed DNA-binding specificity are critical for PML/RARα-induced immortalization, *Proc Natl Acad Sci USA* **103**:9238–9243, 2006.
97. Semple RK, Chatterjee VK, O'Rahilly S, PPARγ and human metabolic disease, *J Clin Invest* **116**:581–589, 2006.
98. Agostini M, Schoenmakers E, Mitchell C, *et al.*, Non-DNA binding dominant-negative, human PPARγ mutations cause lipodystrophic insulin resistance, *Cell Metab* **4**:303–311, 2006.

Chapter 7

Androgen Receptor Coactivators in Prostate Cancer

Nancy L. Weigel and Irina U. Agoulnik

Prostate cancer is an androgen dependent disease. Consequently, the mechanism of action of the androgen receptor (AR), the factors that facilitate androgen receptor action, and means to block AR action are of major interest in combating this disease. Studies in the last dozen years have revealed that a remarkable number of proteins and protein complexes interact with AR, thus facilitating its actions as a regulator of transcription. These proteins, termed coactivators, often have intrinsic enzymatic activity or act to recruit other proteins with enzymatic activities that modify chromatin associated proteins, as well as AR, other coactivators and polymerase. Many candidate coactivators have been identified through interaction and overexpression studies. The contributions of some of these to AR activity and prostate cancer cell growth have been further examined in RNA interference studies. This chapter summarizes the current evidence for the roles for these coactivators in prostate cancer based on their contributions to AR action, cell growth and their expression levels in prostate tumors.

7.1 Introduction

The prostate is an androgen dependent tissue and prostate cancer is an androgen dependent disease.[1] The actions of androgens (testosterone and dihydrotestosterone [DHT]) are mediated by the androgen receptor (AR), a member of the nuclear receptor superfamily. Primary tumors that have not extended beyond the prostate capsule can often be treated successfully by surgical removal of the prostate. However, invasive tumors and metastatic prostate cancer are treated by some form of androgen ablation. Although effective initially, the tumors typically become resistant within a couple of years and they relapse again. There is abundant evidence suggesting that many of these androgen ablation resistant tumors remain AR dependent. Most of the tumors are detected by an elevation in serum prostate specific antigen (PSA), an androgen regulated protease produced by prostate cells. A subset of other androgen regulated genes is also expressed. Remarkably, the most consistent change in mRNA in recurrent tumors compared to androgen dependent tumors is the elevation of AR mRNA.[2] Finally, several investigators have shown that AR positive androgen independent prostate cancer cell lines continue to require AR for growth. The inhibition of AR function by microinjection of an AR antibody inhibits cell growth.[3] Furthermore, the proliferation of androgen independent C4-2 prostate cancer cells grown in androgen depleted medium is abrogated by the elimination of AR using siRNA.[4] Consequently, there is a great deal of interest in understanding the mechanism of action of AR in androgen dependent and androgen independent prostate cancer and also in identifying new therapeutic targets. AR action is dependent upon a number of coactivator complexes and their expression, regulation, and contribution to AR action is an active area of research.

7.2 Mechanism of AR Action

Inactive AR is localized to the cytoplasm and is associated with complexes of chaperone proteins including hsp90 and immunophilins. Hormone binding leads to the dissociation of these complexes, dimerization of the receptor, nuclear localization and binding to target sites in the nucleus (Fig. 7.1). The best characterized binding sites are sequence specific hormone response elements. The agonist bound receptors recruit a series of coactivator complexes containing a variety of enzyme activities including histone acetyltransferases (HAT), methylases, ubiquitin ligases and kinases that modify chromatin and proteins associated with chromatin in order to induce transcription of target genes.[5,6] AR can also induce transcription by binding to other transcription factors rather than binding directly to DNA. Finally, AR also functions as a repressor of transcription. The requirements for coactivators in these latter activities are less well characterized.

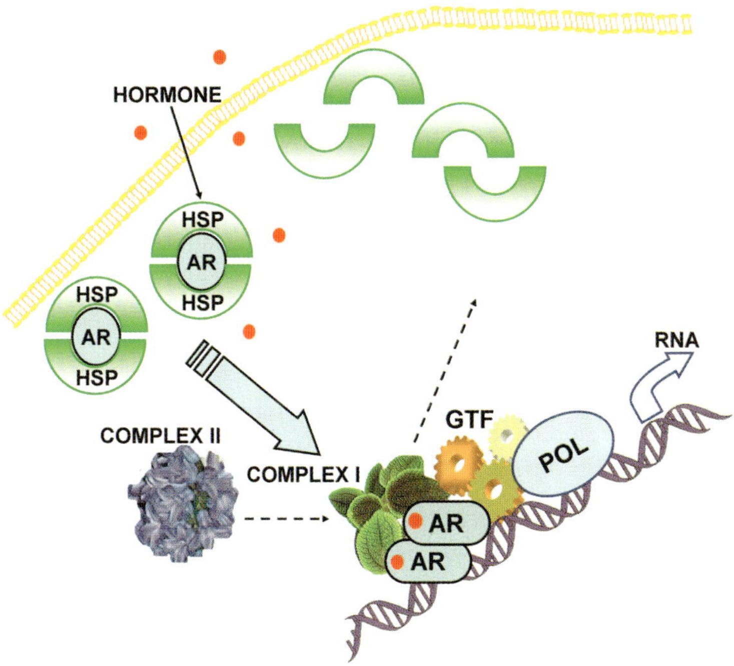

Fig. 7.1. Mechanism of Androgen Receptor Action. In the absence of hormone, androgen receptor (AR) monomers are associated with heat shock protein/chaperone (hsp) complexes. Hormone binding induces conformational changes in the receptor which result in dissociation of heat shock complexes, receptor dimerization, nuclear uptake and binding to DNA. The receptors than recruit a series of coactivator complexes, which modify histones and other proteins, recruit general transcription factors (GTF) and RNA Polymerase II (Pol) to stimulate transcription.

7.3 Structure of Androgen Receptor

Similar to other steroid receptors, AR contains a carboxyl terminal hormone binding domain and a DNA binding domain with two Zn^{++} binding motifs (Fig. 7.2).[7] The amino-terminus of AR is relatively long (>500 aa). Unique to AR, it contains a poly-Gln repeat of variable length (typically 18–24). Receptors with shorter poly-Gln repeats are somewhat more active whereas very large expansions of this repeat (>40) cause spinobulbar muscular atrophy. As a result, the length of individual ARs is variable. The conventional numbering of amino acid residues in AR assumes a total length of 919 amino acids and 21 glutamines in the polyglutamine

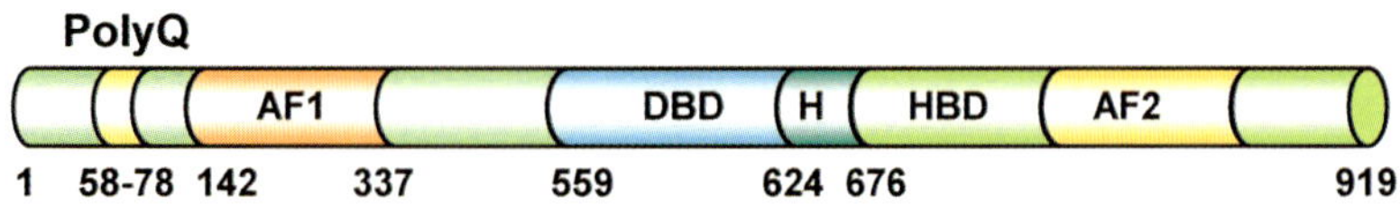

Fig. 7.2. Androgen receptor structure. The androgen receptor is comprised of a carboxyl terminal hormone binding domain (HBD) linked through a hinge region (H) to the DNA binding domain (DBD) that, in turn, is linked to the largely unstructured amino-terminus. The amino terminus contains a variable length polyglutamine tract, which contributes to receptor activity and activation function 1 (AF1), a region that recruits coactivators and is required for optimal transcriptional activity. A weaker activation function, AF2, is located in the carboxyl terminal hormone binding domain. The numbers denote the boundaries of the functional domains assuming that the receptor contains 21 glutamines in its glutamine tract.

stretch. Both the amino terminus and the hormone binding domain contain regions important for transcriptional activation termed AF-1 and AF-2 respectively. The relative importance of these regions is receptor and context dependent. Although AF-2 is the primary activation domain and coactivator interaction region for estrogen receptor α (ERα), AF-1 in AR is the strongest contributor to transcriptional activation. Whereas hormone induces receptor dimerization through the hormone binding domains of many steroid receptors, there is strong evidence that hormone binding induces an interaction between a region in the amino terminus of AR and its hormone binding domain. This interaction may diminish coactivator accessibility to the hormone binding domain.

7.4 Coactivators in AR Action

AR recruits a series of coactivator complexes to regulate the transcription of target genes. Coactivators originally were identified as proteins that interact with receptors and, when overexpressed, enhanced transcription. The primary coactivators often form multi-protein complexes and overexpression of these associated proteins also stimulates receptor activity. Thus, these proteins are also considered to be coactivators. Studies using chromatin immunoprecipitation (ChIP) assays to identify hormone dependent recruitment of proteins to AR regulated genes have shown that a remarkable variety of proteins is recruited to DNA in response to hormone. Many of the proteins have intrinsic enzyme activities whereas others serve as scaffolds to promote interactions between multiple proteins. Some of the coactivators have a relatively limited range of targets (for example, nuclear receptors and a few other transcription factors) whereas others have much broader roles. Several proteins, which have been termed coactivators, have well established roles independent of their regulation of transcription (for example, kinases). Coactivators that are recruited to target genes have been termed type I coregulators; other proteins whose overexpression enhances overall AR activity, but are not necessarily recruited to chromatin are termed type 2 coregulators. They have diverse functions including enhancing AR stability and, consequently, AR expression.

One of the key questions in the field is the identification of the regulatory mechanisms that determine which factor will recruit the coactivator from a limited pool of coactivators. Thus, proteins whose overexpression alters AR function are potential candidate coactivators. Additional studies are required to determine whether and under what circumstances the endogenous protein contributes to AR action.

7.5 Coactivators in Prostate Cancer

Because of the pivotal role of AR in prostate cancer, there is a great deal of interest in measuring coactivator levels in prostate cancer samples and in assessing the contributions of the coactivators in AR action and in prostate cancer. Since none of the candidate coactivators is simply an AR coactivator, caution is needed in interpreting the studies of coactivators in prostate cancer. Many of the coactivators have very important functions independent of their actions in AR signaling. This chapter will highlight what is known about the role of coactivators in AR action in prostate cancer as well as coactivator expression in prostate cancers. A number of early studies sought to study coactivator levels at the mRNA level, in part because of the lack of sensitive, specific antibodies for newly identified coactivators. In most cases, the comparisons have been done on mRNA isolated from normal versus tumor tissue. Because of the different proportions of epithelial and stromal cells, it is difficult to determine whether the observed changes are cancer specific. Coactivator expression is, in some cases, also regulated at the post-transcriptional level. In some cases, the regulation is androgen dependent. For example, androgens reduce the level of p300 protein with no effect on the mRNA level.[8] The agreement between RNA and protein studies has not been particularly good. Because the immunohistochemical studies can measure expression in specific cell populations, they are likely to be more accurate in determining whether an individual coactivator is differentially expressed in tumor versus normal epithelial cells. Consequently, we have summarized the immunohistochemical evidence for differential expression. The coactivators are extensively posttranslationally modified. As a result, there is an additional level of regulation that has not yet been evaluated by immunohistochemical studies. The study of coactivators in steroid hormone action is a relatively new field. The first steroid receptor coactivator, SRC-1 (steroid receptor coactivator-1), was reported in 1995[9] and more than 250 proteins have been identified in the last decade. Accordingly, our understanding of coactivator action is incomplete.

7.5.1 *Histone Acetyl-transferases (HATs) as AR coactivators*

Many of the AR coactivators have intrinsic HAT activity and/or act to recruit other HATs to AR.[10] Originally characterized as proteins that acetylate specific lysines in histones, more recent studies have shown that many other proteins are substrates for these enzymes including other coactivators and AR, itself. Acetylation of AR on lysines in the hinge region of AR enhances AR mediated transactivation of selected target genes.[11] As described in more detailed below, some HATs stimulate AR activity not only by modifying histones, but also by directly acetylating AR.

7.5.2 *The p160 coactivator family*

The first to be identified and the best characterized steroid receptor coactivators, are the p160 coactivators. The proteins were identified and named by multiple groups and alternate names are shown in parentheses. This family is comprised of three structurally related proteins, SRC-1, SRC-2 (GRIP1,TIF2), and SRC-3 (p/CIP,AIB1,ACTR,RAC3,TRAM-1) with molecular weights of approximately 160 kDa.[10] They potentiate the activities of all steroid receptors; SRC-1 and SRC-2 also coactivate a limited set of other transcription factors. In contrast, *in vitro* studies as well as mouse knock out studies show that SRC-3 is a major factor in a wide variety of signaling pathways. All three proteins recruit HATs; SRC-1 and SRC-3 have intrinsic HAT activity. Thus, this family of proteins is important for facilitating the acetylation of histones and other associated proteins. All three members contain multiple LXXLL motifs, which interact with the hormone binding domains of the steroid receptors. However, AR is an exception. The primary interaction is between a glutamine rich region in the carboxyl terminal region of the coactivators and the amino-terminal AF-1 of AR.[12] Their expression and activities in prostate cancer and prostate cancer cells have been the topic of many studies.

7.5.2.1 *SRC-1*

Overexpression of SRC-1 enhances AR transcriptional activity and increases its sensitivity to hormone suggesting that it is an AR coactivator that could potentiate AR activity in the androgen depleted environment produced by androgen deprivation therapy.[4] The first evidence that SRC-1 plays an important physiological role in AR action in prostate was derived from the phenotypic analysis of SRC-1 null mice.[13] Androgen induced prostate growth after castration was reduced compared to that of wild type littermates. Interestingly, there is a compensatory upregulation of SRC-2/TIF2 in these mice. As a result, the role of SRC-1 in prostate may be an underestimate in this model.

SRC-1 is expressed in all prostate cancer cell lines examined, in normal prostate epithelia and in prostate tumors. Using siRNA to deplete SRC-1 expression, Agoulnik *et al.* showed that SRC-1 was required for optimal PSA expression in androgen dependent LNCaP cells.[4] Moreover, the depletion of SRC-1 slowed cell proliferation, but had no effect on the proliferation of AR negative PC-3 prostate cancer cells. Hence, SRC-1 plays a role in androgen dependent proliferation. AR represses transcription of maspin and, surprisingly, SRC-1 was necessary for this repression. Thus, AR coactivators can facilitate both the induction and repression of target genes. The androgen independent C4-2 cells derived from LNCaP cells express PSA in androgen depleted medium. The PSA expression is AR dependent and reducing SRC-1 expression under these conditions reduces both PSA expression and cell proliferation. Thus, SRC-1 also facilitates hormone independent actions of AR.

The expression of SRC-1 in prostate cancer has been examined by a number of groups. In an initial study of a small number of samples, Wilson's group[14] found that SRC-1 was overexpressed in recurrent prostate cancer. SRC-1 expression measured in a tissue microarray consisting of more than 500 cases of primary prostate cancer and normal tissue from the same patients was found to be very variable.[4] There was no overall difference in average staining between normal prostate epithelia and tumor cells. However, higher expression correlated with a number of characteristics of more aggressive disease including clinical stage, Gleason score, extracapsular extension and pelvic lymph node metastases. Moreover, SRC-1 expression was found to be significantly higher in a set of metastatic tumors compared to the primary tumors. One surprising finding was that there was a significant correlation between SRC-1 levels in normal prostate and in prostate tumors. This suggests that men with higher expression of SRC-1 in their normal tissue are more likely to have more aggressive forms of prostate cancer.

The activity of SRC-1 is dependent not only on its expression level, but also on its degree of phosphorylation. Rowan *et al.* have identified several phosphorylation sites in SRC-1 including two sites whose phosphorylation is regulated by p42/p44 MAPK.[15] Ueda *et al.* have found that these sites are important for SRC-1 dependent potentiation of AR activity induced by the cytokine, IL-6;[16] IL-6 dependent induction of AR activity is dependent upon p42/p44 MAPK activity. Gioeli *et al.* reported that p42/p44 MAPK activity frequently is elevated in advanced prostate cancers.[17] As a result, these tumors may have increased SRC-1 activity independent of the absolute level of expression.

7.5.2.2 *SRC-2*

There is also very good evidence that SRC-2/TIF2 is important for AR action in prostate cancer. Eliminating SRC-2 expression using siRNA reduces hormone dependent induction of a number of AR regulated genes including PSA.[18] However, SRC-2 does not appear to play a role in the inhibition of maspin expression by AR. Eliminating SRC-2 also reduces LNCaP cell proliferation; however, SRC-2 is also required for proliferation of AR negative PC-3 cells. Thus, SRC-2 appears to have a broader role in regulating proliferation than does SRC-1. Agoulnik *et al.* found that expression of SRC-2 in LNCaP and C4-2 cells was repressed by androgens. AR binding sites were found both in the promoter and in an intron of SRC-2. Analyses of the cloned promoter revealed that both agonist-bound and antagonist-bound AR repress SRC-2 expression. In this case, an antagonist would have a unique beneficial effect in inhibiting growth relative to simple androgen ablation. Interestingly, in androgen independent CWR-R1 prostate cancer cells, EGF treatment increases the expression of TIF2, induces TIF2 phosphorylation, enhances interactions between AR and TIF2 and potentiates AR transcriptional activity through a p42/p44 MAPK dependent mechanism.[19] Thus, the activity of this coactivator is also stimulated by p42/p44 MAPK.

The studies of SRC-2 expression in prostate cancer are also consistent with a role for androgens in repressing expression. Using small sets of samples, Gregory *et al.*[14] observed that SRC-2 expression in recurrent prostate cancer was very high relative to primary prostate cancer or BPH. Using a tissue microarray with more than 500 cases of prostate cancer, Agoulnik *et al.*[18] found that high levels of TIF2 expression correlated with a shorter time to biochemical recurrence defined as a rising serum PSA level. Interestingly, when AR expression was also considered, tumors with low levels of AR were least likely to recur regardless of the level of TIF2/SRC-2 expression, whereas in samples with high levels of AR expression, those that also had high levels of TIF2 were more likely to recur quickly than those with lower levels of TIF2. This suggests that in primary prostate cancer, TIF2 is facilitating the activity of AR despite its additional ability to potentiate growth of AR independent cells. The analysis of a limited number of recurrent tumors from men who had failed androgen ablation therapy revealed that they all had extremely high levels of TIF2 and that the expression level was higher than in the primary tumors even when samples of the same grade were examined. Thus, it appears that androgen ablation *in vivo* causes an elevation of TIF2 as predicted from the cell based studies.

7.5.2.3 *SRC-3*

In contrast to SRC-1 and SRC-2, SRC-3 has much broader effects on cell function. These actions include induction of overall Akt levels and a corresponding increase in activated Akt.[20,21] Although increasing SRC-3 expression in androgen dependent LNCaP cells increases cell growth and reducing SRC-3 diminishes prostate cancer cell proliferation, this is also true in AR negative cells. SRC-3 is overexpressed in prostate cancer and its expression correlates with increased proliferation and decreased apoptosis. SRC-3 is an AR coactivator and it is recruited to the PSA promoter and enhancer.[22] A more complete discussion of the overall role of SRC-3 in prostate cancer as well as in other cancers can be found in the chapter by Yan, Tsai, and Tsai in this book.

7.5.3 *Additional AR coactivators with HAT activity*

7.5.3.1 *CBP(CREB binding protein)/p300*

These coregulators not only acetylate histones, but have been found to acetylate AR increasing its transcriptional activity. They bind to the p160 coactivators, so their recruitment to AR is primarily through these proteins although they are capable of binding directly to other transcription factors. There are several studies that implicate p300 in the activation of AR by non-steroid mediated pathways. IL-6 mediated induction of AR activity in the absence of androgen requires p300.[23] Treatment with the neuropeptide, bombesin, also activates AR in the absence of ligand.[24,25] This induction is associated with increased AR acetylation; depleting p300 with siRNA reduces AR acetylation. Analyses of p300 expression in prostate cancer show that expression correlates with proliferation, characteristics of more aggressive disease and shorter time to biochemical recurrence. Consequently, it is likely that p300 plays a role in both hormone dependent and hormone independent prostate cancer. Interestingly, androgens reduce expression of p300 at the protein level.[8] Thus, androgen ablation should enhance the expression of p300 facilitating hormone independent activation of AR.

7.5.3.2 *p300/CBP-associated factor (P/CAF)*

P/CAF is a HAT that can directly interact with AR and can also be recruited to AR complexes through binding to proteins such as the p160 coactivators or CBP. P/CAF also acetylates AR and thus may contribute directly to its increased activity through post-translational modification.[26]

7.5.3.3 *Tip60 (Tat-interactive protein, 60 kDa)*

Tip60 is another HAT that increases AR activity and acetylates AR.[26] It also acetylates p53 and this acetylation is required for p53 induced apoptosis in response to DNA damage.[27] Analyses of Tip60 expression in prostate and prostate cancer reveal that nuclear Tip60 is enhanced in androgen refractory cancers.[28] In contrast, there is a substantial amount of cytoplasmic staining in primary tumors and in benign prostate hypertropy (BPH). Studies in cell lines suggest that AR regulates the expression of Tip60; androgen withdrawal increases nuclear expression of Tip60 and treatment with androgen reduces expression.

7.5.4 *Methyltransferases as AR Coactivators*

Methyltransferases methylate specific lysines or arginines in histones. Some of the residues are monomethylated whereas others are multiply methylated. Although many methylations have been associated with the inhibition of transcription (for example, lysine methylation prevents acetylation), others stimulate transcription. One arginine methyl transferase has been well characterized as a nuclear receptor coactivator and there is less complete information supporting roles for two others.

7.5.4.1 *CARM1*

CARM1 is an arginine methyltransferase, which methylates Arg 2, 17, and 26 of histone H3. It acts as an AR coactivator, but requires the presence of a p160 coactivator such as SRC-2/TIF2 in order to potentiate AR activity. CARM1 does not bind directly to AR; rather, it is associated with AR through its direct binding to a p160 coactivator. Reducing CARM-1 expression diminishes AR dependent transcription, slows prostate cancer cell proliferation, and induces apoptosis.[29] CARM1 is not specific to AR signaling, thus some of its actions in inducing apoptosis, for example, may be independent of AR. Analysis of the expression of CARM1 in normal prostate and in prostate cancer revealed that CARM1 expression is higher in tumors than in normal tissue and higher in androgen refractory tumors compared to androgen dependent tumors.

7.5.4.2 *PRMT-1*

PRMT-1 methylates Arg 3 of histone 4. Similar to CARM-1, it interacts with p160 coactivators. CARM-1 and PRMT-1 can synergize in coactivating steroid receptors.[30]

7.5.4.3 *G9a*

G9a methylates Lys9 of histone H3 and often functions as a corepressor. However, in the context of nuclear receptor action, it appears to act as a coactivator. G9a potentiates AR activity in the presence of CARM1.[31] Surprisingly, other methyltransferases do not appear to synergize with G9a. G9a may have a unique as yet uncharacterized function in steroid receptor action. Elimination of its methyltransferase activity does not prevent the coactivation of AR. In contrast, CARM1 coactivation of AR and synergism with G9a depends upon its methyltransferase activity.

7.5.5 *Ubiquitin ligases as AR coactivators*

There is evidence that the activity of AR is regulated by ubiquitin ligases and by ubiquitination. E6-AP is a ubiquitin ligase that interacts with AR and potentiates AR activity when overexpressed. There is evidence that E6-AP plays a role in normal prostate development as E6-AP null mice have smaller prostate glands.[32] Interestingly, the prostate cells express higher levels of AR protein, but reduced levels of the AR regulated target, probasin. This suggests that E6-AP is needed for optimal potentiation of AR activity, but may also facilitate AR degradation. Despite its ability to potentiate AR action, it is expressed at lower levels in prostate cancer compared to normal tissue.[33]

7.5.6 *SWI/SNF Complexes*

Chromatin remodeling is required for the transcription of many target genes. ATPases play a role in positioning of nucleosomes. BRM, but not BRG1 complexes, appear to be important for AR activity. Although BRM does not interact directly with AR, the BAF57 subunit of the complex does interact with AR.[34] BAF57 null cells transfected with an AR expression vector show diminished hormone dependent induction compared to other BAF57 positive lines. Re-expression of BAF57 restores AR activity. In LNCaP cells, dominant negative BAF57 reduces proliferation and AR activity. Similarly, reducing BRM expression with siRNA reduces proliferation. Thus, this complex plays a role in AR action in prostate cancer cells.

7.5.7 *TRAP (thyroid receptor associated proteins)/ DRIP (vitamin D receptor interacting proteins)/ mediator complexes*

These complexes play important roles in facilitating transcription of a variety of transcription factors including the nuclear receptors. TRAP220/Med1 is the subunit in the complex that directly interacts with AR. Reducing TRAP220 expression in LNCaP prostate cancer cells reduced androgen dependent expression of PSA to a greater extent than reducing expression of the p160 coactivator, SRC-1.[35] This suggested that these complexes are critical for AR dependent regulation of transcription. Although reducing TRAP220 expression also reduced cell growth and induced apoptosis, similar experiments in AR negative prostate cancer cells yielded similar responses. Thus, as might be expected for a more general coregulator, many of the responses were AR independent. An immunohistochemical analysis of TRAP220 expression in prostate cancer and in normal prostate from the same patients revealed that TRAP220 is expressed at higher levels and to a greater extent in prostate cancer cells relative to the normal epithelial cells.

7.5.8 *ARA coactivators*

Dr. Chawnshang Chang's laboratory has isolated a series of proteins that interact with AR and enhance AR transcriptional activity. These proteins have been named androgen receptor associated proteins (ARA). They are numbered based on their apparent molecular weights. Most of these proteins have been identified previously. They have alternate names, and in some cases, have known additional functions. The proteins are unrelated structurally. Thus, although the names are similar, their functions may differ significantly.

7.5.8.1 *ARA70 (ELE1,RFG)*

ARA70 is one of the earliest coactivators identified, it was originally thought to be AR specific[36] but subsequent studies revealed that is was capable of coactivating other receptors. ARA70 interacts with the hormone binding domain of AR; one of the most striking findings is that it allows AR to more efficiently utilize a broader range of hormones including adrenal androgens, estradiol, and even antagonists including bicalutamide as agonists.[37] In the case of estradiol, ARA70 reduces the dissociation rate of estradiol from AR. Although ARA70 contains an LXXLL motif, it interacts with AR through an amino-terminal FXXLF and enhances protein expression and stability. A dominant negative ARA70 inhibits AR dependent transcription. Moreover, siRNA for ARA70 reduces both agonist and antagonist dependent induction of PSA in LNCaP cells. ARA70 is predominantly cytoplasmic; it is expressed in prostate epithelial cells and the percentage of ARA70 positive prostate cancer cells is higher than the percentage of ARA70 positive benign prostate cells, suggesting a possible role in prostate cancer.[38]

7.5.8.2 *ARA24 (Ran)*

ARA24 was identified as a candidate AR coactivator through its interaction with the polyglutamine repeat in the amino terminus of AR. ARA24/Ran is a small GTP binding protein that plays a role in nuclear import of proteins. Thus, its ability to enhance AR activity may be a result of enhanced nuclear import of AR. Interestingly, Ran interacts more weakly with AR containing longer polyglutamine repeats and is less capable of potentiating the activity of AR with longer polyglutamine repeats.[39] As a result, this protein may contribute to the enhanced activity of AR with shorter repeats. Analyses of Ran expression in prostate cancer specimens using FISH showed that Ran is expressed more highly than the rest.[40] However, because of Ran's role in general cellular function, it is not possible to determine the precise contribution of Ran to androgen action in prostate cancer.

7.5.8.3 *ARA54*

ARA54 was also detected as an AR interacting protein. Subsequent studies suggested that it is a ubiquitin ligase. It potentiates the activity of AR, but also of progesterone receptor. A dominant negative form of this protein inhibits PSA expression and prostate cancer cell growth.[41]

7.5.8.4 *ARA55/Hic-5*

ARA55 was also identified as an AR interacting protein that is broadly distributed within the cell. It can be associated with focal adhesion complexes. Subsequent studies revealed that it is expressed predominantly in the stromal cells of the prostate. It can also function as an AR coactivator and is recruited to androgen responsive promoters in stromal cells.[42] It is not expressed in LNCaP prostate cancer cells. The limited studies of mRNA expression in prostate suggest reduced expression in tumors, but this may be secondary to the reduced proportion of stromal cells in tumor versus normal tissue.

7.5.9 *Cell signaling molecules that also function as AR coactivators*

A number of molecules that participate in various aspects of cell signaling pathways have also been characterized as AR coactivators. Among these are kinases and phosphatases, proteins that serve as scaffolds for signaling molecules, and effectors of cell signaling pathways. Because of the diverse activities of these proteins, it is often difficult to dissect the contribution of AR dependent activities from more general activities in prostate cancer cells.

7.5.9.1 *β-catenin*

β catenin is best known as a TCF/LEF transcription factor coactivator that is regulated by wnt signaling. A number of studies have shown that β catenin also stimulates AR activity.[43] The contribution of endogenous β catenin to overall AR activity is relatively modest in many circumstances, but is more pronounced in cells stimulated by IGF.[44] Increased Wnt-1 and β catenin have been associated with more advanced prostate cancers.

7.5.9.2 *Caveolin-1*

Caveolin-1 is a scaffold protein that organizes signaling molecules in caveolae, small membrane invaginations enriched in signaling molecules. Caveolin-1 has been found to interact with androgen receptors and to stimulate transcription.[45] More recent studies suggest that a fraction of the androgen receptor may be palmitoylated inducing interactions with caveolae and caveolin-1.[46] Whether it is the signaling induced by these complexes that enhances AR transcriptional activity or some other function of caveolin-1 that stimulates AR activity remains to be determined. Caveolin-1 expression has been correlated with more aggressive prostate cancer.

7.5.9.3 *Cdc25B*

Cdc25B is a dual specificity phosphatase that facilitates cell cycle progression by dephosphorylating cyclin dependent kinases. It has been shown to interact with AR potentiating its activity; Cdc25B synergizes with p300 and P/CAF to enhance AR activity.[47] Surprisingly, the phosphatase activity is not required to potentiate AR action. Analyses of Cdc25B expression reveal that it is overexpressed in prostate cancer, particularly in higher grade tumors.[48]

7.5.9.4 *PELP-1 (Proline-, glutamic acid-, and leucine-rich protein-1)/ /MNAR (modulator of nongenomic activity of estrogen receptor)*

PELP-1 is a recently described protein that interacts with the components of a variety of signaling pathways including multiple kinases, proteins involved in cell cycle progression, and nuclear receptors.[49] Although not fully characterized as yet, PELP-1 is a novel regulator of AR function in that it participates both in the transcriptional activation functions of AR and in the activation of cell signaling pathways by AR. With regard to its coactivator function, it may function primarily through aiding in the recruitment of general coregulators such as CBP and p300. It also interacts with FHL2 (Four-and-a-Half-LIM only Protein 2),[50] a protein described as an androgen receptor specific coactivator. There is also evidence that it participates in hormone dependent activation of src kinase. In androgen dependent LNCaP cells, hormone treatment induces the formation of complexes containing PELP-1/MNAR, AR and src and activation of downstream kinases including p42/p44 MAPK.[51] Interestingly, in a hormone independent derivative of LNCaP cells, this complex is formed in the absence of hormone. In a limited study of expression in prostate cancer samples, PELP1/MNAR was found to be expressed at higher levels in high-grade tumors than in normal tissue or low grade tumors.[50] Thus, there is evidence that PELP1/MNAR may play a role in prostate cancer.

7.6 Summary

The studies of androgen receptor coactivator expression and function in prostate cancer cells suggest that increased expression can contribute to aberrant receptor activation. In normal prostate epithelial cells, the role of AR is to produce prostate specific proteins rather than to stimulate growth. However, in prostate cancer cells, AR stimulates cell growth. One of the surprising findings is that androgens repress the expression of several coactivators important for cell growth and AR function. Thus, there is a feedback loop limiting AR action. Androgen ablation therapy reduces the levels of androgens causing increased expression of coactivators such as SRC-2 and p300. Some of these coactivators not only potentiate AR activity in hormone depleted medium, but also facilitate growth through AR independent pathways.

References

1. Arnold JT, Isaacs JT, Mechanisms involved in the progression of androgen-independent prostate cancers: It is not only the cancer cell's fault, *Endocr Relat Cancer* **9**:61–73, 2002.
2. Holzbeierlein J, Lal P, LaTulippe E, *et al.*, Gene expression analysis of human prostate carcinoma during hormonal therapy identifies androgen-responsive genes and mechanisms of therapy resistance, *Am J Pathol* **164**:217–227, 2004.
3. Zegarra-Moro OL, Schmidt LJ, Huang H, *et al.*, Disruption of androgen receptor function inhibits proliferation of androgen-refractory prostate cancer cells, *Cancer Res* **62**:1008–1013, 2002.
4. Agoulnik IU, Vaid A, Bingman IWE, *et al.*, Role of SRC-1 in the promotion of prostate cancer cell growth and tumor progression, *Cancer Res* **65**:7959–7967, 2005.
5. Heinlein CA, Chang C, Androgen receptor (AR) coregulators: An overview, *Endocr Rev* **23**:175–200, 2002.
6. Edwards J, Bartlett JMS, The androgen receptor and signal-transduction pathways in hormone-refractory prostate cancer, Part 1: Modifications to the androgen receptor, *BJU Int* **95**:1320–1326, 2005.
7. McEwan IJ, Molecular mechanisms of androgen receptor-mediated gene regulation: Structure-function analysis of the AF-1 domain, *Endocr Relat Cancer* **11**:281–293, 2004.
8. Heemers HV, Sebo TJ, Debes JD, *et al.*, Androgen deprivation increases p300 expression in prostate cancer cells, *Cancer Res* **67**:3422–3430, 2007.

9. Onate SA, Tsai SY, Tsai M-J, *et al.*, Sequence and characterization of a coactivator for the steroid hormone receptor superfamily, *Science* **270**:1354–1357, 1995.
10. McKenna NJ, Lanz RB, O'Malley BW, Nuclear receptor coregulators: Cellular and molecular biology, *Endocr Rev* **20**:321–344, 1999.
11. Fu M, Rao M, Wang C, *et al.*, Acetylation of androgen receptor enhances coactivator binding and promotes prostate cancer cell growth, *Mol Cell Biol* **23**:8563–8575, 2003.
12. Bevan CL, Hoare S, Claessens F, *et al.*, The AF1 and AF2 domains of the androgen receptor interact with distinct regions of SRC1, *Mol Cell Biol* **19**:8383–8392, 1999.
13. Xu J, Qiu Y, DeMayo FJ, *et al.*, Partial hormone resistance in mice with disruption of the steroid receptor coactivator-1 (SRC-1) gene, *Science* **279**:1922–1925, 1998.
14. Gregory CW, He B, Johnson RT, *et al.*, A mechanism for androgen receptor-mediated prostate cancer recurrence after androgen deprivation therapy, *Cancer Res* **61**:4315–4319, 2001.
15. Rowan BG, Weigel NL, O'Malley BW, Phosphorylation of steroid receptor coactivator-1: Identification of the phosphorylation sites and phosphorylation through the mitogen-activated protein kinase pathway, *J Biol Chem* **275**:4475–4483, 2000.
16. Ueda T, Mawji NR, Bruchovsky N, *et al.*, Ligand-independent activation of the androgen receptor by interleukin-6 and the role of steroid receptor coactivator-1 in prostate cancer cells, *J Biol Chem* **277**:38087–38094, 2002.
17. Gioeli D, Mandell JW, Petroni GR, *et al.*, Activation of mitogen-activated protein kinase associated with prostate cancer progression, *Cancer Res* **59**:279–284, 1999.
18. Agoulnik IU, Vaid A, Nakka M, *et al.*, Androgens modulate expression of TIF2, an androgen receptor coactivator whose expression level correlates with early biochemical recurrence in prostate cancer, *Cancer Res* **66**:10594–10602, 2006.
19. Gregory CW, Fei X, Ponguta LA, *et al.*, Epidermal growth factor increases coactivation of the androgen receptor in recurrent prostate cancer, *J Biol Chem* **279**:7119–7130, 2004.
20. Zhou HJ, Yan J, Luo W, *et al.*, SRC-3 is required for prostate cancer cell proliferation and survival, *Cancer Res* **65**:7976–7983, 2005.
21. Zhou G, Hashimoto Y, Kwak I, *et al.*, Role of the steroid receptor coactivator SRC-3 in cell growth, *Mol Cell Biol* **23**:7742–7755, 2003.
22. Wang Q, Carroll JS, Brown M, Spatial and temporal recruitment of androgen receptor and its coactivators involves chromosomal looping and polymerase tracking, *Mol Cell* **19**:631–642, 2005.

23. Debes JD, Schmidt LJ, Huang H, *et al.*, p300 mediates androgen-independent transactivation of the androgen receptor by interleukin 6, *Cancer Res* **62**:5632–5636, 2002.
24. Gong J, Zhu J, Goodman OB, *et al.*, Activation of p300 histone acetyltransferase activity and acetylation of the androgen receptor by bombesin in prostate cancer cells, *Oncogene* **25**:2011–2021, 2006.
25. Desai SJ, Ma AH, Tepper CG, *et al.*, Inappropriate activation of the androgen receptor by nonsteroids: Involvement of the Src kinase pathway and its therapeutic implications, *Cancer Res* **66**:10449–10459, 2006.
26. Fu M, Wang C, Zhang X, *et al.*, Acetylation of nuclear receptors in cellular growth and apoptosis, *Biochem Pharmacol* **68**:1199–1208, 2004.
27. Sykes SM, Mellert HS, Holbert MA, *et al.*, Acetylation of the p53 DNA-binding domain regulates apoptosis induction, *Mol Cell* **24**:841–851, 2006.
28. Halkidou K, Gnanapragasam VJ, Mehta PB, *et al.*, Expression of Tip60, an androgen receptor coactivator, and its role in prostate cancer development, *Oncogene* **22**:2466–2477, 2003.
29. Majumder S, Liu Y, Ford OH, *et al.*, Involvement of arginine methyltransferase CARM1 in androgen receptor function and prostate cancer cell viability, *Prostate* **66**:1292–1301, 2006.
30. Wang H, Huang ZQ, Xia L, *et al.*, Methylation of histone H4 at arginine 3 facilitating transcriptional activation by nuclear hormone receptor, *Science* **293**:853–857, 2001.
31. Lee DY, Northrop JP, Kuo MH, *et al.*, Histone H3 lysine 9 methyltransferase G9a is a transcriptional coactivator for nuclear receptors, *J Biol Chem* **281**:8476–8485, 2006.
32. Khan OY, Fu G, Ismail A, *et al.*, Multifunction steroid receptor coactivator, E6-associated protein, is involved in development of the prostate gland, *Mol Endocrinol* **20**:544–559, 2006.
33. Gao X, Mohsin SK, Gatalica Z, *et al.*, Decreased expression of e6-associated protein in breast and prostate carcinomas, *Endocrinology* **146**:1707–1712, 2005.
34. Link KA, Burd CJ, Williams E, *et al.*, BAF57 governs androgen receptor action and androgen-dependent proliferation through SWI/SNF, *Mol Cell Biol* **25**:2200–2215, 2005.
35. Wang Q, Sharma D, Ren Y, *et al.*, A coregulatory role for the TRAP/Mediator complex in androgen receptor mediated gene expression, *J Biol Chem* **277**:42852–42858, 2002.
36. Yeh S, Chang C, Cloning and characterization of a specific coactivator, ARA70, for the androgen receptor in human prostate cells, *Proc Natl Acad Sci USA* **93**:5517–5521, 1996.

37. Yeh SH, Miyamoto H, Shima H, *et al.*, From estrogen to androgen receptor: A new pathway for sex hormones in prostate, *Proc Natl Acad Sci USA* **95**:5527–5532, 1998.
38. Hu YC, Yeh S, Yeh SD, *et al.*, Functional domain and motif analyses of androgen receptor coregulator ARA70 and its differential expression in prostate cancer, *J Biol Chem* **279**:33438–33446, 2004.
39. Hsiao PW, Lin DL, Nakao R, *et al.*, The linkage of Kennedy's neuron disease to ARA24, the first identified androgen receptor polyglutamine region-associated coactivator, *J Biol Chem* **274**:20229–20234, 1999.
40. Li P, Yu X, Ge K, *et al.*, Heterogeneous expression and functions of androgen receptor co-factors in primary prostate cancer, *Am J Pathol* **161**:1467–1474, 2002.
41. Kang HY, Yeh S, Fujimoto N, *et al.*, Cloning and characterization of human prostate coactivator ARA54, a novel protein that associates with the androgen receptor, *J Biol Chem* **274**:8570–8576, 1999.
42. Heitzer MD, DeFranco DB, Hic-5/ARA55, a LIM domain-containing muclear receptor coactivator expressed in prostate stromal cells, *Cancer Res* **66**:7326–7333, 2006.
43. Truica CI, Byers S, Gelmann EP, Beta-catenin affects androgen receptor transcriptional activity and ligand specificity, *Cancer Res* **60**:4709–4713, 2000.
44. Verras M, Sun Z, Beta-catenin is involved in insulin-like growth factor 1-mediated transactivation of the androgen receptor, *Mol Endocrinol* **19**:391–398, 2005.
45. Lu ML, Schneider MC, Zheng Y, *et al.*, Caveolin-1 interacts with androgen receptor. A positive modulator of androgen receptor mediated transactivation, *J Biol Chem* **276**:13442–13451, 2001.
46. Pedram A, Razandi M, Sainson RC, *et al.*, A conserved mechanism for steroid receptor translocation to the plasma membrane, *J Biol Chem* **282**:22278–22288, 2007.
47. Ma Z-Q, Liu Z, Ngan ESW, *et al.*, Cdc25B functions as a novel coactivator for the steroid receptors, *Mol Cell Biol* **21**:8056–8067, 2001.
48. Ngan ES, Hashimoto Y, Ma ZQ, *et al.*, Overexpression of Cdc25B, an androgen receptor coactivator, in prostate cancer, *Oncogene* **22**:734–739, 2003.
49. Vadlamudi RK, Kumar R, Functional and biological properties of the nuclear receptor coregulator PELP1/MNAR, *Nucl Recep Signal* **5**:e004, 2007.
50. Nair SS, Guo Z, Mueller JM, *et al.*, Proline-, glutamic acid-, and leucine-rich protein-1/modulator of nongenomic activity of estrogen receptor enhances androgen receptor functions through LIM-only coactivator, four-and-a-half LIM-only protein 2, *Mol Endocrinol* **21**:613–624, 2007.
51. Unni E, Sun S, Nan B, *et al.*, Changes in androgen receptor nongenotropic signaling correlate with transition of LNCaP cells to androgen independence, *Cancer Res* **64**:7156–7168, 2004.

Chapter 8

PGC-1α and Metabolic Control in Skeletal and Cardiac Muscle

Zolt Arany and Bruce M. Spiegelman

The appropriate regulation of any particular gene in space and time requires the concerted action of DNA-binding transcription factors, the basal transcriptional apparatus and transcriptional coactivator proteins. Whereas most earlier studies of genes that are regulated in development or in altered physiological states focused on transcription factors, it is now very clear that coactivators can be the primary regulators of altered patterns of gene expression. This has probably been best illustrated by the PGC-1 coactivators which is the focus of this review.

8.1 Introduction

PGC-1α (formerly PGC-1) was discovered as a binding partner and coactivator of PPARγ in brown fat. We had previously shown that PPARγ was the dominant regulator of white fat cell differentiation,[1] but later work by others pointed out the key role of PPARγ in brown fat differentiation as well.[2] This raised the paradox as to how a single transcription factor could initiate patterns of gene expression that stimulated lipid storage (white fat) or lipid oxidation (brown fat) in different cells. Hypothesizing the existence of a brown fat-specific coactivator, we cloned PGC-1α using PPARγ as a bait in a yeast two-hybrid screen, using a cDNA library made from brown fat cells.[3]

8.2 Structure of the PGC-1α Complex

A key feature of coactivators is their ability to modify chromatin to allow or cause transcription to proceed. PGC-1α has not been shown to have any of the enzymatic activities present in some coactivators, such as histone acetyl transferase (HAT) activity or ATP-dependent chromatin unwinding activity. On the other hand, PGC-1α recruits several such proteins into a large complex. Murine PGC-1α contains 797 amino acids and most of the N-terminal 150 amino acids encode a transcriptional activation domain that recruits several HAT-containing proteins, including CBP/p300 and SRC-1. The C-terminus of PGC-1α contains an SR domain and RNA-binding domain (so-called RRM), domains that have been implicated in RNA splicing. In fact, these domains do appear to be involved in the processing of many mRNAs whose transcription has been initiated by PGC-1α.[4] This region of PGC-1α also docks the mediator complex, a group of proteins involved in the recruitment of RNA polymerase II.[5] The key to the function of PGC-1α, like all coactivators, is the ability to dock on transcription factors. PGC-1α, in fact, docks on and coactivates most members of the nuclear receptor family and many transcription factors outside this family. The interaction with transcription factors can occur at many places in the PGC-1α structure with most nuclear receptors touching one or more of the LXXLL motifs. The overall architecture of PGC-1α is shared by the two other members of the family, PGC-1β and PRC (Fig. 8.1), but their activities have been less extensively studied. Both PGC-1β and PRC were identified through their sequence homology to PGC-1α.

8.3 Regulatory Proteins in the PGC-1α Complex

Recent studies have shown that several proteins can interact with PGC-1α and regulate its function. SirT1, the mammalian homolog of yeast Sir2, has been shown to bind to PGC-1α.[6] PGC-1α has up to 13 separate acetylation sites which are inhibitory towards its transcriptional functions, and SirT1 has deacetylase activity that can remove each of these acetylations. These acetylations are added by the GCN5 protein, at least in the liver.[7] The notion that a major regulator of aging in many biological systems, SirT1, regulates the activity of a dominant metabolic regulator, PGC-1α, is intriguing. In fact, it appears that a substantial fraction of the activities of SirT1 might be through its actions on PGC-1α.[8,9] One of the most crucial questions that remains is

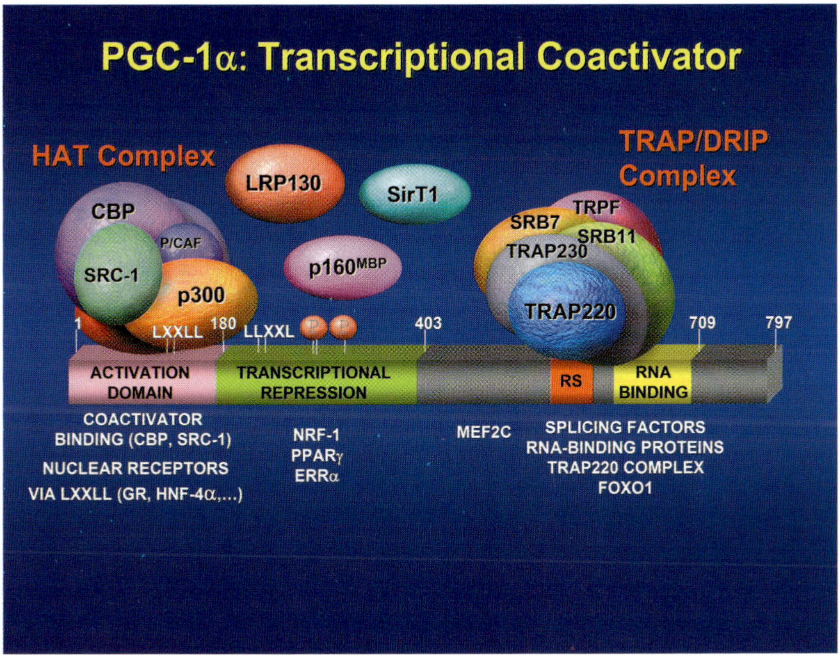

Fig. 8.1. Schematic of PGC-1α and known associated proteins.

whether the acetylation/deacetylation cycle of PGC-1α functions as a simple on/off switch, or whether certain acetylation/deacetylation events affect some biological pathways preferentially.

Other proteins that can bind to and increase PGC-1α's transcriptional and biological action include LRP130,[10] a gene mutated in certain patients with Leigh's Syndrome, and lipin,[11] a gene mutated in a murine model of lipodystrophy. p160^{MBP} binds to PGC-1α and inhibits its activity; interestingly, this interaction is reduced by the p38 MAPK phosphorylation of PGC-1α. CARM1, a protein with methyltransferase activity, binds to and methylates PGC-1α, increasing its activity in transfection studies.[15]

8.4 Overview of Biological Activities of the PGC-1 Coactivators

Many if not most of the known activities of the PGC-1s relate to oxidative metabolism, and mitochondrial biogenesis and function. PGC-1α

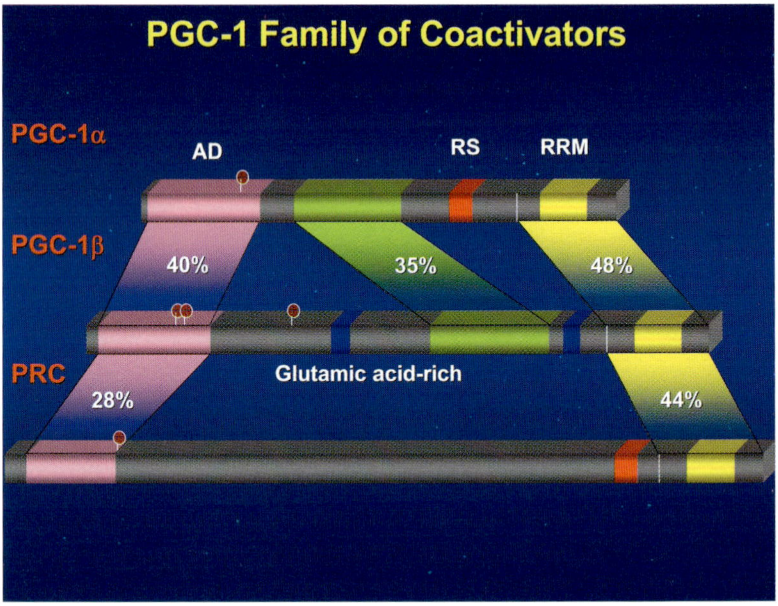

Fig. 8.2. Schematic of the PGC-1 family of transcriptional co-activators. Percentage amino-acid homology is indicated. AD: activation domain. RS: RS-rich region. RRM: RNA-binding domain. Red circles: LXXLL motifs.

and PGC-1β can induce a broad program of mitochondrial gene expression involving those genes encoded in both the nuclear and mitochondrial genomes. Both coactivators induce mitochondrial biogenesis *per se* in virtually all cell types examined (Fig. 8.2). PGC-1α, in particular, is highly inducible in many cell and tissue types, by stimuli such as cold exposure (brown fat) or physical activity (muscle). Thus, it is now clear that the PGC-1s, particularly PGC-1α, mediate the interaction with the extracellular or extra-organismal environment, and thus result in alterations in mitochondrial number and metabolism. The mechanism of control of mitochondrial gene expression by PGC-1α has been studied in some detail, and involves the coactivation of a small group of proteins including ERRα, NRF-1 and NRF-2 (also known as GABP). These three proteins bind directly to, and activate many mitochondrial genes encoded in the nuclear genome. In addition, NRF-1 and PGC-1α activate the nuclear gene encoding TFAM, a protein that is required for the transcription and replication of the mitochondrial genome. PGC-1α also turns on factors B1 and B2, which interact with the mitochondrial DNA polymerase.[16] Genetic studies in mice indicate that the loss of either PGC-1α or PGC-1β causes a quantitative decrease in mitochondrial gene expression and function, but mice are alive and retain substantial mitochondrial activity. The decrease in the levels of PGC-1α *and* PGC-1β, at least in cell culture, causes a collapse of mitochondrial biology, indicating the crucial yet complementary action of these proteins.[17]

One very interesting aspect of the biological function of the PGC-1s is their ability to activate tissue-specific functions that relate to the core program of mitochondrial biogenesis, yet being distinct from this pathway. For example, PGC-1α turns on the genes for UCP-1 and type II deiodinase in brown fat. These proteins are crucial for brown fat-mediated thermogenesis, a process completely dependent on mitochondria. PGC-1α turns on many aspects of the fasting program in liver including the mitochondrial β-oxidation of fatty acids. It also turn on the entire program of gluconeogenesis, which functions in the cytoplasm. As detailed extensively below, PGC-1α and -β control many aspects of oxidative metabolism in skeletal muscle and heart tissues, but also control many non-mitochondiral functions that are linked to these, such as fiber-type switching.

8.4.1 *PGC-1α and reactive oxygen species (ROS)*

One set of recent observations deserving special mention relates to ROS metabolism. ROS are metabolites of oxygen that are highly reactive and can damage lipids, DNA and proteins. ROS include superoxide, hydrogen peroxide and the hydroxyl radical. The majority of ROS in most cell types are derived initially from the "escape" of an unpaired electron from the mitochondrial electron transport chain to molecular oxygen. As PGC-1α activates genes involved in mitochondrial oxidative metabolism, it also turns on a set of genes involved in either suppressing ROS formation (UCP2 and UCP3) or detoxifying ROS species (SOD1, SOD2, GPX1, catalase). The net result of this is that mitochondrial number and respiration can increase without any proportional increase in the levels of ROS within cells (St. Pierre *et al.*, 2006). Importantly, while the PGC-1s can activate mitochondrial genes that are anti-ROS (UCPs and SOD);[18,19] they also increase the expression of cytoplasmic proteins that detoxify ROS, such as SOD1, GPX1 and catalase.

ROS are thought to be causally involved in cell damage and death in many important human diseases, including neurodegenerative disorders such as Huntington's disease and Parkinson's disease, as well as ischemia/reperfusion injury of the heart and brain. The role of PGC-1α in the defense against ROS was investigated genetically in two murine models of neurodegeneration, MPTP-induced Parkinsonism and kainic acid-induced seizures. In both cases, much greater change to the relevant brain areas occurred in the absence of PGC-1α than in controls, and this neuronal death was associated with an excess of oxidative damage to DNA and proteins.[20] The transcription factors on which PGC-1α (and PGC-1β) dock in order to activate the genes of ROS detoxification remain to be determined. However, the importance of this pathway is highlighted by the fact that PGC-1α itself is induced by exposure of cells to an oxidative stressor. Hence, the well known induction of an anti-ROS program by ROS is likely to be mediated, at least in part, by the PGC-1 coactivators. The transcription factor CREB, previously shown to bind to and activate the PGC-1α promoter in the liver,[21] is apparently a major target of H_2O_2 treatment of cells.[20]

8.4.2 *Skeletal muscle energetics*

Skeletal muscle is a heterogeneous tissue with diverse, important physiological functions, and PGC-1's are involved in many of these. Muscle performs a spectrum of types of work, ranging from continuous low-level activities like maintaining posture to sudden bursts of intense activity. To achieve this, fiber types with different biophysical and energetic properties are used. In adult skeletal muscle, there are four main types, named after the myosin heavy chain (MHC) genes primarily expressed: I, IIA, IIX, and IIB. MHC's and their ancillary proteins confer on each fiber its unique biophysical properties. Type I fibers, often called slow-twitch fibers, have a "slow" contraction profile that is also more energetically efficient, while fast-twitch type IIB fibers are more powerful but less efficient. At the same time, type I fibers are rich in mitochondria, while IIB fibers have fewer mitochondria and favor glycolysis as a source of ATP. IIA and IIX fibers fall in between, creating a gradient of options. Hence, muscles enriched for type I and IIA fibers, such as the soleus, can maintain a steady supply of ATP and resist fatigue. This makes them useful for low-grade but constant activities like posture maintenance. On the other hand, type-II rich muscles like the quadriceps are more useful for episodic but more intense bursts of activity.

Although the fiber composition of different muscle beds is determined during development, substantial plasticity remains in the adult.[22] Most notably, chronic training exercise can induce a fast-to-slow transition, to the benefit of, for example, marathon runners. Conversely, disuse causes a slow-to-fast transition. A series of elegant, now classic, experiments demonstrated that the primary determinant of this plasticity is its nerve input: the amplitude and frequency of electrical stimulation dictates directly the identity of fibers. Superimposed on this dominant signal are a number of other hormonal signals, most notably thyroid hormone activity. Importantly, a number of diseases, ranging from dystrophy to sarcopenia, affect fibers differentially. This is discussed in greater detail below.

8.4.3 *PGC-1s and fiber types*

The powerful ability of PGC-1α to drive mitochondrial biogenesis first suggested that it may play a role in determining fiber types. The transgenic expression of PGC-1α in skeletal muscle subsequently demonstrated this quite clearly:[23] normally white muscle throughout the transgenic mouse became red, and rich in oxygen-containing myoglobin and mitochondria. Mitochondrial genes were induced, but genes encoding myofibrillar proteins characteristic of slow fibers, including MHC I and IIA, were also induced. Hence, not only did PGC-1α induce mitochondria, but it activated the program of fiber switching, including both its energetic and biophysical aspects. Consistent with this, explanted muscles from transgenic animals were far more resistant to fatigue than wild type controls.

A number of observations support the fact that PGC-1α is involved in the physiological determination of fiber type switching. First, PGC-1α is more strongly expressed in slow-twitch muscles like the soleus than in fast muscles like the quadriceps.[23] Second, exercise robustly induces PGC-1α expression in both rodents and humans (e.g., Ref. 24 suggesting that its induction contributes to the fast-to-slow transition observed with training). Third, PGC-1α is known to act downstream from a number of signaling cascades involved in fiber type determination. For instance, calcineurin and calmodulin-modulated kinase, both sensors of calcium signaling, impinge on PGC-1α activity.[25] Calcineurin, in particular, is known to respond to low-amplitude low-frequency calcium transients typical of slow-twitch muscles and activate signaling cascades that culminate in the determination of slow-twitch fibers. Transgenic expression of calcineurin in skeletal muscle leads to a fast-to-slow fiber switch,[26] which is likely mediated at least in part via activation of PGC-1α. The adrenergic system, via cyclic AMP and protein kinase A, also probably contributes to PGC-1α expression in the process of fiber-type determination.

It is important to recognize that PGC-1α is probably not the sole determinant of fiber type identity. Transgenic mice expressing PGC-1β in skeletal muscle also undergo a profound fiber type switch. However, in this case type IIX fibers predominated, with an actual decrease in all three of the other fiber types.[27] Animals lacking PGC-1α have apparently normal amounts of type I and IIA fibers, although the mitochondrial activity and function of these fibers does appear diminished.[28] Importantly, however, PGC-1$\alpha^{-/-}$ mice are markedly hyperactive, and this feature may affect skeletal muscle composition.[29] Consistent with this, skeletal muscle-specific PGC-1$\alpha^{-/-}$ mice, which are *not* hyperactive, have marked reductions in MHC I and IIA content (unpublished results). The fiber content of animals lacking PGC-1β has not been reported.

Precisely how PGC-1α activates specific fiber-type genes remains incompletely understood. Activation of the mitochondrial program is likely mediated in large part by the coactivation of NRF's, ERRα, and PPARα and δ, as discussed above. Which transcription factors are involved in the reprogramming of the myofibrillar genes by PGC-1α is less clear. The MEF2 family of transcription factors is activated by exercise[30] and PGC-1α coactivates MEF2's to induce slow-twitch genes like myoglobin and troponin I slow.[23] However, PGC-1β appears to be an equally robust coactivator of MEF2, at least in transfection assays, yet it does not cause the induction of type I and IIA fibers.[27] Clearly, other factors targeted by PGC-1α to induce specific fiber type genes remain to be identified. The complete reprogramming of fibers by a single molecule underscores a recurrent theme in coactivator biology: by simultaneously engaging numerous transcription factors involved in disparate programs, a coactivator like PGC-1α can orchestrate a complex phenotypic shift.

8.4.4 *PGC-1 and muscle wasting*

In addition to its key role in physical movement, skeletal muscle also acts as an important reservoir of calories and amino acids. During periods of food deprivation, this reservoir is tapped by catabolism of the large mass of muscle protein comprising the myofibrillar components. This can lead to sometimes drastic loss of muscle mass and frank emaciation. The amino acids produced are used by the liver for gluconeogenesis and by the entire body as a source of essential amino acids, otherwise not provided by absent food intake. The breakdown of muscle protein is initiated largely due to the induction of particular E3-ubiquitin ligases, murf and atrogin, that target proteins to the proteasomal degradation machinery.[31] Induction of these ligases is mediated by induction and nuclear translocation of the transcription factor Foxo3, which activates the murf and atrogin genes. This process is suppressed by anabolic signals, including the insulin, IGF-1, and Akt kinase pathways.

Although these events are a normal response to prolonged fasting, a number of pathological processes can also lead to profound muscle wasting, at least in part by activating the same degradation machinery. Denervation, chronic renal failure, cancer, sepsis, and diabetes are all associated with profound muscle atrophy, and in experimental animals, they all induce murf and atrogin and activate the proteasomal degradation pathway.[32,33] In these cases muscle breakdown is often maladaptive and contributes to worsening the plight of patients. Indeed, wasting of skeletal muscle predicts a worse outcome for patients with many diseases, including heart failure and cancer.

It has been recognized that different muscle fiber types are catabolized to different extents in response to systemic signals. Type IIB fibers, for instance, are more prone to atrophy than type I fibers in response to fasting, sepsis, or cancer. Moreover, contractile activity itself can protect from atrophy in the face of catabolic signals. Given PGC-1α's role in fiber type determination and processes induced by contraction, these observations suggested that PGC-1α may inhibit atrophy. Supporting this notion, the same catabolic states that induce murf and atrogin in experimental animals also profoundly inhibit PGC-1α expression.[32,34] To address the function of PGC-1α in muscle wasting directly, atrophy was induced by the denervation of hindlimbs in mice, and the response of wild type animals was compared to that of mice transgenically expressing PGC-1α in skeletal muscle.[34] This technique models the atrophy seen in human motor neuron diseases like Lou Gehrig's Disease. Strikingly, little or no loss of skeletal mass was seen in denervated hindlimbs from transgenic animals, whereas up to 40% loss of mass was seen in wild type animals. At the same time, the induction of atrophic genes like murf and atrogin by denervation was blocked in the transgenic animals.

In a similar approach, muscle atrophy was induced by transfecting a plasmid expressing constitutively active Foxo3 directly into skeletal muscle of mice.[34] Co-transfecting a plasmid encoding PGC-1α almost completely abrogated the atrophic response, recapitulating the observations seen with the PGC-1α transgenic animals. It is still unclear precisely how PGC-1α, normally a potent coactivator, can repress the activity of Foxo3. Occupancy of the atrogin promoter by Foxo3 is blocked by the presence of excess PGC-1α, suggesting that the effect may be indirect. One possible indirect mechanism by which PGC-1α may inhibit Foxo3 is through signals emanating from the neuromuscular junction (NMJ), the nerve-muscle interface. By docking the transcription factor GA-binding protein (GABP), PGC-1α can induce a comprehensive program of genes encoding for components of the NMJ, and transgenic animals have higher densities of NMJ's *in vivo* (Handschin *et al.*, 2007). Stabilizing NMJ's even in the face of denervation may thus preserve anti-atrophic pathways. Based on these results, it is clear that PGC-1α has the ability to potently inhibit atrophy in skeletal muscle. Whether this is a direct effect or not remains unclear, but either way PGC-1α presents an exciting and new potential therapeutic avenue in wasting disorders.

8.4.5 *PGC-1α and muscular dystrophy*

Duchenne's muscular dystrophy (DMD) is an X-linked recessive disease caused by a mutation in the gene dystrophin. The absence of dystrophin triggers the degeneration of skeletal and cardiac muscle beginning in childhood, and invariably leads to death by young adulthood. No successful therapy for this terrible disease exists.

Various lines of evidence suggested that PGC-1α might modulate the progression of DMD. Type I fibers appear more resistant to degeneration in DMD than type II fibers.[35] Induction of calcineurin, known to act upstream of PGC-1α, can partially prevent muscle damage seen in mice deficient in the dystrophin gene (mdx mice). Also, certain aspects of DMD are consistent with dysregulated PGC-1α, including reduced mitochondrial activity. To formally test the notion that PGC-1α may protect against, or ameliorate DMD, mdx mice were crossed to mice transgenically expressing increased levels of PGC-1α in skeletal muscle.[36] By five weeks of age, mdx mice show evidence of profound fiber degeneration and regeneration, increased levels in the serum of markers of muscle injury, and diminished maximal exercise capacity. Surprisingly, the introduction of transgenic PGC-1α in muscle of mdx mice significantly improved all of these parameters.

Exactly how PGC-1α protects against degeneration in mdx mice is not yet completely clear. PGC-1α induces utrophin, a homolog of dystrophin that can compensate for the absence of dystrophin.[36,37] PGC-1α also induces a broad program of genes involved in NMJ formation, which might itself compensate in part of dystrophin deficiency.[36] It is also likely that at least some beneficial effects of PGC-1α are indirect, perhaps via induction of the mitochondrial program, changes in calcium signaling, or reduction of ROS-mediated inflammation. It will be interesting to see if PGC-1α can also protect against other types of muscular dystrophy.

8.4.6 *PGC-1α and diabetes mellitus*

Yet another vital function of skeletal muscle is the regulation of serum glucose. Skeletal muscle is the main sink for glucose disposal in response to insulin. Resistance to insulin signaling in skeletal muscle is considered a major defect in type II diabetes mellitus. Recent research has implicated both mitochondria and the PGC-1 coactivators in this process. PGC-1α and β are both reduced in skeletal muscle of diabetic patients, along with many genes of the mitochondrial OXPHOS system.[38,39] Consistent with this, skeletal muscle in diabetic patients also contains fewer slow, oxidative fibers.[40] Importantly, expression of PGC-1α and β and the OXPHOS genes are also reduced in the skeletal muscle of *nondiabetic* patients with only glucose intolerance or with a strong family history of type-2 diabetes.[39] Additionally, aberrant mitochondrial function has been shown in the muscles of elderly patients with insulin resistance.[41] These findings suggest a very early, potentially causal, role in the development of diabetes.

This "mitochondrial/PGC-1" hypothesis of human insulin resistance remains formally unproven but is the subject of intense ongoing study. The hypothesis posits that decreased expression of the PGC-1 coactivators result in the decrease in expression and function of mitochondria in skeletal muscle of affected individuals, resulting in less oxidation of glucose and lipids in muscle, and in turn leading to nefarious results such as increased lipid deposition and/or production of ROS and RNS in muscle. Alternatively, the loss of OXPHOS gene expression could merely be a molecular sentinel for decreased PGC-1α expression, which leads to insulin resistance through mechanisms other than mitochondrial regulation. All of these factors together may contribute importantly to a prediabetic state. Skeletal muscle from mice deficient in either PGC-1α or β have demonstratable deficiencies in mitochondrial gene expression, number, and function.[28,42,43] Surprisingly, the mice do not show whole-body resistance to insulin or glucose.[29,42,43] However, the mice have complex whole-body phenotypes, including important alterations in activity, circadian rhythms, MVO2, and fat content, all of which render changes in serum glucose in response to exogenous glucose or insulin difficult to interpret. Proof or disproof of the "mitochondrial/PGC-1" hypothesis will therefore need to wait for skeletal muscle-specific deletion of PGC-1α and β, or other methods of modulating PGC-1 activity such as pharmaceutical intervention.

8.4.7 *PGC-1α and the heart*

Pumping ten tons of blood daily is a work intensive endeavor, so the human heart consumes energy voraciously. For this, cardiac muscle relies mostly on oxidative phosphorylation, which is far more efficient than the anaerobic generation of ATP. The level of cardiac MVO2 is thus the highest as compared to levels in other tissue in the body, and nearly a third of the volume of each cardiac muscle cell is occupied by mitochondria. The proper function of this machinery is critical to health. Inherited defects in mitochondria in humans, albeit rare, can lead to cardiomyopathy and death.[44] Acquired defects in various components of mitochondria and OXPHOS can also be demonstrated in more generic forms of heart failure.[45]

PGC-1α and β are abundantly expressed in the heart, consistent with the abundance of mitochondria.[3,46] PGC-1α is a powerful inducer of fatty acid oxidation and the mitochondrial program in the heart, as in other tissues.[47,48] Somewhat surprisingly, mice lacking either PGC-1α or β are viable, and cardiac function is largely preserved. However, closer scrutiny reveals important energetic defects. Genes involved in mitochondrial biology are repressed in hearts from PGC-1$\alpha^{-/-}$ mice compared to wild type controls, and mitochondrial enzymatic activities are blunted.[28,42] Explanted beating hearts from these mice have decreased levels of ATP compared to wild type controls, as measured by nuclear magnetic resonance.[28] As a result, the amount of work PGC-1$\alpha^{-/-}$ hearts can perform is significantly diminished, especially in response to adrenergic agents like dobutamine.[28] The cardiac phenotype of PGC-1$\beta^{-/-}$ mice has been less extensively studied, and double knockout mice are unlikely to be forthcoming, given the nearly absolute need for at least one of PGC-1α or PGC-1β for efficient mitochondrial biogenesis.[17]

Following insults like a myocardial infarction or progressive valvular disease, the heart undergoes a series of still poorly understood adaptive and maladaptive changes. Although these changes may be beneficial in the short term by, for example, compensating for hemodynamic instability, in the long run, they can progressively lead to maladaptation and heart failure. A number of metabolic changes occur during this period.[45] The heart abandons lipids as its substrate of choice and turns to carbohydrates. Oxidative phosphorylation is blunted. It has been suggested for many decades that these metabolic derangements may play a causal role in the progression to heart failure, and that the failing heart is an energy-starved heart.

The possibility that PGC-1α may be involved in this energetic maladaptation was first suggested by the observation that levels of PGC-1α diminish during cardiac remodeling in rodent models.[49,50] Since then, a number of other models of heart failure have shown similar, sometimes spectacular reductions in PGC-1α, invariably accompanied by the reductions in target genes of PGC-1α like OXPHOS and fatty acid oxidation.[51,52] Submitting PGC-1$\alpha^{-/-}$ mice to pathological cardiac stress leads to fulminant cardiac failure, demonstrating that reduced levels of PGC-1α predisposes to heart failure.[53] These observations have led to the notion that diminishing levels of PGC-1α during cardiac remodeling is a maladaptive process that leads to diminished reserves of ATP generation and eventual cardiac insufficiency. Blocking this process may therefore be a novel avenue for therapeutic intervention in heart failure. Notably, however, massive overexpression of PGC-1α leads to such powerful induction of mitochondrial biogenesis that myofibers are displaced and heart failure ensues.[47,48] More graded induction of PGC-1α will therefore be needed to test this notion.

8.5 Future Prospects

Research on whether the PGC-1 coactivators represent a target for the development of a new class of therapeutics for human disease is important. Classically, transcription factors and other proteins involved in transcription are not viewed as favorable targets for drug development, with the exception of the nuclear receptors. However, the expression of the PGC-1 coactivators is highly regulated, especially that of PGC-1α, suggesting that drugs that modulate the levels of this protein could be developed. Supporting this concept is the data that PGC-1α is modulated by common signal transduction pathways such as cAMP, AMPK, and the Ca^{++} systems. On the other hand, while elevation of PGC-1α in all tissues could be expected to yield certain benefits in obesity/diabetes, muscle and brain diseases, increased PGC-1α in liver might activate the gluconeogenic gene program. This could result in increased glucose and/or insulin levels, neither of which is desirable chronically.

There are three potential solutions to this problem. The first and most obvious is to apply drugs that elevate PGC-1α expression only to the most serious medical conditions, where elevated blood glucose would not be an unacceptable side effect. Examples of this would be neurodegenerative diseases such as Parkinson's, Alzheimer's and Huntington's. The anti-neurodegenerative effects of PGC-1α seem quite general,[20] at least in mice, so the potential use of agents raising PGC-1α levels in all tissues for these fatal and terrible diseases seems well founded. Similarly, Duchenne's muscular dystrophy is fatal by the late teens or early 20's, and if the elevation of PGC-1α expression can retard this process, certain side effects may be acceptable.

At the other extreme, relatively little in the way of side effects would be tolerable in the context of obesity and type 2 diabetes. Effective treatments exist for type 2 diabetes, and the risk profile of any new agents for this disease must be highly favorable. Few medical treatments for obesity exist, but because of the very chronic nature of this condition, and the fact that it is not necessarily fatal, tolerance for side effects will be low. Treatment of diabetes and obesity with agents that raise PGC-1α would probably depend on the existence of a therapeutic window between levels that will increase energy expenditure in fat and muscle, and those that would increase gluconeogenesis in liver. Whether such a therapeutic window exists is not known.

Lastly, it is possible that tissue-selective activators of PGC-1 expression or action could be developed. Conceivably, agonists for particular nuclear receptors could be synthesized that stimulate the docking of PGC-1α in preference to other coactivators. Such nuclear receptors could be broadly expressed or could be highly tissue-selective. Preferential docking on a tissue-selective receptor could effect a program of energy expenditure in a more selective manner than a drug which activated PGC-1 expression in all tissues.

References

1. Tontonoz P, Hu E, Spiegelman BM, Stimulation of adipogenesis in fibroblasts by PPAR gamma 2, a lipid-activated transcription factor, *Cell* **79**(7):1147–1156, 1994.
2. Tai TA, *et al.*, Activation of the nuclear receptor peroxisome proliferator-activated receptor gamma promotes brown adipocyte differentiation, *J Biol Chem* **271**(47):29909–29914, 1996.
3. Puigserver P, *et al.*, A cold-inducible coactivator of nuclear receptors linked to adaptive thermogenesis, *Cell* **92**(6):829–839, 1998.
4. Monsalve M, *et al.*, Direct coupling of transcription and mRNA processing through the thermogenic coactivator PGC-1, *Mol Cell* **6**(2):307–316, 2000.
5. Wallberg AE, *et al.*, Coordination of p300-mediated chromatin remodeling and TRAP/mediator function through coactivator PGC-1alpha, *Mol Cell* **12**(5):1137–1149, 2003.
6. Rodgers JT, *et al.*, Nutrient control of glucose homeostasis through a complex of PGC-1alpha and SIRT1, *Nature* **434**(7029):113–118, 2005.
7. Lerin C, *et al.*, GCN5 acetyltransferase complex controls glucose metabolism through transcriptional repression of PGC-1alpha, *Cell Metab* **3**(6):429–438, 2006.
8. Baur JA, *et al.*, Resveratrol improves health and survival of mice on a high-calorie diet, *Nature* **444**(7117):337–342, 2006.
9. Lagouge M, *et al.*, Resveratrol improves mitochondrial function and protects against metabolic disease by activating SIRT1 and PGC-1alpha, *Cell* **127**(6):1109–1122, 2006.
10. Cooper MP, *et al.*, Defects in energy homeostasis in Leigh syndrome French Canadian variant through PGC-1alpha/LRP130 complex, *Genes Dev* **20**(21):2996–3009, 2006.

11. Finck BN, *et al.*, Lipin 1 is an inducible amplifier of the hepatic PGC-1alpha/PPARalpha regulatory pathway, *Cell Metab* **4**(3):199–210, 2006.
12. Fan M, *et al.*, Suppression of mitochondrial respiration through recruitment of p160 myb binding protein to PGC-1alpha: Modulation by p38 MAPK, *Genes Dev* **18**(3):278–289, 2004.
13. Knutti D, Kressler D, Kralli A, Regulation of the transcriptional coactivator PGC-1 via MAPK-sensitive interaction with a repressor, *Proc Natl Acad Sci USA*, **98**(17):9713–9718, 2001.
14. Puigserver P, *et al.*, Cytokine stimulation of energy expenditure through p38 MAP kinase activation of PPARgamma coactivator-1, *Mol Cell* **8**(5):971–982, 2001.
15. Teyssier C, *et al.*, Activation of nuclear receptor coactivator PGC-1alpha by arginine methylation, *Genes Dev* **19**(12):1466–1473, 2005.
16. Gleyzer N, Vercauteren K, Scarpulla RC, Control of mitochondrial transcription specificity factors (TFB1M and TFB2M) by nuclear respiratory factors (NRF-1 and NRF-2) and PGC-1 family coactivators, *Mol Cell Biol* **25**(4):1354–1366, 2005.
17. Uldry M, *et al.*, Complementary action of the PGC-1 coactivators in mitochondrial biogenesis and brown fat differentiation, *Cell Metab* **3**(5):333–341, 2006.
18. St-Pierre J, *et al.*, Bioenergetic analysis of peroxisome proliferator-activated receptor gamma coactivators 1alpha and 1beta (PGC-1alpha and PGC-1beta) in muscle cells, *J Biol Chem* **278**(29):26597–26603, 2003.
19. Valle I, *et al.*, PGC-1alpha regulates the mitochondrial antioxidant defense system in vascular endothelial cells, *Cardiovasc Res* **66**(3):562–573, 2005.
20. St-Pierre J, *et al.*, Suppression of reactive oxygen species and neurodegeneration by the PGC-1 transcriptional coactivators, *Cell* **127**(2):397–408, 2006.
21. Herzig S, *et al.*, CREB regulates hepatic gluconeogenesis through the coactivator PGC-1, *Nature* **413**(6852):179–183, 2001.
22. Booth FW, Thomason DB, Molecular and cellular adaptation of muscle in response to exercise: Perspectives of various models, *Physiol Rev* **71**(2):541–585, 1991.
23. Lin J, *et al.*, Transcriptional co-activator PGC-1 alpha drives the formation of slow-twitch muscle fibres, *Nature* **418**(6899):797–801, 2002.
24. Russell AP, *et al.*, Endurance training in humans leads to fiber type-specific increases in levels of peroxisome proliferator-activated receptor-gamma coactivator-1 and peroxisome proliferator-activated receptor-alpha in skeletal muscle, *Diabetes* **52**(12):2874–2881, 2003.
25. Lin J, Handschin C, Spiegelman BM, Metabolic control through the PGC-1 family of transcription coactivators, *Cell Metab* **1**(6):361–370, 2005.

26. Naya FJ, *et al.*, Stimulation of slow skeletal muscle fiber gene expression by calcineurin *in vivo*, *J Biol Chem* **275**(7):4545–4548, 2000.
27. Arany Z, *et al.*, The transcriptional coactivator PGC-1beta drives the formation of oxidative type IIX fibers in skeletal muscle, *Cell Metab* **5**(1):35–46, 2007.
28. Arany Z, *et al.*, Transcriptional coactivator PGC-1 alpha controls the energy state and contractile function of cardiac muscle, *Cell Metab* **1**(4):259–271, 2005.
29. Lin J, *et al.*, Defects in adaptive energy metabolism with CNS-linked hyperactivity in PGC-1alpha null mice, *Cell* **119**(1):121–135, 2004.
30. Wu H, *et al.*, Activation of MEF2 by muscle activity is mediated through a calcineurin-dependent pathway, *EMBO J* **20**(22):6414–6423, 2001.
31. Cao PR, Kim HJ, Lecker SH, Ubiquitin-protein ligases in muscle wasting, *Int J Biochem Cell Biol* **37**(10):2088–2097, 2005.
32. Sacheck JM, *et al.*, Rapid disuse and denervation atrophy involve transcriptional changes similar to those of muscle wasting during systemic diseases, *Faseb J* **21**(1):140–155, 2007.
33. Lecker SH, *et al.*, Multiple types of skeletal muscle atrophy involve a common program of changes in gene expression, *FASEB J* **18**(1):39–51, 2004.
34. Sandri M, *et al.*, PGC-1alpha protects skeletal muscle from atrophy by suppressing FoxO3 action and atrophy-specific gene transcription, *Proc Natl Acad Sci USA* **103**(44):16260–16265, 2006
35. Webster C, *et al.*, Fast muscle fibers are preferentially affected in Duchenne muscular dystrophy, *Cell* **52**(4):503–513, 1988.
36. Handschin C, *et al.*, PGC-1(alpha) regulates the neuromuscular junction program and ameliorates Duchenne muscular dystrophy, *Genes Dev* **21**(7):770–783, 2007.
37. Angus LM, *et al.*, Calcineurin-NFAT signaling, together with GABP and peroxisome PGC-1(alpha), drives utrophin gene expression at the neuromuscular junction, *Am J Physiol Cell Physiol* **289**(4):C908–917, 2005.
38. Mootha VK, *et al.*, PGC-1alpha-responsive genes involved in oxidative phosphorylation are coordinately downregulated in human diabetes, *Nat Genet* **34**(3):267–273, 2003.
39. Patti ME, *et al.*, Coordinated reduction of genes of oxidative metabolism in humans with insulin resistance and diabetes: Potential role of PGC1 and NRF1, *Proc Natl Acad Sci USA* **100**(14):8466–8471, 2003.
40. Hickey MS, *et al.*, Skeletal muscle fiber composition is related to adiposity and *in vitro* glucose transport rate in humans, *Am J Physiol* **268**(3 Pt 1): E453–457, 1995.
41. Petersen KF, *et al.*, Mitochondrial dysfunction in the elderly: Possible role in insulin resistance, *Science* **300**(5622): 1140–1142, 2003.

42. Leone T, *et al.*, PGC-1alpha deficiency causes multi-system energy metabolic derangements: muscle dysfunction, abnormal weight control and hepatic steatosis, *PLOS* **3**:e101, 2005.
43. Lelliott CJ, *et al.*, Ablation of PGC-1beta results in defective mitochondrial activity, thermogenesis, hepatic function, and cardiac performance, *PLoS Biol*, **4**(11):e369, 2006.
44. Kelly DP, Strauss AW, Inherited cardiomyopathies, *N Engl J Med* **330**(13):913–939, 1994.
45. Neubauer S, The failing heart — an engine out of fuel, *N Engl J Med* **356**(11):1140–1151, 2007.
46. Lin J, *et al.*, Peroxisome proliferator-activated receptor gamma coactivator 1beta (PGC-1beta), a novel PGC-1-related transcription coactivator associated with host cell factor, *J Biol Chem* **277**(3):1645–1648, 2002.
47. Lehman JJ, *et al.*, Peroxisome proliferator-activated receptor gamma coactivator-1 promotes cardiac mitochondrial biogenesis, *J Clin Invest* **106**(7):847–856, 2000.
48. Russell LK, *et al.*, Cardiac-specific induction of the transcriptional coactivator peroxisome proliferator-activated receptor gamma coactivator-1alpha promotes mitochondrial biogenesis and reversible cardiomyopathy in a developmental stage-dependent manner, *Circ Res* **94**(4):525–533, 2004.
49. Lehman JJ, Kelly DP, Transcriptional activation of energy metabolic switches in the developing and hypertrophied heart, *Clin Exp Pharmacol Physiol* **29**(4):339–345, 2002.
50. Garnier A, *et al.*, Depressed mitochondrial transcription factors and oxidative capacity in rat failing cardiac and skeletal muscles, *J Physiol* **551**(Pt 2):491–501, 2003.
51. Sano M, *et al.*, Menage-a-trois 1 is critical for the transcriptional function of PPARgamma coactivator 1, *Cell Metab* **5**(2):129–142, 2007.
52. Sano M, *et al.*, Activation of cardiac Cdk9 represses PGC-1 and confers a predisposition to heart failure, *EMBO J* **23**(17):3559–35569, 2004.
53. Arany Z, *et al.*, Transverse aortic constriction leads to accelerated heart failure in mice lacking PPAR-gamma coactivator 1-alpha, *Proc Natl Acad Sci USA*, 2006.

Chapter 9

Coregulators in Metabolic and Neurodegenerative Diseases

Jérôme N. Feige, Hiroyasu Yamamoto and Johan Auwerx

Transcriptional coregulators regulate major aspects of mammalian homeostasis. Among these aspects, metabolic and neuronal regulations are primordial for an organism to achieve higher functions and to adapt to its environment. The prominent role played by coregulators in the signaling pathways controlling these aspects of homeostasis culminates in their implication in major pathologies. Therefore, we review in this chapter how coregulators influence metabolic diseases such as obesity and type 2 diabetes as well as neurodegenerative disorders including Alzheimer's and Huntington's diseases. Interestingly, while metabolic and neuronal regulations have for long been studied distinctly, recent evidence highlights an important cross-talk whereby metabolic homeostasis is central to neuronal function. The dual implication of several coregulators in both of these aspects thus opens some parallels for the pharmacological targeting of metabolic and neurodegenerative disorders.

9.1 Introduction

Transcriptional regulation is vital for homeostasis and enables the adaptation of physiological processes to external cues.[1] In eukaryotes, transcription factors such as nuclear receptors (NRs) are in that respect, key players, which integrate signals coming from dietary, metabolic and endocrine pathways to control target gene expression.[2,3] Transcription factors are assisted by coregulators which can either repress (corepressors) or enhance (coactivators) transcriptional activity by modifying and epigenetically remodeling chromatin structure and by bridging transcription factors to the basal transcriptional machinery.[4–6] Coregulators endowed with enzymatic activities directed towards histone and non-histone targets include acetyltransferases such as CBP, p300, p160 coactivators or GCN-5, deacetylases such as the sirtuins which are activated by NAD^+,[7] as well as many other proteins regulating the methylation, ubiquitylation or sumoylation status.[4] Other coregulators function by creating scaffolds for the assembly of multiproteic transcriptional complexes (p160 and PGC-1 coactivators), by regulating splicing (PGC-1α) or by recruiting the pre-initiation complex to transcriptional start sites (mediator complex anchored to NRs via the Med1/TRAP220/DRIP205/PBP subunit). Given the possibility for an individual coregulator to interact with a large panoply of transcription factors, coregulators can integrate the action of multiple pathways and coordinate different steps of biological programs.[6] They therefore play a predominant role in the tight regulation of the genetic programs crucial to homeostasis.[8] Consistent with this implication, coregulator dysfunction is associated with major diseases including metabolic disorders, neurodegeneration, inflammation and cancer. While the latter themes will be addressed in separate chapters of this book, we will focus on the implication of coregulators in metabolic and neurodegenerative diseases with a strong emphasis on studies both in animal models and in humans.

9.2 Coregulators in Metabolic Diseases

Throughout evolution, the human genome has been selected to maximize energy storage, and the mismatch currently existing between our genes and our lifestyle (reduced activity combined with high calorie western diets) results in a broad range of metabolic disorders. Obesity, type 2 diabetes, hyperlipidemia and hypercholesterolemia are highly prevalent diseases which are commonly referred to as the metabolic syndrome. While a wealth of evidence implicates nuclear receptors in the etiology of such diseases, both in animal models and in human studies, the involvement of their transcriptional coregulators has only started emerging, essentially from studies in animal models. In this section, we will first summarize the implication of coregulators in metabolic diseases and then illustrate how coregulators could be targeted to treat metabolic disorders.

9.2.1 *The implication of coregulators in metabolic diseases: regulating the balance between energy intake, storage and expenditure*

9.2.1.1 *Obesity*

According to the World Health Organization, the pandemic of obesity affects over 400 million people worldwide and this figure is expected to double within the next decade. Obesity is a multifactorial disease resulting from multiple genetic and environmental components. However, this complexity can be reduced to a simple model considering energy intake, storage and expenditure as the three determinants which specify an individual's metabolic state. Coregulators can potentially affect these three aspects of energy homeostasis. Energy intake is modulated by the efficiency of nutrient absorption in the intestine and by the amount of food ingested, which is controlled centrally and involves communication between the brain and peripheral tissues such as the gastrointestinal tract and the adipose tissue.[9] Despite recent progress in the identification of peptides controlling food intake, the cross-talk potentially existing with coregulators remains unexplored. In contrast, coregulators play major roles in regulating the balance between energy storage and expenditure, and deregulations can impede energetic homeostasis and lead to metabolic disorders.

The homologous acetyltransferases CBP and p300, which coactivate nuclear receptors as well as many other transcription factors by acetylating histone tails, are involved in the development and integrity of the adipose tissue. Heterozygous CBP-deficient mice are strongly lipodystrophic and resistant to obesity induced by high fat feeding, a phenotype which has been linked to the involvement of CBP in adipocyte differentiation through the coactivation of PPARγ and C/EBPβ, and in SREBP-mediated lipogenesis.[10] In addition, the related coactivator p300 is also required for adipogenesis.[11] These coactivators most probably also exacerbate fat storage by inhibiting energy expenditure. PPARα and PPARβ/δ target genes controlling fatty acid oxidation and energy uncoupling are induced in the muscle, liver and brown adipose tissue (BAT) of $CBP^{+/-}$ mice,[10] and $CBP^{+/-}$ animals have increased adiponectin levels,[10] further testifying for a catabolic state.[12] Despite the identification of human mutations and the translocations of the CBP and p300 genes responsible for developmental pathologies and certain forms of cancer,[13] no mutations of these coactivators have been described in obese patients.

The steroid receptor coactivators (SRCs) also influence fat storage. The invalidation of SRC-2 (or TIF2 / GRIP1) in mice results in a lean phenotype which correlates with the decreased expression of genes involved in fatty acid uptake and storage and with increased lipolysis.[14] In addition, the expression of genes required for fatty acid synthesis, such as the fatty acid synthase, were decreased in the liver of SRC-2$^{-/-}$ mice.[15] Together with the induction of SRC-2 in WAT upon high-fat feeding and the positive actions of SRC-2 on adipocyte differentiation,[14,16] these results demonstrate that SRC-2 exerts positive effects on fat storage, at least in part by coactivating PPARγ. In addition, SRC-2 also inhibits adaptive thermogenesis in BAT[14] and mitochondrial function in skeletal muscle (Coste and Auwerx, unpublished results), through the repression of the expression of PGC-1α and of its downstream oxidative targets. SRC-3 (or p/CIP/AIB-1/ACTR) also plays an important role in fat storage since it is enriched in the adipose tissue and its inactivation results in decreased adiposity and smaller adipocytes even in mice fed regular chow.[16] Consistently, the absence of SRC-3 totally abolishes adipocyte differentiation by altering PPARγ-dependent transcription of genes important for lipid storage and by inhibiting the capacity of C/EBP α and δ to induce the adipose tissue-specific PPARγ2 isoform. In contrast, SRC-1 protects from obesity by inducing thermogenesis in BAT.[14] Although SRC-1 was proposed to cooperate with SRC-3 because of the lean phenotype of SRC-1/SRC-3 knock-out mice,[17] the absence of a metabolic phenotype in the SRC-1$^{-/-}$ and SRC-3$^{-/-}$ mice in this study strongly contrasts with two reports demonstrating that SRC-1$^{-/-}$ and SRC-3$^{-/-}$ mice are respectively fatter and leaner than their wild-type littermates.[14,16]

The ligand-dependent corepressor RIP140 acts as an inhibitor of energy expenditure in WAT by blocking mitochondrial function in a tissue specialized in fat storage. RIP140$^{-/-}$ mice are leaner than their control littermates even when challenged by high-fat feeding,[18] but this phenotype is not linked to a defect in adipogenesis.[18,19] RIP-140 represses the uncoupling of respiration through a direct inhibition of the UCP-1 promoter.[18,19] In addition, RIP140 also inhibits other aspects of energy expenditure in the adipose tissue as it represses genes implicated in fatty acid oxidation, mitochondrial biogenesis and oxidative phosphorylation, resulting in increased mitochondrial density in adipocytes lacking RIP140.[20]

The PPARγ coactivator-1α (PGC-1α) is the founding member of a family of three related proteins which control major metabolic functions through the coactivation of NRs as well as of other transcription factors.[21,22] While the PGC-1-related coactivator (PRC) is expressed ubiquitously, the expression of PGC-1α and PGC-1β is enriched in mitochondria-rich tissues such as BAT and cardiac and skeletal muscles where PGC-1 family members cooperate to control mitochondrial functions such as oxidative phosphorylation and mitochondrial biogenesis. PGC-1α-mediated mitochondrial control requires the nuclear respiratory factors (NRFs)[23,24] and ERRα,[25,26] transcription factors controlling mitochondrial DNA synthesis and replication, as well as the expression of many subunits of the respiratory chain.[27] In addition, PGC-1α also induces fatty acid oxidation by coactivating PPARα and thyroid hormone receptors.[28–30] The prominent roles of the PGC-1 coactivators in energy expenditure which will be further described in the context of diabetes strongly impact metabolic regulations. PGC-1s can therefore potentially limit fat accretion in the white adipose tissue through an interorgan cross talk where energy is dissipated in tissues such as the skeletal muscle and the BAT. Consistently, PGC-1α expression is reduced in the white adipose tissue from obese patients,[31] and conversely, calorie restriction which induces weight loss in obese subjects, increases PGC-1α expression in skeletal muscle.[32]

The beneficial effects of caloric restriction on longevity and body mass have been associated, at least in part, to the activation of Sirtuin 1 (Sirt1) by elevated levels of NAD^+.[33] In humans, Sirt1 expression is upregulated in calorie-restricted obese patients,[32] and mouse models of obesity have reduced levels of Sirt1 expression.[34] Moreover, Sirt1 activators protect against diet-induced obesity in animal models.[35,36] The protective effects of Sirt1 on weight gain could be linked to the inhibition of the adipogenic actions of PPARγ,[37] but the promotion of energy expenditure most presumably participates to this protection to a large extent. Indeed, Sirt1 activation induces the phosphorylation of the AMP-activated protein kinase (AMPK) and of its downstream oxidative target acetylCoA carboxylase in the liver.[35] Sirt1 also promotes mitochondrial function and fatty acid oxidation in skeletal muscle and BAT by synergizing with and activating PGC-1α.[36,38] In addition, Sirt1 promotes the expression of adiponectin,[34] an adipokine which controls fatty acid oxidation and whose expression is inversely correlated to metabolic disorders. Consistent with these results in mouse models, several single nucleotide polymorphisms associated with variations in energy expenditure in humans were identified in the Sirt1 gene.[36] Furthermore, Sirt3, another member of the sirtuin family, also controls mitochondrial respiration and uncoupling in BAT where it positively regulates thermogenesis.[39] Together with the reduced Sirt3 expression in BAT from obese animal models, this suggests that Sirt3 could participate to the pathogenesis of obesity.

9.2.1.2 *Insulin resistance and type 2 diabetes*

Type 2 diabetes is clinically characterized by a hyperglycemia in fasted patients, resulting primarily from a desensitization to the hypoglycemic actions of insulin. The etiology of type 2 diabetes is, however, a complex issue, and multiple causes can lead to insulin resistance. Nevertheless, obesity and dyslipidaemia constitute two important contributing factors.[40] An excess of circulating free fatty acids (FFA) causes insulin resistance by altering the ability of skeletal muscle to properly take up and metabolize glucose. Type 2 diabetes development is paralleled by a progressive loss of the function and/or mass of pancreatic β-cells, which results in decreased insulin secretion and a further worsening of glucose tolerance. Interestingly, mitochondrial defects, notably in muscle and β-cells, constitute a central factor in the development of insulin resistance and β-cell dysfunction.[41]

Given the tight link between obesity and type 2 diabetes and the absence of clearly established molecular causes to insulin resistance, the action of coregulators on global insulin sensitivity is very often correlated to a lean or an obese phenotype while their precise contribution remains difficult to dissect. For example, SRC-2, RIP140 and CBP all seem to cause insulin resistance since glucose tolerance and insulin sensitivity are enhanced by their genetic invalidation in mice.[10,14,20] The fact that glucose tolerance is correlated with a lean phenotype in these three animal models constitutes a confounding factor which suggests that the effects on glucose homeostasis could be a consequence of differences in adiposity. However, specific mechanisms also probably contribute to insulin sensitivity. The action of RIP140 could relate to its negative effect on glucose uptake in adipocytes,[20] but since this cell type is not the primary site of glucose utilization, other mechanisms probably exist in the skeletal muscle and in the pancreas. The actions of CBP may be linked to its role in gluconeogenesis, as CREB phosphorylated after glucagon-dependent cAMP activation, can recruit CBP to stimulate hepatic glucose production.[42] However, other major sites such as the muscle and the pancreas also deserve investigation.

Given the strong link between mitochondrial function and insulin sensitivity,[41] the prominent role of PGC-1 coactivators in mitochondria potentially implicates them in the pathophysiology of type 2 diabetes. Importantly, the expression of PGC-1α and PGC-1β is decreased in skeletal muscle from diabetic subjects, thereby contributing to impaired oxidative phosphorylation.[24,43] Since mitochondrial dysfunctions have been proposed as a cause of insulin resistance in human skeletal muscle,[44,45] it is likely that the PGC-1 coactivators are instrumental to maintain insulin action and that their deregulation could contribute to the development of insulin resistance. Consistent with this hypothesis, a Gly482Ser mutation of PGC-1α associates with type 2 diabetes in some but not all human populations.[46] Furthermore, since oxidative stress induced by radical oxygen species (ROS) has been implicated in the etiology of insulin resistance,[47] the positive actions of PGC-1α on ROS detoxifying enzymes such as the superoxide dismutase SOD2 and the glutathione peroxidase GPx1 potentially participate in the maintenance of a correct sensitivity to insulin.[48] In contrast, the action of PGC-1α in the liver seems to inhibit insulin sensitivity since coactivation of the hepatocyte nuclear factor 4α (HNF-4α) and of the forkhead box protein O 1 (FOXO-1) stimulates gluconeogenesis.[22] The increased expression of PGC-1α in the liver of diabetic animal models suggests that it can participate in the onset of hyperglycemia by triggering gluconeogenesis.[49] PGC-1α is also increased in the pancreas of diabetic animals where it inhibits insulin secretion.[50] Thus, potential pharmacological intervention targeting PGC-1α for the treatment of type 2 diabetes will require avoiding adverse effects by affecting the beneficial biological programs only. The possibility to uncouple the positive actions of PGC-1α on oxidative phosphorylation from the negative ones on gluconeogenesis by affecting the PGC-1α/ERRα interaction suggests that this goal is feasible.[25] Furthermore, PGC-1β and PRC, which participate with PGC-1α to a common regulatory network to control energy expenditure by promoting oxidative phosphorylation,[51,52] could constitute alternative targets. The absence of gluconeogenic action of PGC-1β is of particular interest in that respect.[53]

Sirtuins are also implicated in the control of glucose homeostasis. Sirt1 positively regulates insulin secretion from pancreatic β cells by repressing UCP2 and thereby increasing the coupling of respiration to ATP synthesis.[54] In addition, β cell specific overexpression of Sirt1 improves glucose tolerance,[55] and Sirt1 activation in skeletal muscle enhances insulin sensitivity by deacetylating and activating PGC-1α.[36] Sirt1 activation could also exert beneficial effects by protecting β cell loss from oxidative stress, via deacetylation of FOXO1.[56] However, the output of hepatic Sirt1 activation is still debatable as on the one hand, Sirt1 induces beneficial effects such as AMPK activation and mitochondrial biogenesis;[35] but on the other, it promotes gluconeogenesis by deacetylating and activating PGC-1α.[57] In contrast, Sirt4 seems to have a negative output on glucose homeostasis by inhibiting amino acid-stimulated insulin secretion from β cells.[58] Sirt4 catalyzes the ADP-ribosylation of glutamate dehydrogenase, which inhibits the conversion of glutamate to α-ketoglutarate and thereby inhibits the stimulation of ATP production and the subsequent insulin secretion regulated by this TCA metabolite. Altogether, the sirtuins are profiling themselves as important players in energy homeostasis, warranting further integrative studies to better define the pleiotropic functions of these coregulators.

9.2.2 *Targeting coregulators for the treatment of metabolic disorders*

Coregulators are clearly emerging as major metabolic actors, which beyond the control of "metabolic" gene expression by transcription factors, provide a second more global level of transcriptional "metabolic adaptation" which integrates various signaling pathways and physiological stimuli. Targeting regulatory nodes under coregulator control to pharmacologically combat metabolic disorders is an ambition which can now be foreseen. PGC-1α is clearly, to date, the prime candidate to achieve this goal since it plays a prominent role in energy expenditure and metabolic diseases, and the pathways controlling its expression, its interaction with transcription factors and its activity can potentially all three be targeted. However, sirtuins are not far down the road given the possibility to pharmacologically modulate their activity.

PGC-1α appears as a unique regulator of metabolic programs because its expression is regulated by physiological challenges such as fasting, cold exposure or exercise. PGC-1α expression is also controlled by several metabolic signaling pathways. Direct transcriptional effects on the PGC-1α promoter involve activation by CREB and its coactivator TORC (transducer of regulated CREB activity).[59–61] Although this signaling pathway promotes the PGC-1α-dependent mitochondrial functions leading to enhanced energy expenditure,[60] targeting PGC-1α through this pathway could eventually coinduce a detrimental hepatic gluconeogenic program.[59,61] PPARβ/δ can also directly activate the PGC-1α promoter,[62] suggesting that the beneficial effects of PPARβ/δ agonists could be linked to enhanced PGC-1α signaling. Furthermore, the activation of the calcium/calmodulin-dependent protein kinase (CaMK)[63] or the inhibition of the S6 kinase[64] control mitochondrial biogenesis at least in part by increasing PGC-1α expression.

The actions of PGC-1α on metabolism are also controlled by post-translational modifications. Such modifications are of major importance since they link intracellular signaling pathways to transcriptional regulation. For example, PGC-1α phosphorylation is most likely involved in translating the effect of cytokine stimulation to oxidative metabolism through p38 MAP kinase-mediated phosphorylation of PGC-1α which enhances its activity.[65,66] In addition, methylation by the protein arginine methyltransferase 1 (PRMT1) can also enhance PGC-1α activity,[67] although the physiological importance of this regulation remains to be determined. The most prominent PGC-1α post-translational modification is most probably acetylation. Indeed, PGC-1α activity was shown to be inhibited by acetylation and the PGC-1α acetylation status is directly regulated by the balanced action of the GCN5 acetyltransferase and the Sirt1 deacetylase.[38,57,68] GCN5 was reported to inhibit PGC-1α-dependent hepatic gluconeogenesis while Sirt1-mediated PGC-1α deacetylation and activation promotes glucose production in hepatocytes and energy expenditure in muscle and BAT. The Sirt-1/PGC-1α connection therefore provides a means to link the energy status of a cell, sensed by Sirt-1 through the NAD^+/NADH ratio,[33] to a transcriptional output on metabolic networks regulated by PGC-1α. The possibility to target Sirt1, by natural or synthetic agonists, and to thereby coaffect PGC-1α signaling therefore opens this regulatory axis to nutraceutical and pharmacological intervention to target metabolic disorders. This possibility has recently been investigated by treating mice with resveratrol, a polyphenol naturally found in red grapes which activates Sirt-1. Resveratrol-treated mice were protected from obesity, essentially because of enhanced energy expenditure which was testified by increased exercise endurance and resistance to cold.[36] This effect was shown to be caused by the Sirt1-dependent and subsequent activation of PGC-1α.[36] In addition, insulin sensitivity was enhanced by resveratrol.[35,36] Interestingly, resveratrol also prolonged lifespan,[35] suggesting that the beneficial effects of Sirt1 activation probably extend beyond the regulation of PGC-1α. The possibility to design more selective and more efficacious Sirt1 activators and to pharmacologically target other sirtuin members therefore opens a new field of drug design aimed at targeting the metabolic syndrome.

9.3 *Coregulators in neurodegenerative diseases*

In higher eukaryotes, neurons ensure a wide range of vital processes including cognitive functions in the brain, the coordination of the activity of different organs and their adaptation to external cues, and the control of motor-sensory information enabling coordinated locomotion. At the cellular level, these processes are largely orchestrated by changes in gene expression which rely to a large extent on the regulation of transcription factor activity by coregulators. Neurons exhibit specific metabolic requirements, and a wealth of evidence implicates metabolic regulations as key features of neuronal function. It is therefore not surprising to uncover a common implication of certain coregulators in neuronal and metabolic physiology and disease. Although we believe that this crosstalk has been understudied in the past years and probably extends far beyond what is currently known, emerging evidence implicates histone acetyltransfrases (HATs), histone deacetylases (HDACs), as well as PGC-1 coactivators in neuronal function and in neurodegenerative disorders.

Neurodegenerative diseases are characterized by the deterioration of neurons which are progressively lost over time. These diseases are classified into pathologies affecting movement and locomotion such as ataxia, and pathologies affecting memory which can ultimately lead to dementia, while some overlap is possible between these two major classes. Major neurodegenerative diseases include Huntington's disease, Alzheimer's disease, Parkinson's disease, and amyotrophic lateral sclerosis. Huntington's disease (HD), also known as Huntington chorea, is an inherited disease characterized by choreiform movements and progressive dementia. HD has been genetically identified as a trinucleotide CAG-repeat mutation on chromosome four which leads to a polyglutamine extension of the huntingtin protein, with the length of the repetition relating to the age of onset and to the rate of disease progression. Alzheimer's disease (AD) is the most common cause of dementia, with a lifetime risk estimated to be as high as 1:4 to 1:2 and more than 14% of individuals older than 65 years suffering from this pathology. AD is characterized by progressive cognitive deterioration which causes declining activities of everyday life, neuropsychiatric symptoms and behavioural changes. While the genetic cause of AD are unknown, the pathology is characterized by the presence of amyloid plaques of aggregated Aβ peptide, by intracellular neurofibrillary tangles and by pronounced cell death.[69] Parkinson's disease (PD) is another common neurological disorders affecting approximately 1% of individuals older than 60 years. The major symptoms of PD are resting tremor, rigidity, bradykinesia and postural instability, which can ultimately culminate in dementia in 15%–30 % of the patients. These symptoms are thought to be related with the loss of pigmented dopaminergic neurons in the substantia nigra pars compacta and with the accumulation of intraneuronal inclusions known as Lewy bodies which are enriched in ubiquitin and α-synuclein.[70] Finally, amyotrophic lateral sclerosis (ALS) is a progressively devastating disorder of the anterior horn cells of the spinal cord and the motor cranial nuclei, leading to progressive muscle weakness and atrophy. Approximately 10% of cases are familial cases out of which a fifth results from mutations in the gene encoding for the ROS detoxifying enzyme SOD1.[71]

9.3.1 *CBP loss of function*

Consistent with the strong implication of CREB in neuronal integrity,[72–74] the CREB coactivator CBP plays a key role in several neurological disorders. Rubinstein-Taybi syndrome, a disease characterized by skeletal malformation with mental retardation, has been associated with CBP loss of function in humans.[75,76] Moreover, heterozygous invalidation of CBP in mice recapitulates the symptoms of this disease.[75,77,78] The disruption of the histone acetyltransferase (HAT) activity of CBP as well as its ability to transactivate CREB is supposed to be the cause of this syndrome.[79]

Decreased expression of CBP is also observed in AD, which probably results from reduced presenilin-dependent Notch signaling.[80] The degradation of CBP by caspase-6 can also participate to the pathogenesis of AD by triggering CBP loss of function.[81] This implication has been further evidenced by the identification of the caspase-6 proenzyme as well as the active caspase-6 fragment in the brain from adult patients suffering from AD.[82] A third mechanism inhibiting CREB-CBP signalling could also be implicated in the etiology of AD. Increased levels of Aβ42 peptides inhibit activity-induced phosphorylation of CREB, and thereby lead to a reduction of CRE-dependent gene expression.[83–85]

CBP loss of function is also associated with polyglutamine (polyQ) diseases such as Huntington's disease (HD). CBP can directly interact with the mutated huntingtin protein, leading to a squelching of CBP which reduces the amount of this coactivator available for transcriptional competence.[86–89] This interaction occurs via the HAT domain of CBP and results in the inhibition of HAT activity, thereby triggering global histone deacetylation and cell death.[90,91] Although CBP and p300 are homologous and share overlapping metabolic functions, mutated huntingtin specifically interacts and inhibits CBP without affecting p300.[92] In addition, the instability of CAG repeats causing polyQ extension associates with decreased CBP activity in a drosophila model of HD and pharmacological intervention to restore acetylation suppresses this instability.[93] Altogether, CBP loss of function and subsequent histone deacetylation are thought to be common features of neurodegeneration.

9.3.2 *Therapeutic strategies to combat neurodegeneration through the modulation of CBP activity*

Based on the notion that CBP loss of function causes neurological disorders by inhibiting histone acetylation, HDAC inhibitors have been tested to limit histone deacetylation and thereby limit the progression of these diseases. Sodium butyrate and suberoylanilide hydroxamic acid (SAHA), were shown to induce a significant decrease in neurodegeneration in both a drosophila[90] and a mouse[94,95] model of HD. These two HDAC inhibitors are now moving to clinical trials for HD. However, HDAC inhibitors have a wide range of action and are in some cases, potent inducers of neuronal apoptosis. A strategy to potentially ensure higher specificity has been to directly overexpress CBP. CBP overexpression reduced polyQ-induced aggregation and neurodegeneration in a drosophila model of HD,[91] and blocked neuronal death in the spinal and bulbar muscular atrophy (SBMA) model.[95,96] However, it is noteworthy that the hyperacetylation induced by CBP-overexpression or by HDAC inhibition can lead to cellular toxicity.[81]

These results demonstrate that the fine tuning of histone acetylation levels plays a key role in neuronal homeostasis. HDAC inhibition and CBP overexpression both seem to be of limited value to counteract neurodegeneration, most probably because they induce non-physiological acetylation levels and/or off-target effects. A mild or a selective CBP activation therefore seems required, and compounds working as CBP activators, such as N-(4-chloro-3-trifluoromethyl-phenyl)-2-ethoxy-6-pentadecyl-benzamide (CTPB),[96] could prove useful in the design of neuroprotective reagents. Although the possibility to target CBP to limit and possibly reverse neurodegeneration is one goal which sounds reasonable to aim for, understanding the molecular determinants which implicate the neuron-specific actions of CBP in neurodegeneration will be required to design specific drugs.

9.3.3 *Sirtuins and neuroprotective effects*

Several reports indicate that Sirt1 and nicotinamide metabolism have a protective role in the neuronal system. Axonal degeneration is a common feature and is observed in the early stage of both peripheral neuropathies and neurodegenerative diseases, such as Alzheimer's disease and amyotrophic lateral sclerosis.[97,98] Wallerian degeneration is a self-destructive degenerative process which begins at the distal portion of a transected axon.[99] Interestingly, Wallerian axonal degeneration is delayed in the *wld*^{s} mouse model,[100–102] resulting from the fusion of the N-terminal domain of the ubiquitylation factor Ube4b to the full-length nicotinamide mononucleotide adenyltransferase 1 (Nmnat 1), an enzyme required for both the *de novo* and salvage pathways of NAD^+ biosynthesis.[103–105] Moreover, overexpression of Nmnat1 alone was recently shown to prevent axonal degeneration.[106] These observations suggest that Sirt1 activation, through increased levels of neuronal NAD^+ by Nmnat1, could have a protective function in axonal degeneration.

In mouse models of AD and Parkinson's disease, it is well known that caloric restriction, a process known to activate Sirt1, can delay the neurodegenerative processes.[107–110] The accumulation of aggregated Aβ peptide in amyloid plaques is hypothesized to initiate a pathological cascade resulting in the onset and progression of AD via

NFκB induction in microglia,[111–114] either by the tumor necrosis factor (TNF)-receptor type 1 or the receptor of advanced glycation end product (AGE).[115,116] Interestingly, the induction of NFκB can be reduced by Sirt1 and by the Sirt1 activator resveratrol,[117] strongly suggesting that Sirt1 can attenuate Aβ-stimulated neurotoxicity and AD-related inflammatory responses. Furthermore, Sirt1 can also prevent Aβ peptide generation by promoting the non-amyloidogenic processing of amyloid precursor protein (APP) through the inhibition of Rho kinase 1 (ROCK1) expression.[118]

Abnormal mitochondrial function has been suspected to be important for the pathogenesis of HD, since mitochondrial dysfunction seems to sensitize neurons to oxidative stress.[119] Patients suffering from HD have reduced glucose metabolism and increased levels of lactate in the basal ganglia, a reduced activity of several key components of the oxidative phosphorylation pathways in the mitochondria of striatal neurons, and pronounced mitochondrial morphological abnormalities including derangement of the matrix and cristae.[120,121] Consistently, Sirt1 and PGC-1α which are major regulators of oxidative mitochondrial metabolism have been implicated in the pathogenesis of HD. HD can result, at least in part, from reduced PGC-1α activity through direct interaction of mutated huntingtin with the PGC-1α promoter or with PGC-1α itself.[122,123] Moreover, PGC-1α$^{-/-}$ mice are more sensitive to chemically-induced neurodegeneration of the substantia nigra and the hippocampus, at least in part because of impaired ROS-scavenging by PGC-1α.[48] PGC-1α also controls the transcription of a neuromuscular genetic program by synergizing with GA-binding protein (GABP; alternatively called nuclear respiratory factor 2), and thereby induces a protective role in animal models of Duchenne muscular dystrophy.[124]

Since PGC-1α activity is stimulated by Sirt1-mediated deacetylation,[36,38,57] and that other sirtuins also control mitochondrial metabolism, the modulation of sirtuin activity could constitute an interesting approach for the therapy of HD. Preliminary results in cellular and animal HD models are encouraging since the polyQ-dependent neuronal dysfunction observed in HD can be rescued by overexpression of Sirt1 or by resveratrol treatment,[125] while this favorable effect is suppressed by sirtuin inhibitors, such as nicotinamide or sirtinol.

Altogether, increasing evidence supports the involvement of coregulators in neurological disorders. The prominent role of acetyltransferases and deacetylases outlines a major role for the acetylation status of both histone and non-histone substrates in neuronal pathophysiology for which pharmacological intervention may not be too far down the road. Furthermore, a picture emerges where the metabolic implication of coregulators in energy expenditure is tightly linked to their neuronal actions because of the strong implication of mitochondrial dysfunctions in neurodegenerative disorders.[119] The implication of other coregulators in neurodegeneration therefore deserves thorough investigation and we predict that the fields of metabolic and neurodegenerative disorders will partially merge in the coming years.

Altogether, coregulators constitute dual targets for drug design in the field of metabolic and neurodegenerative diseases. Combined investigation of the common molecular and cellular pathways and of the organ-specific physiological actions of these key proteins undoubtedly raises hope to translate fundamental research from the bench to the clinic in the near future.

Acknowledgments

We thank the members of the Auwerx laboratory for stimulating interesting discussions. Work in the authors' laboratory was supported by grants from CNRS, INSERM, ULP, Hôpital Universitaire de Strasbourg, FRM, AFM, EU and NIH (DK598204). JNF is supported by a FEBS fellowship.

References

1. Jacob F, Monod J, Genetic regulatory mechanisms in the synthesis of proteins, *J Mol Biol* **3**:318–356, 1961.
2. Desvergne B, Michalik L, Wahli W, Transcriptional regulation of metabolism, *Physiol Rev* **86**:465–514, 2006
3. Francis GA, Fayard E, Picard F, *et al.*, Nuclear receptors and the control of metabolism, *Annu Rev Physiol* **65**:261–311, 2003.
4. Rosenfeld MG, Lunyak VV, Glass CK, Sensors and signals: A coactivator/corepressor/epigenetic code for integrating signal-dependent programs of transcriptional response, *Genes Dev* **20**:1405–1428, 2006
5. Smith CL, O'Malley BW, Coregulator function: A key to understanding tissue specificity of selective receptor modulators, *Endocr Rev* **25**:45–71, 2004
6. Spiegelman BM, Heinrich R, Biological control through regulated transcriptional coactivators, *Cell* **119**:157–167, 2004.

7. Yamamoto H, Schoonjans K, Auwerx J, Sirtuin functions in health and disease, *Mol Endocrinol* **21(8)**:1745–1755, 2007.
8. Feige JN, Auwerx J, Transcriptional coregulators in the control of energy homeostasis, *Trends Cell Biol* **17**:292–301, 2007.
9. Morton GJ, Cummings DE, Baskin DG, *et al.*, Central nervous system control of food intake and body weight, *Nature* **443**:289–295, 2006.
10. Yamauchi T, Oike Y, Kamon J, *et al.*, Increased insulin sensitivity despite lipodystrophy in Crebbp heterozygous mice, *Nat Genet* **30**:221–226, 2002.
11. Takahashi N, Kawada T, Yamamoto T, *et al.*, Overexpression and ribozyme-mediated targeting of transcriptional coactivators CREB-binding protein and p300 revealed their indispensable roles in adipocyte differentiation through the regulation of peroxisome proliferator-activated receptor gamma, *J Biol Chem* **277**:16906–16912, 2002.
12. Kadowaki T, Yamauchi T, Kubota N, *et al.*, Adiponectin and adiponectin receptors in insulin resistance, diabetes, and the metabolic syndrome, *J Clin Invest* **116**:1784–1792, 2006.
13. Kalkhoven E, CBP and p300: HATs for different occasions, *Biochem Pharmacol* **68**:1145–1155, 2004.
14. Picard F, Gehin M, Annicotte J, *et al.*, SRC-1 and TIF2 control energy balance between white and brown adipose tissues, *Cell* **111**:931–941, 2002.
15. Jeong JW, Kwak I, Lee KY, *et al.*, The genomic analysis of the impact of steroid receptor coactivators ablation on hepatic metabolism, *Mol Endocrinol* **20**:1138–1152, 2006.
16. Louet JF, Coste A, Amazit L, *et al.*, Oncogenic steroid receptor coactivator-3 is a key regulator of the white adipogenic program, *Proc Natl Acad Sci USA* **103**:17868–17873, 2006.
17. Wang Z, Qi C, Krones A, *et al.*, Critical roles of the p160 transcriptional coactivators p/CIP and SRC-1 in energy balance, *Cell Metab* **3**:111–122, 2006.
18. Leonardsson G, Steel JH, Christian M, *et al.*, Nuclear receptor corepressor RIP140 regulates fat accumulation, *Proc Natl Acad Sci USA* **101**:8437–8442, 2004.
19. Christian M, Kiskinis E, Debevec D, *et al.*, RIP140-targeted repression of gene expression in adipocytes, *Mol Cell Biol* **25**:9383–9391, 2005.
20. Powelka AM, Seth A, Virbasius JV, *et al.*, Suppression of oxidative metabolism and mitochondrial biogenesis by the transcriptional corepressor RIP140 in mouse adipocytes, *J Clin Invest* **116**:125–136, 2006.
21. Lin J, Handschin C, Spiegelman BM, Metabolic control through the PGC-1 family of transcription coactivators, *Cell Metab* **1**:361–370, 2005.
22. Finck BN, Kelly DP, PGC-1 coactivators: Inducible regulators of energy metabolism in health and disease, *J Clin Invest* **116**:615–622, 2006.

23. Wu Z, Puigserver P, Andersson U, *et al.*, Mechanisms controlling mitochondrial biogenesis and respiration through the thermogenic coactivator PGC-1, *Cell* **98**:115–124, 1999.
24. Patti ME, Butte AJ, Crunkhorn S, *et al.*, Coordinated reduction of genes of oxidative metabolism in humans with insulin resistance and diabetes: Potential role of PGC1 and NRF1, *Proc Natl Acad Sci USA* **100**:8466–8471, 2003.
25. Mootha VK, Handschin C, Arlow D, *et al.*, Erralpha and Gabpa/b specify PGC-1alpha-dependent oxidative phosphorylation gene expression that is altered in diabetic muscle, *Proc Natl Acad Sci USA* **101**:6570–6575, 2004.
26. Schreiber SN, Emter R, Hock MB, *et al.*, The estrogen-related receptor alpha (ERRalpha) functions in PPARgamma coactivator 1alpha (PGC-1alpha)-induced mitochondrial biogenesis, *Proc Natl Acad Sci USA* **101**:6472–6477, 2004.
27. Scarpulla RC, Nuclear control of respiratory gene expression in mammalian cells, *J Cell Biochem* **97**:673–683, 2006.
28. Puigserver P, Wu Z, Park CW, *et al.*, A cold-inducible coactivator of nuclear receptors linked to adaptive thermogenesis, *Cell* **92**:829–839, 1998.
29. Lehman JJ, Barger PM, Kovacs A, *et al.*, Peroxisome proliferator-activated receptor gamma coactivator-1 promotes cardiac mitochondrial biogenesis, *J Clin Invest* **106**:847–856, 2000.
30. Vega RB, Huss JM, Kelly DP, The coactivator PGC-1 cooperates with peroxisome proliferator-activated receptor alpha in transcriptional control of nuclear genes encoding mitochondrial fatty acid oxidation enzymes, *Mol Cell Biol* **20**:1868–1876, 2000.
31. Semple RK, Crowley VC, Sewter CP, *et al.*, Expression of the thermogenic nuclear hormone receptor coactivator PGC-1alpha is reduced in the adipose tissue of morbidly obese subjects, *Int J Obes Relat Metab Disord* **28**:176–179, 2004.
32. Civitarese AE, Carling S, Heilbronn LK, *et al.*, Calorie restriction increases muscle mitochondrial biogenesis in healthy humans, *PLoS Med* **4**:e76, 2007.
33. Guarente L, Sirtuins as potential targets for metabolic syndrome, *Nature* **444**:868–874, 2006.
34. Qiao L, Shao J, SIRT1 regulates adiponectin gene expression through FoxO1-C/enhancer-binding protein alpha transcriptional complex, *J Biol Chem* **281**:39915–39924, 2006.
35. Baur JA, Pearson KJ, Price NL, *et al.*, Resveratrol improves health and survival of mice on a high-calorie diet, *Nature* **444**:337–342, 2006.
36. Lagouge M, Argmann C, Gerhart-Hines Z, *et al.*, Resveratrol improves mitochondrial function and protects against metabolic disease by activating SIRT1 and PGC-1alpha, *Cell* **127**:1109–1122, 2006.

37. Picard F, Kurtev M, Chung N, *et al.*, Sirt1 promotes fat mobilization in white adipocytes by repressing PPAR-gamma, *Nature* **429**:771–776, 2004.
38. Gerhart-Hines Z, Rodgers JT, Bare O, *et al.*, Metabolic control of muscle mitochondrial function and fatty acid oxidation through SIRT1/PGC-1alpha, *EMBO J* **26**:1913–1923, 2007.
39. Shi T, Wang F, Stieren E, *et al.*, SIRT3, a mitochondrial sirtuin deacetylase, regulates mitochondrial function and thermogenesis in brown adipocytes, *J Biol Chem* **280**:13560–13567, 2005.
40. Kahn SE, Hull RL, Utzschneider KM, Mechanisms linking obesity to insulin resistance and type 2 diabetes, *Nature* **444**:840–846, 2006.
41. Lowell BB, Shulman GI, Mitochondrial dysfunction and type 2 diabetes, *Science* **307**:384–387, 2005.
42. Zhou XY, Shibusawa N, Naik K, *et al.*, Insulin regulation of hepatic gluconeogenesis through phosphorylation of CREB-binding protein, *Nat Med* **10**:633–637, 2004.
43. Mootha VK, Lindgren CM, Eriksson KF, *et al.*, PGC-1alpha-responsive genes involved in oxidative phosphorylation are coordinately downregulated in human diabetes, *Nat Genet* **34**:267–273, 2003.
44. Petersen KF, Dufour S, Befroy D, *et al.*, Impaired mitochondrial activity in the insulin-resistant offspring of patients with type 2 diabetes, *N Engl J Med* **350**:664–671, 2004.
45. Petersen KF, Befroy D, Dufour S, *et al.*, Mitochondrial dysfunction in the elderly: Possible role in insulin resistance, *Science* **300**:1140–1142, 2003.
46. Soyal S, Krempler F, Oberkofler H, *et al.*, PGC-1alpha: A potent transcriptional cofactor involved in the pathogenesis of type 2 diabetes, *Diabetologia* **49**:1477–1488, 2006.
47. Houstis N, Rosen ED, Lander ES, Reactive oxygen species have a causal role in multiple forms of insulin resistance, *Nature* **440**:944–948, 2006.
48. St-Pierre J, Drori S, Uldry M, *et al.*, Suppression of reactive oxygen species and neurodegeneration by the PGC-1 transcriptional coactivators, *Cell* **127**:397–408, 2006.
49. Yoon JC, Puigserver P, Chen G, *et al.*, Control of hepatic gluconeogenesis through the transcriptional coactivator PGC-1, *Nature* **413**:131–138, 2001.
50. Yoon JC, Xu G, Deeney JT, *et al.*, Suppression of beta cell energy metabolism and insulin release by PGC-1alpha, *Dev Cell* **5**:73–83, 2003.
51. Andersson U, Scarpulla RC, Pgc-1-related coactivator, a novel, serum-inducible coactivator of nuclear respiratory factor 1-dependent transcription in mammalian cells, *Mol Cell Biol* **21**:3738–3749, 2001.
52. St-Pierre J, Lin J, Krauss S, *et al.*, Bioenergetic analysis of peroxisome proliferator-activated receptor gamma coactivators 1alpha and 1beta

(PGC-1alpha and PGC-1beta) in muscle cells, *J Biol Chem* **278**:26597–26603, 2003.
53. Lin J, Tarr PT, Yang R, *et al.*, PGC-1beta in the regulation of hepatic glucose and energy metabolism, *J Biol Chem* **278**:30843–30848, 2003.
54. Bordone L, Motta MC, Picard F, *et al.*, Sirt1 regulates insulin secretion by repressing UCP2 in pancreatic beta cells, *PLoS Biol* **4**:e31, 2006.
55. Moynihan KA, Grimm AA, Plueger MM, *et al.*, Increased dosage of mammalian Sir2 in pancreatic beta cells enhances glucose-stimulated insulin secretion in mice, *Cell Metab* **2**:105–117, 2005.
56. Kitamura YI, Kitamura T, Kruse JP, *et al.*, FoxO1 protects against pancreatic beta cell failure through NeuroD and MafA induction, *Cell Metab* **2**:153–163, 2005.
57. Rodgers JT, Lerin C, Haas W, *et al.*, Nutrient control of glucose homeostasis through a complex of PGC-1alpha and SIRT1, *Nature* **434**:113–118, 2005.
58. Haigis MC, Mostoslavsky R, Haigis KM, F *et al.*, SIRT4 inhibits glutamate dehydrogenase and opposes the effects of calorie restriction in pancreatic beta cells, *Cell* **126**:941–954, 2006.
59. Koo SH, Flechner L, Qi L, *et al.*, The CREB coactivator TORC2 is a key regulator of fasting glucose metabolism, *Nature* **437**:1109–1111, 2005.
60. Wu Z, Huang X, Feng Y, *et al.*, Transducer of regulated CREB-binding proteins (TORCs) induce PGC-1alpha transcription and mitochondrial biogenesis in muscle cells, *Proc Natl Acad Sci USA* **103**:14379–14384, 2006.
61. Herzig S, Long F, Jhala US, *et al.*, CREB regulates hepatic gluconeogenesis through the coactivator PGC-1, *Nature* **413**:179–183, 2001.
62. Schuler M, Ali F, Chambon C, *et al.*, PGC1alpha expression is controlled in skeletal muscles by PPARbeta, whose ablation results in fiber-type switching, obesity, and type 2 diabetes, *Cell Metab* **4**:407–414, 2006.
63. Wu H, Kanatous SB, Thurmond FA, *et al.*, Regulation of mitochondrial biogenesis in skeletal muscle by CaMK, *Science* **296**:349–352, 2002.
64. Um SH, Frigerio F, Watanabe M, *et al.*, Absence of S6K1 protects against age- and diet-induced obesity while enhancing insulin sensitivity, *Nature* **431**:200–205, 2004.
65. Knutti D, Kressler D, Kralli A, Regulation of the transcriptional coactivator PGC-1 via MAPK-sensitive interaction with a repressor, *Proc Natl Acad Sci USA* **98**:9713–9718, 2001.
66. Puigserver P, Rhee J, Lin J, *et al.*, Cytokine stimulation of energy expenditure through p38 MAP kinase activation of PPARgamma coactivator-1, *Mol Cell* **8**:971–982, 2001.

67. Teyssier C, Ma H, Emter R, *et al.*, Activation of nuclear receptor coactivator PGC-1alpha by arginine methylation, *Genes Dev* **19**:1466–1473, 2005.
68. Lerin C, Rodgers JT, Kalume DE, *et al.*, GCN5 acetyltransferase complex controls glucose metabolism through transcriptional repression of PGC-1alpha, *Cell Metab* **3**:429–438, 2006.
69. Hardy JA, Higgins GA, Alzheimer's disease: The amyloid cascade hypothesis, *Science* **256**:184–185, 1992.
70. Maguire-Zeiss KA, Federoff HJ, Convergent pathobiologic model of Parkinson's disease, *Ann NY Acad Sci* **991**:152–166, 2003.
71. Rosen DR, Siddique T, Patterson D, *et al.*, Mutations in Cu/Zn superoxide dismutase gene are associated with familial amyotrophic lateral sclerosis, *Nature* **362**:59–62, 1993.
72. Lonze BE, Riccio A, Cohen S, *et al.*, Apoptosis, axonal growth defects, and degeneration of peripheral neurons in mice lacking CREB, *Neuron* **34**:371–385, 2002.
73. Mantamadiotis T, Lemberger T, Bleckmann SC, *et al.*, Disruption of CREB function in brain leads to neurodegeneration, *Nat Genet* **31**:47–54, 2002.
74. Riccio A, Ahn S, Davenport CM, *et al.*, Mediation by a CREB family transcription factor of NGF-dependent survival of sympathetic neurons, *Science* **286**:2358–2361, 1999.
75. Petrij F, Giles RH, Dauwerse HG, *et al.*, Rubinstein-Taybi syndrome caused by mutations in the transcriptional co-activator CBP, *Nature* **376**:348–351, 1995.
76. Bartsch O, Schmidt S, Richter M, *et al.*, DNA sequencing of CREBBP demonstrates mutations in 56% of patients with Rubinstein-Taybi syndrome (RSTS) and in another patient with incomplete RSTS, *Hum Genet* **117**:485–493, 2005.
77. Oike Y, Hata A, Mamiya T, *et al.*, Truncated CBP protein leads to classical Rubinstein-Taybi syndrome phenotypes in mice: implications for a dominant-negative mechanism, *Hum Mol Genet* **8**:387–396, 1999.
78. Tanaka Y, Naruse I, Maekawa T, *et al.*, Abnormal skeletal patterning in embryos lacking a single Cbp allele: A partial similarity with Rubinstein-Taybi syndrome, *Proc Natl Acad Sci USA* **94**:10215–10220, 1997.
79. Bourtchouladze R, Lidge R, Catapano R, *et al.*, A mouse model of Rubinstein-Taybi syndrome: Defective long-term memory is ameliorated by inhibitors of phosphodiesterase 4, *Proc Natl Acad Sci USA* **100**:10518–10522, 2003.
80. Saura CA, Choi SY, Beglopoulos V, *et al.*, Loss of presenilin function causes impairments of memory and synaptic plasticity followed by age-dependent neurodegeneration, *Neuron* **42**:23–36, 2004.

81. Rouaux C, Jokic N, Mbebi C, *et al.*, Critical loss of CBP/p300 histone acetylase activity by caspase-6 during neurodegeneration, *EMBO J* **22**:6537–6549, 2003.
82. LeBlanc A, Liu H, Goodyer C, *et al.*, Caspase-6 role in apoptosis of human neurons, amyloidogenesis, and Alzheimer's disease, *J Biol Chem* **274**: 23426–23436, 1999.
83. Gong B, Vitolo OV, Trinchese F, *et al.*, Persistent improvement in synaptic and cognitive functions in an Alzheimer's mouse model after rolipram treatment, *J Clin Invest* **114**:1624–1634, 2004.
84. Tong L, Thornton PL, Balazs R, *et al.*, Beta-amyloid-(1-42) impairs activity-dependent cAMP-response element-binding protein signaling in neurons at concentrations in which cell survival Is not compromised, *J Biol Chem* **276**:17301–17306, 2001.
85. Vitolo OV, Sant'Angelo A, Costanzo V, *et al.*, Amyloid beta -peptide inhibition of the PKA/CREB pathway and long-term potentiation: Reversibility by drugs that enhance cAMP signaling, *Proc Natl Acad Sci USA* **99**:13217–13221, 2002.
86. Kazantsev A, Preisinger E, Dranovsky A, *et al.*, Insoluble detergent-resistant aggregates form between pathological and nonpathological lengths of polyglutamine in mammalian cells, *Proc Natl Acad Sci USA* **96**:11404–11409, 1999.
87. Nucifora FC Jr, Sasaki M, Peters MF, *et al.*, Interference by huntingtin and atrophin-1 with cbp-mediated transcription leading to cellular toxicity, *Science* **291**:2423–2428, 2001.
88. Steffan JS, Kazantsev A, Spasic-Boskovic O, *et al.*, The Huntington's disease protein interacts with p53 and CREB-binding protein and represses transcription, *Proc Natl Acad Sci USA* **97**:6763–6768, 2000.
89. Chai Y, Shao J, Miller VM, *et al.*, Live-cell imaging reveals divergent intracellular dynamics of polyglutamine disease proteins and supports a sequestration model of pathogenesis, *Proc Natl Acad Sci USA* **99**.9310–9315, 2002.
90. Steffan JS, Bodai L, Pallos J, *et al.*, Histone deacetylase inhibitors arrest polyglutamine-dependent neurodegeneration in Drosophila, *Nature* **413**:739–743, 2001.
91. Taylor JP, Taye AA, Campbell C, *et al.*, Aberrant histone acetylation, altered transcription, and retinal degeneration in a Drosophila model of polyglutamine disease are rescued by CREB-binding protein, *Genes Dev* **17**:1463–1468, 2003.
92. Cong SY, Pepers BA, Evert BO, *et al.*, Mutant huntingtin represses CBP, but not p300, by binding and protein degradation, *Mol Cell Neurosci* **30**:12–23, 2005.

93. Jung J, Bonini N, CREB-binding protein modulates repeat instability in a Drosophila model for polyQ disease, *Science* **315**:1857–1859, 2007.
94. Hockly E, Richon VM, Woodman B, *et al.*, Suberoylanilide hydroxamic acid, a histone deacetylase inhibitor, ameliorates motor deficits in a mouse model of Huntington's disease, *Proc Natl Acad Sci USA* **100**:2041–2046, 2003.
95. Ferrante RJ, Kubilus JK, Lee J, *et al.*, Histone deacetylase inhibition by sodium butyrate chemotherapy ameliorates the neurodegenerative phenotype in Huntington's disease mice, *J Neurosci* **23**:9418–9427, 2003.
96. Balasubramanyam K, Swaminathan V, Ranganathan A, *et al.*, Small molecule modulators of histone acetyltransferase p300, *J Biol Chem* **278**:19134–19140, 2003.
97. Fischer LR, Culver DG, Tennant P, *et al.*, Amyotrophic lateral sclerosis is a distal axonopathy: Evidence in mice and man, *Exp Neurol* **185**:232–240, 2004.
98. Stokin GB, Lillo C, Falzone TL, *et al.*, Axonopathy and transport deficits early in the pathogenesis of Alzheimer's disease, *Science* **307**:1282–1288, 2005.
99. Raff MC, Whitmore AV, Finn JT, Axonal self-destruction and neurodegeneration, *Science* **296**:868–871, 2002.
100. Conforti L, Tarlton A, Mack TG, *et al.*, A Ufd2/D4Cole1e chimeric protein and overexpression of Rbp7 in the slow Wallerian degeneration (WldS) mouse, *Proc Natl Acad Sci USA* **97**:11377–11382, 2000.
101. Lunn ER, Perry VH, Brown MC, *et al.*, Absence of Wallerian degeneration does not hinder regeneration in peripheral nerve, *Eur J Neurosci* **1**:27–33, 1989.
102. Mack TG, Reiner M, Beirowski B, *et al.*, Wallerian degeneration of injured axons and synapses is delayed by a Ube4b/Nmnat chimeric gene, *Nat Neurosci* **4**:1199–1206, 2001.
103. Bieganowski P, Brenner C, Discoveries of nicotinamide riboside as a nutrient and conserved NRK genes establish a Preiss-Handler independent route to NAD+ in fungi and humans, *Cell* **117**:495–502, 2004.
104. Llorente B, Dujon B, Transcriptional regulation of the Saccharomyces cerevisiae DAL5 gene family and identification of the high affinity nicotinic acid permease TNA1 (YGR260w), *FEBS Lett* **475**:237–241, 2000.
105. Grubisha O, Smith BC, Denu JM, Small molecule regulation of Sir2 protein deacetylases, *FEBS J* **272**:4607–4616, 2005.
106. Araki T, Sasaki Y, Milbrandt J, Increased nuclear NAD biosynthesis and SIRT1 activation prevent axonal degeneration, *Science* **305**:1010–1013, 2004.
107. Luchsinger JA, Tang MX, Shea S, *et al.*, Caloric intake and the risk of Alzheimer's disease, *Arch Neurol* **59**:1258–1263, 2002.

108. Maswood N, Young J, Tilmont E, *et al.*, Caloric restriction increases neurotrophic factor levels and attenuates neurochemical and behavioral deficits in a primate model of Parkinson's disease, *Proc Natl Acad Sci USA* **101**:18171–18176, 2004.
109. Patel NV, Gordon MN, Connor KE, *et al.*, Caloric restriction attenuates Abeta-deposition in Alzheimer transgenic models, *Neurobiol Aging* **26**:995–1000, 2005.
110. Wang J, Ho L, Qin W, *et al.*, Caloric restriction attenuates beta-amyloid neuropathology in a mouse model of Alzheimer's disease, *FASEB J* **19**:659–661, 2005.
111. Haass C, Take five — BACE and the gamma-secretase quartet conduct Alzheimer's amyloid beta-peptide generation, *EMBO J* **23**:483–488, 2004.
112. Masters CL, Simms G, Weinman NA, *et al.*, Amyloid plaque core protein in Alzheimer's disease and Down syndrome, *Proc Natl Acad Sci USA* **82**:4245–4249, 1985.
113. Selkoe DJ, Amyloid beta-protein and the genetics of Alzheimer's disease, *J Biol Chem* **271**:18295–18298, 1996.
114. Selkoe DJ, Alzheimer's disease: Genes, proteins, and therapy, *Physiol Rev* **81**:741–766, 2001.
115. Li R, Yang L, Lindholm K, *et al.*, Tumor necrosis factor death receptor signaling cascade is required for amyloid-beta protein-induced neuron death, *J Neurosci* **24**:1760–1771, 2004.
116. Yan SD, Chen X, Fu J, *et al.*, RAGE and amyloid-beta peptide neurotoxicity in Alzheimer's disease, *Nature* **382**:685–691, 1996.
117. Chen J, Zhou Y, Mueller-Steiner S, *et al.*, SIRT1 protects against microglia-dependent amyloid-beta toxicity through inhibiting NF-kappaB signaling, *J Biol Chem* **280**:40364–40374, 2005.
118. Qin W, Yang T, Ho L, *et al.*, Neuronal SIRT1 activation as a novel mechanism underlying the prevention of Alzheimer's disease amyloid neuropathology by calorie restriction, *J Biol Chem* **281**:21745–21754, 2006.
119. Chan DC, Mitochondria: Dynamic organelles in disease, aging, and development, *Cell* **125**:1241–1252, 2006.
120. Browne SE, Beal MF, The energetics of Huntington's disease, *Neurochem Res* **29**:531–546, 2004.
121. Squitieri F, Cannella M, Sgarbi G, *et al.*, Severe ultrastructural mitochondrial changes in lymphoblasts homozygous for Huntington disease mutation, *Mech Ageing Dev* **127**:217–220, 2006.
122. Cui L, Jeong H, Borovecki F, *et al.*, Transcriptional repression of PGC-1alpha by mutant huntingtin leads to mitochondrial dysfunction and neurodegeneration, *Cell* **127**:59–69, 2006.

123. Weydt P, Pineda VV, Torrence AE, *et al.*, Thermoregulatory and metabolic defects in Huntington's disease transgenic mice implicate PGC-1alpha in Huntington's disease neurodegeneration, *Cell Metab* **4**:349–362, 2006.
124. Handschin C, Kobayashi YM, Chin S, *et al.*, PGC-1alpha regulates the neuromuscular junction program and ameliorates Duchenne muscular dystrophy, *Genes Dev* **21**:770–783, 2007.
125. Parker JA, Arango M, Abderrahmane S, *et al.*, Resveratrol rescues mutant polyglutamine cytotoxicity in nematode and mammalian neurons, *Nat Genet* **37**:349–350, 2005.

Chapter 10

Role of the RIP140 Corepressor in Metabolic Regulation

Malcolm G. Parker, Mark Christian, Evangelos Kiskinis, Asha Seth, Donna Nichol and Roger White

Energy homeostasis depends on the expression of metabolic gene network that control the activity and anabolic and catabolic pathways. A number of nuclear receptors regulate transcription from these gene networks in metabolic tissues. Their transcriptional activity varies according to the recruitment of coactivators or corepressors. The function of the corepressor RIP140 in suppressing the expression of catabolic gene networks in adipose and muscle is the focus of this review.

10.1 Introduction

Nuclear receptors play a central role in energy homeostasis by regulating the expression of gene networks that control metabolic functions in response to changing environmental conditions. Their ability to control metabolic pathways is largely determined by cofactors that are capable of remodelling the state of chromatin in the vicinity of target genes or modulating the function of the transcription machinery. The expression of cofactors, their binding to nuclear receptors and their activity is determined by a combination of intrinsic factors such as intracellular metabolites that act directly as ligands and extrinsic stimuli that bind cell surface receptors and trigger downstream signalling pathways. Fluctuations in, for example, nutritional status, physical activity, stress and temperature result in the activation of distinct kinase cascades that phosphorylate nuclear receptors or their cofactors. Many nuclear receptors are involved in the regulation of energy homeostasis either through their direct actions in metabolic tissues or indirectly via systemic control pathways involving peripheral tissues or the central nervous system.[1] Of these, peroxisome proliferator activated receptors (PPARs)[2] and estrogen-related receptors (ERRs)[3] are emerging as crucial regulators in metabolic tissues such as adipose, liver and muscle.

The ability of nuclear receptors to activate or repress transcription from gene networks depends on the recruitment of a repertoire of cofactors that function as coactivators or corepressors. The approach we took to identify cofactors was to screen for receptor interacting proteins (RIPs) that interact with the ligand binding domain of the estrogen receptor in a ligand-dependent manner. Accordingly, we discovered proteins we referred to as RIP140 and RIP160 with molecular weights in the region of 140 kDa and 160 kDa respectively.[4] It emerged that RIP140 functions as a corepressor for most, if not all, nuclear receptors and RIP160 corresponds to the steroid receptor coactivator (SRC) family that Bert O'Malley had discovered using yeast two hybrid approaches.[5] SRC1 was the prototype for a family of coactivators that also includes TIF2 and p/CIP.[6–8] They seem to play a role in a vast number of biological responses including the control of energy homeostasis.[9,10] More recently, the PPARγ coactivator (PGC) family of coactivators, PGC1α and PGC1β, have been found to serve a crucial role in metabolic pathways in adipose, liver and muscle tissues by mediating the transcriptional activity of PPARs and ERRs.[11,12] Several of these pathways are also subject to regulation by RIP140. In this review, we focus on the role of RIP140 as a global regulator of metabolic gene networks and its relationship to other transcriptional cofactors that control energy balance.

10.2 Distribution and Expression of RIP140

The RIP140 gene is widely expressed in tissues but is localized to specific cell types. High levels are detected in metabolic tissues including adipose tissue, liver and muscle tissues.[13] RIP140 mRNA is transcribed from multiple promoters although the open reading frame is confined to a single exon (Fig. 10.1). Expression is regulated by a number of hormones including estrogen,[14] retinoic acid,[15] progestin,[16] and vitamin D.[17] In adipocytes, the gene is transcribed predominantly from a promoter we refer to as P2 which seems to be activated during the process of

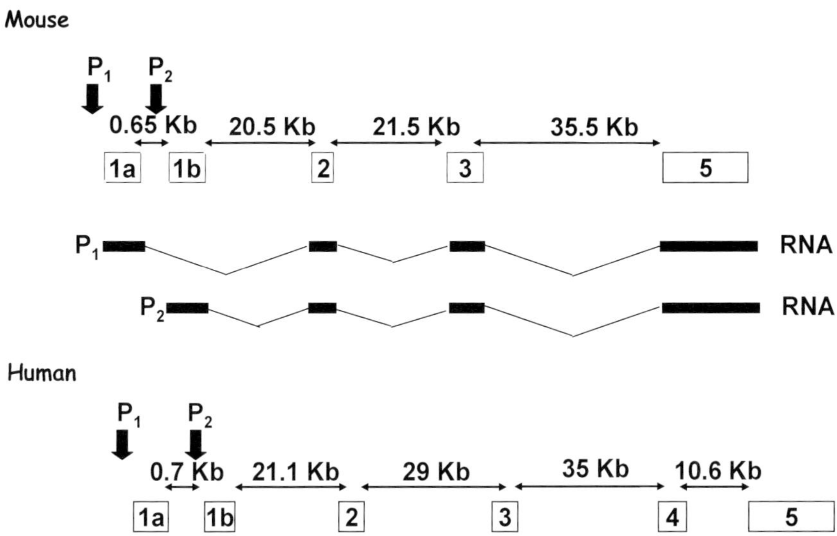

Fig. 10.1. Organization of the mouse and human RIP140 gene. RIP140 mRNA transcripts were analyzed by 5′RACE using RNA extracted from differentiated mouse 3T3-L1 adipocyte cells and human ZR-75 breast cancer cells. Exons were identified by comparison of mRNA transcripts with the sequences of chromosome 16 (mouse) or chromosome 21 (human). Alternative promoters P_1 and P_2 are indicated. Exons 1a, 1b, 2, 3, and 5 are conserved between species. mRNA transcripts comprising exons 1a or 1b, 2, 3, and 5 were detected in mouse 3T3-L1 adipocytes, and transcripts containing exon 1a or 1b, 2, 3, 4b, and 5 were identified in ZR-75 cells.

adipogenesis as exemplified in 3T3-L1 cells.[18] Expression studies and chromatin immunoprecipitation experiments indicate that ERRα stimulates transcription from the P2 promoter. Further analysis indicates that ERRα is capable of activating RIP140 gene transcription by two mechanisms. They does so directly by binding to an ERE/ERRE at -650/-633 and indirectly through Sp1 binding sites in the proximal promoter. The analysis of skeletal muscle indicates that RIP140 is expressed in a fibre type specific manner with relatively high levels of expression in muscles that are rich in glycolytic "fast-twitch" fibres such as gastrocnemius and extensor digitorum longus and little expression in those rich in oxidative "slow-twitch" fibres such as soleus or diaphragm. However, the molecular basis for this difference has yet to be determined.

10.3 Biological Roles of RIP140

The phenotype of RIP140 null mice indicates that the corepressor plays a crucial role in energy homeostasis. Both male and female mice devoid of RIP140 exhibit a lean phenotype with a 70% reduction in total body fat reflecting almost a complete absence of subcutaneous fat and a marked decrease in other fat depots [Fig. 10.2(A)].[13] Histology indicates that the size of adipocytes in white adipose tissue (WAT) is decreased with a corresponding 70% reduction in their volume indicating that their storage capacity is impaired. There is no evidence of lipodystrophy and high fat feeding indicates that RIP140 null mice are resistant to diet-induced obesity with no evidence of hepatic steatosis. There is also a reduction in circulating triglycerides and free fatty acids as well as reduced leptin levels. The alteration in fat accumulation cannot be explained by a reduction in food intake. However, it seems to reflect an increase in energy expenditure since RIP140 null mice consume approximately 30% more oxygen. The major organ responsible for energy expenditure is skeletal muscle, which varies in fibre type content according to the type of physical activity performed. Muscles involved in sustained contractile activity comprise a large proportion of slow fibres rich in mitochondria that generate energy by oxidative phosphorylation while muscles required for transient bursts of contractility comprise fast-switch fibres that generate energy by reductive glycolysis. The deletion of RIP140 leads to a morphological change in appearance with muscles rich in reductive fibres becoming redder than controls suggesting an increase in myoglobin content [Fig. 10.2(A)]. This is typical of slow fibres with increased mitochondrial biogenesis and increased oxidative capacity, consistent with the observed increase in oxygen consumption. The reduced storage of triglyceride in the adipose cells of RIP140 null mice, together with the alterations observed in muscle, predict that these changes would be accompanied by beneficial effects on glucose clearance or insulin sensitivity. We were therefore surprised when there was no apparent improvement when glucose tolerance and insulin sensitivity tests were performed on relatively young mice (approx 8–10 weeks;[13]). However, the benefits were clearly observed after feeding the mice a high fat diet or as mice age (six to nine months) when we found that RIP140 null mice show a marked improvement in glucose tolerance and enhanced responsiveness to insulin.[19] The relative contribution of changes in adipose and muscle tissues to the beneficial affects on insulin sensitivity have yet to be determined but it is

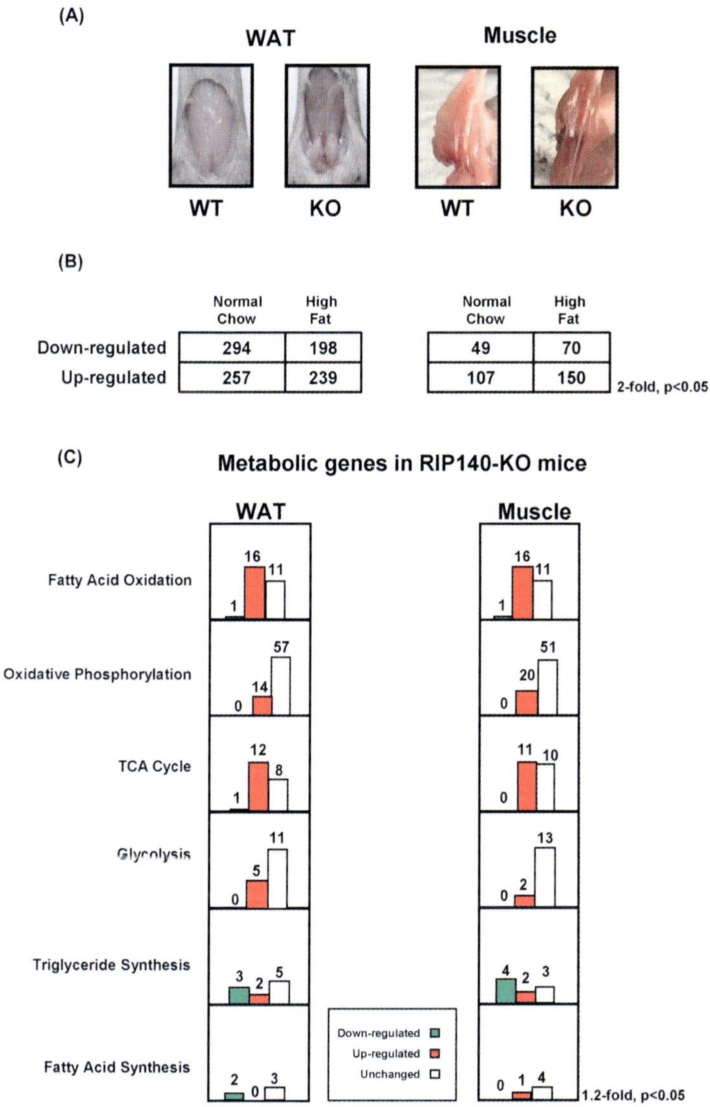

Fig. 10.2. RIP140 is a global repressor of metabolic gene expression. (A) RIP140 null mice exhibit a metabolic phenotype with an almost complete absence of subcutaneous white adipose tissue and a switch from reductive to oxidative (redder) fibres. (B) Summary of number of genes down- and up-regulated in RIP140 null mice in WAT and muscle tissues reflecting the corresponding morphological changes. Differences in numbers of genes in mice fed a normal or a high-fat diet are shown. (C) Analysis of the expression of genes in metabolic pathways shows a significant proportion are upregulated in the tissues of RIP140 null mice.

clear that RIP140 plays a key role in regulating metabolic activity with implications for obesity and type II diabetes.

10.4 RIP140 is a Global Repressor of Metabolic Gene Expression

To identify genes that were responsible for the metabolic changes observed in adipose and muscle tissues, we compared their gene expression profiles in wild type and RIP140 null mice using Affymetrix DNA microarrays. Given that RIP140 functions as a corepressor for nuclear receptors, we were initially surprised that the overall number of genes that were upregulated in the absence of the corepressor was similar to the number of downregulated genes [Fig. 10.2(B)]. However, this was not the case for metabolic genes of which 33% were upregulated and only 4% were downregulated, consistent with the function of RIP140 as a corepressor. The vast majority of upregulated genes in both adipose and muscle tissues[59] were involved in catabolic pathways including fatty acid oxidation, oxidative phosphorylation, glycolysis, and the tricarboxylic acid cycle [Fig. 10.2(C)]. Many of the upregulated genes were common to both tissue types. Interestingly, in the metabolic pathway gene cluster analysis, the expression of only seven genes in WAT and five genes in muscle were downregulated in the absence of RIP140, the majority of which were involved in anabolic pathways, namely fatty acid and triglyceride synthesis [Fig. 10.2(C)]. It is unclear how many of these genes are direct target genes for RIP140 but ChIP-on-chip experiments should address this point. We speculate that many of them will be targets for nuclear receptors with PPARs and ERRs being good candidates. It is conceivable that the downregulated genes might be targets for other transcription factors for which RIP140 might function as a coactivator, but this remains to be determined.

10.5 Molecular Mechanism of Repression by RIP140

RIP140 seems to function as a scaffold protein that bridges nuclear receptors to chromatin remodelling enzymes involved in chromatin condensation and transcriptional repression. The recruitment to nuclear receptors is mediated by LXXLL motifs (where L is leucine and X is any amino acid) of which there are nine in RIP140[20] and a tenth motif (LYYML) (where Y is tyrosine and M is methionine) at the C-terminus which seems to bind selectively to retinoid receptors[21] (Fig. 10.3). The presence of ten different NR-interaction motifs in RIP140 may suggest some functional redundancy. Studies have shown that particular NRs have a clear preference for specific RIP140 NR boxes[22–24] and isolated fragments of RIP140 can interact either constitutively with a NR, or only in the presence of ligand.[25] Although the interaction of RIP140 with many NRs is ligand-dependent, certain NRs have no known ligands or show interaction in the absence of ligand (Table 10.1). It is conceivable that alternative mechanisms, such as its expression level or post-translational modifications, may also modulate recruitment of RIP140 to nuclear receptors.

The identification of four distinct repression domains (RDs) in RIP140[26] that are conserved across vertebrate species (Fig. 10.3) predicts that these act as binding sites for different repressive enzymatic complexes. The mechanism of repression for both RD1 and RD2

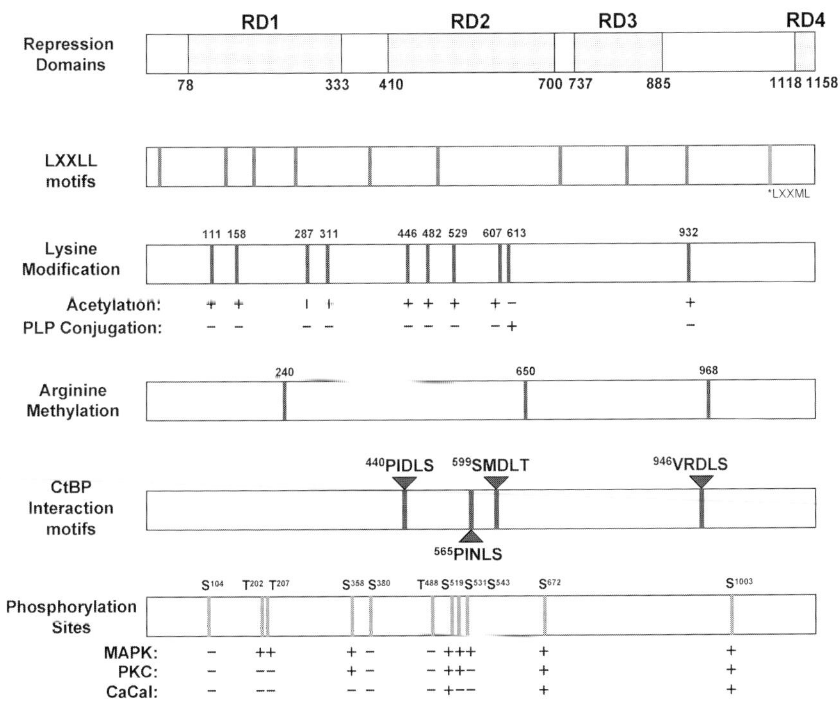

Fig. 10.3. Schematic representation of the repression domains, interaction motifs, and post-translational modifications of RIP140.

Table 10.1. Nuclear Receptors Known to Interact with RIP140 *in vitro*[a,b]

Nuclear Receptor	Ligand-Enhanced Interaction	Ligand-Independent Interaction	Reference
TRα (*NR1A1*)	√		[40]
TRβ (*NR1A2*)	√		[40–42]
RARα (*NR1B1*)	√		[43]
RARβ (*NR1B2*)	√		[41]
PPARα (*NR1C1*)	√	√	[40, 44–46]
PPARδ (*NR1C2*)		√	[46]
PPARγ (*NR1C3*)	√		[40]
RORβ (*NR1F2*)			[42]
LXRβ (*NR1H2*)		√	[42, 47]
LXRα (*NR1H3*)		√	[42, 44, 47]
VDR (*NR1I1*)	√		[42, 48]
PXR (*NR1I2*)	√		[49]
HNF4α (*NR2A1*)			[42]
RXRα (*NR2B1*)	√		[40, 41, 44, 50]
RXRβ (*NR2B2*)	√		[43]
TR2 (*NR2C1*)		√	[51]
TR4 (*NR2C2*)		√	[45]
ERα (*NR3A1*)	√		[41, 52]
ERβ (*NR3A2*)			[42]
ERRα (*NR3B1*)		√	[42, 53]
ERRβ (*NR3B2*)		√	[53]
ERRγ (*NR3B3*)		√	[53, 54]
GR (*NR3C1*)			[55, 56]
AR (*NR3C4*)		√	[57]
SF-1 (*NR5A1*)		√	[58]
DAX-1 (*NR0B1*)		√	[58]

[a]Ligand-dependent or -independent interaction is indicated if reported.
[b]Abbreviations: AR, androgen receptor; DAX-1, dosage-sensitive sex reversal-adrenal hypoplasia congenita critical region on the X chromosome, gene 1; HNF4, hepatocyte nuclear factor 4; LXR, liver X receptor; PXR, pregnane X receptor; RAR, retinoic acid receptor; ROR, RAR-related orphan receptor; RXR, retinoid X receptor; SF-1, steroidogenic factor 1; TR2, testicular orphan receptor 2; TR4 testicular orphan receptor 4; VDR, vitamin D receptor.

involves the recruitment of HDAC modifying enzymes,[27] but that of RDs 3 and 4 is yet to be fully characterized. In the case of RD2, HDAC recruitment is mediated by the binding of CtBPs (C-terminal binding proteins), which is partially relieved by the HDAC inhibitor TSA.[26]

This suggests that CtBP is acting via HDAC-dependent and independent mechanisms. CtBP transcriptional corepressors were initially identified through the interaction with a motif in the C-terminus of the adenoviral transforming protein E1A. There are two family members, CtBP1 and CtBP2, with both distinct and overlapping functions. Four motifs that facilitate CtBP recruitment have been identified in RIP140 (Fig. 10.3). Two of these, PIDLS[26,28,29] and PINLS,[26,28] are required for repression by RD-2.

CtBP is a dehydrogenase that can act as a redox sensor through changes in NADH/NAD$^+$ ratio affecting its repressive function.[30–32] Large changes in cellular redox state occur in certain metabolic abnormalities, such as diabetes, and alterations in the intracellular redox status by increasing the concentration of NAD(P)H are reported to reduce the amount of abdominal adipose tissue.[33] Thus, the repressive activity of RIP140 may be subject to fluctuations in cellular redox potential mediated by CtBP.

The repressive function of RIP140 can be modulated by post-translational modifications. It is subject to phosphorylation on up to 11 different residues, with functional consequences including enhanced repression by increased HDAC3 recruitment.[34] Conversely, the biological activity of the corepressor is inhibited by arginine methylation.[35] Additionally, there are nine acetylated lysines identified RIP140, one of which (Lys446) prevents CtBP recruitment when acetylated[29] a further lysine modification of RIP140 is the novel Lys613 conjugation of pyridoxal 5′-phosphate, the biologically active form of vitamin B6, resulting in enhanced repressive activity.[36] Thus, the multiple modifications of RIP140 provide a mechanism for the dynamic control of gene expression in response to external stimuli. PGC-1α is similarly subject to regulation by second messenger signaling pathways. Phosphorylation by p38MAPK is required for full induction of Ucp1 expression in response to β-adrenergic stimulation of brown adipocytes.[37]

10.6 Opposing Roles of RIP140 and PGC-1α

Interestingly, RIP140 and PGC-1α, in metabolic tissues, appear to fulfill opposing roles in metabolic tissues. In muscle, the switch from reductive to oxidative fibres that occurs in RIP140-null mice[59] also occurs following PGC-lα overexpression.[38] In white adipocytes, similar phenotypes manifest with increased mitochondrial biogenesis and Ucp1 expression when either RIP140 is ablated or PGC-1α is exogenously expressed.[13,19,39] The opposing metabolic roles of RIP140 and PGC-1α are likely due to shared common target genes that are regulated in opposite directions due to the recruitment of additional cofactors and enzymes including HDACs and HATs, respectively. In addition, the relative expression of RIP140 and PGC-1α, in conjunction with their functional regulation controlled by signaling pathways, is likely to be crucial for the regulation of metabolic gene expression by nuclear receptors. Thus it appears that the primary role of RIP140 in energy homeostasis is in modulating signaling pathways that determine metabolic processes. This is supported by the broad range of phenotypes generated on altering the level of expression of RIP140. It is evident that an understanding of the relative contribution of the function of cofactors in specific tissues will enhance our knowledge of the action of nuclear receptors in metabolism, while the potential to target specific receptor-cofactor interactions *in vivo* may allow intervention in metabolic-related diseases.

References

1. Francis GA, *et al.*, Nuclear receptors and the control of metabolism, *Annu Rev Physiol* **65**:261–311, 2003.
2. Lee CH, Olson, P, Evans, RM, Minireview: Lipid metabolism, metabolic diseases, and peroxisome proliferator-activated receptors, *Endocrinology* **144**(6):2201–2207, 2003.
3. Luo J, *et al.*, Reduced fat mass in mice lacking orphan nuclear receptor estrogen-related receptor alpha, *Mol Cell Biol* **23**(22):7947–7956, 2003.
4. Cavailles V, *et al.*, Interaction of proteins with transcriptionally active estrogen receptors, *Proc Natl Acad Sci USA*, **91**(21):10009–10013, 1994.
5. Onate SA, *et al.*, Sequence and characterization of a coactivator for the steroid hormone receptor superfamily, *Science* **270**(5240):1354–1357, 1995.
6. Glass CK, Rose DW and Rosenfeld MG, Nuclear receptor coactivators, *Curr Opin Cell Biol* **9**(2):222–232, 1997.
7. Xu J, Li Q, Review of the *in vivo* functions of the p160 steroid receptor coactivator family, *Mol Endocrinol* **17**(9):1681–1692, 2003.
8. Xu J, O'Malley BW, Molecular mechanisms and cellular biology of the steroid receptor coactivator (SRC) family in steroid receptor function, *Rev Endocr Metab Disord* **3**(3):185–192, 2002.
9. Picard F, *et al.*, SRC-1 and TIF2 control energy balance between white and brown adipose tissues, *Cell* **111**(7):931–941, 2002.

10. Wang Z, *et al.*, Critical roles of the p160 transcriptional coactivators p/CIP and SRC-1 in energy balance, *Cell Metab* **3**(2):111–122, 2006.
11. Puigserver P, Tissue-specific regulation of metabolic pathways through the transcriptional coactivator PGC1-alpha, *Int J Obes (Lond)* **29**(Suppl 1): S5–9, 2005.
12. Puigserver P, Spiegelman BM, Peroxisome proliferator-activated receptor-gamma coactivator 1alpha (PGC-1alpha): Transcriptional coactivator and metabolic regulator, *Endocr Rev* **24**(1):78–90, 2003.
13. Leonardsson G, *et al.*, Nuclear receptor corepressor RIP140 regulates fat accumulation, *Proc Natl Acad Sci USA* **101**(22):8437–8442, 2004.
14. Thenot S, *et al.*, Estrogen receptor cofactors expression in breast and endometrial human cancer cells, *Mol Cell Endocrinol* **156**(1–2):85–93, 1999.
15. Kerley JS, *et al.*, Transcriptional activation of the nuclear receptor corepressor RIP140 by retinoic acid: A potential negative-feedback regulatory mechanism, *Biochem Biophys Res Commun* **285**(4):969–975, 2001.
16. Graham JD, *et al.*, Altered progesterone receptor isoform expression remodels progestin responsiveness of breast cancer cells, *Mol Endocrinol* **19**(11):2713–2735, 2005.
17. Lin R, *et al.*, Expression profiling in squamous carcinoma cells reveals pleiotropic effects of vitamin D3 analog EB1089 signaling on cell proliferation, differentiation, and immune system regulation, *Mol Endocrinol* **16**(6):1243–1256, 2002.
18. Nichol D, *et al.*, RIP140 expression is stimulated by estrogen-related receptor alpha during adipogenesis, *J Biol Chem* **281**(43):32140–32147, 2006.
19. Powelka AM, *et al.*, Suppression of oxidative metabolism and mitochondrial biogenesis by the transcriptional corepressor RIP140 in mouse adipocytes, *J Clin Invest* **116**(1):125–136, 2006.
20. Heery DM, *et al.*, A signature motif in transcriptional co-activators mediates binding to nuclear receptors, *Nature* **387**(6634):733–736, 1997.
21. Wei LN, Farooqui M, Hu X, Ligand-dependent formation of retinoid receptors, receptor-interacting protein 140 (RIP140), and histone deacetylase complex is mediated by a novel receptor-interacting motif of RIP140, *J Biol Chem* **276**(19):16107–16112, 2001.
22. Moore JM, *et al.*, Quantitative proteomics of the thyroid hormone receptor-coregulator interactions, *J Biol Chem* **279**(26):27584–27590, 2004.
23. Heery DM, *et al.*, Core LXXLL motif sequences in CREB-binding protein, SRC1, and RIP140 define affinity and selectivity for steroid and retinoid receptors, *J Biol Chem* **276**(9):6695–6702, 2001.
24. Tazawa H, *et al.*, Regulation of subnuclear localization is associated with a mechanism for nuclear receptor corepression by RIP140, *Mol Cell Biol*, **23**(12):4187–4198, 2003.

25. Wei LN, Hu X, Receptor interacting protein 140 as a thyroid hormone-dependent, negative co-regulator for the induction of cellular retinoic acid binding protein I gene, *Mol Cell Endocrinol* **218**(1–2):39–48, 2004.
26. Christian M, Tullet JM, Parker MG, Characterization of four autonomous repression domains in the corepressor receptor interacting protein 140, *J Biol Chem* **279**(15):15645-15651, 2004.
27. Wei LN, *et al.*, Receptor-interacting protein 140 directly recruits histone deacetylases for gene silencing, *J Biol Chem* **275**(52):40782–40787, 2000.
28. Castet A, *et al.*, Multiple domains of the receptor-interacting protein 140 contribute to transcription inhibition, *Nucleic Acids Res* **32**(6):1957–1966, 2004.
29. Vo N, Fjeld C, Goodman RH, Acetylation of nuclear hormone receptor-interacting protein RIP140 regulates binding of the transcriptional corepressor CtBP, *Mol Cell Biol* **21**(18):6181–6188, 2001.
30. Fjeld CC, Birdsong WT, Goodman RH, Differential binding of NAD+ and NADH allows the transcriptional corepressor carboxyl-terminal binding protein to serve as a metabolic sensor, *Proc Natl Acad Sci USA* **100**(16):9202–9207, 2003.
31. Kumar V, *et al.*, Transcription corepressor CtBP is an NAD(+)-regulated dehydrogenase, *Mol Cell* **10**(4):857–869, 2002.
32. Zhang Q, Piston, DW, Goodman, RH, Regulation of corepressor function by nuclear NADH, *Science* **295**(5561):1895–1897, 2002.
33. Gaikwad A, *et al.*, *In vivo* role of NAD(P)H:quinone oxidoreductase 1 (NQO1) in the regulation of intracellular redox state and accumulation of abdominal adipose tissue, *J Biol Chem* **276**(25):22559–22564, 2001.
34. Gupta P, *et al.*, Regulation of co-repressive activity of, and HDAC recruitment to, RIP140 by site-specific phosphorylation, *Mol Cell Proteomics* **4**(11):1776–1784, 2005.
35. Mostaqul Huq MD, *et al.*, Suppression of receptor interacting protein 140 repressive activity by protein arginine methylation, *EMBO J* **25**(21):5094–5104, 2006.
36. Huq MD, *et al.*, Vitamin B6 conjugation to nuclear corepressor RIP140 and its role in gene regulation, *Nat Chem Biol* **3**(3):161–165, 2007.
37. Cao W, *et al.*, p38 mitogen-activated protein kinase is the central regulator of cyclic AMP-dependent transcription of the brown fat uncoupling protein 1 gene, *Mol Cell Biol* **24**(7):3057–3067, 2004.
38. Lin J, *et al.*, Transcriptional co-activator PGC-1 alpha drives the formation of slow-twitch muscle fibres, *Nature* **418**(6899):797–801, 2002.
39. Orci L, *et al.*, Rapid transformation of white adipocytes into fat-oxidizing machines, *Proc Natl Acad Sci USA* **101**(7):2058–2063, 2004.
40. Treuter E, *et al.*, A regulatory role for RIP140 in nuclear receptor activation, *Mol Endocrinol* **12**(6):864–881, 1998.

41. L'Horset F, *et al.*, RIP-140 interacts with multiple nuclear receptors by means of two distinct sites, *Mol Cell Biol* **16**(11):6029–6036, 1996.
42. Albers M, *et al.*, Automated yeast two-hybrid screening for nuclear receptor-interacting proteins, *Mol Cell Proteomics* **4**(2):205–213, 2005.
43. Lee CH, Wei LN, Characterization of receptor-interacting protein 140 in retinoid receptor activities, *J Biol Chem* **274**(44):31320–31326, 1999.
44. Miyata KS, *et al.*, Receptor-interacting protein 140 interacts with and inhibits transactivation by, peroxisome proliferator-activated receptor alpha and liver-X-receptor alpha, *Mol Cell Endocrinol* **146**(1–2):69–76, 1998.
45. Yan ZH, *et al.*, Regulation of peroxisome proliferator-activated receptor alpha-induced transactivation by the nuclear orphan receptor TAK1/TR4, *J Biol Chem* **273**(18):10948–10957, 1998.
46. Lim HJ, Moon I, Han K, Transcriptional cofactors exhibit differential preference toward peroxisome proliferator-activated receptors alpha and delta in uterine cells, *Endocrinology* **145**(6):2886–2895, 2004.
47. Jakobsson T, *et al.*, Molecular basis for repression of liver X receptor mediated gene transcription by the receptor-interacting protein 140, *Biochem J* **405**:31–39, 2007.
48. Masuyama H, *et al.*, Evidence for ligand-dependent intramolecular folding of the AF-2 domain in vitamin D receptor-activated transcription and coactivator interaction, *Mol Endocrinol* **11**(10):1507–1517, 1997.
49. Masuyama H, *et al.*, The expression of pregnane X receptor and its target gene, cytochrome P450 3A1, in perinatal mouse, *Mol Cell Endocrinol* **172**(1–2):47–56, 2001.
50. Wiebel FF, *et al.*, Ligand-independent coregulator recruitment by the triply activatable OR1/retinoid X receptor-alpha nuclear receptor heterodimer, *Mol Endocrinol* **13**(7):1105–1118, 1999.
51. Lee CH, Chinpaisal, C, Wei, LN, Cloning and characterization of mouse RIP140, a corepressor for nuclear orphan receptor TR2, *Mol Cell Biol* **18**(11):6745–6755, 1998.
52. Cavailles V, *et al.*, Nuclear factor RIP140 modulates transcriptional activation by the estrogen receptor, *EMBO J*, **14**(15):3741–3751, 1995.
53. Castet A, *et al.*, Rip140 differentially regulates estrogen receptor-related receptor transactivation depending on target genes, *Mol Endocrinol* **20**:1035–1047, 2006.
54. Sanyal S, *et al.*, Deoxyribonucleic acid response element-dependent regulation of transcription by orphan nuclear receptor estrogen receptor-related receptor gamma, *Mol Endocrinol* **18**(2):312–325, 2004.
55. Subramaniam N, Treuter E, Okret S, Receptor interacting protein RIP140 inhibits both positive and negative gene regulation by glucocorticoids, *J Biol Chem* **274**(25):18121–18127, 1999.

56. Zilliacus J, *et al.*, Regulation of glucocorticoid receptor activity by 14 — 3-3-dependent intracellular relocalization of the corepressor RIP140, *Mol Endocrinol* **15**(4):501–511, 2001.
57. Carascossa S, *et al.*, Rip140 is a repressor of the androgen receptor activity, *Mol Endocrinol* **20**:1506–1518, 2006.
58. Sugawara T, *et al.*, RIP 140 modulates transcription of the steroidogenic acute regulatory protein gene through interactions with both SF-1 and DAX-1, *Endocrinology* **142**(8):3570–3577, 2001.
59. Seth A, Steel JH, Nichol D, *et al.*, The transcriptional corepressor RIP140 regulates oxidative metabolism in skeletal muscle, *Cell Metab* **6(3)**:236–245, 2007.

Chapter 11

Nuclear Receptor Corepressors and Metabolism

Theresa Alenghat and Mitchell A. Lazar

Nuclear receptors (NRs) are critical to nearly all aspects of metabolism. NRs that are particularly important in regulating gene expression in metabolic tissues such as adipose tissue, muscles, liver include thyroid hormone receptors, liver X receptors and peroxisome proliferator-activated receptors. NRs activate gene transcription in the presence of their cognate ligands. However, in the absence of ligand, corepressor proteins are recruited by these NRs to target genes and function to repress transcription. By mediating transcriptional repression at metabolically important NR target genes, corepressors play a critical regulatory role in carbohydrate and lipid metabolism. Therefore, understanding which NR target genes are selectively regulated by corepressors and the mechanisms of how these proteins function to repress gene transcription will provide great insight into metabolic regulation and will enable the development of appropriate treatments for prevalent diseases such as obesity, diabetes, and atherosclerosis. This chapter will focus on two of the most extensively examined corepressors; nuclear receptor corepressor (N-CoR) and its homologue; silencing mediator for retinoid and thyroid receptor (SMRT); the mechanism of NR repression; and the role of corepressors in metabolism.

11.1 Introduction

Nuclear hormone receptors (NRs) function as transcription factors that are critical in development, differentiation, and metabolism.[1] NRs alter gene regulation of target genes depending on the presence or absence of small hydrophobic molecules, termed ligands. NR ligands bind directly to the NR and they include steroid hormones (estrogen, androgen, glucocorticoids), non-steroid hormones (thyroid hormone, retinoic acid), as well as products of metabolism (fatty acids, bile acids, oxysterols).[2] NRs provide powerful targets for designing new and effective drugs. In fact, many currently used pharmaceutical agents, such as tamoxifen in breast cancer, thiazolidinediones in diabetes, and dexamethasone in inflammatory conditions, ameliorate disease by altering NR activity.[3]

11.1.1 *Nuclear receptor structure*

NRs are modular proteins that consist of six conserved domains (A–F).[4] The A/B domain in the N-terminal region varies significantly between different NRs and comprises activation function 1 (AF-1), which can activate transcription in a ligand-independent manner.[5,6] The C region is composed of two type II zinc fingers which encompass the highly conserved DNA binding domain (DBD).[7] This region is responsible for recognizing and specifically binding to DNA consensus sequences known as response elements (RE). The P box, which is found in a helical region of the first zinc finger, directly contacts the major groove nucleotides and mediates DNA sequence recognition.[8] Adjacent to region C is a variable hinge domain (D region) which may permit the NR to alter its conformation and has been shown to contain a nuclear localization region and/or activation domain in some receptors.[9]

The largest domain, region E, also termed the ligand binding domain (LBD) contains ligand-dependent activation function 2 (AF-2) and is essential for ligand recognition and binding, receptor dimerization, as well as ligand-dependent transcriptional regulation.[4] Although there is significant sequence variation in the LBDs of different NRs, the overall structural folding of this domain is highly similar.[10] Ligand binding within the hydrophobic pocket of the LBD induces a conformational change of the NR. More specifically, helix 12, which is located within the LBD C-terminus and is required for the ligand dependent transactivation function (AF-2), undergoes the most significant movement in response to ligand binding.[11–13] This ligand induced conformational change is often referred to as the molecular switch that permits the binding of coactivators to the LBD and simultaneously inhibits association with corepressors, resulting in their release (discussed in more detail below). At the C-terminus, some NRs contain a variable F region, the function of which has yet to be determined.

11.1.2 *Nuclear receptor classifications*

The NR superfamily is most commonly subdivided into four groups based on dimerization as well as DNA-binding properties.[1] Class I NRs bind to inverted DNA REs as homodimers and include the steroid hormone receptors such as glucococorticoid receptor (GR), mineralocorticoid receptor (MR), progesterone receptor, androgen receptor (AR) and estrogen receptor (ER). These NRs are found in the cytoplasm and are induced through ligand association to form homodimers then translocate to the nucleus to function as transcription factors through direct DNA binding.[9]

The Class II NRs heterodimerize with retinoid X receptor (RXR) and most commonly bind to direct repeat REs. Unlike Class I receptors, Class II receptors are found in the nucleus and bind DNA regardless of ligand levels, permitting these receptors to activate transcription in the presence of ligand as well as repress transcription in the absence of ligand.[14] These NRs include the non-steroid ligand binding NRs such as thyroid hormone receptor (TR), retinoic acid receptor (RAR), vitamin D receptors (VDR), and peroxisome proliferator-activated receptor (PPAR).[1] Since these NRs each bind to direct repeats of AGGTCA, the binding specificity is generally dependent on the number of nucleotide spacing between repeats.[9] For instance, TR preferentially binds DNA at direct repeats of AGGTCA with four nucleotide spacers (DR4), whereas RAR optimally binds to direct repeats of AGGTCA with five nucleotide spacers (DR5). This binding specificity is broadly defined by the "3-4-5 rule" in which DR3, DR4, or DR5 response elements are bound by VDR, TR, or RAR, respectively.[15]

Many NRs bind direct repeat REs as homodimers and are classified as Class III NRs, whereas Class IV NRs bind DNA as monomers.[1] These two classes contain several "orphan" NRs such as RAR-related orphan receptor (ROR), nerve growth factor IB-like receptor (NGF1-B), and Rev-erb, which were identified based on homology to other NRs, but have no known ligand.[16] Orphan receptors present a challenging and exciting area of research in which novel ligands and modes of regulation may be identified. However, while many of these orphan receptors could have ligands yet to be discovered, it is quite possible that the activity of some orphan receptors may not be ligand-dependent.[17]

11.2 Corepressors

Prior to the identification of NR corepressor proteins, the C-terminal region of TR and RAR were found to be required for transcriptional silencing in the absence of ligand.[18] The LBDs of these NRs were later shown to squelch the repressive function of DNA bound NRs, suggesting the association of inhibitory factors with the unliganded NR LBD.[19–21] Soon after this discovery, the first NR corepressors, nuclear receptor corepressor (N-CoR), silencing mediator of retinoid and thyroid receptors (SMRT), and RIP-140 were identified.[22–24] Of these proteins, N-CoR and SMRT will be discussed in greater detail in this chapter whereas RIP-140 will be covered in a later chapter. In the past ten years, other potential NR corepressors have been discovered such as Alien,[25] Hairless,[26] L-CoR,[27] SUN-CoR,[28] SHARP,[29] and SLIRP.[30]

11.2.1 *N-CoR/SMRT*

Yeast two-hybrid screens for proteins that associate with TR and RAR in the absence of ligand led to the critical discoveries of the corepressors N-CoR and SMRT.[22,23] Although this chapter will focus on the recruitment of N-CoR/SMRT by NRs, it is important to note that these proteins have since been identified as corepressors for various non-receptor transcription factors.[31] N-CoR and SMRT are highly homologous proteins that are encoded by distinct genes and are expressed fairly ubiquitously, although at varying levels depending on the tissue and stage of development.[32–34]

11.2.1.1 *Structure*

N-CoR and SMRT are both large, 270 kDa proteins that share several highly conserved domains (Fig. 11.1). The C-terminus of SMRT and N-CoR contain distinct NR interactions domains (IDs), which are required for tethering the corepressor to the NR LBD.[35] Each ID contains a critical L/I- X-X-I/V-I motif, referred to as the CoRNR box, that is required for NR interaction.[36–38] The CoRNR box, which parallels the L-X-X-L-L recognition motif found in NR coactivator proteins, is predicted to form

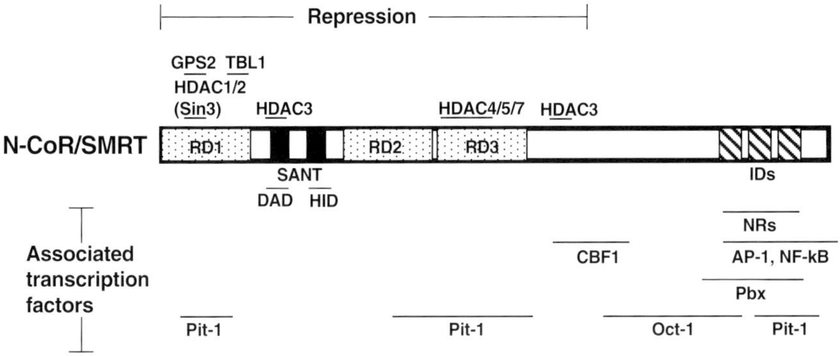

Fig. 11.1. Schematic of N-CoR/SMRT domains and associated proteins.

an extended α-helical domain that docks in a complementary groove formed by NR helices 3, 4, and 5. Although N-CoR and SMRT are highly homologous, specific amino acid sequences in the C-terminus of the corepressors permit the preferential association of certain NRs with either N-CoR or SMRT.[39–41]

The domains responsible for transcriptional repression are contained primarily within the amino terminus of N-CoR and SMRT. Multiple, non-redundant, repressive regions were originally characterized due to their ability to function as autonomous repression domains (RD1, RD2, and RD3) when fused to DNA binding proteins.[22,42] These RDs, which are critical for protein associations, enable the formation of the various corepressor complexes that are discussed in more detail below.

Between RD1 and RD2 are two distinct SANT (SWI3, ADA2, N-CoR, TFIIB) motifs which are also critical for corepressor function. Both N-CoR and SMRT stably associate with and activate histone deacetylase 3 (HDAC3) through the deacetylase activation domain (DAD) which is primarily composed of the first SANT motif.[43–45] This domain alone can function to strongly repress transcription when fused to DNA binding proteins and deacetylation of local histones by HDAC3 results in repression of gene transcription.[41] The second SANT motif defines the histone interaction domain (HID) because it mediates direct interactions with unacetylated histone H4 N-terminus tails, suggesting a feed-forward model for corepressor binding to nucleosomal DNA.[46,47]

11.2.1.2 *N-CoR/SMRT complexes*

A variety of proteins have been reported to associate with N-CoR and SMRT (Fig. 11.1). Biochemical purification of the corepressors has repeatedly identified a core complex containing either N-CoR or SMRT as well as HDAC3, transducin (beta)-like protein (TBL-1) or TBL-1 related protein (TBLR1), and G-protein pathway suppressor 2 (GPS2).[48–51] The critical role of HDAC3 in repression by N-CoR/SMRT will be discussed below. Both TBL1 and GPS2 associate with RD1 of N-CoR and SMRT.[50] TBL-1, a WD40-repeat protein, also mediates repression, potentially through its ability to associate with histones H2B and H4.[51] More recently, TBL1 and TBLR1 have been suggested to play a critical role in the switch from gene repression to activation by mediating the dissociation and degradation of N-CoR and SMRT.[52] The role of GPS2 is not well understood, but this protein may further stabilize the corepressor complex through its association with TBL1.[50]

Yeast two-hybrid and *in vitro* studies originally found that N-CoR and SMRT interact with the mammalian switch-independent 3 (mSin3)-HDAC1/2 complex.[53,54] The association of N-CoR and SMRT with HDAC1/2 occurs via an interaction of mSin3 with RD1. However, the physiologic relevance for this association is controversial. Even though the mSin3 protein functions as an important corepressor for several transcription factors, N-CoR or SMRT complex purification from cells often fails to identify the mSin3-HDAC1/2 components as members of the core N-CoR/SMRT corepressor complex.[48–51] Furthermore, only HDAC3, not HDAC1 or HDAC2, have been found to be recruited by NRs to repressed genes.[41,55] Nevertheless, biochemical fractionation of complexes containing N-CoR from Xenopus oocyte extracts revealed the existence of distinct Sin3-dependent and Sin3-independent complexes which suggests that N-CoR may associate in multiple corepressor complexes.[56]

In addition to the class I HDACs, class II HDACs, HDAC4, HDAC5, and HDAC7, can interact directly with RD3 of N-CoR and SMRT *in vitro*, suggesting that these HDACs may also contribute to repression.[57,58] In fact, a complex containing Sin3A, N-CoR and HDAC4 was recently suggested to be induced by the binding of apoptotic compounds to the orphan NR small heterodimer partner (SHP).[59] It has also been reported that the N-CoR complex contains components of the SWI/SNF complex, such as BRG1, BAF170, BAF155, and BAF47 as well as KAP-1, a heterochromatin silencing protein.[60] A direct interaction between this SWI/SNF ATPase and the N-CoR complex suggests that the corepressor itself may contain chromatin remodeling activity that could play a role in repression. However, intrinsic remodeling activity of the complex or a role for BRG1 in NR-mediated repression has not been reported.

11.3 Mechanisms of Repression by Corepressors

Ligand binding to the LBD of NRs stimulates a conformational change in the receptor that stimulates the release of corepressor proteins and the recruitment of coactivators, several of which possess histone acetyltransferase (HAT) activity. The ligand stimulated recruitment of these HATs as well as the association of HDACs in recruited NR corepressor complexes suggest the importance of dynamic chromatin acetylation and deacetylation in NR mediated gene regulation.

11.3.1 *Chromatin and the histone code*

Eukaryotic cells package their DNA around histone proteins to form a higher order structure termed chromatin. The repetitive element within chromatin, called the nucleosome, is composed of 147 bp of DNA tightly wound around a histone octamer. Histone H1 functions as a linker between nucleosomes that permits further condensation of the chromatin structure. Chromatin, itself, is generally repressive by limiting access of transcriptional machinery to gene promoters. However, covalent nucleosomal modifications as well as ATP dependent chromatin remodeling ATPases enable chromatin flexibility in response to specific cellular signals. The chromatin structure, therefore, undergoes local condensation or decondensation which facilitates various processes such as DNA replication, repair, or transcription.[61] Histone N-terminus tails extend from the nucleosomal core and provide a template for various covalent modifications such as acetylation, phosphorylation, methylation, SUMOylation, as well as ubiquitination. The pattern of modifications on these histone tails establish a "histone code" which dictates recruitment of specific cofactors or itself stimulates structural condensing or opening of the chromatin structure.[62,63]

Due to its early correlation with transcriptional activity, histone acetylation has been the most extensively studied covalent nucleosomal modification.[64,65] Acetylation of histone tails is believed to disrupt the DNA-histone interaction, causing local decondensing of the chromatin and permitting access for transcription machinery. Furthermore, histone acetylated-lysines are the preferred substrate for bromodomain containing proteins which include several coactivators with HAT activity.[66,67] Therefore, histone acetylation propagates increased acetylation by recruiting HATs. Conversely, removal of the acetyl groups by HDACs promotes tighter DNA-histone associations and inhibits transcriptional activity. Similar to HATs, the deacetylation of histone tails by HDACs establishes a preferred template for core components of the N-CoR/SMRT complex and likely enables more stable interactions with the corepressor complex.[46,47,51]

11.3.2 *Ligand independent repression*

In the absence of ligand binding to class II NRs, such as TR and RAR, the NR assumes a conformation that interacts strongly with N-CoR and SMRT.[22,23,68] The prevailing mechanistic model by which these corepressors mediate gene repression are through the co-recruitment of directly bound HDACs (HDAC3, HDAC4, HDAC5, HDAC7) or indirectly bound HDACs (HDAC1/2 via mSin3) as discussed previously. Currently, the association between N-CoR/SMRT and HDAC3 seems to be functionally critical with regards to NR mediated repression (Fig. 11.2). HDAC3, which must be presented to the corepressor by the multiprotein TCP-1 ring complex (TRiC) chaperone, directly binds in a tight complex with SMRT and N-CoR *in vivo*.[69] In fact, HDAC3's association with the DAD of N-CoR or SMRT is required for the enzymatic deacetylase activity of HDAC3. Correspondingly, recruitment of the N-CoR/HDAC3 complex to NR target genes induces deacetylation of local histones, resulting in repression of TR mediated transcription.[41,44,45] Histone deacetylation also seems to create a preferred template for N-CoR and SMRT as well as TBL1 and, therefore, HDAC3 activity likely leads to further stabilization of the corepressor complex at NR repressed genes.[46,47,51] Additionally, the ATP dependent chromatin remodeling ATPase SNF2H can secondarily be recruited to unacetylated histone H4 tails and enhance repression of TR target genes.[70]

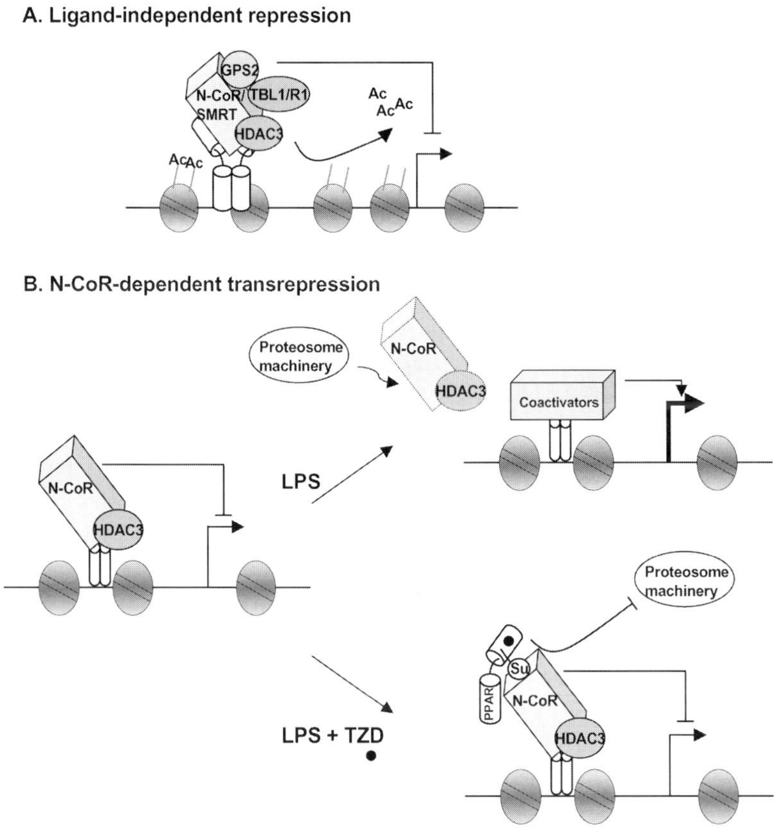

Fig. 11.2. (A) Mechanism of NR-directed N-CoR/SMRT dependent transcriptional repression. (B) Model of N-CoR dependent transrepression by TZD stimulated PPARγ SUMOylation.

As discussed earlier, biochemical complex purification suggests that multiple corepressor complexes may exist and that these complexes may form or associate with NRs differently depending on cell type, cellular signals, corepressor levels, or NR type.[31] Although the inhibition of HDAC3 binding to N-CoR's DAD depletes a significant amount of TR directed repression on a transicntly transfected reporter gene, it is probable that the mechanism of repression is more complex such that other regions of N-CoR/SRMT, particularly the RDs, play a critical role in N-CoR/SMRT mediated repression at endogenous NR target genes. RD1, RD2, and RD3 can each repress transcription when fused to DNA binding proteins. Co-recruitment of other proteins, particularly HDACs, likely contributes to the mechanism by which these domains repress.[54,57] TBL1 is also required for TR mediated repression, possibly through a HDAC independent mechanism.[51] Additionally, there is evidence that NR repression may also be partly mediated by interactions of N-CoR and SMRT with basal transcription components, such as TFIIB, which may cause direct interference with formation of the preinitiation complex.[71]

Furthermore, different NRs differentially recruit either N-CoR or SMRT complexes.[40] For instance, TR repression is almost completely lost with N-CoR depletion in 293T cells, whereas both N-CoR and SMRT contribute to Rev-erb directed repression.[41] Deletion of either N-CoR or SMRT in mice results in embryonic lethality even though SMRT expression is unaffected, demonstrating that these corepressors cannot fully compensate for each other.[34,72a] However, while it is evident that N-CoR and SMRT are differentially recruited and that these proteins are not functionally redundant, it is not clear whether these highly homologous corepressors differ in their mechanism of repression. With regards to complex formation, both N-CoR and SMRT generally associate with the same proteins, except for the reported interactions of IR10 and TAB2 exclusively with N-CoR.[51,72]

In addition to ligand binding to the NR, post-translational modifications have been suggested to further modify the association of N-CoR/SRMT with NRs. SMRT phosphorylation by MAPK-extracellular signal-regulated kinase 1 (MEK-1) and MEK-1 kinase (MEKK-1) inhibits the association of SMRT with NRs and may also lead to the redistribution of SMRT out of the nucleus.[73] On the other hand, phosphorylation of SMRT by casein kinase II (CK2) is believed to stabilize the interaction of SMRT with TR.[74] IL-1β has also been suggested to stimulate MEKK-1 phosphorylation of TAB2, a N-CoR associated protein identified in a yeast two-hybrid screen. This covalent modification of TAB2 has been shown to stimulate nuclear export of the N-CoR/TAB2/HDAC3 complex.[72] Furthermore, a model has been proposed in which TBL1 and TBLR1 function as corepressor/coactivator exchange factors by mediating N-CoR degradation in the presence of ligand binding through recruitment of a ubiquitin/19S proteasome complex.[52] Lastly, direct phosphorylation of the NR itself appears to further modulate corepressor association.[75,76]

11.3.3 *Ligand dependent repression*

Some corepressors, such as RIP140 and L-CoR, function to repress transcription in a ligand-dependent manner. Unlike N-CoR and SMRT, these corepressors are recruited by DNA bound NRs in the presence of ligand. Analogous to coactivator proteins, RIP140 utilizes a LxxLL motif to interact with ligand bound NR. However, similar to N-CoR and SMRT, RIP140 serves as a platform for HDACs and functions to repress transcription in the presence of ligand. RIP140, which plays an important role in metabolism, will be covered in greater detail in a later chapter.[77]

11.3.3.1 *Transrepression*

A subset of ligand dependent repression, termed transrepression, involves protein-protein interactions between NRs and promoter bound transcription factors rather than direct binding of the NR to DNA.[78] In general, transrepression refers to all mechanisms of ligand dependent repression that do not involve direct binding of the NR to the promoter. For example, the anti-inflammatory effect of glucocorticoids rely on GR mediated repression but, often, do not require DNA-binding of GR to the promoter of active inflammatory genes.[79] Mechanisms of NR-mediated transrepression include coactivator competition, inhibition of coactivator recruitment, disruption of coactivator complexes, and blocking of RNA polymerase hyperphosphorylation.[78]

With regards to corepressors, inhibition of HDAC activity has been shown to relieve transrepression and, consistently, GR has recently been found to recruit HDAC2 to the IL-1β promoter in a ligand dependent manner.[80] Furthermore, a N-CoR dependent model of transrepression has also emerged to explain ligand induced PPARγ and LXR-mediated inhibition of inflammatory response genes in macrophages[81,82] (Fig. 11.2). A subset of nuclear factor-κB (NF-κB) regulated inflammatory genes require N-CoR/SMRT for basal repression. In the presence of an inflammatory stimulus, such as lipopolysaccharide (LPS), these genes are activated by corepressor dissociation from the transcription factor and coactivator recruitment.[52,72] However, PPARγ and LXR agonists have been found to antagonize the activation of NF-κB target genes.[81,82] The mechanism for this transrepressive response results from ligand dependent SUMOylation of the NR and subsequent targeting of the modified NR to N-CoR/HDAC3 complexes bound at inflammatory gene promoters. This association prevents dissociation of the N-CoR/HDAC3 complex from the gene and, therefore, maintains repression of the promoter even in the presence of LPS. Furthermore, depletion of N-CoR or HDAC3 from the cells results in a loss of transrepression, suggesting that these core components of the complex are required.[81] The implications for corepressor dependent transrepression in relation to inflammation and obesity will be discussed in the next section.

11.4 Corepressors and Metabolism

Animal models have recently been used to determine the biological importance of NR mediated repression and, specifically, the corepressor proteins. N-CoR was shown by Jepsen *et al*. to be essential for normal development and postnatal survival. Mice lacking this corepressor protein died *in utero* by day 15 due to severe anemia and secondary edema. Furthermore, the N-CoR knockout embryos were defective in erythrocyte, thymocyte, as well as neural development.[34] In *Xenopus*, expression of a dominant negative form of SMRT that blocks RAR mediated repression was found to inhibit embryonic development of anterior structures, demonstrating that RAR directed repression was critical for proper head development. Furthermore, SMRT-/- mice also die *in utero* and demonstrate significant defects in retinoic acid dependent forebrain development.[72a]

Knockout studies also suggest a physiologic role for TR directed repression. This is most evident by the finding that mice lacking thyroid hormone have a more severe phenotype than mice with all TR genes deleted, suggesting that release from the repressed state is required for normal development and that chronic repression by TR carries deleterious consequences.[84] In the TR knockout mice, basal expression of genes normally repressed by TR is likely sufficient to compesate for the absence of thyroid hormone (TH) stimulated release from repression. This finding was further supported by microarray analyses showing significant differences in gene expression between TR knockouts and hypothyroid mice. More specifically, a subset of genes was found to be repressed in hypothyroid mice relative to the basal level detected in TR knockouts.[85] Dominant negative strategies have been utilized to further demonstrate a biological role of TR mediated repression. Mutant forms of both TRα and TRβ that are unable to bind TH such that the receptor fails to release corepressors causes various CNS abnormalities when introduced into mice, emphasizing the importance of unliganded TR in development.[86,87] Furthermore, liver specific expression of a dominant negative form of N-CoR, which lacks repression domains but retains the NR interactions regions, leads to de-repression of TR target genes and promotes hepatocyte proliferation.[88]

While animal models have demonstrated the importance of corepressors in normal development and physiology, abnormal corepressor function has been implicated in various human diseases. For instance, a genetic disease termed resistance to thyroid hormone (RTH) results from mutations in the ligand binding domain of TRβ that inhibit its binding to TH and, therefore, cause the receptor to maintain corepressor interactions and transcriptional repression in the presence of TH.[89] Furthermore, several forms of leukemia have been found to carry genetic rearrangements which lead to inappropriate recruitment of N-CoR or SMRT. Lastly, mislocation of N-CoR exclusively to the cytoplasm through an interaction with the Huntington's disease gene product, huntingtin, has been implicated in progression of this disease.[31]

11.4.1 *Metabolism and inflammation*

NRs are critical transcriptional regulators of several metabolic genes and pathways. Complex coordination of signaling by NRs as well as other transcription factors is required for normal metabolic processes to occur within the cell. Malfunctions in NR signaling, specifically PPARs and LXRs, contribute to metabolic diseases such as insulin resistance, type II diabetes, obesity, and atherosclerosis such that drugs targeting NRs are currently being used as well as continually being investigated as therapeutics.[3] Furthermore, recent evidence has shown a significant correlation between inflammation in metabolic organs with the development of insulin resistance and obesity.[90] Interestingly, PPARs and LXRs are also important regulators of gene expression within inflammatory cells, suggesting a critical integrated relationship between NR mediated regulation of inflammation and metabolism.[91]

Presently, corepressor proteins, the mediators of NR directed transcriptional repression, are being studied with regards to metabolism and potential therapeutic targets. Leonardsson *et al.* found that mice lacking the corepressor RIP140 are lean and resistant to diet-induced obesity, emphasizing a critical role for RIP140 in metabolism.[92] Unfortunately, the embryonic lethality of deleting N-CoR in mice has limited *in vivo* analysis of this corepressor in adult metabolic regulation. This section highlights significant contributions from recent cellular studies that have elucidated the importance of N-CoR and SMRT in NR-directed regulation of both metabolic and inflammatory genes.

11.4.1.1 *PPARγ*

Although the specific roles of N-CoR and SMRT in metabolism are not fully understood, the most substantial evidence that these corepressors play a critical role in metabolic and inflammatory gene regulation comes from studies of the peroxisome proliferator-activated receptor (PPAR) family. The PPARs, consisting of PPARα, PPARδ (also known as PPARβ), and PPARγ, have been implicated in various biological processes but are most well known for their key regulatory role in lipid metabolism. These NRs activate in response to dietary fatty acids as well as their metabolic derivatives, suggesting that they function as lipid sensors in the body.[93] PPARα and PPARγ are predominantly expressed in liver and adipose tissue, respectively, whereas PPARδ is expressed fairly ubiquitously.[94] Although all three PPARs play important roles in lipid homeostasis and metabolism, here we focus on PPARγ, the most extensively studied PPAR in relation to corepressor function.

PPARγ, particularly the PPARγ2 isoform, is a critical regulator of adipocyte differentiation.[95,96] In fact, the antidiabetic drugs termed thiazolidinediones (TZDs) that are commonly used to increase insulin sensitivity were discovered to be selective activators of PPARγ, thus linking PPARγ directed regulation and insulin sensitivity.[97] Consistently, dominant negative mutations in PPARγ found in humans have been associated with insulin resistance, diabetes mellitus, and hypertension.[98,99]

Mouse models have been integral to defining the *in vivo* role of PPARγ in adipogenesis and insulin sensitivity. Homozygous deletion of PPARγ is embryonic lethal due to placental defects, however, rescue of a PPARγ knockout mouse by placental reconstitution as well as examination of cells harvested from chimeric mice demonstrated that PPARγ is required for the proper development of adipose tissue.[100,101] Consistently, mice engineered with an adipose specific hypomorphic PPARγ allele develop lipodystrophy and hyperlipidemia, whereas mice that have PPARγ deleted from adipose in adulthood show a progressive loss of adipocytes, indicating that PPARγ expression in adipose is necessary for both adipose development and maintenance.[102,103] Liver specific PPARγ knockouts develop increased adiposity, hyperlipidemia, as well as insulin resistance.[104] Similarly, specific deletion of PPARγ in muscle results in whole body insulin resistance and increased adiposity.[105,106] Since tissue specific deletions of PPARγ lead to abnormal fat distribution and dysfunction in glucose and insulin homeostasis, PPARγ is further characterized as a critical regulator of cross talk between adipose tissue, the liver and muscle. Interestingly, mice that are heterozygous for PPARγ deficiency demonstrate improved insulin sensitivity compared to their wild type counterparts.[107,108] Although the mechanism underlying this phenotype is not clearly defined, the genetic reduction in unliganded receptor would be expected to de-repress the expression of PPARγ target genes and thus partially mimic the activating effects of PPARγ agonists.

11.4.1.2 *PPARγ and N-CoR/SMRT*

Early studies examining the interaction between PPARγ and N-CoR/SMRT suggested that PPARγ could interact *in vitro* with corepressors and that it may preferentially associate with SMRT.[75,109] More recent studies have supported a model in which PPARγ recruits either of these corepressors *in vivo* and that N-CoR/SMRT likely play an important role in PPARγ directed transcriptional regulation as well as adipogenesis.[99,110,111]

Molecular studies in 3T3-L1 adipocytes have demonstrated the recruitment of N-CoR/SMRT to the promoters of specific PPARγ target genes during repression. Guan *et al.* compared PPARγ and coregulator recruitment to two different PPARγ target genes, aP2, an adipocyte specific gene that is expressed at high levels in adipocytes, and glycerol kinase (GyK), a gene that is expressed at low levels in adipocytes but is significantly induced by TZDs. In the absence of the exogenous ligand, PPARγ and coactivator proteins are recruited to the aP2 promoter, consistent with the high level of aP2 expression in untreated adipocytes. Under these same conditions, however, PPARγ along with the corepressor proteins N-CoR, SMRT, and HDAC3 are recruited to the GyK promoter. Furthermore, treatment with HDAC inhibitors increases endogenous GyK expression and the depletion of both corepressors in adipocytes using small interfering RNA (siRNA) increases basal expression of a GyK reporter gene, indicating that the N-CoR/SMRT-HDAC containing complex is critical for maintaining repression of this PPARγ target gene. In the presence of TZDs, corepressor proteins dissociate from the PPARγ bound GyK gene and coactivator proteins are recruited, inducing gene activation.[111] This same pattern of cofactor recruitment has been shown for oxidized LDL receptor 1 (OLR1), another PPARγ target gene that is repressed in 3T3-L1 adipocytes but activated by TZDs.[112] Therefore, PPARγ selectively recruits corepressors to specific promoters such that some targets are repressed and some are activated in the absence of exogenous ligands, suggesting the existence of separate populations of corepressor dependent and independent PPARγ target genes (Fig. 11.3).

Both GyK and OLR1 are critical regulators of metabolites in the body. For instance, GyK catalyzes the phosphorylation of glycerol to glycerol 3-phosphate, the backbone of triglycerides (TGs). In fact, TZD induction of GyK stimulates glycerol incorporation into TGs and,

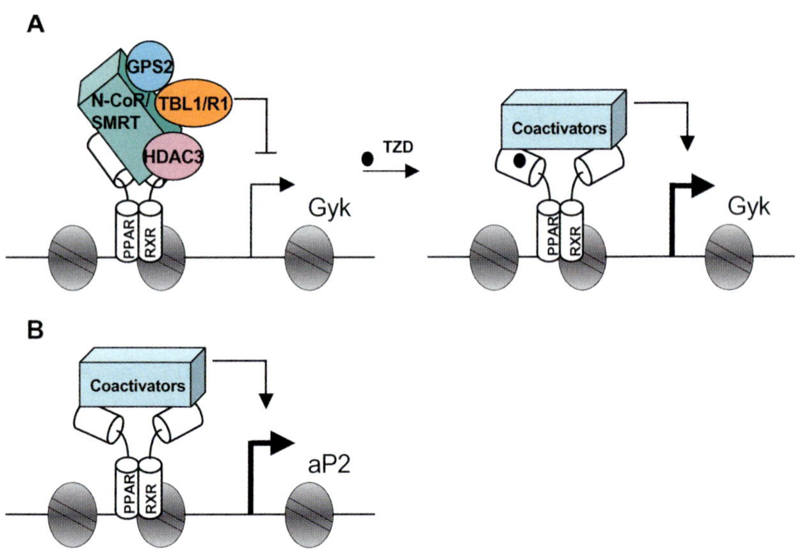

Fig. 11.3. Molecular mechanism underlying differential ligand response of PPARγ target genes in adipocytes. In the absence of exogenous ligand, PPARγ recruits either corepressor N-CoR/SMRT complex (A) or coactivators (B) to PPARγ target genes in adipocytes. TZDs stimulate displacement of the corepressor complex from the GyK promoter and recruitment of coactivators to the GyK promoter.

therefore, contributes to lowering of serum free fatty acids.[113] OLR1 increases oxLDL uptake as well as total cholesterol levels in adipocytes.[112] Given the potential regulatory role of GyK and OLR1 in adipocyte metabolism, corepressor dependent repression of these genes is likely to be critical to the proper maintenance of lipid homeostasis. Moreover, it is likely that corepressors may regulate additional metabolic genes that are TZD-responsive in adipocytes.

Consistent with their role in PPARγ directed transcriptional regulation, N-CoR and SMRT have been found to be physiologically critical during adipogenesis. Gurnell *et al.* utilized a dominant-negative mutant receptor to inhibit the activity of wildtype PPARγ *in vivo*. This mutant PPARγ maintains a stronger association with N-CoR/SMRT in the absence as well as in the presence of ligand and represses basal transcription relative to wild type PPARγ. Interestingly, the expression of this dominant negative PPARγ mutant in primary human preadipocytes blocks TZD-induced differentiation, suggesting that the PPARγ-corepressor association plays a role in inhibiting adipocyte development.[114] This finding was further supported by Yu *et al.* who found that PPARγ can recruit both N-CoR and SMRT in 3T3-L1 cells. Furthermore, 3T3-L1 fibroblast cells depleted of N-CoR or SMRT by siRNA that are induced to differentiate demonstrate increased expression of adipocyte specific proteins as well as lipid accumulation compared to control cells, consistent with enhanced adipocyte differentiation.[110] Furthermore, SIRT1, an NAD-dependent protein deacetylase, has been shown to act similarly to N-CoR/SMRT such that it inhibits expression of PPARγ target genes and restricts adipogenesis. Surprisingly, the mechanism by which SIRT1 functions in this context appears to be through an association with N-CoR/SMRT at the promoters of PPARγ target genes in adipocytes.[115] Taken together, these studies strongly implicate N-CoR and SMRT function in the inhibition of adipogenesis, likely through PPARγ directed repression.

11.4.1.3 *Inflammation and N-CoR/SMRT*

It has become evident over the past decade that chronic inflammation, particularly in adipose tissue, occurs with obesity. Chronic inflammation plays a critical role in the development of obesity-linked diseases such as type 2 diabetes and atherosclerosis.[90] More specifically, macrophages have emerged as a potential link between inflammation and metabolism. Recently, macrophage accumulation in adipose tissue was found to characterize obesity in mice and humans, suggesting that macrophage infiltration of adipose tissue may play a role in the development of inflammatory conditions associated with obesity.[116,117] With regards to corepressors, a subset of nuclear factor-κB (NF-κB) and activator protein(AP)-1 regulated inflammatory genes require N-CoR/SMRT for basal repression.[118,119] Furthermore, macrophages derived from N-CoR knockout embryos exhibit a partially activated phenotype and de-repression of AP-1 target genes, suggesting that N-CoR functions in repressing macrophage activation.[118]

Interestingly, two families of NRs, the PPARs and the liver X receptors (LXRs), are key regulators of both lipid metabolism and inflammatory gene expression in macrophages. As discussed earlier, PPARs are critical to adipogenesis and the maintenance of glucose and triglyceride homeostasis. LXRs, which consist of two isoforms LXRα and LXRβ, are regulated by derivatives of cholesterol, termed oxysterols, and play an essential role in cholesterol and fatty acid metabolism.[120] However, in macrophages, both PPAR and LXR agonists inhibit the lipopolysaccharide (LPS) induction of several proinflammatory genes.[120] Recent reports from Pascual *et al.* and Ghisletti *et al.* indicate that this ligand induced inhibition occurs through a N-CoR dependent transrepression mechanism (discussed in previous section), therefore suggesting a critical role for corepressors in integrating inflammatory and metabolic transcriptional signals.

11.5 Conclusions and Future Directions

In the past decade, corepressors have emerged as critical regulators of NR mediated transcriptional repression. This chapter focused on N-CoR and SMRT and initial evidence that these proteins play a key role in metabolic regulation, particularly through PPARγ and LXRs. However, these studies are just the beginning and future *in vivo* studies are necessary to determine the precise physiological roles for N-CoR/SMRT directed repression. Conditional knockout and mutant knock-in studies will provide insight into the tissue specific roles of N-CoR and SMRT as well as the biological significance of conserved domains and protein associations, potentially directing the development of new therapeutics. Importantly, other corepressors have already been identified and it is likely that new corepressor proteins will be uncovered in the future years. The challenge will not only be to determine the role of these corepressors in NR mediated transcriptional regulation but to also integrate the functions of multiple corepressors and NRs into our understanding of metabolic regulation.

References

1. Mangelsdorf DJ, *et al.*, The nuclear receptor superfamily: The second decade, *Cell* **83**(6):835–839, 1995.
2. Francis GA, *et al.*, Nuclear receptors and the control of metabolism, *Annu Rev Physiol* **65**:261–311, 2003.
3. Gronemeyer H, Gustafsson JA, Laudet V, Principles for modulation of the nuclear receptor superfamily, *Nat Rev Drug Discov* **3**(11):950–964, 2004.
4. Bourguet W, Germain P, Gronemeyer H, Nuclear receptor ligand-binding domains: Three-dimensional structures, molecular interactions and pharmacological implications, *Trends Pharmacol Sci* **21**(10):381–388, 2000.
5. Giguere V, *et al.*, Functional domains of the human glucocorticoid receptor, *Cell* **46**(5):645–652, 1986.
6. Kumar V, *et al.*, Functional domains of the human estrogen receptor, *Cell* **51**(6):941–951, 1987.
7. Luisi BF, *et al.*, Crystallographic analysis of the interaction of the glucocorticoid receptor with DNA, *Nature* **352**(6335):497–505, 1991.
8. Umesono K, Evans RM, Determinants of target gene specificity for steroid/thyroid hormone receptors, *Cell* **57**(7):1139–1146, 1989.
9. Tsai MJ, O'Malley BW, Molecular mechanisms of action of steroid/thyroid receptor superfamily members, *Annu Rev Biochem* **63**:451–486, 1994.
10. Wurtz JM, *et al.*, A canonical structure for the ligand-binding domain of nuclear receptors, *Nat Struct Biol* **3**(1):87–94, 1996.
11. Durand B, *et al.*, Activation function 2 (AF-2) of retinoic acid receptor and 9-cis retinoic acid receptor: Presence of a conserved autonomous constitutive activating domain and influence of the nature of the response element on AF-2 activity, *EMBO J* **13**(22):5370–5382, 1994.
12. Barettino D, Vivanco Ruiz MM, Stunnenberg HG, Characterization of the ligand-dependent transactivation domain of thyroid hormone receptor, *EMBO J* **13**(13):3039–3049, 1994.
13. Danielian PS, *et al.*, Identification of a conserved region required for hormone dependent transcriptional activation by steroid hormone receptors, *EMBO J* **11**(3):1025–1033, 1992
14. Glass CK, Rosenfeld MG, The coregulator exchange in transcriptional functions of nuclear receptors, *Genes Dev* **14**(2):121–141, 2000.

15. Umesono K, *et al.*, Direct repeats as selective response elements for the thyroid hormone, retinoic acid, and vitamin D3 receptors, *Cell* **65**(7):1255–1266, 1991.
16. Laudet V, Adelmant G, Nuclear receptors, Lonesome orphans, *Curr Biol* **5**(2):124–127, 1995.
17. Benoit G, Malewicz M, Perlmann T, Digging deep into the pockets of orphan nuclear receptors: Insights from structural studies, *Trends Cell Biol* **14**(7):369–376, 2004.
18. Baniahmad A, Kohne AC, Renkawitz R, A transferable silencing domain is present in the thyroid hormone receptor, in the v-erbA oncogene product and in the retinoic acid receptor, *EMBO J* **11**(3):1015–1023, 1992.
19. Tong GX, *et al.*, Transcriptional silencing by unliganded thyroid hormone receptor beta requires a soluble corepressor that interacts with the ligand-binding domain of the receptor, *Mol Cell Biol* **16**(5):1909–1920, 1996.
20. Baniahmad A, *et al.*, The tau 4 activation domain of the thyroid hormone receptor is required for release of a putative corepressor(s) necessary for transcriptional silencing, *Mol Cell Biol* **15**(1):76–86, 1995.
21. Casanova J, *et al.*, Functional evidence for ligand-dependent dissociation of thyroid hormone and retinoic acid receptors from an inhibitory cellular factor, *Mol Cell Biol* **14**(9):5756–5765, 1994.
22. Horlein AJ, *et al.*, Ligand-independent repression by the thyroid hormone receptor mediated by a nuclear receptor co-repressor, *Nature* **377**(6548): 397–404, 1995.
23. Chen JD, Evans RM, A transcriptional co-repressor that interacts with nuclear hormone receptors, *Nature* **377**(6548):454–457, 1995.
24. Cavailles V, *et al.*, Nuclear factor RIP140 modulates transcriptional activation by the estrogen receptor, *EMBO J* **14**(15):3741–3751, 1995.
25. Dressel U, *et al.*, Alien, a highly conserved protein with characteristics of a corepressor for members of the nuclear hormone receptor superfamily, *Mol Cell Biol* **19**(5):3383–3394, 1999.
26. Potter GB, *et al.*, The hairless gene mutated in congenital hair loss disorders encodes a novel nuclear receptor corepressor, *Genes Dev* **15**(20): 2687–2701, 2001.
27. Fernandes I, *et al.*, Ligand-dependent nuclear receptor corepressor LCoR functions by histone deacetylase-dependent and -independent mechanisms, *Mol Cell* **11**(1):139–150, 2003.
28. Zamir I, *et al.*, Cloning and characterization of a corepressor and potential component of the nuclear hormone receptor repression complex, *Proc Natl Acad Sci USA* **94**(26):14400–14405, 1997.
29. Shi Y, *et al.*, Sharp, an inducible cofactor that integrates nuclear receptor repression and activation, *Genes Dev* **15**(9):1140–1151, 2001.

30. Hatchell EC, *et al.*, SLIRP, a small SRA binding protein, is a nuclear receptor corepressor, *Mol Cell* **22**(5):657–668, 2006.
31. Jepsen K, Rosenfeld MG, Biological roles and mechanistic actions of corepressor complexes, *J Cell Sci* **115**(Pt 4):689–698, 2002.
32. Martinez de Arrieta C, Koibuchi N, Chin WW, Coactivator and corepressor gene expression in rat cerebellum during postnatal development and the effect of altered thyroid status, *Endocrinology* **141**(5):1693–1698, 2000.
33. Chen JD, Umesono K, Evans RM, SMRT isoforms mediate repression and anti-repression of nuclear receptor heterodimers, *Proc Natl Acad Sci USA* **93**(15):7567–7571, 1996.
34. Jepsen K, *et al.*, Combinatorial roles of the nuclear receptor corepressor in transcription and development, *Cell* **102**(6):753–763, 2000.
35. Privalsky ML, The role of corepressors in transcriptional regulation by nuclear hormone receptors, *Annu Rev Physiol* **66**:315–360, 2004.
36. Hu X, Lazar MA, The CoRNR motif controls the recruitment of corepressors by nuclear hormone receptors, *Nature* **402**(6757):93–96, 1999.
37. Nagy L, *et al.*, Mechanism of corepressor binding and release from nuclear hormone receptors, *Genes Dev*, **13**(24):3209–3216, 1999.
38. Perissi V, *et al.*, Molecular determinants of nuclear receptor-corepressor interaction, *Genes Dev* **13**(24):3198–3208, 1999.
39. Hu X, Li Y, Lazar MA, Determinants of CoRNR-dependent repression complex assembly on nuclear hormone receptors, *Mol Cell Biol* **21**(5): 1747–1758, 2001.
40. Cohen RN, *et al.*, The specificity of interactions between nuclear hormone receptors and corepressors is mediated by distinct amino acid sequences within the interacting domains, *Mol Endocrinol* **15**(7):1049–1061, 2001.
41. Ishizuka T, Lazar MA, The N-CoR/histone deacetylase 3 complex is required for repression by thyroid hormone receptor *Mol Cell Biol* **23**(15): 5122–5131, 2003.
42. Ordentlich P, *et al.*, Unique forms of human and mouse nuclear receptor corepressor SMRT, *Proc Natl Acad Sci USA* **96**(6):2639–2644, 1999.
43. Webb P, *et al.*, The nuclear receptor corepressor (N-CoR) contains three isoleucine motifs (I/LXXII) that serve as receptor interaction domains (IDs), *Mol Endocrinol* **14**(12):1976–1985, 2000.
44. Guenther MG, Barak O, Lazar MA, The SMRT and N-CoR corepressors are activating cofactors for histone deacetylase 3, *Mol Cell Biol* **21**(18): 6091–6101, 2001.
45. Ishizuka T, Lazar MA, The nuclear receptor corepressor deacetylase activating domain is essential for repression by thyroid hormone receptor, *Mol Endocrinol* **19**(6):1443–1451, 2005.

46. Yu J, *et al.*, A SANT motif in the SMRT corepressor interprets the histone code and promotes histone deacetylation, *EMBO J* **22**(13):3403–3410, 2003.
47. Hartman HB, *et al.*, The histone-binding code of nuclear receptor corepressors matches the substrate specificity of histone deacetylase 3, *EMBO Rep* **6**(5):445–451, 2005.
48. Guenther MG, *et al.*, A core SMRT corepressor complex containing HDAC3 and TBL1, a WD40-repeat protein linked to deafness, *Genes Dev* **14**(9):1048–1057, 2000.
49. Li J, *et al.*, Both corepressor proteins SMRT and N-CoR exist in large protein complexes containing HDAC3, *EMBO J* **19**(16):4342–4350, 2000.
50. Zhang J, *et al.*, The N-CoR-HDAC3 nuclear receptor corepressor complex inhibits the JNK pathway through the integral subunit GPS2, *Mol Cell* **9**(3):611–623, 2002.
51. Yoon HG, *et al.*, Purification and functional characterization of the human N-CoR complex: The roles of HDAC3, TBL1 and TBLR1, *EMBO J* **22**(6):1336–1346, 2003.
52. Perissi V, *et al.*, A corepressor/coactivator exchange complex required for transcriptional activation by nuclear receptors and other regulated transcription factors, *Cell* **116**(4):511–526, 2004.
53. Nagy L, *et al.*, Nuclear receptor repression mediated by a complex containing SMRT, mSin3A, and histone deacetylase, *Cell* **89**(3):373–380, 1997.
54. Heinzel T, *et al.*, A complex containing N-CoR, mSin3 and histone deacetylase mediates transcriptional repression, *Nature* **387**(6628):43–48, 1997.
55. Li J, *et al.*, Specific targeting and constitutive association of histone deacetylase complexes during transcriptional repression, *Genes Dev* **16**(6):687–692, 2002.
56. Jones PL, *et al.*, Multiple N-CoR complexes contain distinct histone deacetylases, *J Biol Chem* **276**(12):8807–8811, 2001.
57. Huang EY, *et al.*, Nuclear receptor corepressors partner with class II histone deacetylases in a Sin3-independent repression pathway, *Genes Dev* **14**(1):45–54, 2000.
58. Kao HY, *et al.*, Isolation of a novel histone deacetylase reveals that class I and class II deacetylases promote SMRT-mediated repression, *Genes Dev* **14**(1):55–66, 2000.
59. Farhana L, *et al.*, Adamantyl-substituted retinoid-related molecules bind small heterodimer partner and modulate the Sin3A repressor, *Cancer Res* **67**(1):318–325, 2007.
60. Underhill C, *et al.*, A novel nuclear receptor corepressor complex, N-CoR, contains components of the mammalian SWI/SNF complex and the corepressor KAP-1, *J Biol Chem* **275**(51):40463–40470, 2000.

61. Richmond TJ, Davey CA, The structure of DNA in the nucleosome core, *Nature* **423**(6936):145–150, 2003.
62. Jenuwein T, Allis CD, Translating the histone code, *Science* **293**(5532):1074–1080, 2001.
63. Strahl BD, Allis CD, The language of covalent histone modifications, *Nature* **403**(6765):41–45, 2000.
64. Grunstein M, Histone acetylation in chromatin structure and transcription, *Nature* **389**(6649):349–352, 1997.
65. Allfrey VG, Faulkner R, Mirsky AE, Acetylation and methylation of histones and their possible role in the regulation of rna synthesis, *Proc Natl Acad Sci USA* **51**:786–794, 1964.
66. Dhalluin C, *et al.*, Structure and ligand of a histone acetyltransferase bromodomain, *Nature* **399**(6735):491–496, 1999.
67. Owen DJ, *et al.*, The structural basis for the recognition of acetylated histone H4 by the bromodomain of histone acetyltransferase gcn5p, *EMBO J* **19**(22):6141–6149, 2000.
68. Hu X, Lazar MA, Transcriptional repression by nuclear hormone receptors, *Trends Endocrinol Metab* **11**(1):6–10, 2000.
69. Guenther MG, *et al.*, Assembly of the SMRT-histone deacetylase 3 repression complex requires the TCP-1 ring complex, *Genes Dev* **16**(24):3130–3135, 2002.
70. Alenghat T, Yu J, Lazar MA, The N-CoR complex enables chromatin remodeler SNF2H to enhance repression by thyroid hormone receptor, *EMBO J* **25**(17):3966–3974, 2006.
71. Wong CW, Privalsky ML, Transcriptional repression by the SMRT-mSin3 corepressor: Multiple interactions, multiple mechanisms, and a potential role for TFIIB, *Mol Cell Biol* **18**(9):5500–5510, 1998.
72. Baek SH, *et al.*, Exchange of N-CoR corepressor and Tip60 coactivator complexes links gene expression by NF-kappaB and beta-amyloid precursor protein, *Cell* **110**(1):55–67, 2002.

72a. Jepsen K, *et al.*, SMRT-mediated repression of an H3K27 demethylase in progression from neural stem cell to neuron, *Nature* **1**:6, 2007.

73. Hong SH, Privalsky ML, The SMRT corepressor is regulated by a MEK-1 kinase pathway: Inhibition of corepressor function is associated with SMRT phosphorylation and nuclear export, *Mol Cell Biol* **20**(17):6612–6625, 2000.
74. Zhou Y, *et al.*, The SMRT corepressor is a target of phosphorylation by protein kinase CK2 (casein kinase II), *Mol Cell Biochem* **220**(1–2):1–13, 2001.
75. Lavinsky RM, *et al.*, Diverse signaling pathways modulate nuclear receptor recruitment of N-CoR and SMRT complexes, *Proc Natl Acad Sci USA* **95**(6):2920–2925, 1998.

76. Davis PJ, *et al.*, Thyroxine promotes association of mitogen-activated protein kinase and nuclear thyroid hormone receptor (TR) and causes serine phosphorylation of TR, *J Biol Chem* **275**(48):38032–38039, 2000.
77. Christian M, White R, Parker MG, Metabolic regulation by the nuclear receptor corepressor RIP140, *Trends Endocrinol Metab* **17**(6):243–250, 2006.
78. Pascual G, Glass CK, Nuclear receptors versus inflammation: Mechanisms of transrepression, *Trends Endocrinol Metab* **17**(8):321–327, 2006.
79. Reichardt HM, *et al.*, Repression of inflammatory responses in the absence of DNA binding by the glucocorticoid receptor, *EMBO J* **20**(24):7168–7173, 2001.
80. Ito K, *et al.*, Histone deacetylase 2-mediated deacetylation of the glucocorticoid receptor enables NF-kappaB suppression, *J Exp Med* **203**(1):7–13, 2006.
81. Pascual G, *et al.*, A SUMOylation-dependent pathway mediates transrepression of inflammatory response genes by PPAR-gamma, *Nature* **437**(7059):759–763, 2005.
82. Ghisletti S, *et al.*, Parallel SUMOylation-dependent pathways mediate gene- and signal-specific transrepression by LXRs and PPAR-gamma, *Mol Cell* **25**(1):57–70, 2007.
83. Koide T, *et al.*, Active repression of RAR signaling is required for head formation, *Genes Dev* **15**(16):2111–2121, 2001.
84. Gothe S, *et al.*, Mice devoid of all known thyroid hormone receptors are viable but exhibit disorders of the pituitary-thyroid axis, growth, and bone maturation, *Genes Dev* **13**(10):1329–1341, 1999.
85. Yen PM, *et al.*, Effects of ligand and thyroid hormone receptor isoforms on hepatic gene expression profiles of thyroid hormone receptor knockout mice, *EMBO Rep* **4**(6):581–587, 2003.
86. Venero C, *et al.*, Anxiety, memory impairment, and locomotor dysfunction caused by a mutant thyroid hormone receptor alpha1 can be ameliorated by T3 treatment, *Genes Dev*, **19**(18):2152–2163, 2005.
87. Hashimoto K, *et al.*, An unliganded thyroid hormone receptor causes severe neurological dysfunction, *Proc Natl Acad Sci USA*, **98**(7):3998–4003, 2001.
88. Feng X, *et al.*, Transgenic targeting of a dominant negative corepressor to liver blocks basal repression by thyroid hormone receptor and increases cell proliferation, *J Biol Chem* **276**(18):15066–15072, 2001.
89. Tagami T, Jameson JL, Nuclear corepressors enhance the dominant negative activity of mutant receptors that cause resistance to thyroid hormone, *Endocrinology* **139**(2):640–650, 1998.
90. Wellen KE, Hotamisligil GS, Inflammation, stress, and diabetes, *J Clin Invest* **115**(5):1111–1119, 2005.

91. Glass CK, Ogawa S, Combinatorial roles of nuclear receptors in inflammation and immunity, *Nat Rev Immunol* **6**(1):44–55, 2006.
92. Leonardsson G, *et al.*, Nuclear receptor corepressor RIP140 regulates fat accumulation, *Proc Natl Acad Sci USA* **101**(22):8437–8442, 2004.
93. Evans RM, Barish GD, Wang YX, PPARs and the complex journey to obesity, *Nat Med* **10**(4):355–361, 2004.
94. Ferre P, The biology of peroxisome proliferator-activated receptors: Relationship with lipid metabolism and insulin sensitivity, *Diabetes* **53** Suppl 1:S43–50, 2004.
95. Tontonoz P, Hu E, Spiegelman BM, Stimulation of adipogenesis in fibroblasts by PPAR-gamma 2, a lipid-activated transcription factor, *Cell* **79**(7): 1147–1156, 1994.
96. Rosen ED, *et al.*, C/EBPalpha induces adipogenesis through PPAR-gamma: A unified pathway, *Genes Dev* **16**(1):22–26, 2002.
97. Lehmann JM, *et al.*, An antidiabetic thiazolidinedione is a high affinity ligand for peroxisome proliferator-activated receptor gamma (PPAR-gamma), *J Biol Chem* **270**(22):12953–12956, 1995.
98. Barroso I, *et al.*, Dominant negative mutations in human PPAR-gamma associated with severe insulin resistance, diabetes mellitus and hypertension, *Nature* **402**(6764):880–883, 1999.
99. Savage DB, *et al.*, Human metabolic syndrome resulting from dominant-negative mutations in the nuclear receptor peroxisome proliferator-activated receptor-gamma, *Diabetes* **52**(4):910–917, 2003.
100. Barak Y, *et al.*, PPAR-gamma is required for placental, cardiac, and adipose tissue development, *Mol Cell* **4**(4):585–595, 1999.
101. Rosen ED, *et al.*, PPAR-gamma is required for the differentiation of adipose tissue *in vivo* and *in vitro*, *Mol Cell* **4**(4):611–617, 1999.
102. Koutnikova H, *et al.*, Compensation by the muscle limits the metabolic consequences of lipodystrophy in PPAR-gamma hypomorphic mice, *Proc Natl Acad Sci USA* **100**(24):14457–14462, 2003.
103. He W, *et al.*, Adipose-specific peroxisome proliferator-activated receptor gamma knockout causes insulin resistance in fat and liver but not in muscle, *Proc Natl Acad Sci USA* **100**(26):15712–15717, 2003.
104. Gavrilova O, *et al.*, Liver peroxisome proliferator-activated receptor gamma contributes to hepatic steatosis, triglyceride clearance, and regulation of body fat mass, *J Biol Chem* **278**(36):34268–34276, 2003.
105. Hevener AL, *et al.*, Muscle-specific Pparg deletion causes insulin resistance, *Nat Med* **9**(12):1491–1497, 2003.
106. Norris AW, *et al.*, Muscle-specific PPAR-gamma-deficient mice develop increased adiposity and insulin resistance but respond to thiazolidinediones, *J Clin Invest* **112**(4):608–618, 2003.

107. Miles PD, *et al.*, Improved insulin-sensitivity in mice heterozygous for PPAR-gamma deficiency, *J Clin Invest* **105**(3):287–292, 2000.
108. Kubota N, *et al.*, PPAR-gamma mediates high-fat diet-induced adipocyte hypertrophy and insulin resistance, *Mol Cell* **4**(4):597–609, 1999.
109. Zamir I, Zhang J, Lazar MA, Stoichiometric and steric principles governing repression by nuclear hormone receptors, *Genes Dev* **11**(7):835–846, 1997.
110. Yu C, *et al.*, The nuclear receptor corepressors NCoR and SMRT decrease peroxisome proliferator-activated receptor gamma transcriptional activity and repress 3T3-L1 adipogenesis, *J Biol Chem* **280**(14):13600–13605, 2005.
111. Guan HP, *et al.*, Corepressors selectively control the transcriptional activity of PPAR-gamma in adipocytes, *Genes Dev* **19**(4):453–461, 2005.
112. Chui PC, *et al.*, PPAR-gamma regulates adipocyte cholesterol metabolism via oxidized LDL receptor 1, *J Clin Invest* **115**(8):2244–2256, 2005.
113. Guan HP, *et al.*, A futile metabolic cycle activated in adipocytes by antidiabetic agents, *Nat Med* **8**(10):1122–1128, 2002.
114. Gurnell M, *et al.*, A dominant-negative peroxisome proliferator-activated receptor gamma (PPAR-gamma) mutant is a constitutive repressor and inhibits PPAR-gamma-mediated adipogenesis, *J Biol Chem* **275**(8): 5754–5759, 2000.
115. Picard F, *et al.*, Sirt1 promotes fat mobilization in white adipocytes by repressing PPAR-gamma, *Nature* **429**(6993):771–776, 2004.
116. Weisberg SP, *et al.*, Obesity is associated with macrophage accumulation in adipose tissue, *J Clin Invest* **112**(12):1796–1808, 2003.
117. Xu H, *et al.*, Chronic inflammation in fat plays a crucial role in the development of obesity-related insulin resistance, *J Clin Invest* **112**(12): 1821–1830, 2003.
118. Ogawa S, *et al.*, A nuclear receptor corepressor transcriptional checkpoint controlling activator protein 1 dependent gene networks required for macrophage activation, *Proc Natl Acad Sci USA* **101**(40):14461–14466, 2004.
119. Hoberg JE, Yeung F, Mayo MW, SMRT derepression by the IkappaB kinase alpha: A prerequisite to NF-kappaB transcription and survival, *Mol Cell* **16**(2):245–255, 2004.
120. Li AC, Glass CK, PPAR- and LXR-dependent pathways controlling lipid metabolism and the development of atherosclerosis, *J Lipid Res* **45**(12): 2161–2173, 2004.

Chapter 12

Coregulators in CNS Function and Disease

O.C. Meijer and E.R. de Kloet

Nuclear receptor coregulators are broadly present in the brain, and relevant for neuronal development, maintenance of neuronal integrity and responsiveness to steroid hormones. Coregulators integrate multiple signaling pathways at the level of the transcription machinery, and may be necessary for neurons to adequately respond to a great diversity of hormonal signals under different circumstances. Their expression patterns suggest widespread functions in the brain, both globally and concerning processes linked with very specific brain nuclei. Corepressor NCoR and coactivator SRC-1 have been shown to be relevant to the proper development of particular neuronal populations. SRC-1 modulates the responses to steroid hormones in the adult brain in relation to reproductive or stress state. The integrative nature of coregulators is possibly best exemplified in the brain by the phenotype of PGC-1α knockout mice that show enhanced neuronal cell death during aging and after excitotoxic challenges, due to failure to mount a coordinate neuroprotective genetic program. These mice are the first example of the relevance of coregulator function in neurodegerative disease. Brain pathophysiology may be linked with organism-wide changes in the metabolic or immune system. The role of coregulators in brain disease might best be viewed in relation to such general states that form contexts in which many nuclear receptor mediated signals change, and in which the role of coregulators as orchestrators of adaptation at the cellular level becomes crucial.

12.1 Introduction

The brain links the environment with endocrine secretions that in turn feedback on the brain as potent modulators of metabolism and neuronal excitability. This is illustrated by the action of gonadal and adrenal steroid hormones that target precisely those brain circuits that responded to environmental stimulation and initially triggered steroid secretion through the activation of hypothalamic-pituitary hormonal cascades. This action exerted by steroids serves to modulate behavioral responses to environmental stimuli. The response of these nerve cells to the steroid hormones depends not only on the expression of their cognate receptors but also on the genomic signal transduction machinery. The repertoire of nuclear receptor coregulators is a key component of this gene regulatory mechanism.[1,2] The coregulators determine the nature and magnitude of a hormonal signal mediated via nuclear receptors. Moreover, because coregulators are capable to interact with several activated nuclear receptors, they may integrate multiple intracellular signalling pathways into a coordinated genomic response. This function of coregulators is reminiscent to that of the steroid hormones themselves, which on another level of complexity, coordinate the function of cells, tissues and organs into an organismal response.

The importance of coregulators for brain function has become apparent from two types of research approaches. The first involves the analysis of specific steroid hormone signalling pathways in the brain in relation to the expression of coregulators. The second approach has relied on phenotypes of coregulator knockout mouse models, often irrespective of a specific nuclear receptor pathway. By observing the development of several coregulator knockout mice, it has become clear that ablation of various coregulators has important consequences for neuronal development. Coregulators also are crucial for neuronal integrity during the aging process: recently, the knockout mouse for the coactivator PGC-1α has revealed a dramatic phenotype in terms of neuronal integrity in the adult brain. Despite their pivotal roles in these aspects of brain function, the particular nuclear receptors (or other transcription factors) involved in the developmental phenotypes of coregulator knockout mouse models often have not been identified.

Brain coregulator function has been studied in relation to three main classical steroid signalling pathways. First, coregulators modulate sex steroid actions underlying female reproductive cycles. These are exemplified by estrogen and progesterone receptor mediated responses, which proceed in a cyclic and highly cell type specific manner in the regulation of female reproductive behavior. Together with the initial *in vivo* findings of partial estrogen resistance in knockout mice for steroid receptor coactivator-1,[3] these findings have triggered substantial interest in the role of coregulators as mediators of sex steroid effects in the brain.

Second, glucocorticoid action during adaptation to stress also depends on coregulators of the two types of brain corticosteroid receptors i.e. glucocorticoid (GR) and mineralocorticoid (MR) receptors. Glucocorticoids act via these receptors in many brain areas to control stress reactions, and to promote behavioral adaptation. However, inadequate or excessive action of the glucocorticoids may play a role in the development of stress related disease.[4] While it is obvious that glucocorticoids influence many aspects of higher brain functions, including emotional responses and cognitive processing of information, at present, the main focus of research on coregulator function *in vivo* has been on the regulation of the hypothalamus-pituitary-adrenal (HPA) axis.

Third, thyroid hormone receptors (TR) and retinoic acid receptors (RAR) have important functions in brain,[5] particularly for brain development. Coregulators have been studied in this context, but mainly in a correlative fashion. Because TR and RAR are DNA-bound in the absence of hormone, the role of corepressors is of more obvious importance than it has appeared so far for the classical steroid receptors.

Since the initial discovery of the first coactivator SRC-1 in 1995,[6] over a hundred different coregulators have been identified. In relation to brain function, most attention has been on the p160 family of SRCs. Therefore, we will focus in this chapter on these coregulators in this chapter. However, we will cover also the available information on some of the numerous other coregulators of equal or potentially even more interest that are left understudied at this time. We will first describe the expression of several coregulators in the brain, which is then followed by sections on the more studied coregulators and steroid pathways mentioned above. In a final discussion, a section on the relevance of these findings is addressed from the perspective of a conceptual framework assigning to coregulators a crucial function in the integration of nuclear receptor signals.

12.2 Presence of Nuclear Receptor Coregulators in the Brain

The brain contains numerous different cell types and nuclei that are dedicated to specific tasks ranging from vegetative control to emotional and cognitive functions. Many coregulators are expressed in the brain, at times exclusively and at times in a highly cell specific manner. Most known coregulators have rather widespread expression in the body, but the literature show two cases of coregulators that are highly enriched in the brain: ERAP140, a coactivator for ER and the thyroid hormone receptor (TR),[7] and nrip-2/NIX1, a corepressor of retinoic acid receptor (RAR) and TR, but not of steroid hormone receptors. Transcripts for both coregulators are expressed neuronal, and Nrip-2 shows a rather discrete expression pattern in the dentate gyrus of the hippocampus, the amygdala, thalamic, and hypothalamic regions of the brain.[8]

Inspection of large scale expression mapping studies, such as the Allen Brain Atlas, readily reveals a number of striking examples of cell specific expression of other coregulator mRNAs in the brain.[9] According to this resource, the well studied ER corepressor RIP140 [or nuclear receptor interacting protein 1, or nrip-1, involved in energy metabolism and reproduction],[10] is highly expressed in a restricted number of brain nuclei, including the dopaminergic substantia nigra, and brain stem (Fig. 12.1). Its family member nrip-3 shows a distinct but equally specific expression pattern. Two members of the coregulator family of thyroid receptor interacting proteins, trip-4 and trip-10,[11] are expressed in a highly specific fashion in mouse hypothalamus (Fig. 12.1). The high and relatively selective expression in areas like the paraventricular nucleus, the suprachiasmatic nucleus, the subtantia nigra, and particular hippocampal subareas gives clues to the involvement of these proteins in respectively regulation of stress, circadian rhythms, goal directed behavior and spatial memory processes. However, the hormones, nuclear receptors or other interacting proteins relevant for signalling in these particular areas remain, as yet, unknown.

If only for historic reasons, the most widely studied coregulators are probably the p160 SRCs, and the corepressors NCoR and SMRT. The latter corepressors are both expressed widely in the brain. There are however clear regional differences in expression, and in particular, neuroendocrine cell groups have been reported to be devoid of either

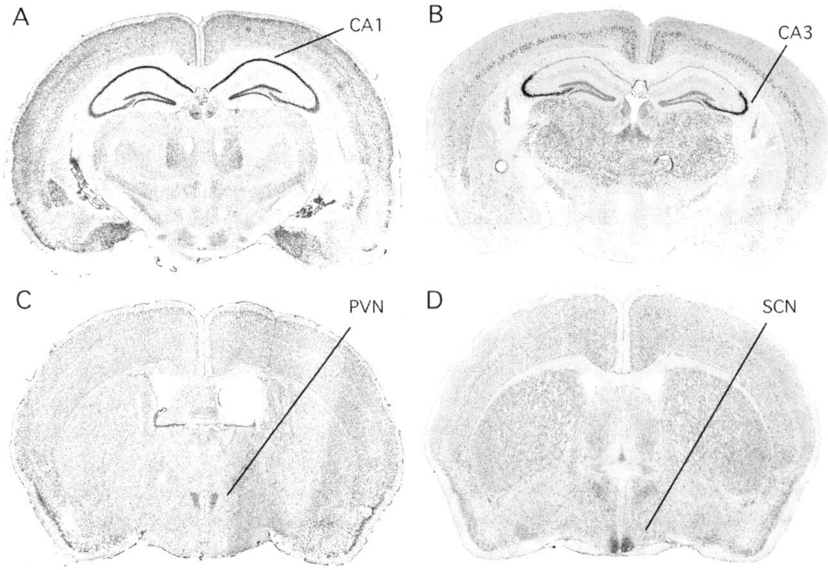

Fig. 12.1. Specific expression of coregulator mRNAs in the brain. (A) Nrip1 (RIP140) is expressed highly in.the hippocampal CA1 areas, cortex and amygdala. (B) Nrip3 shows high levels of mRNA expression in the hippocampal CA3, layer 4 of the cortex and the thalamus. (C) Thyroid hormone receptor interacting protein 4 is expressed specifically in hypothalamic paraventricular nucleus. (D) Trip10 expression is particularly high in the suprachiasmatic nucleus, the seat of the biological clock. Adapted from Allen Brain Atlas [Internet]. Seattle (WA): Allen Institute for Brain Science. © 2004 [2007]. Available from: http://www.brain-map.org.

corepressor.[12–14] Of the p160 coactivators, SRC-1 is expressed ubiquitously in the rodent brain.[15–17] Interestingly, mRNA for splice variants SRC-1a and 1e show highly specific differences in expression, particularly in hypothalamus and brain stem.[16] SRC-2 mRNA is also expressed widely throughout the brain but is generally found at lower levels (at least in the male brain), while expression of the more distant family member SRC-3 is very low, with the highest levels in the hippocampus, cerebellum and olfactory bulb.[17–20] The anterior pituitary, which is of importance for many of the neuroendocrine processes regulated by brain steroid receptors, contains appreciable levels of all three p160 SRCs.

12.3 Nuclear Receptor Coregulators in Brain Development & Aging

A number of nuclear receptors are crucially involved in brain development, such as retinoic acid receptors (general brain patterning), thyroid hormone receptors (cerebellar development), and the orphan nuclear receptors nurr1 and nur77 (dopamine neurons).[5,21,22] Some coregulators have specifically been implicated in brain development, as revealed by knockout mice models. In general, these show brain region specific defects in neurodevelopment, emphasizing the cell specific importance of particular coregulator repertoires. The corepressor NCoR is necessary for normal thalamic and cortical development,[14] and in neuronal stem cells is important in the determination of glial versus neuronal cell fate.[23] SRC-1 is expressed at high levels in the developing brain, particularly in neuronal lineage, and may therefore be involved as a coregulator of for example RAR-mediated developmental signalling.[24,25] SRC-1 knockout mice show retarded development of cerebellar Purkinje cells, and accordingly, have a slight motor dysfunction.[17] As is the case for other phenotypes of this mouse, the absence of SRC-1 is possibly compensated for by increased SRC-2 expression in some cells.[3,17] Compensatory expression in knockout mice may be circumvented by techniques resulting in knockdown using inducible promoters in gene targetting or siRNA models, and/or viral targetting techniques in adult animals. For the SRC-1 mice, this is discussed in some detail in the section of sex steroid responses and coregulators.

12.3.1 *PGC-1α and neurodegeneration*

The most striking brain phenotypes of coregulator mouse models to date are those of the PPARγ coactivator 1α (PGC-1α) knockout mice lines, which display widespread neurodegeneration that is particularly marked in the striatum.[26,27] PGC-1α has a central role in the control of energy metabolism, mitochondrial biogenesis and adaptive thermogenesis. It turns out that PGC-1α is not only responsible for the genesis of energy, but also a central component of mounting the cellular defense mechanism against the damaging by products of oxydative phosphorylation, the reactive oxygen species (ROS).[28]

PGC-1α knockout mice show a pronounced hyperactivity and other motor aberrations that "seem to model Huntington's disease better than the actual Huntington's disease mouse models." The mice display spongiform lesions in the brain that are caused by excessive exposure to ROS.[26] The link between Huntington's disease and PGC-1α was substantiated in several ways: PGC-1α target genes are reduced in the striatum of HD patients and of mice expressing the disease version of the huntingtin gene. In addition, PGC-1α expression can reverse cellular mitochondrial deficits in "Huntington mice" and transfer resistance to neurotoxic lesions.[29] Yet another link implies involvement of the transcription factor CREB, which stimulates PGC-1α expression after neurotoxic simuli; CREB knockout mice also show neurodegeneration, possibly because of a lack of PGC-1 induction.[28]

The PGC-1α link to neurodegeneration is compelling, and seems to extend beyond one disease model, as the PGC-1α null mice are highly susceptible to a number of neurodegenerative challenges, including neurotoxic challenges of dopamine neurons and kainic acid endangerment of hippocampal CA1 neurons. The nuclear receptor PPARγ has been proposed as a target for neuroprotective drugs.[30] On the other hand, PGC-1α is in many tissues, highly inducible. An interesting question that reflects the integrative nature of coregulators would be what constitutes a better neuroprotective strategy: boost the expression of the coregulator or providing ligand to activate one of its recruiting nuclear receptors?

12.3.2 *Thyroid hormone & regulation of coregulator expression*

The importance of coregulator "make up" for nuclear receptor signalling has led to the question whether expression of the coregulators themselves is subject to regulation. The specific nature of thyroid hormone signalling in brain development in the early postnatal period in rodents has triggered the hypothesis that coregulator expression is specifically modulated during that "critical period." Indeed, expression of SRC1, 2 and NCoR mRNA in cerebellum remains dynamic during the first weeks of life.[18] While analysis of knockout mice has revealed that the gene dose of SRCs is important for TR signalling,[31] natural variations of thyroid hormone responsiveness are as yet not unequivocally explained in terms of coregulator variability.

A related question is whether, as part of positive or negative regulatory loops, hormonal signals regulate the machinery of their own signal transduction. Interestingly, in spite of the rather ubiquitous expression of copressors NCoR and SMRT in the brain, and the clear established role in TR mediated signalling, these corepressors seem to be absent in the cells in the hypothalamus that are responsible for TH negative feedback in the adult hypothalamus.[12] However, the mRNA levels of several coregulators was shown to be rapidly regulated by estrogen and thyroid hormones in the pituitary,[15] an organ in which coregulator function has been extensively described.

In the neonate brain, attention has focussed on regulation of coregulator expression by thyroid hormones, apparently in an attempt to probe this system for autoregulatory mechanisms. Conflicting data have been reported with respect to the regulation of SRCs and NCoR/SMRT by thyroid hormone in neonate rat and mouse brain at times that TR activation is crucial for neuronal development.[18,32] The apparent lack of a robust effect seems to argue against a major role of TR-mediated coregulator expression in normal brain development.

12.4 Sex Steroid Signalling

Both the earliest experiments in endocrinology, as well as the earliest described effects of nuclear receptor coregulators involve the effects of the sex steroids, mediated via androgen, estrogen and progesterone receptors.[6] The brain plays a pivotal role in reproduction, in its role as a master gland and the secretion of e.g. gonadotropin releasing hormone, and through regulation of sexual behavior, both in female and male animals. SRC-1 was identified as a coactivator of the ER, and the SRC-1 knockout mouse displays a partial resistance to estrogens.[3] It is also the coregulator that has been studied most (and almost exclusively) in relation to both male and female reproductive behavior. Indeed, there is evidence that SRC-1 is important for aspects of ER, PR and AR signalling to the brain, but possibly in a receptor- and developmental stage dependent manner.

12.4.1 *Estrogen receptor mediated signalling*

The issue of whether the amount of SRCs is rate-limiting at a given a steroid signal, and the related question on the importance of regulation of SRC expression for steroid responsiveness has been addressed extensively for the estrogen receptor signalling in the brain.

The extent of exposure to androgens in early life can program sexual behavior in rodents via defeminizing effects (mediated by ER, after aromatization of the ligand) and masculinizing effects (mediated by AR). Knockdown of SRC-1 protein during ontogeny by infusion of antisense oligonucleotides led to clear effects on the organization of rat sexual behavior in adulthood, particularly those aspects that depend on ER, but not AR, function. These effects in adult rats treated as pups included the size of the sexual dimorphic nucleus and lordosis behavior, which could no longer be programmed by testosterone treatment during the first days of life after knockdown of SRC-1.[33]

In the adult female rat brain, estrogens have activational effects via induction of progesterone receptors in diverse hypothalamic nuclei. PR induction is a dramatic effect that has provided a robust readout for the modulation of ER-mediated signalling. Colocalization studies have shown that SRC-1 is expressed in the neuronal cell types relevant for this effect.[34] Moreover, SRC-1 knockdown interferes with ER-mediated activational effects, namely induction of PR in the ventromedial nucleus of the hypothalamus (VMH) and the related PR-dependent intensity of lordosis behavior. In these experiments, the effects in the adult rat with reduced SRC-1 seem to be weaker, and in fact, required concomitant knockdown of the cointegrator CBP.[35,36]

A study using a similar strategy in mice[37] evaluated the role of all three p160 SRCs as mediators of the activational effects of estrogens on female reproductive behavior. Here, local infusion of SRC-1 antisense oligonucleotides in the VMH resulted in a dramatic loss of induction of PR by estrogens, and of lordosis behavior. Interestingly, also treatment with SRC-2, but not SRC-3, antisense oligonucleotides led to a complete reduction in ER responsiveness in the adult female mouse. From these data, it follows that both SRC-1 and 2 are part of transcriptional complex(es) necessary for ER(alpha) mediated transactivation of the PR gene. This suggests that the amount of both SRCs is rate limiting and that in this case, the amount of ER is such that all SRC-1 and 2 molecules that are available have to be recruited for the full induction of the PR gene to occur. It remains unclear whether the contrast in magnitude of the SRC-1 knockdown effect between the above mentioned studies in adult rats and mice is caused by an unexpected large species difference, or due to methodological issues such as the extent of knockdown, which indeed was much larger in the study in mice.[35,37]

An additional line of evidence shows strong effects of SRC-1 on the action of sex steroids in the brain of the male Japanese quail.[38,39] Adult male birdbrains are very sensitive to changes in circulating testosterone, which like in mammals, can act directly on AR, or after conversion to estrogen, on ER. Activation of these receptors leads to distinct changes in the hypothalamus, including an increase in size of the preoptic area with enhanced vasotocin expression. These changes become manifest in a few days as changes in reproductive behavior that can be either AR or ER-dependent. Much like in the rodent studies, infusion of antisense (LNA) oligonucleotides against SRC-1 led to a pronounced loss of steroid responsiveness, but in this case, both ER and AR dependent phenomena could be blocked. Interestingly, a rebound effect was observed after cessation of antisense infusion: the testosterone-dependent aromatase expression in the hypothalamus after withdrawal exceeded that in control male animals. This is a rarely observed phenomenon, and was interpreted as based on a transiently higher SRC-1 expression. Thus, from these studies it appears that both ER and AR mediated activational effects of testosterone on the brain depend on SRC-1, and that SRC-1 is in fact rate limiting given a sufficient level of hormone and receptor expression.

Despite the strong evidence that SRC-1 is necessary for programming as well as activational effects mediated by ER in the brain, SRC-1 knockout mice have normal fertility and estrogen-induced brain changes. This may be due to the reported compensatory upregulation of SRC-2 in brain.[3] Indeed, in these mice, ER-mediated effects on reproductive behavior have been reported to depend completely on SRC-2.[37] To what extent SRC-1 and 2 are redundant is an intriguing question. First, there are dramatically different roles of the two coactivators in other tissues, suggesting substantial differences between their "target genes" or "target transcription factors."[40] Second, SRC-1 has specific splice variants that have not been reported for SRC-2. Not only do these splice variants differ in their expression in for example hypothalamus,[16] they also have rather large differences in the way they function as coactivators for estrogen receptors.[41] In any case, it is clear that the answers that come closest to physiology will have to come from knockdown that is induced shortly before the steroidal challenge is given, be it by advanced transgenic models, or by transient knockdown using antisense techniques.

12.4.2 *Androgen and progesterone receptor signalling*

Not only ER function, but also AR and PR mediated signalling can be affected by knockdown of SRC-1 in the brain. This is to be expected as SRC-1 is capable to interact with a broad spectrum of nuclear receptors,[6] and therefore may function as a more general sensitizer of hormone responsiveness. The ER-induced PR expression in the female hypothalamus provides an example of a cell population that depends on proper responsiveness to at least two types of steroids. Indeed, by using differential timing of SRC-1 antisense administration, the point has been made that SRC-1 (and CBP) is necessary for adequate ER as well as PR signalling.[36] Likewise, the testosterone effects on male reproductive behavior in the quail depend on coordinated ER and AR mediated effects, both of which can be attenuated by SRC-1 knockdown.[38,39] However, the earlier mentioned study on programming during rat ontogeny could not find the effects of SRC-1 knockdown of AR-mediated signalling,[33] which may indicate that the nature of integrative signalling by SRC-1 varies between species or developmental stages.

12.4.3 *Regulation of coactivators during reproductive stages*

The effects in knockdown studies suggest that at a given hormone concentration and receptor presence, the amount of coactivator (SRC-1) can indeed be a rate limiting factor for steroid signalling to the brain. The logical next question is then: does the activity of this rate limiting factor in specific brain areas vary between individual animals in order to direct physiological and behavioral responsiveness? Misiti *et al.*[15] showed quite early on that mRNA levels of coregulators can be very sensitive to hormonal stimuli, at least in pituitary cells, where estrogen was able to strongly downregulate SRC-1 mRNA. However, the answer to the question of regulation of expression in brain is not as clear as one would hope for. The activity of a given coregulator protein may depend on many factors, including expression levels, post-translational modifications and the presence of interacting proteins. The only approach that is feasible in specific brain areas has been to measure expression levels, either at the mRNA levels (high spatial resolution, but indirect measure), or at the protein level using Western blot (actual protein but loss of resolution), or immunohistochemistry (protein at high resolution but problematic quantification).

A straightforward hypothesis is that in the relevant brain nuclei just before periods of reproductive behavior, there is an upregulation of coregulators that stimulates sex steroid responsiveness. In the seasonally breeding siberian hamster this seems to be the case, as shorter day length is associated with testicular regression and lower reproductive activity, and accompanied by a modest decrease in SRC-1 positive cells in part of the bed nucleus of the stria terminalis.[42] However, there was no clear effect of androgen treatment on SRC-1 expression, and the plasticity of AR expression as a function of day length and hormone milieu was far greater than that of SRC-1. In the quail, testosterone is able to upregulate SRC-1 mRNA and protein specifically in the preoptic area.[43]

An equivalent question has been addressed by several groups in relation to the female ovarian cycle in rodents, and the increased ER and PR mediated signalling that occurs in particular in the ventromedial nucleus. While some groups have reported an increase in SRC-1 expression during proestus in rat, and after estrogen treatment of ovariectomy,[37,44,45] other reports are negative.[15,34] While the idea of regulating the regulator is attractive, the jury is still out as to how significant this mechanism is in relation to SRC-1 and sex steroid signalling to the brain. In the case of female reproductive behavior in the rat, the induction of the PR is in any case a more prominent adjustment of steroid responsiveness than coregulator regulation.

SRC-1 does seem to constitute an important modulator of ER, AR and PR mediated signalling in the hypothalamus of birds and mammals. Much of the evidence from knockdown studies suggests that SRC-1 expression constitutes a rate limiting step in the transduction of the steroid receptor activation to transcriptional changes. The extent to which SRC-1 itself is regulated is not clear. Finally, as was also mentioned in a previous section, there are many more, presently understudied coregulators that are potentially relevant for sex steroid signalling in the brain.

12.5 Glucocorticoids and Stress

Glucocorticoid hormones secreted by the adrenal gland as part of the stress response have important actions on the brain. These actions serve to promote adaptation to the stressor by modulating a broad range of processes ranging from autonomic and endocrine regulation to cognitive and emotional processing. They are mediated by two types of nuclear receptors, mineralo and glucocorticoid receptors (MR and GR). Both can bind the endogenous hormones cortisol and corticosterone, but MR has a ten-fold higher affinity and is occupied significantly at basal levels of hormone, while GR is occupied mainly after exposure to stress levels of corticosteroids. GR is present in almost all brain areas, but MR has a much more restricted expression pattern with particular enrichment in limbic brain regions. Interestingly, besides their classic genomic actions, these receptors also play a role in the immediate response to stressors by mediating non-genomic effects on neurotransmission in the brain. Thus, brain MR and GR bind the same ligand, but under different conditions, and in concert, act to promote the responsiveness to and the recovery from stressors over a timescale ranging from minutes to hours and days.[4,46]

The actions of corticosteroid hormones are in principle adaptive in the context of most stressors. Consequently, excessive or inadequate MR and GR mediated signalling is thought to be a relevant factor for the development of stress related psychopathology. The response to the dexamethasone suppression test, in which the balance between stimulatory drive and glucocorticoid negative feedback on the pituitary is measured, is different in many psychiatric diseases. Despite a certain lack of disease specificity, this points to a GR mediated signal that is inadequate in relation to the drive from hypothalamus to pituitary gland. In a more general sense, the extent of MR and GR mediated signalling and the balance in the activation of the two receptor types is thought to underlie successful or inadequate coping with stress.[4]

MR and GR are nuclear receptors and accordingly can act as transcription factors by binding to response elements or interaction with other transcription factors. Although the receptors have a high degree of homology, and in fact share the same response elements, they can mediate opposite genomic effects, even within a single neuronal cell type. Receptor specific effects have been difficult to understand. In most contexts, GR is a stronger transcription factor, both as transactivator and as transrepressor. However, MR activation has intrinsic effects via genomic mechanisms and therefore has its own specific sets of target genes in brain.[46]

Glucocorticoids can have pleiotropic effects on cellular metabolism, but also induce changes in neuronal excitability that are highly specific to a particular area. This makes sense as there will be local differences in the demand of modulation of neuronal activity. Thus, transcriptional responses to glucocorticoid stimulation within the brain are in part shared, and in part specific to particular neurons. For example, effects of GR activation on a single gene can be opposite depending on the brain region: CRH expression is potently downregulated via GR in the hypothalamic paraventricular nucleus, but upregulated in other brain areas such as the amygdala. Repression of the 5-HT (serotonin) 1A receptor is specific to certain subfields of the hippocampus, and absent in the serotonergic cell bodies in the raphe nuclei, where the 5-HT1A receptor is also expressed.[19] The discovery of the coregulators has provided an excellent opportunity to test a number of hypotheses on the cause of receptor and cell specific effects of MR and GR in the brain, and of possible changes in corticosteroid sensitivity.

12.5.1 *Cell specificity*

Because the nature of the glucocorticoid signal is so diverse in different brain areas, coregulators that are expressed in specific brain areas may shape the cellular response to activation of MR and GR. In this respect, there is a large number of unexplored candidate glucocorticoid signalling modifiers that are expressed in a brain region specific manner, and could account for cell specific responses (Fig. 12.1). We have observed that splice variants SRC-1a and 1e have a strikingly different distribution in the hypothalamus and brain stem compared to the rest of the brain.[16] Although what ties these nuclei together is not obvious, their common coregulator status argues for a particular steroid responsiveness for neuroendocrine and motor output systems of the brain. We also observed that these splice variants have pronounced functional differences in coactivation properties of corticosteroid receptors, that possibly relate to the amount of response elements present in a promoter.[47] Likewise, the ratio between NCoR and SMRT differs between brain areas, as measured by immunohistochemistry and mRNA *in situ* hybridisation.[13] These corepressors not only affect the efficacy of antagonist action on classical steroid receptors, but may also influence transactivtion by the endogenous agonists.[48] While the availability of proper neuronal cell lines to study glucocorticoid effects remains problematic, studies in different cell types of glial lineage have shown that the recruitment of SRCs by GR on a synthetic promoter differs between cell types.[49] The consequences of cell specific coregulator expression and for specific gene regulation by glucocorticoids in relation to *in vivo* fluctuations of hormones are as yet unknown.

12.5.2 *Receptor specificity*

The differences in transcriptional activity of MR and GR raise the question whether there are coregulators that are specific for one of the two receptor types. Most attention has been dedicated to proteins that specifically interact with the low homology N-terminal part of the receptors. We have found slight differences for the coactivation of MR and GR by SRC-1e, possibly due to an interaction between the MR N-terminal part and a non-NR box domain of the coactivator.[47] The most striking receptor-specific coregulator in this respect reported to date is the elongation factor ELL, which acts as a coactivator for MR but a repressor for GR. The issue of corticosteroid receptor specificity is covered in a recent review article.[50]

12.5.3 *Transcriptional effects of MR and GR coregulators in brain cells*

As to the actual relevance of coregulators for glucocorticoid signalling in brain cells, most work has focussed on SRC-1 and SRC-2, mainly because of the availability of knockout mice. There are unfortunately no known genes in neurons in the brain that are regulated by glucocorticoids as robustly as in the case of PR induction by ER. Although genomic responses mediated by MR and GR are necessary for many of the neuromodulatory effects of glucocorticoids,[46] the fold change observed for mRNA induction is generally too small to provide a good window for modulatory factors of glucocorticoid action.[51]

Evidence for actual involvement of SRCs on gene expression in the brain comes not from neurons, but from glial cells. In primary cultures of brain astrocytes, knockdown of SRC-1 and 2, but not SRC-3 led to a reduced transactivation via GR. In myelinating cells from the peripheral nervous system, SRC-1 and SRC-3 act as coactivators.[49,52] We have corroborated these findings by studying induction of mRNA of serum and glucocorticoid regulated kinase 1 (sgk-1) in SRC-1 knockout mice

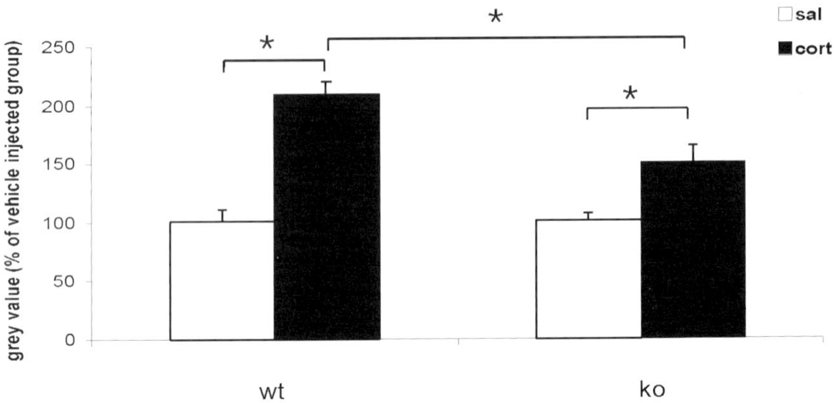

Fig. 12.2. Induction of sgk-1 mRNA in the corpus callosum of the mouse brain 3 hours after treatment with a high dose of corticosterone (300 μg orally). The induction, liekly mediated via GR, is less in the SRC-1 knockout mice. Data by OC Meijer, J Xu, E Apostolakis and ER de Kloet.

brain. Sgk-1 mRNA provides a robust read out for glucocorticoid action in rodent brain white matter through its striking upregulation within an hour of a single dose of corticosterone, presumably reflecting transcriptional regulation in oligodendrocytes.[53] In SRC-1 knockout mice, we measured this induction in the corpus callosum, and found it to be significantly lower than in wild type littermates (Fig. 12.2).

12.5.4 *Stress & the hypothalamus pituitary adrenal axis*

The hormonal cascade brought about by the hypothalamus pituitary adrenal (HPA) axis is of pivotal importance for the control of circulating glucocortioids, and can be used as a read out of the brain activity in relation to stress. It is sensitive to glucocorticoid negative feedback, both at the level of the pituitary, and the hypothalamic paraventricular nucleus (PVN). The negative feedback involves inhibition of secretion of ACTH and corticosterone, as well as transcriptional downregulation of the secretagog mRNAs that are involved. Thus, POMC and CRH mRNA can be potently downregulated by glucocorticoids via GR in anterior pituitary and PVN, respectively. In addition, MR also exerts inhibitory control over the HPA axis, but this involves receptor sites in brain areas outside the "core" of the HPA axis, such as the hippocampus.

The straightforward hypothesis with regard to coregulator function might be that lower levels of GR coactivators would lead to lower GR sensitivity and escape of negative feedback on the HPA axis, i.e. higher basal and/or stress-induced ACTH and corticosterone levels. An immediate caveat of this hypothesis is that part of the negative feedback mechanisms involve the repression of gene activity, and that little is know about the role of coregulators in transrepression via protein–protein interaction and via negative GREs. The HPA axis has been studied both in SRC-1 and SRC-2 knockout mice.[54,55] Both lines show changed activity of the HPA axis, but also have a major phenotype at the level of the adrenal cortex, which complicates the interpretation of the data. For example, SRC-2 knockouts have lowered basal and stress induced corticosterone levels. The higher ACTH content of the pituitary and increased hypothalamic CRH mRNA expression likely are the consequence of decreased steroid levels, and may only be seen as evidence for changed efficacy of GR under conditions in which corticosterone levels are (experimentally made) equal. On the other hand, part of the complex phenotype of these animals may well reflect defective gene expression in the brain, but the contribution of MR and GR in these processes is unresolved.[55]

SRC-1 knockout mice also exhibit an altered HPA axis. Their phenotype involves adrenal cortex, which if challenged, displays a greater ACTH sensitivity. However, the increased corticosterone output is not readily compensated for by negative feedback, suggesting a level of corticosteroid receptor insensitivity. In fact, there is a strong transcriptional phenotype for GR signalling, as POMC mRNA repression by the synthetic glucocorticoid dexamethasone is markedly reduced in SRC-1 knockout mice.[54] This observation suggests that in fact SRC-1 can also act as a corepressor, and is involved in signal transduction from the negative GRE that is present in the POMC promoter.

Given the clear GR resistance, at least at the level of the pituitary POMC gene, the HPA-axis phenotype of these mice is remarkably mild. Our own unpublished observations in the brains of SRC-1 knockout mice suggest that there are also differences in central CRH expression, but also only a very mild hyperactivation of the HPA axis. This is possibly due to compensatory changes during development, such as the

SRC-2 upregulation that has been reported in several tissues.[3,37,54] However, the discrepancy between clearly disturbed transcription regulation (POMC) and relatively mild phenotype indicates that compensation not only occurs by substitution of SRC-1 by SRC-2, but also involves more complex adaptations.

12.5.5 *Regulation of expression*

There is published evidence that SRC-1 can act *in vivo* as a coregulator of GR signalling in the pituitary and in glial cells. Based on *in vitro* studies, and *in vivo* distribution patterns in the brain, it is to be expected that SRC-1 also plays a role in neuronal signalling via MR and GR. However, all available data for corticosteroid signalling *in vivo* have relied on knockout models rather than knockdown, and so although a necessary factor, SRC-1 is not a proven rate limiting factor. Nonetheless, the question has been addressed whether the regulation of expression of SRC-1 and other coregulators may explain changes in corticosteroid responsiveness.

As is the case with the regulation of coactivators by sex steroids, the reports on regulation by glucocorticoids are mixed. High doses of glucocorticoids by themselves are able to downregulate expression of SRC-1 in many tissues including the brain.[56] More physiological elevations of corticosterone in the rat however had no consequences for hippocampal SRC-1 mRNA expression, but did lower SRC-1e mRNA levels in (unidentified cells in) the anterior pituitary.[47] However, long lasting programming effects of prenatal dexamethasone on the HPA axis do no seem to involve changed expression of SRC-1, or 2.[57]

A probably principal aspect of coregulator function is that they modulate diverse hormonal signals. The conceptual consequences of this issue will be addressed in more detail below, but with respect to the regulation of coactivator levels, it is interesting to note that stress blocks stimulatory effect of testosterone on SRC-1 expression in the quail brain.[43] This emphasizes that coregulator function should also be approached from the viewpoint of coordinate regulation of several steroid (or other nuclear) receptors.

In summary, the available data do suggest a role for SRC-1 in at least GR sensitivity in the brain, and the HPA axis. For both corticosteroid receptor types, genetic variations have pointed to a role of glucocorticoid signalling in human stress responsiveness, as well as stress-related psychopathology.[4,58] Possibly, data from human genetics may reveal the involvement of particular coactivators in human stress responsiveness, and stress related disease, analogous to what has been shown for MR and GR.

12.6 Perspectives Beyond the Receptor

What is the role of brain nuclear receptor coregulators for intracellular steroid signalling?

To address this question, we would like to consider the coregulator as integrator of multiple intracellular signalling pathways in brain cells. In such a model, the brain would link the environment with hormone secretions. The hormones in turn coordinate cell, tissue and organ function in order to cope with environmental demands and to facilitate reproductive, adaptive and vegetative functions. For this purpose, the hormones feed back on precisely those brain circuits that initially have led to their own secretion. These circuits underlie emotional arousal, cognitive processes, ingestive behavior and aspects of sexual behavior which are modified by the hormones. Coregulators operate in these complex processes as integrators of many nuclear receptor signals within neurons and glia cells.

Viewed from the perspective of a single nuclear receptor type, there are many factors that regulate the extent of hormonal signalling. Steroids are secreted in regular pulses or premature secretory bursts, as is the case after stress. Cells respond to these pulses in a process called frequency encoding of which very little is understood. The magnitude of the pulses depends on binding to plasma proteins, the penetration in the target cell and local enzymatic conversion. Receptor diversity can be considerable at the level of genes, splice variants, protein variants and post-translational modification. Next are differences in the signal transduction machinery, which are by definition less specific than the hormone and its cognate receptor, but serve as an integrating role and bias the cellular responses towards current needs.

Given the specific interactions of coregulators with subsets of nuclear receptors, in a promoter specific manner, it is likely that they are (de)sensitizers of particular sets of genes that are responsive to several stimuli. For example, opposite effects of stress and reproductive hormones may be enhanced by particular coregulator stoichiometry. In the case of cellular differentiation, such processes definitely play a role, for example in the determination of white or brown fat cell fate.[40] The PGC-1 story is a nice example of coordinate regulation of energy metabolism with protection against damage associated with increased cellular catabolism. In general, the regulation of the repertoire of active coregulators should direct a cell towards states of different functionality, defined as the combination of the signals it responds to and the signals it produces (secretory, structurally, electrically). For the brain, these changes underlie *a.o.* degrees of neuronal plasticity.

It is still early days with regard to in depth understanding of the roles nuclear receptor coregulators have in the functioning of the brain. Global expression data are available for many coregulators,[9] and their analysis may give clues for the function of specific protein patterns. Functional analysis of coregulator-nuclear receptor interactions in the brain have received a one to one approach for a limited number of factors. This analysis has been hampered by complex phenotypes and compensatory mechanisms of knockout models as well as the labor intensive nature of local knockdown approaches. As a consequence, the integrative nature of coregulator function has received scarce attention. SRC-1 may be viewed as a factor that coordinates the actions of ER, AR and PR in the hypothalamus. However, the p160 SRCs can interact with a broad range of nuclear receptors, and their potential roles as coordinators of nuclear receptor function in the interactions of for example stress and reproduction, metabolism and reproduction, or stress and metabolism remain to be established.

Coregulator relevance for particular brain related diseases might range from neurodevelopmental processes to activational effects of steroids, but remains mostly unknown at this stage. The notable exception is PGC-1α, given the dramatically increased neuronal vulnerability of PGC-1α knockout mice that is relevant for a broad range of neurodegenerative disorders. Future clues to the role of other coregulators in brain disease await data from human association studies, and further insight in the physiology of coregulator function. However, it has become clear that brain pathophysiology may be linked with organism-wide changes such as the metabolic syndrome and activated immune status. Such general states form contexts in which many nuclear receptor mediated signals change, and in which the role of coregulators as orchestrators of adaptation at the cellular level becomes crucial for the adaptation and the health of the organism as a whole.

Acknowledgment

The support by the Netherlands Organization for Scientific Research (VIDI nr 917.36.381.to OCM) and the Royal Netherlands Academy for Arts and Sciences is gratefully acknowledged.

References

1. Metivier R, Reid G, Gannon F, Transcription in four dimensions: Nuclear receptor-directed initiation of gene expression, *EMBO Rep* **7**:161–167, 2006.
2. Lonard DM, O'Malley BW, The expanding cosmos of nuclear receptor coactivators, *Cell* **125**:411–414, 2006.
3. Xu J, Qiu Y, DeMayo FJ, *et al.*, Partial hormone resistance in mice with disruption of the steroid receptor coactivator-1 (SRC-1) gene, *Science* **279**:1922–1925, 1998.
4. de Kloet ER, Derijk RH, Meijer OC, Therapy insight: Is there an imbalanced response of mineralocorticoid and glucocorticoid receptors in depression? *Nat Clin Pract Endocrinol Metab* **3**:168–179, 2007.
5. Oppenheimer JH, Schwartz HL, Molecular basis of thyroid hormone-dependent brain development, *Endocr Rev* **18**:462–475, 1997.
6. Onate SA, Tsai SY, Tsai MJ, *et al.*, Sequence and characterization of a coactivator for the steroid hormone receptor superfamily, *Science* **270**:1354–1357, 1995.
7. Shao W, Halachmi S, Brown M, ERAP140, a conserved tissue-specific nuclear receptor coactivator, *Mol Cell Biol* **22**:3358–3372, 2002.
8. Greiner EF, Kirfel J, Greschik H, *et al.*, Differential ligand-dependent protein-protein interactions between nuclear receptors and a neuronal-specific cofactor, *Proc Natl Acad Sci USA* **97**:7160–7165, 2000.
9. Lein ES, Hawrylycz MJ, Ao N, *et al.*, Genome-wide atlas of gene expression in the adult mouse brain, *Nature* **445**:168–176, 2007.
10. Steel JH, White R, Parker MG, Role of the RIP140 corepressor in ovulation and adipose biology, *J Endocrinol* **185**:1–9, 2005.
11. Lee JW, Choi HS, Gyuris J, *et al.*, Two classes of proteins dependent on either the presence or absence of thyroid hormone for interaction with the thyroid hormone receptor, *Mol Endocrinol* **9**:243–254, 1995.
12. Becker N, Seugnet I, Guissouma H, *et al.*, Nuclear corepressor and silencing mediator of retinoic and thyroid hormone receptors corepressor expression is incompatible with T(3)-dependent TRH regulation, *Endocrinology* **142**: 5321–5331, 2001.
13. van der Laan S, Lachize SB, Schouten TG, *et al.*, Neuroanatomical distribution and colocalisation of nuclear receptor corepressor (N-CoR) and

silencing mediator of retinoid and thyroid receptors (SMRT) in rat brain, *Brain Res* **1059**:113–121, 2005.
14. Jepsen K, Hermanson O, Onami TM, *et al.*, Combinatorial roles of the nuclear receptor corepressor in transcription and development, *Cell* **102**:753–763, 2000.
15. Misiti S, Schomburg L, Yen PM, *et al.*, Expression and hormonal regulation of coactivator and corepressor genes, *Endocrinology* **139**:2493–2500, 1998.
16. Meijer OC, Steenbergen PJ, De Kloet ER, Differential expression and regional distribution of steroid receptor coactivators SRC-1 and SRC-2 in brain and pituitary, *Endocrinology* **141**:2192–2199, 2000.
17. Nishihara E, Yoshida-Komiya H, Chan CS, *et al.*, SRC-1 Null mice exhibit moderate motor dysfunction and delayed development of cerebellar purkinje cells, *J Neurosci* **23**:213–222, 2003.
18. Martinez de Arrieta C, Koibuchi N, Chin WW, Coactivator and corepressor gene expression in rat cerebellum during postnatal development and the effect of altered thyroid status, *Endocrinology* **141**:1693–1698, 2000.
19. Meijer OC, van der Laan S, Lachize S, *et al.*, Steroid receptor coregulator diversity: What can it mean for the stressed brain? *Neuroscience* **138**:891–899, 2006.
20. Xu J, Liao L, Ning G, *et al.*, The steroid receptor coactivator SRC-3 (p/CIP/RAC3/AIB1/ACTR/TRAM-1) is required for normal growth, puberty, female reproductive function, and mammary gland development, *Proc Natl Acad Sci USA* **97**:6379–6384, 2000.
21. Gavalas A, Krumlauf R, Retinoid signalling and hindbrain patterning, *Curr Opin Genet Dev* **10**:380–386, 2000.
22. Levesque D, Rouillard C, Nur77 and retinoid X receptors: Crucial factors in dopamine-related neuroadaptation, *Trends Neurosci* **30**:22–30, 2007.
23. Hermanson O, Jepsen K, Rosenfeld MG, N-CoR controls differentiation of neural stem cells into astrocytes, *Nature* **419**:934–939, 2002.
24. Nishihara E, Moriya T, Shinohara K, Expression of steroid receptor coactivator-1 is elevated during neuronal differentiation of murine neural stem cells, *Brain Res* **1135**:22–30, 2007.
25. Misiti S, Koibuchi N, Bei M, *et al.*, Expression of steroid receptor coactivator-1 mRNA in the developing mouse embryo: a possible role in olfactory epithelium development, *Endocrinology* **140**:1957–1960, 1999.
26. Lin J, Wu PH, Tarr PT, *et al.*, Defects in adaptive energy metabolism with CNS-linked hyperactivity in PGC-1alpha null mice, *Cell* **119**:121–135, 2004.
27. Leone TC, Lehman JJ, Finck BN, *et al.*, PGC-1alpha deficiency causes multi-system energy metabolic derangements: Muscle dysfunction, abnormal weight control and hepatic steatosis, *PLoS Biol* **3**:e101, 2005.

28. St-Pierre J, Drori S, Uldry M, *et al.*, Suppression of reactive oxygen species and neurodegeneration by the PGC-1 transcriptional coactivators, *Cell* **127**:397–408, 2006.
29. Weydt P, Pineda VV, Torrence AE, *et al.*, Thermoregulatory and metabolic defects in Huntington's disease transgenic mice implicate PGC-1alpha in Huntington's disease neurodegeneration, *Cell Metab* **4**:349–362, 2006.
30. Sundararajan S, Jiang Q, Heneka M, *et al.*, PPARgamma as a therapeutic target in central nervous system diseases, *Neurochem Int* **49**:136–144, 2006.
31. Weiss RE, Gehin M, Xu J, *et al.*, Thyroid function in mice with compound heterozygous and homozygous disruptions of SRC-1 and TIF-2 coactivators: Evidence for haploinsufficiency, *Endocrinology* **143**:1554–1557, 2002.
32. Ramos HE, Weiss RE, Regulation of nuclear coactivator and corepressor expression in mouse cerebellum by thyroid hormone, *Thyroid* **16**:211–216, 2006.
33. Auger AP, Tetel MJ, McCarthy MM, Steroid receptor coactivator-1 (SRC-1) mediates the development of sex-specific brain morphology and behavior, *Proc Natl Acad Sci USA* **97**:7551–7555, 2000.
34. Tetel MJ, Siegal NK, Murphy SD, Cells in behaviourally relevant brain regions coexpress nuclear receptor coactivators and ovarian steroid receptors, *J Neuroendocrinol*, 2007.
35. Molenda HA, Griffin AL, Auger AP, *et al.*, Nuclear receptor coactivators modulate hormone-dependent gene expression in brain and female reproductive behavior in rats, *Endocrinology* **143**:436–444, 2002.
36. Molenda-Figueira HA, Williams CA, Griffin AL, *et al.*, Nuclear receptor coactivators function in estrogen receptor- and progestin receptor-dependent aspects of sexual behavior in female rats, *Horm Behav* **50**:383–392, 2006.
37. Apostolakis EM, Ramamurphy M, Zhou D, *et al.*, Acute disruption of select steroid receptor coactivators prevents reproductive behavior in rats and unmasks genetic adaptation in knockout mice, *Molecular Endocrinology* **16**:1511–1523, 2002.
38. Charlier TD, Ball GF, Balthazart J, Inhibition of steroid receptor coactivator-1 blocks estrogen and androgen action on male sex behavior and associated brain plasticity, *J Neurosci* **25**:906–913, 2005.
39. Charlier TD, Harada N, Ball GF, *et al.*, Targeting steroid receptor coactivator-1 expression with locked nucleic acids antisense reveals different thresholds for the hormonal regulation of male sexual behavior in relation to aromatase activity and protein expression, *Behav Brain Res* **172**:333–343, 2006.
40. Picard F, Gehin M, Annicotte JS, *et al.*, SRC-1 and TIF2 control energy balance between white and brown adipose tissues, *Cell* **111**:931–941, 2002.

41. Kalkhoven E, Valentine JE, Heery DM, *et al.*, Isoforms of steroid receptor co-activator 1 differ in their ability to potentiate transcription by the oestrogen receptor, *EMBO J* **17**:232–243, 1998.
42. Tetel MJ, Ungar TC, Hassan B, *et al.*, Photoperiodic regulation of androgen receptor and steroid receptor coactivator-1 in Siberian hamster brain, *Brain Res Mol Brain Res* **131**:79–87, 2004.
43. Charlier TD, Ball GF, Balthazart J, Plasticity in the expression of the steroid receptor coactivator 1 in the Japanese quail brain: Effect of sex, testosterone, stress and time of the day, *Neuroscience* **140**:1381–1394, 2006.
44. Mitev YA, Wolf SS, Almeida OF, *et al.*, Developmental expression profiles and distinct regional estrogen responsiveness suggest a novel role for the steroid receptor coactivator SRC-1 as a discriminative amplifier of estrogen signalling in the rat brain, *FASEB J*, 2003.
45. Camacho-Arroyo I, Neri-Gomez T, Gonzalez-Arenas A, *et al.*, Changes in the content of steroid receptor coactivator-1 and silencing mediator for retinoid and thyroid hormone receptors in the rat brain during the estrous cycle, *J Steroid Biochem Mol Biol* **94**:267–272, 2005.
46. Joels M, Corticosteroid effects in the brain: U-shape it, *Trends Pharmacol Sci* **27**:244–250, 2006.
47. Meijer OC, Kalkhoven E, van der Laan S, *et al.*, Steroid receptor coactivator-1 splice variants differentially affect corticosteroid receptor signalling, *Endocrinology* **146**:1438–1448, 2005.
48. Wang Q, Blackford JA Jr, Song LN, *et al.*, Equilibrium interactions of corepressors and coactivators with agonist and antagonist complexes of glucocorticoid receptors, *Mol Endocrinol* **18**:1376–1395, 2004.
49. Grenier J, Trousson A, Chauchereau A, *et al.*, Differential recruitment of p160 coactivators by glucocorticoid receptor between Schwann cells and astrocytes, *Mol Endocrinol* **20**:254–267, 2006.
50. Pascual-Le Tallec L, Lombes M, The mineralocorticoid receptor: A journey exploring its diversity and specificity of action, *Mol Endocrinol* **19**:2211–2221, 2005.
51. Morsink MC, Steenbergen PJ, Vos JB, *et al.*, Acute activation of hippocampal glucocorticoid receptors results in different waves of gene expression throughout time, *J Neuroendocrinol* **18**:239–252, 2006.
52. Cavarretta IT, Martini L, Motta M, *et al.*, SRC-1 is involved in the control of the gene expression of myelin protein Po, *J Mol Neurosci* **24**:217–226, 2004.
53. van Gemert NG, Meijer OC, Morsink MC, *et al.*, Effect of brief corticosterone administration on SGK1 and RGS4 mRNA expression in rat hippocampus, *Stress* **9**:165–170, 2006.

54. Winnay JN, Xu J, O'Malley BW, *et al.*, Steroid receptor coactivator-1-deficient mice exhibit altered hypothalamic-pituitary-adrenal axis function, *Endocrinology* **147**:1322–1332, 2006.
55. Patchev AV, Fischer D, Wolf SS, *et al.*, Insidious adrenocortical insufficiency underlies neuroendocrine dysregulation in TIF-2 deficient mice, *FASEB J* **21**:231–238, 2007.
56. Kurihara I, Shibata H, Suzuki T, *et al.*, Expression and regulation of nuclear receptor coactivators in glucocorticoid action, *Mol Cell Endocrinol* **189**:181–189, 2002.
57. Setiawan E, Owen D, McCabe L, *et al.*, Glucocorticoids do not alter developmental expression of hippocampal or pituitary steroid receptor coactivator-1 and -2 in the late gestation fetal guinea pig, *Endocrinology* **145**:3796–3803, 2004.
58. van Rossum EF, Lamberts SW, Polymorphisms in the glucocorticoid receptor gene and their associations with metabolic parameters and body composition, *Recent Prog Horm Res* **59**:333–357, 2004.

Chapter 13

Tissue Repair and Cancer Control through PPARs and Their Coregulators

Liliane Michalik and Walter Wahli

Tissue response to injury involves a coordinated series of events, which include inflammation/immune cell recruitment and oxidative stress, as well as cell survival, proliferation, migration, differentiation, and extracellular matrix (ECM) production/remodeling. Loss of control of the repair processes can be deleterious, and defense mechanisms that are beneficial may become harmful if they do not initiate and resolve in time. Peroxisome proliferator-activated receptors (PPARs) are transcription factors which control the repair program by modulating gene expression. Their activity depends on their binding to ligands produced after injury, followed by the recruitment of subsets of cofactors. In this chapter, we review our knowledge of PPARs and their cofactor functions in tissue repair and cancer, and discuss the potential for these nuclear receptors as therapeutic targets for organ protection against damage and for tissue repair.

13.1 Introduction

Organ injury triggers complex processes so that repair of the wound takes place. These complex processes involve various cell types and biochemical pathways. Damage occurs in response to chemical, biological, and mechanical stress or insufficient oxygen supply (hypoxia) that is usually caused by blood vessel constriction or obstruction (ischemia). Depending on their position in the body, organs have different susceptibility to attack. The skin is particularly exposed to mechanical damage, UV radiation, microorganisms and xenobiotics, whereas the liver is exposed to any xenobiotic present in the portal blood arriving from the digestive tract. In addition, oxygenation levels and sensitivity to hypoxia differ among the organs, although in most cases short periods of ischemia and reperfusion (I/R) cause extensive damage.

Irrespective of the organ or of the type of injury, the goal of tissue repair is to maintain tissue and organ viability, via processes that require tight control over cell survival and cell death mechanisms, cell growth and differentiation, and extracellular matrix (ECM) production and remodeling. The cells in the wounded organ change their intercellular contacts, proliferate, modify their production of ECM, and new blood vessels form rapidly. In all organs, moderate inflammation is a major component of efficient tissue repair. Inflammatory cytokines and chemokines are produced soon after the injury and these recruit immune cells to the wounded organ. These cells amplify the initial response by producing additional inflammatory mediators. Among these, some factors promote cell proliferation, cell migration, or neovascularization, whereas others increase pain, or even delay healing. The loss of control of the repair processes is deleterious. One striking example is excessive fibrosis, which is the consequence of overproduction of fibrous connective tissue in damaged organs such as the liver and lung. Fibrosis is involved in several life-threatening diseases for which no effective therapy is available.

The behavior of cells during healing, such as proliferation, migration, or participation in blood vessel formation, is the same as that occuring during primary tumor development and metastasis. This suggests that the same repair mechanisms, that are activated in response to injury may promote cancer if uncontrolled in a permissive context. Thus, cancer is often described as "a wound that never heals." In favor of such a definition is the observation that the epithelia, which heal very efficiently after injury, are also responsible for 95% of all cancer deaths. In addition to similar cell behavior and histological characteristics, the gene expression profiles of healing wounds and developing tumors also bear a resemblance to one other.[1,2] For example, the response of fibroblasts to blood serum exposure includes the activation of genes used to classify breast cancers[1] and other tumors, especially those with poor prognosis which are prone to metastasis.[2]

Establishing the molecular characteristics of an injured tissue, and understanding the repair process that follows, should aid the design of novel therapeutic interventions that favor organ healing and inhibit or reverse fibrosis and cancer. During the recent years, several key growth factors and cytokines were identified, but many questions remain unanswered concerning the transcriptional control of wound healing.[3] In fact, transcription factors, which control the repair program by changing gene expression, are attractive pharmacological targets. The peroxisome proliferator-activated receptors, PPARα, PPARβ/δ (PPARβ below) and PPARγ, also known as NR1C1, NR1C2 and NR1C3, respectively[4] are examples of such transcription factors. PPARs are members of the nuclear hormone receptor (NHR) family. They are organized into four structural domains, two of which are highly conserved in all members of the NHR family, namely the DNA binding domain (DBD) and the ligand-binding domain (LBD). The DBD consists of a two zinc-finger motif that is the hallmark of the NHR superfamily. The LBD comprises the hydrophobic cavity where ligands are buried, and harbors the ligand-dependent transcriptional activation function of the receptor called the activation function 2 (AF-2). In addition, the LBD provides an interface for the dimerization of PPARs with their obligate partner, the retinoid X receptor (RXR, NR2B), and for the interaction with regulatory proteins called cofactors. The regulation of gene expression by PPARs results from the binding of the receptor to response elements in the regulatory region of their target genes, ligand binding, and corepressor/coactivator exchange. The central role of PPARs and coregulators is the regulation of energy metabolism. However, in the present review, the focus is on the participation of PPARs in the pathways of tissue repair and cancer. Firstly, the consequences of PPAR activation with regard to healing in several organs are reviewed.[5,6] Secondly, the functions of some of the NHR cofactors in tissue injury and repair are described. However, in most cases so far, the nuclear receptors involved in the action of these cofactors in these processes have not been identified.

13.2 PPARs in Tissue Repair and Cancer

13.2.1 *PPARs and skin wound healing*

The protective function of the skin against diverse attacks and water loss resides in the epidermis. During development, the epidermis evolves from a single layer of epithelial cells to a fully stratified and differentiated epithelium. The outermost epidermal layer, called the stratum corneum, is the product of keratinocytes in their final differentiation stage. It consists of a layer of cross-linked proteins and lipids, which functions as an efficient barrier to dehydration and protection against diverse aggressions. Several studies suggest that epidermal maturation is PPAR-independent, but it is well established that PPARs can promote keratinocyte differentiation.[7–9] The expression of the three PPAR isotypes is undetectable in the adult mouse interfollicular epidermis, but the upregulation of the expression of both PPARα and PPARβ in the keratinocytes at the edges of mechanically-induced injuries[7] suggests that PPARs participate in the transcriptional control of skin healing.[3] Indeed, the transiently stimulated PPARα expression allows for a better control of inflammation soon after wounding.[7] Following skin damage, pro-inflammatory cytokines released by the injured cells stimulate the expression of PPARβ and the production of an unidentified PPARβ natural agonist.[10] PPARβ then activates the protein kinase Bα (PKBα, also known as Akt-1)-dependent prosurvival pathway, through transcriptional upregulation of 3-phosphoinositide-dependent kinase (PDK1) and indirect repression of the phosphatase and tensin homologue 10 (PTEN).[11] The increased resistance of keratinocytes to cell death contributes to faster re-epithelialization. After the completion of the new epithelium, activation of the TGF-β pathway in keratinocytes decreases the PPARβ expression levels.[7,12] Binding of CCAAT/enhancer-binding protein α (C/EBPα) and C/EBPβ to the promoter of PPARβ, and histone deacetylation via histone deacetylase-1 (HDAC-1) also repress PPARβ promoter activity in interfollicular and hair follicle keratinocytes.[13] The importance of PPARα and PPARβ for efficient skin wound repair is illustrated by the phenotype of PPARα- and PPARβ-null mice, which is characterized by delayed wound closure after injury.[7]

13.2.2 *PPARs and kidney ischemia/reperfusion (I/R)*

Acute renal failure (ARF) is a devastating syndrome which affects about 5% of hospitalized patients. It has a high mortality rate due to the lack of an effective treatment. Kidney damage is mostly due to toxins, infections, immune reactions, or ischemia. Damage to renal tubuli provokes epithelial cell apoptosis and necrosis, and is accompanied by the back-leakage of glomerular filtrate.[14] PPARα-mediated induction of FA oxidation in mice is thought to protect renal structure and function in situations of kidney I/R.[15] In support of this hypothesis, renal damage was enhanced in PPARα-null mice subjected to I/R injury[15] or after long term exposure to the plasticizer, Di-(2-Ethylhexyl)Phtalate, which induced severe glomerulonephritis accompanied by high oxidative stress.[16] Ligand-dependent PPARα activation also reduced ARF caused by treatment with cisplatin, an antitumor agent responsible for nephrotoxicity in humans.[17–19] As for PPARα, PPARβ activation protects mice from I/R-induced renal damage, with reduced necrosis, apoptosis and inflammation. Conversely, PPARβ-null animals are more susceptible to I/R injury and exhibit greater kidney dysfunction than wild-type mice. The protective role of PPARβ is due to the activation of the PKB/Akt survival pathway, similar to that observed in keratinocytes at skin wounds, and also similar to the increased spreading of epithelial cells.[20] Finally, the activation of PPARγ preserved the kidney not only from I/R, but also from diabetic nephropathy, hypertensive nephropathy, experimental glomerulonephritis, and cyclosporine- or glycerol-induced renal injury.[21–23] It protects the kidney through improved glucose metabolism and insulin resistance as well as anti-inflammatory, anti-fibrotic, anti-apoptotic effects, and reduced oxidative stress.[24–29] Together, these observations underscore the protective power of all three PPAR isotypes in several kidney dysfunctions.

13.2.3 *PPARs and lung I/R and fibrosis*

As is the case of kidney failure, no treatment except lung transplantation exists for patients with end stage pulmonary diseases following I/R lung injury or fibrosis. Furthermore, I/R lung injury remains the principal cause of death during the first month after transplantation.[30] In mice and rats, PPARγ activation significantly reduces I/R-induced lung damage, mostly via its anti-inflammatory functions, resulting in reduced pulmonary edema and increased survival.[22,31,32] Pulmonary fibrosis, such as that provoked by anticancer treatment with bleomycin, is characterized by inflammation, proliferation and accumulation of myofibroblasts, and excessive deposition of ECM proteins in the lung connective tissue, leading to shortness of breath.[33,34] In mice, bleomycin-induced pulmonary fibrosis was reduced by PPARγ ligands.[35] This beneficial effect of PPARγ activation may also apply to humans, since

PPARγ agonists interrupted the profibrotic effects of TGF-β in human pulmonary fibroblasts in culture.[36]

13.2.4 *PPARs and digestive tract I/R*

Intestinal I/R causes severe damage characterized by endothelial cell swelling, capillary plugging, and mucosal barrier dysfunction.[37] It occurs as a consequence of acute mesenteric ischemia, traumatic, hemorrhagic or septic shock, abdominal aortic aneurysm, severe burns or small bowel transplantation. Activation of PPARα or PPARγ had a beneficial action in several rodent models of intestinal I/R injury. PPARγ activation has anti-inflammatory effects, most likely due to the inhibition of NF-κB[38,39] and the prevention of neutrophil infiltration.[40] This PPARγ protective effect is reduced in PPAR$\gamma^{+/-}$ mice or by PPARγ inhibition with antagonists.[38,41,42] Targeted deletion of the PPARγ gene in macrophages increased the susceptibility of mice to dextran sulfate sodium (DSS)-induced colitis, suggesting that the anti-inflammatory role of PPARγ in this model depends on its activity in these immune cells.[43] In addition, the beneficial consequences of enteral nutrition administered soon after severe gut I/R injury was associated with PPARγ induction and abrogated by a PPARγ antagonist.[44] Similarly, activation of PPARα also attenuated I/R injury in the intestine, mostly via an anti-inflammatory action.[45] Recent evidence also suggests that the protective function of PPARγ in the digestive tract is not limited to the intestines, but extends to the gastric mucosa, via anti-inflammatory effects, attenuation of lipid peroxidation and inhibition of apoptosis.[46–49] Taken together, all these observations, in addition to the demonstration that PPARγ is the target that mediates the anti-inflammatory effect of 5-aminosalicylic acid, a drug widely used to treate inflammatory bowel diseases (IBD),[50] make this receptor an interesting therapeutic target for the treatment of intestinal injuries.

13.2.5 *PPARs and liver injury (cirrhosis and fibrosis)*

Uncontrolled inflammation and excess extracellular matrix deposition in the liver mesenchyme, due to chronic liver disease, lead to severe fibrosis and end stage cirrhosis, for which the only treatment is liver transplantation.[51] The causes of liver fibrosis and cirrhosis include genetic abnormalities; viral hepatitis forms B and C; toxic, alcoholic and autoimmune-mediated damage; and non-alcoholic steato-hepatitis associated with the metabolic syndrome. Cytokines and oxidative stress activate quiescent hepatic stellate cells (HSC, also known as Ito cells or lipocytes) in the damaged areas. In response to activation, HSC transdifferentiate into proliferative myofibroblasts that produce ECM in the liver mesenchyme. Normal deposition of ECM results from a balance between ECM production and ECM degradation by matrix metalloproteinases (MMPs), whose activity is restricted by the tissue inhibitors of metalloproteinases (TIMPs). In liver fibrosis, this balance is broken, leading to excess ECM production and deposition by the HSC. The most effective therapeutic treatment to reverse fibrosis would be the inactivation of HSC/myofibroblasts, the reduction/reversal of ECM deposition, or the modification of the MMP-TIMP balance in favor of ECM degradation. In fact, the improvement of liver architecture in experimental models indeed suggests that fibrosis is attenuated either by enhancing HSC apoptosis, blocking HSC transdifferentiation, or stimulating ECM degradation.[52]

The activation of PPARα appears to have an antifibrotic action via reduction of inflammation and oxidative stress in several experimental models of liver injury, such as the rat thioacetamide model of cirrhosis,[53] and the mouse model of liver I/R.[54] PPARα may also be involved in the control of liver repair after partial hepatectomy, although its functions in this context are still unclear. Although PPARα regulates genes involved in cell-cycle progression, cytokine signaling, and metabolic changes, it is not required for the compensatory hyperplasia induced by partial hepatectomy.[55–58]

Quiescent HSC express relatively high levels of PPARγ, which rapidly declines during their transdifferentiation into myofibroblasts. Stimulating PPARγ in activated HSC decreases their proliferation and fibrotic activity.[59–62] An increase in PPARγ expression and ectopic overexpression of PPARγ in myofibroblasts reverses their phenotype to more quiescent cells. These latter express lower levels of activation markers and are able to store retinyl esters, which is the hallmark of quiescent HSC.[63–65] Experimental data also suggest that the agonists of the farnesoid X receptor (FXR) increase and maintain high PPARγ expression in HSC, thereby enhancing the antifibrotic action of PPARγ agonists.[66] The COX-2 inhibitor SC-236 attenuates liver inflammation and fibrosis through several mechanisms, including PPARγ activation.[67] Finally, the beneficial action of PPARγ on liver fibrosis is not restricted to HSC, since PPARγ agonists reduce ductal proliferation in a model of bile duct ligation in rats.[68]

Not much is known about the functions of PPARβ in liver fibrosis. Contrary to PPARγ, the expression of this isotype is strongly induced after HSC activation, and it appears to play a profibrogenic role. In a model of carbon tetrachloride-induced acute liver damage, its activation stimulated the proliferation of HSC and the expression of fibrotic markers.[69] Therefore, because PPARγ and PPARβ appear to have opposite effects, manipulating the balance of PPARγ and PPARβ activities to reverse fibrosis represents an interesting therapeutic challenge.

13.2.6 *PPARs in ischemic brain injury and neurodegenerative diseases*

The main causes of brain damage are transient (from syncope or ischemic attack) and permanent (from infarct or irreversible stroke) ischemia, and neurodegenerative diseases. The latter two are major causes of disability and death in developed countries. Very few therapeutic strategies exist to stimulate brain damage repair, although prevention may reduce the severity of ischemic stroke in humans.[70] The fibrate gemfibrozil, a hypolipidemic compound that activates PPARα, reduces the incidence of stroke in men with low HDL and LDL cholesterol, who suffer from coronary heart disease.[71] Studies in mice suggest that this beneficial effect of gemfibrozil is independent of lipid metabolism, being most likely due to improved endothelial relaxation, reduced oxidative stress, and decreased VCAM-1 and ICAM-1 expression in the brain.[72] In experimental brain stroke in mice, the activation of PPARα was required to mediate the protective effect of resveratrol, a polyphenol found in grapes.[73] Activation of PPARα and γ may also be beneficial in neurodegenerative diseases. The PPARα agonist Wy-14,463 decreased oxidative stress and β-amyloid peptide-dependent neurotoxicity in rat hippocampal neurons.[74] Similarly, PPARγ agonists have neuroprotective effects in several models of brain injury, such as ischemia, Alzheimer's disease, multiple sclerosis, and autoimmune encephalomyelitis.[75–79] Activating PPARγ may also be an efficient way to reduce motor neuron cell death, as suggested by a mouse model of amyotrophic lateral sclerosis (ALS). In this model, pioglitazone improved muscle strength and body weight and delayed disease onset, thereby increasing life span.[80] Finally, the PPARγ agonist, pioglitazone, improved motor neuron survival, tissue repair and the recovery of locomotor functions after spinal cord injury in rats.[81]

13.2.7 *PPARs and cardiac I/R*

Cardiac I/R is associated with myocardial infarction, but also with the re-establishment of blood flow by coronary bypass surgery, thrombolysis and angioplasty. Healthy cardiac tissue uses fatty acids (FAs) as the major energy source, whereas utilization of glucose rather than FAs is thought to improve heart recovery.[82] In fact, pharmacological treatments that favor this switch in the utilization of energy sources appear beneficial for cardiac recovery.[83] Decreased PPARα expression and activity in situations of hypoxia or pressure overload is an important molecular determinant of this switch,[84] whereas experimental overexpression of PPARα in the heart impairs recovery following ischemia.[85] Selective overexpression of TNF-α in the heart in a transgenic mouse model suggests that this cytokine participates in the regulation of PPARα expression in cardiac cells. In this model, TNF-α production indirectly decreased PPARα expression, most likely through interaction with TGF-β signaling.[86] Taken together, these data suggest that inhibition of cardiac PPARα by an antagonist could be a therapeutic approach in treating cardiac damage.[85] However, the final outcome of PPARα activation in the heart remains debatable, as beneficial effects of PPARα activation on I/R damage have also been reported.[87–89] Thus, depending on the context, the anti-inflammatory or antifibrotic actions of PPARα could be beneficial, whereas the inhibition of the switch towards glucose utilization would be deleterious when circulating FA levels are high.

Activation of PPARγ reduced myocardial infarction size in several animal models, probably through increasing glucose uptake, improving insulin sensitivity and reducing post-ischemic myocardial apoptosis.[88,90–95] However, activating PPARγ may also be deleterious, since the PPARγ activators, thiazolidinediones (TZDs), were associated with increased susceptibility to ventricular fibrillation during myocardial I/R in pigs[96] and increased mortality after myocardial infarction in rats.[97] The role of PPARγ in heart failure in humans is also debatable. A retrospective cohort study suggested that TZDs may increase the risk of heart failure in patients suffering from type 2 diabetes,[98,99] while the PROactive study concluded that pioglitazone may improve cardiovascular outcome.[100] Finally, activation of PPARα or PPARγ may attenuate cardiac damage through antifibrotic and anti-inflammatory actions in anoxia-reoxygenation and pressure-overloaded hearts.[101–104] However, because beneficial and deleterious effects have both been reported, additional research is necessary before using PPARs as therapeutic targets to improve heart repair.[105]

13.3 PPARs in Shock and Sepsis

Multiple organ failure resulting from septicemic and hemorrhagic shocks is a major cause of death in intensive care units, with mortality reaching over 80% when four or more organs dysfunction simultaneously.[106] In several rodent models of endotoxic shock, treatment with the prostaglandin 15d-PGJ2 was protective,[107,108] probably through the activation of a PPARγ-dependent anti-inflammatory response and stimulation of the heat shock response,[109] a beneficial effect that was reduced by the PPARγ antagonist GW9662.[107] Similarly, the preventive activation of PPARα with fenofibrate protected the endothelia in a rabbit model of endotoxin-induced shock.[110] The protective role of 15d-PGJ2 was also seen in a rat model with hemorrhagic shock, in which 15d-PGJ2 attenuated liver injury and kidney dysfunction.[111] Similarly, in a rodent model of skin burn-induced injury, another PPARγ activator, rosiglitazone, decreased systemic inflammation and prevented damage in remote organs such as the liver, lung and kidney.[112] Further studies using these animal models of shocks will certainly help the design of future therapeutic trials.[113]

13.4 PPARs in Cancer

As mentioned above, the processes involved in the healing of an injury, such as cell proliferation, migration, ECM remodeling and angiogenesis, are also important features of tumor development and spreading. As such, the pathways regulating these processes are classic targets for cancer therapy design. PPARs participate in the regulation of many of these pathways, although their involvement in cancer development remains highly debated. In fact, the involvement of PPARs in the regulation of cancer cell proliferation, differentiation and apoptosis, and their potential as pharmacological targets in this context, have been reviewed extensively.[6,114]

In the last few years, the inhibition of angiogenesis in order to slow down tumor growth has become a major therapeutic goal.[15,16] PPARγ agonists have received much attention in this context.[117] Surprisingly, despite upregulation of the vascular endothelial growth factor (VEGF) expression levels in cultured cells, PPARγ agonists have overall anti-angiogenesis activities, *in vitro* and *in vivo*,[118–123] through several direct and indirect mechanisms.[124,125,127] The roles of PPARα and PPARβ in angiogenesis are less well documented. The PPARα agonist fenofibrate has angiostatic properties *in vitro* and *in vivo*,[126,127] and like PPARγ, it controls the expression of FIAF, an angiopoietin-like protein also known as angiopoietin-like protein 4.[128] Interestingly, the lack of PPARα expression in the stroma of PPARα-null mice inhibited tumor growth, due to exacerbated inflammation and the production of an angiostatic factor in the tumor bed.[129] PPARβ and VEGF signaling appear to be associated in head and neck squamous cell carcinomas (SCCs), as was suggested by the presence of higher microvessel density in these tumors when expressing PPARβ.[130,131] In models of endothelial cell (EC) cultures, the activation of PPARβ stimulated EC proliferation and migration, suggesting that PPARβ favors angiogenesis.[132] Because angiogenesis is necessary for organ repair, although it also supports tumor growth, developing compounds that activate or suppress this process is of interest.

13.5 Cofactors of PPARs in Repair and Cancer

As detailed above, PPARs play an important role in injured organs. In most cases, they are associated with the protection against damage or improvement of repair. Like other nuclear hormone receptors or transcription factors, the regulation of transcription by PPARs depends on their interactions with regulatory proteins called cofactors. Cofactors are proteins that interact with a transcription factor and repress (corepressor) or enhance (coactivator) its transcriptional activity, through modification of the chromatin structure and/or interaction with the basal transcriptional machinery.[133–135] More than 200 cofactors have been identified, which usually bind to and regulate the activity of several of the 49 NHR, in addition to other transcription factors.[133] The PPAR coactivators include CBP/p300, the p160/SRC family (SRC-1/NCoA; SRC-2/GRIP1/TIF2; SRC-3/pCIP/RAC3/ACTR/AIB1/TRAM-1), the PGC-1 family (PGC-1α; PGC-1β), PBP/TRAP220/DRIP205/MED1,

NRC/PRIP/ASC2/AIB3/RAP250/TRBP, and the PRIC DNA helicases (PRIC285; PRIC320). In addition to these proteins that directly bind to PPARs, coactivator-associated proteins have also been described.[133] These proteins do not directly bind to PPARs but are present in complexes regulating PPAR activity, via binding to coactivators.

Corepressors called SMRT, N-CoR and RIP140 were also identified. Corepressors are thought to bind to the apo-NHRs, while binding of a ligand to the NHR induces the release of the corepressors and the recruitment of a complex of coactivators, which initiates and facilitates the transcription of target genes.[136] Recently, experiments performed in cell cultures using fluorescence-labelled proteins demonstrated that PPARs associate with their obligate partner RXR even in the absence of a ligand. The recruitment of cofactors by PPARs is most likely independent of DNA binding, and the formation of a transcriptional complex precedes binding of the PPAR:RXR heterodimer to the promoter of target genes. Furthermore, PPARs exhibit a reduced mobility when ligand-activated, which confirms that binding of an agonist induces the recruitment of large complexes of cofactors. However, the recruitment of cofactors is also possible, although less efficient, in the absence of a ligand, which provides a molecular basis for PPAR constitutive activity.[137,138] The recruitment of cofactors by the estrogen receptor (ER) on the promoter of a target gene was studied in detail. Chromatin immunoprecipitations showed that the interaction between ER, chromatin, and NHR cofactors, which leads to transcriptional activity, is a highly dynamic process, with successive waves of recruitment and dissociation of ER and cofactors from the promoter.[139] Although no such study was performed with PPARs and PPAR target genes, it seems reasonable to consider this dynamic process as a general mode of function.

In summary, the transcriptional activity of PPARs relies on their recruitment of corepressors and coactivators, an interaction which is regulated by their binding to a ligand. Although the importance of PPARs in repair on the one hand, and the necessity for cofactor recruitment for transcriptional activity by the NHRs on the other hand, are well documented, the contribution of cofactors in the regulation of repair or cancer pathways by PPARs has not been explored. The following discussion therefore focuses on the roles of some PPAR cofactors with regard to tissue repair, although the link with specific PPAR actions in this context is unknown (Table 13.1).

Table 13.1. PPARs and cofactors in tissue repair. This table summarizes the interactions of PPARs with some of their coactivators in the liver, heart, skin and corepressor in macrophages, with regard to organ protection against damage and tissue repair.?, denotes a likely but not demonstrated cooperation between the PPAR and the cofactor.

	Liver	Heart	Skin	Macrophages
PPARα	PBP Peroxisome proliferation Hepatocyte proliferation Activation of PPARα target genes	PGC-1 (?) Switch from FAs to glucose utilization		
PPARβ			NRC (?) Keratinocyte migration Cell survival	BCL6 Repression of MCP-1 expression
PPARγ				NCoR Repression of NF-κB target genes SMRT Repression of STAT activity

Several cofactors may be associated with PPAR functions in the liver (Table 13.1). The PPAR binding protein (PBP) is a central coactivator of PPARα and γ. When challenged with a peroxisome proliferator, liver-specific PBP-null mice closely mimicked PPARα-null mice, with almost no peroxisome proliferation, modest hepatocyte proliferation, no induction of PPARα target genes and, finally, no PPARα-induced carcinogenesis in the liver, demonstrating that PBP is a central partner of PPARα activation and stimulation of these processes.[140,141] Interestingly, PBP is also required for CAR-dependent functions, hepatotoxicity,[142] and liver regeneration after hepatectomy.[141] Unlike PBP, SRC1 and PRIP/NRC are not necessary for PPARα activity in the liver. The SRC1-null mice responded to PPARα activation like their wild-type counterparts, showing hepatomegaly, peroxisome proliferation and PPARα target gene regulation.[143] Similarly, liver-specific disruption of the gene encoding PRIP/NRC did not abolish PPARα activity.[144] In an hepatic cell line, treatment with the inflammatory cytokines TNF-α and IL-1 reduced the expression of the NHRs RXRα, PPARα and γ, and of the coactivators SRC-1, PGC-1α and PGC-β.[145] *In vivo*, decreased expression of PPARα was reported in LPS-induced liver inflammation[146] and in liver biopsies of patients suffering from hepatitis C virus chronic infection.[147] Such a decrease in the expression of PPARα, if accompanied by a decreased expression of several of its coactivators, would most likely enhance inflammatory-induced damage, because of decreased anti-inflammatory functions.

The anti-inflammatory functions of PPARs make a large contribution to their protective role against organ damage (Table 13.1). In mouse macrophages, the corepressor NCoR appears to collaborate with PPARγ to repress the transcriptional activation of inflammatory genes. Following sumoylation of its LBD, PPARγ binds to NCoR/HDAC3 complexes, thereby stabilizing the corepressor complexes on chromatin and repressing NF-κB target genes.[148] PPARs also indirectly repress the expression of inflammatory genes in macrophages.[149] For example, the liganded PPARγ represses STAT3 target genes through dissociation from the corepressor SMRT, which is then available for repressing STAT3 activity.[150] Similarly, PPARβ indirectly inhibits the pro-inflammatory gene MCP-1, through the ligand-dependent release of the transcriptional repressor BCL6.[151] Activated PPARγ is certainly beneficial to tissue repair too, inhibiting IFN-γ and/or LPS-induced iNOS expression in macrophages, most likely via titration of the coactivators CBP/p300, and thereby inhibiting the activity of AP-1, STAT and NF-κB.[152,153]

The PPARγ co-activator 1α (PGC-1α) was first identified as a PPARγ cofactor, hence its name.[154] It is now known that PGC-1α also co-activates other NHRs, including PPARα and β.[155] Although PGC-1α is best known as an important factor of mitochondrial biogenesis, where its activity is due to co-activation of the NHR ERR, NRF-1 and NRF-2,[156] it also participates in the regulation of several pathways with beneficial protective consequences in the heart, the brain and muscle. As suggested by the invalidation of the PGC-1α gene in mice, PGC-1α protects the heart and brain against reactive oxygen species (ROS)-induced damage. PGC-1α is necessary for the activation of the expression of ROS-detoxifying enzymes in these tissues, whose importance is highlighted in neurodegenerative diseases.[157] In a rodent model of muscle atrophy, the over-expression of PGC-1α in transgenic mice induced resistance against muscle atrophy. The protective functions of PGC-1α result from counteracting an increase of the expression of genes promoting atrophy (atrogenes) and from preventing the decrease of genes involved in muscle functions.[158] PGC-1α expression in muscle is increased by exercise[159] whereas its decreased expression in situations of muscle atrophy most likely enhances the progression of damage. Finally, as mentioned above, myocytes switch from FAs towards glucose utilization in the damaged heart, and a decrease in the activity of PPARα appears to be an important molecular basis for this metabolic behavior. In addition, a down-regulation of PGC-1 contributes in this shift of cardiac cells towards a more glycolytic metabolism in situations of transient ischemia, which would favor repair.[160] Consistent with this proposition, LPS-induced sepsis in the mouse heart decreased PPARα and β expression, as well as that of the PPAR partner RXR, and of the NHR coactivators CBP, SRC-1, SRC-3, TRAP220 and PGC-1. In this model, the level of expression of several genes involved in lipid metabolism, including the fatty acyl-CoA synthase, the fatty acid transporter CD36 and LPL, is also decreased, consistent with a decrease in FA utilization.[161]

PGC-1β is also thought to participate in mitochondrial energy metabolism, but its functions are less well known than those of PGC-1α. The PGC-1β-null mice show impaired expression of genes involved in mitochondrial functions in the heart, the brain, muscle, the liver and the brown adipose tissue. On a high fat diet, these animals develop hepatic steatosis,[162] which may favor liver fibrosis in the long term. Thus, addressing the expression and activity of PGC-1β in non-alcoholic steato-hepatitis associated with metabolic syndrome would certainly be of interest.[163]

Finally, although several other cofactors are expressed in the skin,[164] the Nuclear Hormone Receptor Coregulator (NRC, also known as ASC-2/PRIP/RAP250/TRBP) is of particular interest with regard to the functions of PPARs in skin wound healing. NRC-null mice are embryonic lethal due to placental defect, liver dysfunction, cardiac hypoplasia and general growth deficiency. NRC$^{+/-}$ animals are viable, but develop chronic skin wounds as a result of grooming. These wounds heal poorly and show no migrating epithelial tongue. The migration of keratinocytes from cultures of NRC$^{+/-}$ skin explants is severely impaired, and apoptosis is strongly increased in NRC$^{+/-}$ embryonic fibroblasts.[165,166]

Table 13.2. PPARs and coactivators in cancer. This table summarizes the interactions of PPARs with some of their cofactors in the liver, ovarian cells, and mammary gland, with regard to carcinogenesis.

	Liver	Ovarian cells	Mammary gland
PPARα	PBP Peroxisome proliferation Hepatocyte proliferation		
PPARγ		PGC-1 Apoptosis stimulation	NRC Proliferation inhibition

Taken together, these observations are reminiscent of the phenotype of PPARβ-null mice.[7,10,11,167] These striking similarities between PPARβ-null and NRC$^{+/-}$ mice suggest that NRC is a central PPARβ coactivator in the keratinocyte response to injury.

PPARs regulate many processes involved in tumor development, such as cell proliferation, apoptosis and differentiation (Table 13.2). Although the safety of activating compounds is debated, PPARs remain potential pharmacological targets in this context.[6,114] Very few data are available with regard to the collaboration between PPARs and their cofactors in tumor development (Table 13.2). Interestingly, the coactivator SRC-3/NCoA3 was identified among genes that are co-regulated in fibroblasts as part of a wound response program, as well as in many human cancers.[1] This transcriptional signature, characteristic of a wound-like phenotype in tumors, is associated with poor prognosis in breast, lung and gastric carcinomas, and in cutaneous melanomas.[1,168,169] In SRC-3-null mice, the development of the mammary gland was retarded, and these mice were more resistant to DMBA-induced mammary epithelial tumors.[170] The absence of SRC-3 protected the animal from mammary tumors, but not from skin tumor development. This is indicative of a tissue selective role of SRC-3. This selectivity most likely reflects lower expression levels of SRC-3 in the skin than in the mammary gland. It also reflects different profiles of NHR expression. Although high SCR-3 is associated with increased proliferation and carcinogenesis in most reported cases, this coactivator had an antiproliferative activity in lymphocytes, its absence resulting in the development of lymphomas in aging SRC-3-null mice.[171] Whereas the pro-tumorigenesis action of SRC-3 was associated with its action on the PI3K/Akt-1 pathway,[170–172] its antiproliferative role correlated with NF-κB constitutive activation in the lymphocytes.[171] Thus, the dual role of SRC-3 is clearly cell context dependent. This example of SRC-3 illustrates the complexity of NHR cofactor functions, which depends on the presence of different partners in different cell types. Although PPARs are among the SRC-3 partners, the functional outcome of these associations remains to be determined. PPARγ has anticarcinogenic activities in many cell types, through stimulation of apoptosis and cell differentiation, and inhibition of proliferation.[6] In many cases, the *in vivo* relevance of these actions is not proven, and the cofactors involved have not been explored. However, PPARγ was reported to be linked to actions of the cofactors PGC-1α and NRC in ovarian and mammary cancer cells, respectively. In ovarian cancer cells, PGC-1α had pro-apoptotic functions, which were inhibited by a PPARγ antagonist and by the repression of PPARγ expression, suggesting that in this cell type, PGC-1α action requires association with PPARγ.[173] In a mammary tumorigenesis model *in vivo*, the haploid inactivation of the gene encoding NRC promoted the development of mammary tumors, and reduced the antiproliferative effect of troglitazone, a PPARγ agonist.[174] These two examples illustrate that PPARγ may interact with various cofactors in a cell type dependent manner, leading to various outcomes. This cell type selectivity certainly partially explains the apparently conflicting and debated data that have been reported with regard to PPAR involvement in cancer.[6] Finally, as described above, PBP is a central partner of PPARα activity in the liver, and is therefore most probably involved in PPARα-dependent carcinogenicity in rodent liver.[140,141]

13.6 Perspectives

PPARs are best known as major regulators of lipid, glucose, and amino acid metabolism. Several other articles in this issue review the role of cofactors as coregulators of PPARs in these metabolic functions. Accumulating knowledge underscores the importance of PPARs as regulators of tissue protection and repair, through systemic (anti-inflammatory, anti-oxidant and metabolic) and cellular (apoptosis, migration, proliferation) actions. In most of the situations described above, the activation of PPARs is expected to reduce or prevent damage, or to improve healing. In some other cases, as in the heart, antagonizing PPAR action may be beneficial, whereas in the liver, manipulating the balance between the activity of PPARβ and γ may efficiently reverse fibrosis. Therefore, PPARs are attractive targets for the development of new strategies to promote healing or to reverse fibrosis, especially for syndromes with high morbidity and mortality rates. In addition, short term preconditioning strategies with PPAR agonists could help in preventing damage before surgery for instance, or when there is a likelihood of kidney ischemia. However, although activating PPARs has proven beneficial for the treatment of metabolic diseases, adverse side effects of PPAR-activating compounds may slow down the development of new strategies. These side effects include edema, weight gain, or liver toxicity.[114] Safety concerns also include potential carcinogenic consequences of PPAR activation, most particularly in the liver, bladder and colon.

These adverse effects, so far associated with full PPAR agonists, have to be considered when developing new compounds. Therefore, new strategies have emerged, aiming at designing selective PPAR modulators (SPPARMs) rather than full PPAR agonists or antagonists. Instead of stimulating the recruitment of a whole set of cofactors by PPARs, these compounds are expected to induce the recruitment of different sets of cofactors. Various PPAR-associated transcriptional complexes would induce the transcription of different sets of target genes in a cell-type selective manner. This approach is very promising and will possibly allow the development of PPAR modulating compounds that will retain beneficial actions, but with lower or no adverse effects.[114]

However, the precise nature and function of endogenous activators of PPARs, and of the cofactors they recruit to modulate cell behavior during tissue repair is unknown. Further studies of PPAR-cofactor interactions in the repair of damaged organs would certainly speed up the design of selective compounds that promote or antagonize PPAR action according to the desired cellular response. Although redundancy is frequent among cofactors, in cell or *in vivo* approaches in which PPAR-dependent activity is monitored in the absence of a given cofactor should help in elucidating the nature of the transcriptional complexes recruited, the corresponding transcriptional response and the physiological output. Recent data suggest that an additional dimension should be added to these complex regulatory mechanisms. The observation that cytoplasmic ligand-binding proteins play an active role in guiding a ligand, retinoic acid (RA), either to RAR or to PPARβ, depending on the cell context is intriguing.[175,176] Remarkably, the activation of RAR or PPARβ by RA commits the cell to opposite fates, apoptosis or survival, respectively. Thus, the transcriptional response of a cell to a given agonist not only depends on the nature of the NHRs and cofactors that are present, but may also be the consequence of the guiding of the compound to its cognate receptor by a cytoplasmic transporter, which translocates to the cell nucleus.

Finally, increasing our knowledge about the roles of PPARs and their cofactors in the regulation of cell cycle, cell death, cell migration and angiogenesis is also a central issue for improving our understanding of the mechanisms of cancer development. Developing this field in the near future is of importance, not only in order to better understand the potential beneficial and adverse effects of novel PPAR-activating compounds, but also to develop new tissue repair and anticarcinogenic therapies.

Acknowledgments

The authors acknowledge grant support from the Swiss National Science Foundation and the Etat de Vaud. The authors thank Nathalie Constantin for her help in preparing the manuscript.

References

1. Chang HY, Sneddon JB, Alizadeh AA, *et al.*, Gene expression signature of fibroblast serum response predicts human cancer progression: Similarities between tumors and wounds, *PLoS Biol* **2**:E7, 2004.
2. Chang HY, Nuyten DS, Sneddon, JB, *et al.*, Robustness, scalability, and integration of a wound-response gene expression signature in predicting breast cancer survival. *Proc Natl Acad Sci USA* **102**:3738–3743, 2005.
3. Schafer M, Werner S, Transcriptional control of wound repair, *Annu Rev Cell Dev Biol* **23**:69–92, 2007.
4. *Nuclear Receptors Nomenclature Committee*, 1999. A unified nomenclature system for the nuclear receptor superfamily. *Cell* **97**:161–163.
5. Michalik L, Wahli W, Involvement of PPAR nuclear receptors in tissue injury and wound repair, *J Clin Invest* **116**:598–606, 2006.

6. Michalik L, Desvergne B, Wahli W, Peroxisome-proliferator-activated receptors and cancers: Complex stories, *Nat Rev Cancer* **4**:61–70, 2004.
7. Michalik L, Desvergne B, Tan NS, *et al.*, Impaired skin wound healing in peroxisome proliferator-activated receptor (PPAR)alpha and PPARbeta mutant mice, *J Cell Biol* **154**:799–814, 2001.
8. Barak Y, Liao D, He W, *et al.*, Effects of peroxisome proliferator-activated receptor delta on placentation, adiposity, and colorectal cancer, *Proc Natl Acad Sci USA* **99**:303–308, 2002.
9. Mao-Qiang M, Fowler AJ, Schmuth M, *et al.*, Peroxisome-proliferator-activated receptor (PPAR)-gamma activation stimulates keratinocyte differentiation, *J Invest Dermatol* **123**:305–312, 2004.
10. Tan NS, Michalik L, Noy N, *et al.*, Critical roles of PPAR beta/delta in keratinocyte response to inflammation, *Genes Dev* **15**:3263–3277, 2001.
11. Di-Poi N, Tan NS, Michalik L, *et al.*, Antiapoptotic role of PPARbeta in keratinocytes via transcriptional control of the Akt1 signaling pathway, *Mol Cell* **10**:721–733, 2002.
12. Tan NS, Michalik L, Di-Poi N, *et al.*, Essential role of Smad3 in the inhibition of inflammation-induced PPARbeta/delta expression, *EMBO J* **23**: 4211–4221, 2004.
13. Di-Poi N, Desvergne B, Michalik L, *et al.*, Transcriptional repression of peroxisome proliferator-activated receptor beta/delta in murine keratinocytes by CCAAT/enhancer-binding proteins, *J Biol Chem* **280**: 38700–38710, 2005.
14. Molitoris BA, Sutton TA, Endothelial injury and dysfunction: Role in the extension phase of acute renal failure, *Kidney Int* **66**:496–499, 2004.
15. Portilla D, Dai G, Peters JM, *et al.*, Etomoxir-induced PPARalpha-modulated enzymes protect during acute renal failure, *Am J Physiol Renal Physiol* **278**:F667–675, 2000.
16. Kamijo Y, Hora K, Nakajima T, *et al.*, Peroxisome proliferator-activated receptor alpha protects against glomerulonephritis induced by long-term exposure to the plasticizer di-(2-ethylhexyl)phthalate, *J Am Soc Nephrol* **18**:176–188, 2007.
17. Arany I, Safirstein RL, Cisplatin nephrotoxicity, *Semin Nephrol* **23**:460–464, 2003.
18. Li S, Gokden N, Okusa MD, *et al.*, Anti-inflammatory effect of fibrate protects from cisplatin-induced ARF, *Am J Physiol Renal Physiol* **289**: F469–480, 2005.
19. Li S, Basnakian A, Bhatt R, *et al.*, PPAR-alpha ligand ameliorates acute renal failure by reducing cisplatin-induced increased expression of renal endonuclease G, *Am J Physiol Renal Physiol* **287**:F990–998, 2004.

20. Letavernier E, Perez J, Joye E, *et al.*, Peroxisome proliferator-activated receptor beta/delta exerts a strong protection from ischemic acute renal failure, *J Am Soc Nephrol* **16**:2395–2402, 2005.
21. Chung BH, Lim SW, Ahn KO, *et al.*, Protective effect of peroxisome proliferator activated receptor gamma agonists on diabetic and non-diabetic renal diseases, *Nephrology (Carlton)* **10** Suppl 2:S40–43, 2005.
22. Cuzzocrea S, Peroxisome proliferator-activated receptors gamma ligands and ischemia and reperfusion injury, *Vascul Pharmacol* **41**:187–195, 2004.
23. Yousefipour Z, Hercule H, Truong L, *et al.*, Ciglitazone, a PPAR{gamma} inducer, ameliorates renal pre-glomerular production and activity of angiotensin II and TxA2 in glycerol-induced acute renal failure, *J Pharmacol Exp Ther* **322**:461–468, 2007.
24. Guan Y, Peroxisome proliferator-activated receptor family and its relationship to renal complications of the metabolic syndrome, *J Am Soc Nephrol* **15**:2801–2815, 2004.
25. Panchapakesan U, Sumual S, Pollock CA, *et al.*, PPARgamma agonists exert antifibrotic effects in renal tubular cells exposed to high glucose, *Am J Physiol Renal Physiol* **289**:F1153–1158, 2005.
26. Zafiriou S, Stanners SR, Polhill TS, *et al.*, Pioglitazone increases renal tubular cell albumin uptake but limits proinflammatory and fibrotic responses, *Kidney Int* **65**:1647–1653, 2004.
27. Sivarajah A, Chatterjee PK, Patel NS, *et al.*, Agonists of peroxisome-proliferator activated receptor-gamma reduce renal ischemia/reperfusion injury, *Am J Nephrol* **23**:267–276, 2003.
28. Zheng F, Fornoni A, Elliot SJ, *et al.*, Upregulation of type I collagen by TGF-beta in mesangial cells is blocked by PPARgamma activation, *Am J Physiol Renal Physiol* **282**:F639–648, 2002.
29. Chatterjee PK, Patel NS, Cuzzocrea S, *et al.*, The cyclopentenone prostaglandin 15-deoxy-Delta(12,14)-prostaglandin J2 ameliorates ischemic acute renal failure, *Cardiovasc Res* **61**:630–643, 2004.
30. King RC, Binns OA, Rodriguez F, *et al.*, Reperfusion injury significantly impacts clinical outcome after pulmonary transplantation, *Ann Thorac Surg* **69**:1681–1685, 2000.
31. Ito K, Shimada J, Kato D, *et al.*, Protective effects of preischemic treatment with pioglitazone, a peroxisome proliferator-activated receptor-gamma ligand, on lung ischemia-reperfusion injury in rats, *Eur J Cardiothorac Surg* **25**:530–536, 2004.
32. Okada M, Yan SF, Pinsky DJ, Peroxisome proliferator-activated receptor-gamma (PPAR-gamma) activation suppresses ischemic induction of Egr-1 and its inflammatory gene targets, *FASEB J* **16**:1861–1868, 2002.
33. Gross TJ, Hunninghake GW, Idiopathic pulmonary fibrosis, *N Engl J Med* **345**:517–525, 2001.

34. Sleijfer S, Bleomycin-induced pneumonitis, *Chest* **120**:617–624, 2001.
35. Genovese T, Cuzzocrea S, Di Paola R, *et al.*, Effect of rosiglitazone and 15-deoxy-Delta12,14-prostaglandin J2 on bleomycin-induced lung injury, *Eur Respir J* **25**:225–234, 2005.
36. Burgess HA, Daugherty LE, Thatcher TH, *et al.*, PPARgamma agonists inhibit TGF-beta induced pulmonary myofibroblast differentiation and collagen production: Implications for therapy of lung fibrosis, *Am J Physiol Lung Cell Mol Physiol* **288**:L1146–1153, 2005.
37. Cappell MS, Intestinal (mesenteric) vasculopathy. II. Ischemic colitis and chronic mesenteric ischemia, *Gastroenterol Clin North Am* **27**:827–860, 1998.
38. Nakajima A, Wada K, Miki H, *et al.*, Endogenous PPAR gamma mediates anti-inflammatory activity in murine ischemia-reperfusion injury, *Gastroenterology* **120**:460–469, 2001.
39. Han X, Benight N, Osuntokun B, *et al.*, Tumour necrosis factor alpha blockade induces an anti-inflammatory growth hormone signalling pathway in experimental colitis, *Gut* **56**:73–81, 2007.
40. Naito Y, Takagi T, Uchiyama K, *et al.*, Suppression of intestinal ischemia-reperfusion injury by a specific peroxisome proliferator-activated receptor-gamma ligand, pioglitazone, in rats, *Redox Rep* **7**:294–299, 2002.
41. Cuzzocrea S, Pisano B, Dugo L, *et al.*, Rosiglitazone and 15-deoxy-Delta12,14-prostaglandin J2, ligands of the peroxisome proliferator-activated receptor-gamma (PPAR-gamma), reduce ischaemia/reperfusion injury of the gut, *Br J Pharmacol* **140**:366–376, 2003.
42. Desreumaux P, Dubuquoy L, Nutten S, *et al.*, Attenuation of colon inflammation through activators of the retinoid X receptor (RXR)/peroxisome proliferator-activated receptor gamma (PPARgamma) heterodimer. A basis for new therapeutic strategies, *J Exp Med* **193**:827–838, 2001.
43. Shah YM, Morimura K, Gonzalez FJ, Expression of peroxisome proliferator-activated receptor-gamma in macrophage suppresses experimentally induced colitis, *Am J Physiol Gastrointest Liver Physiol* **292**:G657–666, 2007.
44. Sato N, Moore FA, Kone BC, *et al.*, Diffential induction of PPAR gamma by luminal glutamine and iNOS by luminal arginine in the rodent post ischemic small bowel, *Am J Physiol Gastrointest Liver Physiol* **292**:G616–623, 2006.
45. Cuzzocrea S, Di Paola R, Mazzon E, *et al.*, WY 14643, a potent exogenous PPAR-alpha ligand, reduces intestinal injury associated with splanchnic artery occlusion shock, *Shock* **22**:340–346, 2004.
46. Ichikawa H, Naito Y, Takagi T, *et al.*, A specific peroxisome proliferator-activated receptor-gamma (PPAR-gamma) ligand, pioglitazone, ameliorates gastric mucosal damage induced by ischemia and reperfusion in rats, *Redox Rep* **7**:343–346, 2002.

47. Konturek PC, Brzozowski T, Kania J, *et al.*, Pioglitazone, a specific ligand of the peroxisome proliferator-activated receptor gamma reduces gastric mucosal injury induced by ischaemia/reperfusion in rat, *Scand J Gastroenterol* **38**:468–476, 2003.
48. Wada K, Nakajima A, Takahashi H, *et al.*, Protective effect of endogenous PPARgamma against acute gastric mucosal lesions associated with ischemia-reperfusion, *Am J Physiol Gastrointest Liver Physiol* **287**:G452–458, 2004.
49. Villegas I, Martin AR, Toma W, *et al.*, Rosiglitazone, an agonist of peroxisome proliferator-activated receptor gamma, protects against gastric ischemia-reperfusion damage in rats: Role of oxygen free radicals generation, *Eur J Pharmacol* **505**:195–203, 2004.
50. Rousseaux C, Lefebvre B, Dubuquoy L, *et al.*, Intestinal anti-inflammatory effect of 5-aminosalicylic acid is dependent on peroxisome proliferator-activated receptor-gamma, *J Exp Med* **201**:1205–1215, 2005.
51. Elsharkawy AM, Oakley F, Mann DA, The role and regulation of hepatic stellate cell apoptosis in reversal of liver fibrosis, *Apoptosis* **10**:927–939, 2005.
52. Iredale JP, Cirrhosis: New research provides a basis for rational and targeted treatments, *BMJ* **327**:143–147, 2003.
53. Toyama T, Nakamura H, Harano Y, *et al.*, PPARalpha ligands activate antioxidant enzymes and suppress hepatic fibrosis in rats, *Biochem Biophys Res Commun* **324**:697–704, 2004.
54. Okaya T, Lentsch AB, Peroxisome proliferator-activated receptor-alpha regulates postischemic liver injury, *Am J Physiol Gastrointest Liver Physiol* **286**:G606–612, 2004.
55. Rao MS, Peters JM, Gonzalez FJ, *et al.*, Hepatic regeneration in peroxisome proliferator-activated receptor alpha-null mice after partial hepatectomy, *Hepatol Res* **22**:52–57, 2002.
56. Wheeler MD, Smutney OM, Check JF, *et al.*, Impaired Ras membrane association and activation in PPARalpha knockout mice after partial hepatectomy, *Am J Physiol Gastrointest Liver Physiol* **284**:G302–312, 2003.
57. Anderson SP, Yoon L, Richard EB, *et al.*, Delayed liver regeneration in peroxisome proliferator-activated receptor-alpha-null mice, *Hepatology* **36**:544–554, 2002.
58. Skrtic S, Carlsson L, Ljungberg A, *et al.*, Decreased expression of peroxisome proliferator-activated receptor alpha and liver fatty acid binding protein after partial hepatectomy of rats and mice, *Liver Int* **25**:33–40, 2005.
59. Xu J, Fu Y, Chen A, Activation of peroxisome proliferator-activated receptor-gamma contributes to the inhibitory effects of curcumin on rat hepatic

stellate cell growth, *Am J Physiol Gastrointest Liver Physiol* **285**:G20–30, 2003.

60. Miyahara T, Schrum L, Rippe R, *et al.*, Peroxisome proliferator-activated receptors and hepatic stellate cell activation, *J Biol Chem* **275**:35715–35722, 2000.
61. Marra F, Efsen E, Romanelli RG, *et al.*, Ligands of peroxisome proliferator-activated receptor gamma modulate profibrogenic and proinflammatory actions in hepatic stellate cells, *Gastroenterology* **119**:466–478, 2000.
62. Galli A, Crabb DW, Ceni E, *et al.*, Antidiabetic thiazolidinediones inhibit collagen synthesis and hepatic stellate cell activation *in vivo* and *in vitro*, *Gastroenterology* **122**:1924–1940, 2002.
63. Hazra S, Xiong S, Wang J, *et al.*, Peroxisome proliferator-activated receptor gamma induces a phenotypic switch from activated to quiescent hepatic stellate cells, *J Biol Chem* **279**:11392–11401, 2004.
64. Yavrom S, Chen L, Xiong S, *et al.*, PPARgamma suppresses proximal alpha 1(I) collagen promoter via inhibition of p300-facilitated NF-I binding to DNA in hepatic stellate cells, *J Biol Chem* **280**:40650–40659, 2005.
65. She H, Xiong S, Hazra S, *et al.*, Adipogenic transcriptional regulation of hepatic stellate cells, *J Biol Chem* **280**:4959–4967, 2005.
66. Fiorucci S, Rizzo G, Antonelli E, *et al.*, Cross-talk between farnesoid-X-receptor (FXR) and peroxisome proliferator-activated receptor gamma contributes to the antifibrotic activity of FXR ligands in rodent models of liver cirrhosis, *J Pharmacol Exp Ther* **315**:58–68, 2005.
67. Planaguma A, Claria J, Miquel R, *et al.*, The selective cyclooxygenase-2 inhibitor SC-236 reduces liver fibrosis by mechanisms involving non-parenchymal cell apoptosis and PPARgamma activation, *FASEB J* **19**:1120–1122, 2005.
68. Marra F, DeFranco R, Robino G, *et al.*, Thiazolidinedione treatment inhibits bile duct proliferation and fibrosis in a rat model of chronic cholestasis, *World J Gastroenterol* **11**:4931–4938, 2005.
69. Hellemans K, Michalik L, Dittie A, *et al.*, Peroxisome proliferator-activated receptor-beta signaling contributes to enhanced proliferation of hepatic stellate cells, *Gastroenterology* **124**:184–201, 2003.
70. Jonas S, Prophylactic pharmacologic neuroprotection against focal cerebral ischemia, *Ann N Y Acad Sci* **765**:21–25, 1995.
71. Bloomfield Rubins H, Davenport J, Babikian V, *et al.*, Reduction in stroke with gemfibrozil in men with coronary heart disease and low HDL cholesterol: The Veterans Affairs HDL Intervention Trial (VA-HIT), *Circulation* **103**:2828–2833, 2001.
72. Deplanque D, Gele P, Petrault O, *et al.*, Peroxisome proliferator-activated receptor-alpha activation as a mechanism of preventive neuroprotection induced by chronic fenofibrate treatment, *J Neurosci* **23**:6264–6271, 2003.

73. Inoue H, Jiang XF, Katayama T, *et al.*, Brain protection by resveratrol and fenofibrate against stroke requires peroxisome proliferator-activated receptor alpha in mice, *Neurosci Lett* **352**:203–206, 2003.
74. Santos MJ, Quintanilla RA, Toro A, *et al.*, Peroxisomal proliferation protects from {beta}-Amyloid Neurodegeneration, *J Biol Chem* **280**: 41057–41068, 2005.
75. Chu K, Lee ST, Koo JS, *et al.*, Peroxisome proliferator-activated receptor-gamma-agonist, rosiglitazone, promotes angiogenesis after focal cerebral ischemia, *Brain Res* **1093**:208–218, 2006.
76. Shimazu T, Inoue I, Araki N, *et al.*, A peroxisome proliferator-activated receptor-gamma agonist reduces infarct size in transient but not in permanent ischemia, *Stroke* **36**:353–359, 2005.
77. Sundararajan S, Gamboa JL, Victor NA, *et al.*, Peroxisome proliferator-activated receptor-gamma ligands reduce inflammation and infarction size in transient focal ischemia, *Neuroscience* **130**:685–696, 2005.
78. Zhao Y, Patzer A, Gohlke P, *et al.*, The intracerebral application of the PPARgamma-ligand pioglitazone confers neuroprotection against focal ischaemia in the rat brain, *Eur J Neurosci* **22**:278–282, 2005.
79. Kielian T, Drew PD, Effects of peroxisome proliferator-activated receptor-gamma agonists on central nervous system inflammation, *J Neurosci Res* **71**:315–325, 2003.
80. Schutz B, Reimann J, Dumitrescu-Ozimek L, *et al.*, The oral antidiabetic pioglitazone protects from neurodegeneration and amyotrophic lateral sclerosis-like symptoms in superoxide dismutase-G93A transgenic mice, *J Neurosci* **25**:7805–7812, 2005.
81. McTigue DM, Tripathi R, Wei P, *et al.*, The PPAR gamma agonist Pioglitazone improves anatomical and locomotor recovery after rodent spinal cord injury, *Exp Neurol* **205**:396–406, 2007.
82. Fragasso G, Piatti Md PM, Monti L, *et al.*, Short- and long-term beneficial effects of trimetazidine in patients with diabetes and ischemic cardiomyopathy, *Am Heart J* **146**:E18, 2003.
83. Kantor PF, Lucien A, Kozak R, *et al.*, The antianginal drug trimetazidine shifts cardiac energy metabolism from fatty acid oxidation to glucose oxidation by inhibiting mitochondrial long-chain 3-ketoacyl coenzyme A thiolase, *Circ Res* **86**:580–588, 2000.
84. Barger PM, Kelly DP, PPAR signaling in the control of cardiac energy metabolism, *Trends Cardiovasc Med* **10**:238–245, 2000.
85. Sambandam N, Morabito D, Wagg C, *et al.*, Chronic activation of PPAR{alpha} is detrimental to cardiac recovery following ischemia, *Am J Physiol Heart Circ Physiol* **290**:H87–95, 2006.
86. Sekiguchi K, Tian Q, Ishiyama M, *et al.*, Inhibition of PPAR-alpha activity in mice with cardiac-restricted expression of tumor necrosis factor: Potential

role of TGF-beta/Smad3, *Am J Physiol Heart Circ Physiol* **292**:H1443–1451, 2007.

87. Yue TL, Bao W, Jucker BM, *et al.*, Activation of peroxisome proliferator-activated receptor-alpha protects the heart from ischemia/reperfusion injury, *Circulation* **108**:2393–2399, 2003.
88. Wayman NS, Hattori Y, McDonald MC, *et al.*, Ligands of the peroxisome proliferator-activated receptors (PPAR-gamma and PPAR-alpha) reduce myocardial infarct size, *FASEB J* **16**:1027–1040, 2002.
89. Tabernero A, Schoonjans K, Jesel L, *et al.*, Activation of the peroxisome proliferator-activated receptor alpha protects against myocardial ischaemic injury and improves endothelial vasodilatation, *BMC Pharmacol* **2**:10, 2002.
90. Lee TM, Chou TF, Troglitazone administration limits infarct size by reduced phosphorylation of canine myocardial connexin43 proteins, *Am J Physiol Heart Circ Physiol* **285**:H1650–1659, 2003.
91. Shiomi T, Tsutsui H, Hayashidani S, *et al.*, Pioglitazone, a peroxisome proliferator-activated receptor-gamma agonist, attenuates left ventricular remodeling and failure after experimental myocardial infarction, *Circulation* **106**:3126–3132, 2002.
92. Yue TL, Chen J, Bao W, *et al.*, *In vivo* myocardial protection from ischemia/reperfusion injury by the peroxisome proliferator-activated receptor-gamma agonist rosiglitazone, *Circulation* **104**:2588–2594, 2001.
93. Khandoudi N, Delerive P, Berrebi-Bertrand I, *et al.*, Rosiglitazone, a peroxisome proliferator-activated receptor-gamma, inhibits the Jun NH(2)-terminal kinase/activating protein 1 pathway and protects the heart from ischemia/reperfusion injury, *Diabetes* **51**:1507–1514, 2002.
94. Sidell RJ, Cole MA, Draper NJ, *et al.*, Thiazolidinedione treatment normalizes insulin resistance and ischemic injury in the zucker Fatty rat heart, *Diabetes* **51**:1110–1117, 2002.
95. Liu HR, Tao L, Gao E, *et al.*, Anti-apoptotic effects of rosiglitazone in hypercholesterolemic rabbits subjected to myocardial ischemia and reperfusion, *Cardiovasc Res* **62**:135–144, 2004.
96. Xu Y, Lu L, Greyson C, *et al.*, Deleterious effects of acute treatment with a peroxisome proliferator-activated receptor-gamma activator in myocardial ischemia and reperfusion in pigs, *Diabetes* **52**:1187–1194, 2003.
97. Lygate CA, Hulbert K, Monfared M, *et al.*, The PPARgamma-activator rosiglitazone does not alter remodeling but increases mortality in rats post-myocardial infarction, *Cardiovasc Res* **58**:632–637, 2003.
98. Delea TE, Edelsberg JS, Hagiwara M, *et al.*, Use of thiazolidinediones and risk of heart failure in people with type 2 diabetes: A retrospective cohort study, *Diabetes Care* **26**:2983–2989, 2003.

99. Nikolaidis LA, Levine TB, Peroxisome proliferator activator receptors (PPAR), insulin resistance, and cardiomyopathy: Friends or foes for the diabetic patient with heart failure? *Cardiol Rev* **12**:158–170, 2004.
100. Dormandy JA, Charbonnel B, Eckland DJ, *et al.*, Secondary prevention of macrovascular events in patients with type 2 diabetes in the PROactive Study (PROspective pioglitAzone Clinical Trial In macroVascular Events): A randomised controlled trial, *Lancet* **366**:1279–1289, 2005.
101. Iglarz M, Touyz RM, Viel EC, *et al.*, Peroxisome proliferator-activated receptor-alpha and receptor-gamma activators prevent cardiac fibrosis in mineralocorticoid-dependent hypertension, *Hypertension* **42**:737–743, 2003.
102. Ogata T, Miyauchi T, Sakai S, *et al.*, Myocardial fibrosis and diastolic dysfunction in deoxycorticosterone acetate-salt hypertensive rats is ameliorated by the peroxisome proliferator-activated receptor-alpha activator fenofibrate, partly by suppressing inflammatory responses associated with the nuclear factor-kappa-B pathway, *J Am Coll Cardiol* **43**: 1481–1488, 2004.
103. Diep QN, Benkirane K, Amiri F, *et al.*, PPAR alpha activator fenofibrate inhibits myocardial inflammation and fibrosis in angiotensin II-infused rats, *J Mol Cell Cardiol* **36**:295–304, 2004.
104. Chen K, Li D, Zhang X, *et al.*, Anoxia-reoxygenation stimulates collagen type-I and MMP-1 expression in cardiac fibroblasts: Modulation by the PPAR-gamma ligand pioglitazone, *J Cardiovasc Pharmacol* **44**:682–687, 2004.
105. Rizza R, Henry R, Kahn R, Commentary on the results and clinical implications of the PROactive study, *Diabetes Care* **28**:2965–2967, 2005.
106. Baue AE, Durham R, Faist E, Systemic inflammatory response syndrome (SIRS), multiple organ dysfunction syndrome (MODS), multiple organ failure (MOF): Are we winning the battle? *Shock* **10**:79–89, 1998.
107. Abdelrahman M, Sivarajah A, Thiemermann C, Beneficial effects of PPAR-gamma ligands in ischemia-reperfusion injury, inflammation and shock, *Cardiovasc Res* **65**:772–781, 2005.
108. Zingarelli B, Cook JA, Peroxisome proliferator-activated receptor-gamma is a new therapeutic target in sepsis and inflammation, *Shock* **23**:393–399, 2005.
109. Kaplan JM, Cook JA, Hake PW, *et al.*, 15-Deoxy-Delta(12,14)-prostaglandin J(2) (15D-PGJ(2)), a peroxisome proliferator activated receptor gamma ligand, reduces tissue leukosequestration and mortality in endotoxic shock, *Shock* **24**:59–65, 2005.
110. Wiel E, Lebuffe G, Robin E, *et al.*, Pretreatment with peroxysome proliferator-activated receptor alpha agonist fenofibrate protects endothelium

in rabbit *Escherichia coli* endotoxin-induced shock, *Intensive Care Med* **31**:1269–1279, 2005.

111. Abdelrahman M, Collin M, Thiemermann C, The peroxisome proliferator-activated receptor-gamma ligand 15-deoxyDelta12, 14 prostaglandin J2 reduces the organ injury in hemorrhagic shock, *Shock* **22**:555–561, 2004.
112. Sener G, Sehirli AO, Gedik N, *et al.*, Rosiglitazone, a PPAR-gamma ligand, protects against burn-induced oxidative injury of remote organs, *Burns* **33**:587–593, 2007.
113. Marshall JC, Deitch E, Moldawer LL, *et al.*, Preclinical models of shock and sepsis: What can they tell us? *Shock* **24** Suppl 1:1–6, 2005.
114. Rubenstrunk A, Hanf R, Hum DW, *et al.*, Safety issues and prospects for future generations of PPAR modulators, *Biochim Biophys Acta* **1771**:1065–1081, 2007.
115. Zakarija A, Soff G, Update on angiogenesis inhibitors, *Curr Opin Oncol* **17**:578–583, 2005.
116. Carmeliet P, Angiogenesis in life, disease and medicine, *Nature* **438**:932–936, 2005.
117. Margeli A, Kouraklis G, Theocharis S, Peroxisome proliferator activated receptor-gamma (PPAR-gamma) ligands and angiogenesis, *Angiogenesis* **6**:165–169, 2003.
118. Keshamouni VG, Arenberg DA, Reddy RC, *et al.*, PPAR-gamma activation inhibits angiogenesis by blocking ELR+CXC chemokine production in non-small cell lung cancer, *Neoplasia* **7**:294–301, 2005.
119. Sarayba MA, Li L, Tungsiripat T, *et al.*, Inhibition of corneal neovascularization by a peroxisome proliferator-activated receptor-gamma ligand, *Exp Eye Res* **80**:435–442.
120. Panigrahy D, Singer S, Shen LQ, *et al.*, PPAR-gamma ligands inhibit primary tumor growth and metastasis by inhibiting angiogenesis, *J Clin Invest* **110**:923–932, 2002.
121. Murata T, He S, Hangai M, *et al.*, Peroxisome proliferator-activated receptor-gamma ligands inhibit choroidal neovascularization, *Invest Ophthalmol Vis Sci* **41**:2309–2317, 2000.
122. Xin X, Yang S, Kowalski J, *et al.*, Peroxisome proliferator-activated receptor gamma ligands are potent inhibitors of angiogenesis *in vitro* and *in vivo*, *J Biol Chem* **274**:9116–9121, 1999.
123. Fauconnet S, Lascombe I, Chabannes E, *et al.*, Differential regulation of vascular endothelial growth factor expression by peroxisome proliferator-activated receptors in bladder cancer cells, *J Biol Chem* **277**:23534–23543, 2002.
124. Goetze S, Bungenstock A, Czupalla C, *et al.*, Leptin induces endothelial cell migration through Akt, which is inhibited by PPARgamma-ligands, *Hypertension* **40**:748–754, 2002.

125. Rieusset J, Auwerx J, Vidal H, Regulation of gene expression by activation of the peroxisome proliferator-activated receptor gamma with rosiglitazone (BRL 49653) in human adipocytes, *Biochem Biophys Res Commun* **265**:265–271, 1999.
126. Varet J, Vincent L, Mirshahi P, *et al.*, Fenofibrate inhibits angiogenesis *in vitro* and *in vivo*, *Cell Mol Life Sci* **60**:810–819, 2003.
127. Kasai T, Miyauchi K, Yokoyama T, *et al.*, Efficacy of peroxisome proliferative activated receptor (PPAR)-alpha ligands, fenofibrate, on intimal hyperplasia and constrictive remodeling after coronary angioplasty in porcine models, *Atherosclerosis* **188**:274–280, 2006.
128. Kersten S, Mandard S, Tan NS, *et al.*, Characterization of the fasting-induced adipose factor FIAF, a novel peroxisome proliferator-activated receptor target gene, *J Biol Chem* **275**:28488–28493, 2000.
129. Kaipainen A, Kieran MW, Huang S, *et al.*, PPARalpha deficiency in inflammatory cells suppresses tumor growth, *PLoS ONE* **2**:e260, 2007.
130. Nijsten T, Geluyckens E, Colpaert C, *et al.*, Peroxisome proliferator-activated receptors in squamous cell carcinoma and its precursors, *J Cutan Pathol* **32**:340–347, 2005.
131. Jaeckel EC, Raja S, Tan J, *et al.*, Correlation of expression of cyclooxygenase-2, vascular endothelial growth factor, and peroxisome proliferator-activated receptor delta with head and neck squamous cell carcinoma, *Arch Otolaryngol Head Neck Surg* **127**:1253–1259, 2001.
132. Piqueras L, Reynolds AR, Hodivala-Dilke KM, *et al.*, Activation of PPARbeta/delta induces endothelial cell proliferation and angiogenesis, *Arterioscler Thromb Vasc Biol* **27**:63–69, 2007.
133. Yu S, Reddy JK, Transcription coactivators for peroxisome proliferator-activated receptors, *Biochim Biophys Acta* **1771**:936–951, 2007.
134. Feige JN, Gelman L, Michalik L, *et al.*, From molecular action to physiological outputs: Peroxisome proliferator-activated receptors are nuclear receptors at the crossroads of key cellular functions, *Prog Lipid Res* **45**:120–159, 2006.
135. Michalik L, Auwerx J, Berger JP, *et al.*, International Union of Pharmacology. LXI. Peroxisome proliferator-activated receptors, *Pharmacol Rev* **58**:726–741, 2006.
136. Rosenfeld MG, Lunyak VV, Glass CK, Sensors and signals: A coactivator/corepressor/epigenetic code for integrating signal-dependent programs of transcriptional response, *Genes Dev* **20**:1405–1428, 2006.
137. Tudor C, Feige JN, Pingali H, *et al.*, Association with coregulators is the major determinant governing peroxisome proliferator-activated receptor mobility in living cells, *J Biol Chem* **282**:4417–4426, 2007.
138. Feige JN, Gelman L, Tudor C, *et al.*, Fluorescence imaging reveals the nuclear behavior of peroxisome proliferator-activated receptor/retinoid X

receptor heterodimers in the absence and presence of ligand, *J Biol Chem* **280**:17880–17890. 2005.
139. Metivier R, Penot G, Hubner MR, *et al.*, Estrogen receptor-alpha directs ordered, cyclical, and combinatorial recruitment of cofactors on a natural target promoter, *Cell* **115**:751–763, 2003.
140. Jia Y, Qi C, Kashireddi P, *et al.*, Transcription coactivator PBP, the peroxisome proliferator-activated receptor (PPAR)-binding protein, is required for PPARalpha-regulated gene expression in liver, *J Biol Chem* **279**:24427–24434, 2004.
141. Matsumoto K, Yu S, Jia Y, *et al.*, Critical role for transcription coactivator PBP/TRAP220 in liver regeneration, and PPARalpha ligand-induced liver tumor development, *J Biol Chem* **282**:17053–17060, 2007.
142. Jia Y, Guo GL, Surapureddi S, *et al.*, Transcription coactivator peroxisome proliferator-activated receptor-binding protein/mediator 1 deficiency abrogates acetaminophen hepatotoxicity, *Proc Natl Acad Sci USA* **102**:12531–12536, 2005.
143. Qi C, Zhu Y, Pan J, *et al.*, Mouse steroid receptor coactivator-1 is not essential for peroxisome proliferator-activated receptor alpha-regulated gene expression, *Proc Natl Acad Sci USA* **96**:1585–1590, 1999.
144. Sarkar J, Qi C, Guo D, *et al.*, Transcription coactivator PRIP, the peroxisome proliferator-activated receptor (PPAR)-interacting protein, is redundant for function of nuclear receptors PPAR and CARm, the constitutive androstane receptor, in mouse liver, *Gene expression* **13**(4–5):255–269, 2007.
145. Kim MS, Sweeney TR, Shigenaga JK, *et al.*, Tumor necrosis factor and interleukin 1 decrease RXRalpha, PPARalpha, PPARgamma, LXRalpha, and the coactivators SRC-1, PGC-1alpha, and PGC-1beta in liver cells, *Metabolism* **56**:267–279, 2007.
146. Tai ES, bin Ali A, Zhang Q, *et al.*, Hepatic expression of PPARalpha, a molecular target of fibrates, is regulated during inflammation in a gender-specific manner, *FEBS Lett* **546**:237–240, 2003.
147. Dharancy S, Malapel M, Perlemuter G, *et al.*, Impaired expression of the peroxisome proliferator-activated receptor alpha during hepatitis C virus infection, *Gastroenterology* **128**:334–342, 2005.
148. Pascual G, Fong AL, Ogawa S, *et al.*, A SUMOylation-dependent pathway mediates transrepression of inflammatory response genes by PPAR-gamma, *Nature* **437**:759–763, 2005.
149. Ricote M, Glass CK, PPARs and molecular mechanisms of transrepression, *Biochim Biophys Acta* **1771**:926–935, 2007.
150. Wang LH, Yang XY, Zhang X, *et al.*, Transcriptional inactivation of STAT3 by PPARgamma suppresses IL-6-responsive multiple myeloma cells, *Immunity* **20**:205–218, 2004.

151. Lee CH, Chawla A, Urbiztondo N, *et al.*, Transcriptional repression of atherogenic inflammation: Modulation by PPARdelta, *Science* **302**:453–457, 2003.
152. Ricote M, Li AC, Willson TM, *et al.*, The peroxisome proliferator-activated receptor-gamma is a negative regulator of macrophage activation, *Nature* **391**:79–82, 1998.
153. Li M, Pascual G, Glass CK, Peroxisome proliferator-activated receptor gamma-dependent repression of the inducible nitric oxide synthase gene, *Mol Cell Biol* **20**:4699–4707, 2000.
154. Puigserver P, Wu Z, Park CW, *et al.*, A cold-inducible coactivator of nuclear receptors linked to adaptive thermogenesis, *Cell* **92**:829–839, 1998.
155. Wang YX, Lee CH, Tiep S, *et al.*, Peroxisome-proliferator-activated receptor delta activates fat metabolism to prevent obesity, *Cell* **113**:159–170, 2003.
156. Mootha VK, Handschin C, Arlow D, *et al.*, Erralpha and Gabpa/b specify PGC-1alpha-dependent oxidative phosphorylation gene expression that is altered in diabetic muscle, *Proc Natl Acad Sci USA* **101**:6570–6575, 2004.
157. St-Pierre J, Drori S, Uldry M, *et al.*, Suppression of reactive oxygen species and neurodegeneration by the PGC-1 transcriptional coactivators, *Cell* **127**:397–408, 2006.
158. Sandri M, Lin J, Handschin C, *et al.*, PGC-1alpha protects skeletal muscle from atrophy by suppressing FoxO3 action and atrophy-specific gene transcription, *Proc Natl Acad Sci USA* **103**:16260–16265, 2006.
159. Schuler M, Ali F, Chambon C, *et al.*, PGC1alpha expression is controlled in skeletal muscles by PPARbeta, whose ablation results in fiber-type switching, obesity, and type 2 diabetes, *Cell Metab* **4**:407–414, 2006.
160. Cook SA, Matsui T, Li L, *et al.*, Transcriptional effects of chronic Akt activation in the heart, *J Biol Chem* **277**:22528–22533, 2002.
161. Feingold K, Kim MS, Shigenaga J, *et al.*, Altered expression of nuclear hormone receptors and coactivators in mouse heart during the acute-phase response, *Am J Physiol Endocrinol Metab* **286**:E201–207, 2004.
162. Sonoda J, Mehl IR, Chong LW, *et al.*, PGC-1beta controls mitochondrial metabolism to modulate circadian activity, adaptive thermogenesis, and hepatic steatosis, *Proc Natl Acad Sci USA* **104**:5223–5228, 2007.
163. Brunt EM, Pathology of fatty liver diseasel, *Mod Pathol* **20** Suppl 1:S40–48, 2007.
164. Westergaard M, Henningsen J, Svendsen ML, *et al.*, Modulation of keratinocyte gene expression and differentiation by PPAR-selective ligands and tetradecylthioacetic acid, *J Invest Dermatol* **116**:702–712, 2001.
165. Mahajan MA, Das S, Zhu H, *et al.*, The nuclear hormone receptor coactivator NRC is a pleiotropic modulator affecting growth, development,

apoptosis, reproduction, and wound repair, *Mol Cell Biol* **24**:4994–5004, 2004.
166. Mahajan MA, Samuels HH, Nuclear hormone receptor coregulator: Role in hormone action, metabolism, growth, and development, *Endocr Rev* **26**:583-597, 2005.
167. Michalik L, Wahli W, Peroxisome proliferator-activated receptors (PPARs) in skin health, repair and disease, *Biochim Biophys Acta* **1771**:991–998, 2007.
168. Rangel J, Torabian S, Shaikh L, *et al.*, Prognostic significance of nuclear receptor coactivator-3 overexpression in primary cutaneous melanoma, *J Clin Oncol* **24**:4565–4569, 2006.
169. Haqq C, Nosrati M, Sudilovsky D, *et al.*, The gene expression signatures of melanoma progression, *Proc Natl Acad Sci USA* **102**:6092–6097, 2005.
170. Kuang SQ, Liao L, Wang S, *et al.*, Mice lacking the amplified in breast cancer 1/steroid receptor coactivator-3 are resistant to chemical carcinogen-induced mammary tumorigenesis, *Cancer Res* **65**:7993–8002, 2005.
171. Coste A, Antal MC, Chan S, *et al.*, Absence of the steroid receptor coactivator-3 induces B-cell lymphoma, *EMBO J* **25**:2453–2464, 2006.
172. Torres-Arzayus MI, Font de Mora J, Yuan J, *et al.*, High tumor incidence and activation of the PI3K/AKT pathway in transgenic mice define AIB1 as an oncogene, *Cancer Cell* **6**:263–274, 2004.
173. Zhang Y, Ba Y, Liu C, *et al.*, PGC-1alpha induces apoptosis in human epithelial ovarian cancer cells through a PPARgamma-dependent pathway, *Cell Res* **17**:363–373, 2007.
174. Zhang H, Kuang SQ, Liao L, *et al.*, Haploid inactivation of the amplified-in-breast cancer 3 coactivator reduces the inhibitory effect of peroxisome proliferator-activated receptor gamma and retinoid X receptor on cell proliferation and accelerates polyoma middle-T antigen-induced mammary tumorigenesis in mice, *Cancer Res* **64**:7169–7177, 2004.
175. Michalik L, Wahli W, Guiding ligands to nuclear receptors, *Cell* **129**:649–651, 2007.
176. Schug TT, Berry DC, Shaw NS, *et al.*, Opposing effects of retinoic Acid on cell growth result from alternate activation of two different nuclear receptors, *Cell* **129**:723–733, 2007.

Chapter 14

Coregulators and Inflammation

Serena Ghisletti, Wendy Huang and Christopher K. Glass

Inflammation is an essential host response that serves to direct the immune system to sites of infection, but when excessive, it contributes to the pathogenesis of numerous diseases that include atherosclerosis, diabetes and cancer. Responses to microbial pathogens are initially mediated by pattern-recognition receptors that regulate the expression and activities of transcription factors that include nuclear factor κB (NF-κB) and interferon-regulatory factors (IRFs). These transcription factors in turn bind to specific DNA response elements on inflammatory target genes and stimulate transcription by recruiting coactivators. In addition to this activation step, recent studies indicate that the transcriptional activation of many inflammatory genes also requires a signal-dependent derepression step, involving active removal of corepressor complexes. Several members of the nuclear receptor superfamily, including the glucocorticoid receptor, peroxisome proliferator-activated receptors, and the liver X receptors, can counter-regulate inflammatory responses in a ligand-dependent manner by inhibiting the formation of activator/coactivator complexes or by preventing the signal-dependent clearance of corepressor complexes. These activities influence the development of innate and adaptive immune responses to microbial infections and represent significant targets for therapeutic intervention in inflammatory diseases.

14.1 Introduction

Inflammation is a complex physiological response to infection and tissue injury involving the actions of many immune and non-immune cell types. At the tissue level, local inflammatory responses are characterized by swelling, redness, heat, pain and loss of function. These physical signs and symptoms are brought about by the stimulus-dependent accumulation of inflammatory cells and their release of potent inducers of inflammation that include bioactive lipid mediators (e.g. prostaglandins), cytokines, chemokines, and reactive oxygen species. This response is an essential component of innate immunity to viral and bacterial pathogens, and sets the stage for development of acquired immunity. At a cellular level, genes that mediate inflammatory responses must remain tightly repressed under normal conditions, but must also be rapidly and highly induced in the setting of infection or injury. Too little response results in overwhelming infection, while inappropriate inflammation can result in the induction or amplification of chronic disease states. The signal-dependent switch from repression to activation requires the coordinated dismissal of transcriptional repressors and corepressors in exchange for activators and coactivators. In this chapter, we provide a brief overview of pathogen sensors and the biological responses to their activation. We also discuss the roles of specific transcription factors and coregulators in positive and negative regulation of the inflammatory responses generated by these and other inflammatory signaling pathways.

14.2 Sensing Invasion: Pattern-Recognition Receptors

Inflammatory responses are initiated by signals that convey evidence of infection or tissue injury. In the past decade, the knowledge of host-pathogen recognition has been greatly enhanced with the discovery of pattern-recognition receptors (PPRs). Four main families of PPRs have been described: Toll-like receptors (TLRs);[1] C-type lectin-like molecules (including the mannose receptor and the β-glucan receptors);[2,3] the nucleotide-binding oligomerization domain-like receptor family (NOD receptors and NALPs); and a family of receptors that have an RNA-helicase domain joined to two caspase-recruitment domains (RIG-1 and MDA5). Members from the same or different PPR families cooperate in the innate immune response to pathogens.[4]

TLR family members recognize lipid, carbohydrate, peptide and nucleic-acid structures that are broadly expressed by different groups of microorganisms. In this chapter, we will focus on three well characterized TLRs and their signaling components that intersect with nuclear receptor and their coregulators: TLR4 recognizes lipopolysaccharides (LPS), a major component of gram-negative bacteria cell wall; TLR2 can dimerize with either TLR1 or TLR6 to recognize bacterial surface tri-acyl lipopeptides and di-acyl lipopeptides; TLR3 recognizes double-stranded RNA as a signature for viral infection. TLRs contain the Toll/interleukin-1 receptor (TIR) domain required for interaction with TIR-recognition domain-containing adaptor molecules. All TLRs (except TLR3) use myeloid differentiation primary-response gene 88 adaptor molecule (MyD88) to initiate downstream signaling, which eventually leads to the activation of NF-κB and the induction of inflammatory gene expression.[5] TLR3 and TLR4 use a MyD88-independent signaling pathway that involves the adaptor molecule TIR-domain-containing adaptor protein inducing interferon-β(TRIF). Activation of TLR3 or TLR4 allows TRIF to associate with the kinase TANK-binding kinase 1 (TBK1) and induces the phosphorylation and nuclear translocation of IFN-regulatory factor 3 (IRF3) and the induction of inflammatory gene expressions.[1]

14.3 Initiation of Inflammatory Responses

Many cell types are now recognized to express TLRs, enabling localized responses to microbial pathogens. Resident macrophages in particular express a broad variety of TLRs and serve as sentinel cells in many tissues. The initial cellular responses to TLR activation includes the secretion of local inflammatory mediators that promote vascular permeability, induce expression of cell adhesion molecules on vascular endothelium, and lead to the recruitment of leukocytes. Neutrophils are recruited to the site of infection or injury as early as three hours and play an important role during the early phase of acute inflammation. Activated neutrophils upregulate their expression of Fc receptors for binding antibody and complement complexes that recognize foreign antigens, enhancing their phagocytic ability and release reactive oxygen and nitrogen intermediates to facilitate the killing of microorganisms. In addition, neutrophils also play key roles in recruiting other immune cells to the site.[6] Monocytes arrive at the scene at five or six hours following infection/injury, and differentiate into macrohages. Upon recognition of microorganisms and activation by local inflammatory mediators, activated monocytes releases additional inflammatory mediators and cytokines that can further enhance macrophage and neutrophil phagocytic activity, induce coagulation, increase vascular permeability, and stimulate recruitment of additional lymphocytes and monocytes from the circulation to facilitate the clearance of infection and initiate repair of tissue damage.

The normal inflammatory response is an acute process that resolves after removal of the inciting stimulus. It has also become increasingly apparent that many of the mechanisms that underlie acute inflammatory responses to infection also contribute to the initiation or amplification of chronic inflammatory disease states that include atherosclerosis, diabetes mellitus, arthritis, inflammatory bowel diseases, and neurodegenerative diseases. A challenge of therapeutic interventions aimed at reducing inflammation is to inhibit inflammatory programs that promote chronic disease processes without disarming the ability of the immune system to respond to infection. One potential level of intervention is at the level of coactivator and/or corepressor function, as illustrated by the anti-inflammatory activities of nuclear receptors.

14.4 Coactivators and Transcription Factors Inducing the Inflammatory Response

Signals generated by an inflammatory stimulus are communicated to changes in gene transcription, leading to enhanced expression of the effector molecules previously described. This process depends on activation of various inducible transcription factors. In this context, the prototypical transcription factor that has been extensively studied is nuclear factor-κB (NF-κB), which pivotally controls the expression of multiple genes involved in inflammatory responses. More recently, members of the INF-regulatory factor (IRF) family of transcription factors have also been shown to play key roles in the regulation of the inflammatory response. The activation of a wide variety of inflammatory genes by these transcription factors is achieved by the interaction with coactivators that allow the recruitment of various components of the basal transcription apparatus to promote gene expression (Fig. 14.1).

14.4.1 *NF-κB*

The mammalian NF-κB transcription-factor family consists of REL-A (p65), NF-κB1 (p50 and its precursor p105), NF-κB2 (p52 and its precursor p100), c-REL and REL-B. The p50 and p52 proteins are generated by proteolytic processing of precursor p105 and p100 respectively.

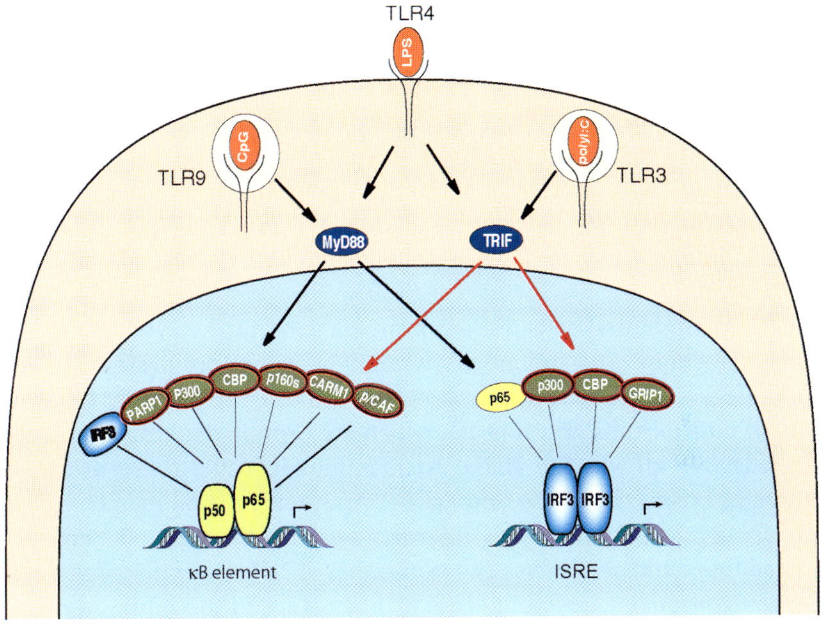

Fig. 14.1. Coactivators and transcription factors inducing the inflammatory response. TLR family members initiate inflammatory response signaling, which leads to the activation of transcription factors (such as NF-κB and IRFs), recruitment of coactivators and induction of inflammatory gene expression. TLR4 recognizes LPS, TLR3 recognizes double-strand RNA (poly I:C) and TLR9 recognizes CpG oligonucleotide sequences. While TLR4 signals through both MyD88-dependent and TRIF-dependent pathways, TLR3 signals exclusively trough the TRIF-dependent pathway and TLR9 (as well as other TLRs not discussed here) signals exclusively through the MyD88-dependent pathway. The activated NF-κB transcription factors and IRFs bind to the responsive elements on the inflammatory target genes and activate transcription by recruiting coactivators such as CBP/p300, p160s, PARP1, CARM1, p/CAF, and GRIP1. IRF3 is a promoter-specific co-activator of NF-κB, while IRF3-mediated activation of ISRE by TLR4 and TLR9 through MyD88 requires that p65 functions as a signal-specific co-activator. TLR3-specific activation of IRF3 through the TRIF signaling pathway is p65-independent.

Genes that encode all five members of the NF-κB family have been deleted in mice. These gene knock-out animal models indicate the distinct role of the NF-κB proteins in the regulation of innate and adaptive immune responses. These proteins have a structurally conserved amino-terminal REL-homology domain (RHD), which contains a motif for dimerization, a motif for binding specific DNA sequences and a motif for nuclear localization. The p50 and p52 proteins contain only an RHD, whereas c-REL, REL-A and REL-B also contain a carboxy-terminal transactivation domain. Each members of the NF-κB family can form homodimers as well as heterodimers with one another, with the exception of REL-B, which interacts only with p50 or p52. The best studied and most abundant of these complexes is the p50-p65 heterodimer.[7,8]

In unstimulated cells, NF-κB proteins exist in the cytoplasm in an inactive form by association with a family of inhibitory proteins known as inhibitors of NF-κB (IκBs). The IκBs are characterized by the presence of multiple ankyrin repeats, which are protein domains that interact with NF-κB via the RHD. In mammalian cells, there are three principal IκBs, IκBα, IκBβ and IκBε.[7,8] Another unusual member of the IκB family is BCL-3, which interacts specifically with p50 and p52 homodimers and can induce the expression of NF-κB-regulated genes as a transcriptional coactivator. This is in contrast to the inhibitory function of the other IκB proteins. The prevailing view is that IκB retains NF-κB in the cytoplasm by masking the conserved nuclear localization sequence (NLS) of the NF-κB subunits. However, recent studies have indicated that the cytoplasmic localization of the inactive NF-κB complexes is achieved by balancing continuous movement between the nuclear and cytoplasmic compartments. Only one of the two NLSs in an NF-κB dimer is masked by IκBα and the NF-κB-IκBα complexes shuttle into the nucleus even in the absence of cellular stimulation. At the same time, the nuclear export sequence (NES) that is located at the amino terminus of the IκBα protein causes the rapid export of such complexes back to the cytoplasm. The dynamic balance between cytosolic and nuclear localization is altered upon IκBα degradation, because it removes the contribution of the IκB NES and exposes the masked NLS of p65, resulting in predominantly nuclear localization of NF-κB. By contrast, NF-κB-IκBβ complexes are retained in the cytoplasm owing to the masking of both NLSs on the NF-κB dimer by IκBβ. It has been well documented that IκBα regulates transient NF-κB activation and that IκBβ maintains persistent NF-κB activation.[7]

There are at least two distinct NF-κB activation pathways.[8] The most frequently observed is the classical pathway, which is induced in response to various inflammatory stimuli previously described. This pathway is typified by the rapid phosphorylation of IκBα on two conserved serine residues by IκB kinase (IKK), a complex that is composed of a regulatory subunit, IKKγ (also known as NEMO), and two catalytic subunits, IKKα and IKKβ. The phosphorylated IκBα is then ubiquitylated by the SCF-E3-ubiquitin ligase complex, which targets it for degradation by the 26S proteasome. In many cells types, IκBβ and IκBε are also subject to phosphorylation and degradation, but with slower kinetics. After they have been freed, NF-κB dimers translocate to the nucleus, bind their cognate DNA-binding sites and induce transcription of inflammatory target genes.

A subset of stimuli, such as the stimulation of the CD40 and lymphotoxin-βreceptors, B-cell-activating factor of the TNF family (BAFF), receptor activator of NF-κB (RANK) ligand and latent membrane protein-1 (LMP1) of Epstein-Barr virus activate the non-canonical pathway. Here, activation of IKKα by the NF-κB inducing kinase (NIK) results in the processing of p100 into p52 by the activation of the 26S proteasome. p52-REL-B heterodimers, which are frequently activated as a consequence of non-canonical pathway activation, have a higher affinity for distinct kB elements and might therefore regulate a distinct subset of NF-κB target genes such as those mediating cell-cell interactions that are critical for lymphoid organ development and adaptive immunity.[9]

As with other sequence-specific transcription factors, NF-κB-dependent gene expression requires interactions with multiple transcriptional coactivators that function to remodel nucleosomes, modify histone tails to control interactions with chromatin-remodeling factors and influence chromatin architecture, and ultimately recruit core transcriptional machinery and RNA polymerase II (Fig. 14.1). The two key coactivators of NF-κB, are the widely required histone acetyltransferases p300 and its homolog, the cAMP-response element-binding protein (CREB)-binding protein (CBP), which directly associate with the NF-κB subunits p50 and p65.[10] These coactivators are thought to promote the rapid formation of the pre-initiation complexes by bridging NF-κB to the basal transcription machinery, thereby facilitating the transcription. Additionally, the histone acetyltransferases p300 and CBP can modify the amino-terminal tails of nucleosome histones, thereby altering the local chromatin structure.

Although the recruitment of p300 or CBP to NF-κB dependent enhanceosomes is required for synergistic activation, tethering p300/CBP alone to the promoter through NF-κB is not sufficient for full activity of NF-κB in the context of chromatin. Several reports indicated that the combined actions and interactions of distinct transcriptional coactivators complexes and cofactors seem to be attributable to the strong transcriptional activity of NF-κB, depending on the stimuli and cell type. Inactivation of either CBP, members of the p160 family of coactivators, or the CBP-associated factor (p/CAF) by nuclear antibody microinjection prevents NF-κB-dependent transactivation.[11] It has also been shown that steroid receptor-coactivator-1 (SRC-1), or nuclear receptor coactivator-1 (NCoA-1), interacts with p50 and potentiates NF-κB-mediated transactivation. This coactivator is a member of a group of related coactivators (the p160 family) that includes SRC-1/NCoA-1, SRC-2/TIF2/GRIP1/NCoA-2, and SRC-3/p-CIP/ACTR/AIB1.[12] More recently, Poly(ADP-ribose) polymerase-1 (PARP-1), a nuclear chromatin-associated protein, has been reported to act as a promoter-specific coactivator of NF-κB. PARP-1 directly interacts with p300 and both subunits of NF-κB (p65 and p50) and synergistically coactivates NF-κB-dependent transcription. In addition, the coactivator-associated arginine methyltransferase CARM1/PRMT4 has been demonstrated as a novel transcriptional coactivator of NF-κB and functions as a promoter-specific regulator of NF-κB recruitment to chromatin. CARM1 directly binds to p65 and synergistically coactivates NF-κB-mediated transactivation, in concert with p300/CBP and the p160 family of coactivators.[13] Also the coactivator Tip60 has been reported to enhance BCL-3-p50 activated transcription, suggesting the role of BCL-3 as a bridging factor between NF-κB and nuclear coregulators.[14] Tip60 is also recruited with BCL-3-p50 on the IL-1β-induced NF-κB target gene *KAI1/CD82*.[15]

The ability of NF-κB family members to interact with coactivators is regulated by covalent modification such as phosphorylation and acetylation. Several groups have shown that inflammatory stimuli induce the activation of p65 by pathways that are distinct from those that lead to IκB degradation and NF-κB nuclear translocation. For example, phosphorylation of p65 Ser276 by PKA is essential for the efficient binding of p65 to CBP. Several additional p65 phosphorylation events have been reported and they can all be generally described as stimulatory modifications that enhance the transcriptional activity of p65 and its ability to interact with coactivators.[16] RELA is also acetylated at several sites by p300 and CBP. The p300/CBP and p300/CBP-associated factor (p/CAF) coactivators represent acetyltransferases that have been implicated as the effectors of this RELA acetylation. Furthermore, acetylated RELA, like the histone proteins, is subject to deacetylation by HDACs, and HDAC3 is important for RELA deacetylation. In addition, the association of p65 with HDAC1 and HDAC2 has been reported to inhibit the expression of NF-κB regulated genes at both basal and induced levels. Intriguingly, recent work has shown that the association of p65 with CBP/p300 or HADC1 is determined by its phosphorylation status.[10]

14.4.2 *IRFs*

The IRFs are a family of transcription factors that are involved both in the induction of type I INFs and in the reponse to INFs. So far, of the nine known members of the IRF family, IRF1, IRF3, IRF5 and IRF7 have been shown to function as direct transducers of virus-mediated signaling. IRF3 and IRF7, which are highly homologous, are the key regulators of type I INF gene expression elicited by viruses. IRF3 is constitutively expressed and, in response to viral infection, its C-terminal regulatory domain is activated by phosphorylation, which allows the formation of IRF3 dimers (either a homodimer or a heterodimer with IRF7).[17] After dimerization, IRF3 translocates rapidly to the nucleus and activates transcription of the type I INFs genes through the interaction with coactivators CBP or p300 (Fig. 14.1). IRF3 and CBP/p300 form a complex that is able to bind to the target sequence ISRE (INF-stimulated response element) and activates the transcription of type I INFs genes and certain chemokine genes such as *ccl5*, *cxcl9* and *cxcl10*.[18]

In addition to their role in type I INFs response modulation, several IRF-family members are also activated by the MyD88-dependent and MyD88-independent pathways and contribute to the specific gene expression programs induced by TLRs. The infβ, *cxcl9* and *cxcl10* gene induction in response to activation of TLR4 by LPS is mediated by IRF3 trough a TRIF-dependent pathway. Moreover, activation of inducible nitric-oxide synthase gene (*inos*) depends on IRF1 in LPS stimulated macrophages.[19] The activation of IRF3 by signaling trough TLR3 is more rapid and potent than that triggered by TLR4 signaling.[1] Two IKK-related proteins, IKKε and TBK1, have recently been identified as the kinases that phosphorylate IRF3 in reponse to stimulation of TLR3.[20] Similarly to TLR4, the activation of TLR3 can induce type I

INF expression by a MyD88-independent, TRIF and TBK1-dependent signaling pathway.[21] The TLR9-subfamily members exclusively use MyD88 as the signaling adaptor, which interacts directly with IRF7 at the endosomal compartment.[22] In addition to IRF7, IRF5 also interacts with MyD88. Activation of the MyD88-dependent pathway by a TLR7 or TLR9 ligand leads to the translocation of IRF5 to the nucleus.[23]

It is well established that members of the NF-κB and IRF families cooperatively control the transcription of several cytokine genes. The classical interaction between IRFs and NF-κB occurs through independent binding of these transcription factors to adjacent sites. By contrast, much evidence indicates that IRF3 can form a complex directly with p65. In this way, IRF3 can function as a promoter-specific and signal-specific cofactor to activate transcription of a set of NF-κB-dependent genes without directly binding to an ISRE.[24–26] Conversely, mouse embryonic fibroblasts deficient in p65 show markedly impaired induction of ISRE-containing genes following ligation of TLR4, indicating that IRF3-p65 complexes (with IRF3 binding the ISRE) lead to the full induction of these genes.[24] Therefore, depending on the nature of the signaling, either IRF3 function as an essential coactivator of p65 for the transcription of NF-κB-dependent genes, or p65 functions as a cofactor of IRF3 for the transcription of IRF-dependent genes.

14.5 Active Repression of Inflammatory Response Genes

Recent studies have begun to provide evidence that in addition to a signal-dependent "activation" step, transcriptional activation of many inflammatory response genes also requires a signal-dependent "de-repression" step (Fig. 14.2). The nuclear receptor corepressor NCoR and the related factor SMRT have recently been established to play essential roles in mediating active repression of a subset of inflammatory response genes in the absence of a stimulus. NCoR and SMRT are key corepressors for a large assortment of different transcription factors. They are encoded by distinct genetic loci but share a common molecular architecture, form similar complexes with other corepressor proteins and exert overlapping biological functions. They are able to interact with unliganded nuclear receptors through an elongated helix of sequence LXX I/H IXXX I/L, alternatively referred as the Cornr-box. SMRT is subject to negative regulation by MAPK signaling pathways.

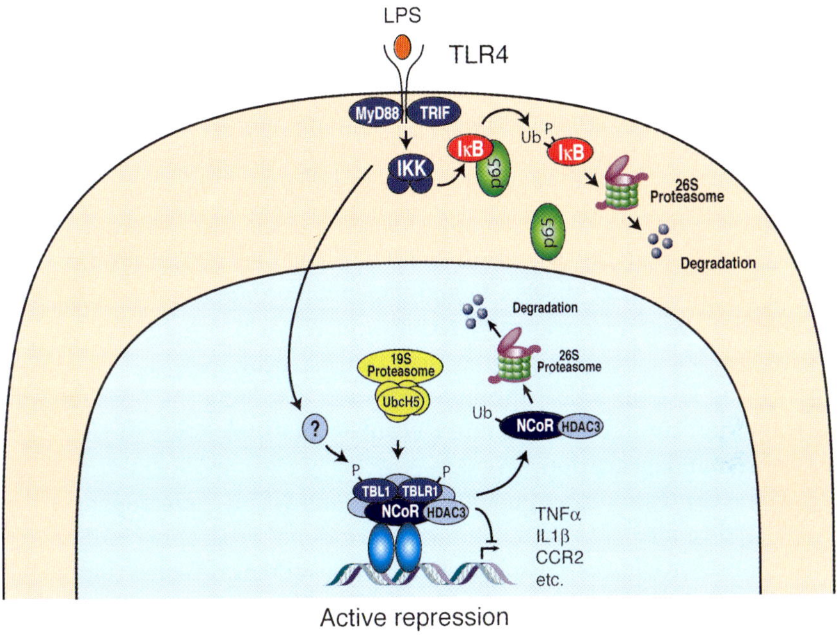

Fig. 14.2. Derepression is a prerequisite to activation of inflammatory response genes. In the absence of signal, many inflammatory target genes are occupied by various corepressor complexes and are actively repressed. One well-characterized corepressor complex contains nuclear-receptor-co-repressor (NCoR), histone deacetylase 3 (HDAC3), trasducin-beta-like1 (TBL1) and TBL1-related protein (TBLR1). After LPS activation of TLR4, activation of the E3 ligase activities of TBL1 and TBLR1 leads to the recruitment of UBCH5-19S-ubiquitin-proteasome complexes, ubiquitylation of the NCoR complex and subsequent proteasomal degradation. TLR4 stimulation also activates the IKK complex, which results in IκB phosphorylation, ubiquitylation, and degradation by the 26S proteasome. Release of the NF-κB complex (here shown as p65) allows it to relocate to the nucleus, where it binds to inflammatory target genes to initiate transcription.

Whereas activation of MEKK1 leads to phosphorylation of SMRT, dissociation from its transcription factor partners, and its redistribution from the cell nucleus to a cytoplasmic compartment, NCoR is refractory to this form of regulation.[27]

Biochemical purification of NCoR complex has defined HDAC3, GPS2 and the transducin beta-like factors TBL1 and TBLR1 are core components of the larger NCoR/SMRT holocomplexes.[28] Additional low-affinity components, including Sin3A, HDAC1, 2 and the Bgr1 complex also contribute to the NCoR-SMRT-dependent transrepression.[29] Whereas ligand binding is sufficient to cause dissociation of NCoR and SMRT from retinoic acid and thyroid hormone receptors *in vitro*, recent studies indicate that an additional ubiquitylation-mediated proteolytic step is required for the removal of NCoR complexes from nuclear receptor target genes *in vivo*.[30] In the case of retinoic acid receptor target genes, the TBLR1 component of the NCoR complex appears to function as an E3-ligase that directs ligand-dependent ubiquitylation and proteosome-mediated clearance of NCoR and HDAC3.[30] A similar mechanism has been described by which the estrogen receptor markedly reduce the level of NCoR through a process involving the upregulation of the ubiquitin ligase Siah2 and the subsequent targeting of NCoR for proteasomal degradation.[31]

Gene expression profiling of wild type and $NCoR^{-/-}$ macrophages has been used to identify endogenous transcriptional programs regulated by NCoR in macrophages.[32] Unexpectedly, these studies revealed that NCoR acts as a transcriptional checkpoint for NF-κB and AP-1-dependent gene networks that regulate diverse biological processes including inflammation, cell migration and collagen catabolism. Deletion of the NCoR gene in macrophages resulted in de-repression of genes encoding matrix metalloproteinases, chemokines and chemokine receptors. Consistent with this pattern of altered gene expression, $NCoR^{-/-}$ macrophages exhibited a partially activated phenotype, exemplified by enhanced ability to migrate through a gelatin barrier.[32] Macrophage expression of matrix metalloproteinases has been suggested to contribute to the instability of complex atherosclerotic lesions, increasing the risk of plaque rupture and acute myocardial infarction.[33] These findings suggest that NCoR complexes play key roles in controlling inflammatory transcriptional programs that contribute to the development of atherosclerosis and its clinical consequences. A similar role of SMRT has been suggested in the negative regulation of NF-κB target genes involved in regulation of anti-apoptotic programs following matrix attachment.[34]

Chromatin immunoprecipitation studies demonstrated that NCoR/HDAC3/TBL1/TBLR1 corepressor complexes occupy numerous NF-κB and AP-1 target genes under basal conditions, and are rapidly cleared in response to TLR4 activation and/or activation of protein kinase C. NCoR complexes were recruited to several AP1 target genes through interaction with c-Jun dimer. Here the NCoR corepressor complex imposes an active block of exchange of c-Jun for c-Jun/c-Fos heterodimers. The checkpoint function of NCoR is relieved by signal-dependent phosphorylation of c-jun by the inflammatory stimulus, which directs the removal of NCoR/HDAC3/TBL1/TBLR1 complexes through recruitment of the ubiquytilation complex Ubc5/19S.[32] NF-κB activated genes that are target of NCoR/SMRT/HDAC3/TBL1/TBLR1 complexes under basal conditions include *inos*. Activation of *inos* by TLR4 agonist LPS resulted in the clearance of the NCoR complex, dependent on the actions of TBL1, TBLR1 and Ubc5.[35] Moreover, IKKα has been proven to phosphorylate SMRT and to induce de-repression of NFκB target genes, stimulating the exchange of corepressor for coactivator complexes.[34] IKKα coordinates the simultaneous phosphorylation of REL-A/p65 and SMRT, displacing SMRT-HDAC3 repressor and allowing p300 to acetylate REL-A/p65.[36]

14.6 Counter-Regulation of the Inflammatory Response by Nuclear Receptors

Given the importance of preventing excessive inflammatory responses, and the need to resolve inflammation following eradication of an inciting stimulus, it is not surprising that inflammation is subject to counter-regulation at multiple levels. Here, we will focus on counter-regulation mediated by members of the nuclear receptor superfamily of ligand-dependent transcription factors, taking advantage of recent evidence indicating that much of the crosstalk between nuclear receptor and transcription factors that drive inflammatory responses occurs at the levels of transcriptional coregulators. Nuclear receptors play diverse roles in the regulation of development, homeostasis and immune responses by both positively and negatively regulating gene expression.[37–39] Recent profiling studies indicate that many nuclear receptors are expressed in immune cells and tissues. For example, 30 of the 48 known mammalian nuclear receptors are expressed in macrophages, and many of these receptors exhibit significant changes in expression in response to inflammatory stimuli.

The glucocorticoid receptor (GR) is prototypic of a subset of ligand-dependent nuclear receptors that integrate host immune responses with physiological circuits that are required for the maintenance of necessary organ functions. The ability of GR to repress transcriptional responses to inflammatory signals is an essential component of its homeostatic functions and a primary mechanism by which natural and synthetic GR agonists exert anti-inflammatory effects in a variety of disease settings.[40] Negative regulation of inflammatory responses is thought to result, at least in part, from the ability of GR to interfere by transrepression, with the activities of other signal-dependent transcription factors that include NF-κB and activator protein-1 (AP-1) family members. Numerous models have been proposed for GR-mediated transrepression, including direct interactions with NF-κB components reviewed in;[41] regulation of components of signal transduction pathways involved in NF-κB and AP-1 activation;[42,43] competition for essential coactivators;[44,45] alternative utilization of coactivators;[46,47] recruitment of corepressors;[48] and modifications of core transcription factors.[48,49]

Anti-inflammatory activities have also been documented *in vivo* and/or *in vitro* for several other members of the nuclear receptor family, including estrogen receptors (ERs),[50] vitamin D receptors (VDRs),[51] peroxisome proliferator-activated receptors (PPARs)[52,53] and LXRs.[54–56] PPARs and LXRs are regulated by fatty acid and cholesterol metabolites, respectively, and were initially characterized as nuclear receptors that play critical roles in lipid homeostasis.[57,58] Emerging evidence suggests that their ability to counter-regulate inflammatory responses plays important roles in both immunity and metabolic control.[52,55] For example, PPARγ is thought to improve insulin resistance by both positively and negatively regulating gene expression, with PPARγ agonists suppressing the expression of inflammatory genes in adipocytes and adipose tissue-associated macrophages that are correlated with impaired insulin signaling.[59,60]

Recent genome-wide studies have revealed that nuclear receptors antagonize inflammatory responses in a gene and signal-specific manner.[25] For example, glucocorticoid receptor agonists such as dexamethasone inhibit only about 50% of the genes that are strongly induced by TLR4 signaling in macrophages, indicating that the GR regulates selective aspects of the inflammatory response. Similarly, ligands for PPARγ and LXRs also inhibited overlapping, but distinct, subsets of LPS-responsive genes, suggesting receptor-specific effects on innate immune responses. These findings support the concept that nuclear receptors are able to utilize a number of different mechanisms to negatively regulate the expression of specific subsets of genes that drive inflammation and immunity.[61] While a complete discussion of the broad range of mechanisms utilized by nuclear receptors lies beyond the scope of this chapter, several of the major concepts are illustrated by recent studies of the GR, PPARγ and liver X receptors (LXRs).

14.6.1 *Glucocorticoid Receptor*

GR agonists provide effective anti-inflammatory therapy for many chronic inflammatory diseases, including rheumatoid arthritis, systemic lupus erythematosus, inflammatory bowel disease, psoriasis, eczema, asthma and transplant rejection.[62] In general, glucocorticoids (GCs) suppress maturation, differentiation and proliferation of immune cells.[63] GCs upregulate anti-inflammatory genes, such as *il10*, and at the same time downregulate proinflammatory gene expression, including *il12p40*, *il1β* and *tnfα*, MHC class II and co-stimulatory molecules in lymphocytes, macrophages, neutrophils, mast cells, eosinophils, endothelial cells, parenchymal cells of many organs and fibroblasts.[64] Clinically, the common side effects include increased susceptibility to infection, osteoporosis, insulin resistance increased appetite and weight gain. Thus, a better understanding of the molecular mechanisms used by GR to suppress inflammation may facilitate better drug designs in the future.

14.6.2 *Transcriptional activation of negative regulators*

One general mechanism by which GR has been established to negatively regulate inflammatory responses is to induce expression of IκBα, which binds and sequesters NF-κB, leading to the decreased production of pro-inflammatory cytokines (such as IL-1, IL-6 and TNFα).[65] This mechanism functions in a cell-type specific manner, enabling broad inhibition of NF-κB-dependent gene expression in cells exhibiting this response, but not in other cell types, such as endothelial cells[66] or macrophages.[25] In addition, GCs can induce glucocorticoid-induced leucine zipper protein (GILZ), which inhibits both NF-κB and AP-1 transcriptional activity.[67] Furthermore, GCs upregulate the expression of mitogen-activated protein kinase phosphatase 1 (MKP-1), which dephosphorylates and thus prevents activation of ERK and p38.[68] As p38 MAPK is also implicated in the stabilization of pro-inflammatory gene mRNAs (e.g. *cox2*, *tnfα*, *il6*, and *il8*), MKP-1 induction by GCs consequently results in mRNA destabilization of these transcripts.[68]

14.6.3 *Disruption of activator/coactivator complexes*

A major question is how GR can inhibit inflammatory programs of gene expression in a gene-specific manner. One mechanism appears to relate to the ability of GR to disrupt transcriptional activator/coactivator complexes that are required for the activation of subsets of NF-κB and IRF3 target genes. For example, studies of TLR3 and TLR4-dependent gene expression have demonstrated that NF-κB and IRF3 can utilize each other as context-dependent coactivators.[26] p65-IRF3 complexes are required for MyD88-dependent, but not TRIF-dependent, transcriptional activation of IRF target genes. An unexpected consequence of this interaction was suggested by a large scale gene expression study, in which a cohort of functionally related inflammatory-response genes that were sensitive to glucocorticoid receptor repression, when activated by TLR4 or TLR9 agonists (utilizing MyD88), were found to be resistant to glucocorticoid receptor repression when activated by the TLR3 agonist poly I:C (dependent on TRIF). Biochemical and molecular studies suggested that IRF3 target genes were sensitive to glucocorticoid-mediated repression in response to TLR4 or TLR9 stimulation because of the ability of the glucocorticoid receptor to bind to the REL-homology domain of REL-A/p65 and prevent its interaction with IRF3. Because p65 is not required for the transcriptional activity of IRF3 when cells are stimulated by TLR3 agonists, IRF3-dependent gene expression is resistant to glucocorticoid-receptor-mediated repression. Conversely, p65-IRF3 complexes are required for the activation of a subset of NF-κB target genes which are dependent on the specific sequence of κB elements.[26] By binding to p65 on target genes and preventing this interaction, GR inhibited the IRF3-dependent subset of NF-κB genes induced by TLR4 or TLR9 agonists in macrophages.[25] These findings have potential clinical implications. Steroid use is generally contraindicated in the setting of HSV infections due to their ability to exacerbate the severity of infection, consistent with glucocorticoids inactivating the MyD88-dependent pathway required for TLR9 signaling.

A second example of the ability of GR to repress inflammatory responses in a gene specific manner by preventing coregulator interactions is provided by studies of the *il8* and *IκBα* genes in A549 cells.[48] TNFα induction of *il8* is sensitive to GR-mediated repression, while the *IκBα* gene is not, due to the ability of GR to specifically block the recruitment of P-TEFb to the *il8* promoter. Since P-TEFb functions as a transcription elongation factor by mediating phosphorylation of serine

2 in the CTD of RNA Polymerase II, these studies provide the evidence for a transrepression mechanism occurring at a post-transcriptional initiation step.

A third example of the ability of GR to influence inflammatory responses is provided by studies of its interactions with GR interacting protein-1 (GRIP-1/TIF2). GRIP1 is a member of the p160 family and functions as a nuclear receptor coactivator for a subset of the NR family.[69] Intriguingly, GRIP1 can also function as a GR corepressor when it is tethered to AP-1 sites dependent on a GRIP1-specific corepression domain (RD). The ability of GRIP1 to interact with both GR and IRF3 suggests that it is a shared subunit in their respective transcription complexes. It has been proposed that activated GR may sequester GRIP1 away from IRF3, thereby antagonizing IRF3-dependent expression induction.[70] Overexpression of GRIP1 alleviated GC inhibition of IRF3-dependent gene expression, whereas GR-mediated repression of AP1 and NF-κB targets was relieved by the activation of IRF3. Local "redistribution" of GRIP1 model suggests that GR can displace GRIP1 from IRF3 activation complex and recruit it as a corepressor to an AP1 or NF-κB tethering GRE on the same promoter.

14.6.4 *Negative regulation of inflammatory responses at the level of corepressors*

Emerging evidence suggests that regulation of corepressor complexes represents an additional level of control of the inflammatory responses by nuclear receptors. In the case of GR, a role for the recruitment of histone deacetylases has been suggested. Upon ligand binding, GR is acetylated. While GR acetylation is important for the regulation of the *SLPI* gene, acetylated GR must be deacetylated by HDAC2 before GR is able to downregulate inflammatory NF-κB target gene expression. In addition, activated GR recruits HDAC2 to the activated transcriptional complex, resulting in deacetylation of hyperacetylated histones, and thus a decrease in inflammatory gene expression.[71] Interestingly, patients with chronic obstructive pulmonary disease (COPD) with reduced HDAC2 expression become resistant to the anti-inflammatory actions of GCs.

Recently, studies of PPARγ and the liver X receptors (LXRs) have led to the identification of transrepression pathways that operate by preventing the signal-dependent clearance of corepressor complexes from inflammatory gene promoters as a prerequisite for full gene activation. PPARγ is the molecular target of the thiazolidinedione class of drugs that act as insulin sensitizers and are used in the treatment of type 2 diabetes. PPARγ ligands have also been shown to inhibit the development of atherosclerosis in animal models and to reduce the expression of inflammatory mediators within atherosclerotic lesions.[72] Previous studies have shown that PPARs can inhibit inflammatory gene expression by several mechanisms, including forming complexes with AP1 and NF-κB family members.[73] However, recent studies indicating that only a subset of NF-κB target genes are subject to repression by PPAR agonists in macrophages are inconsistent with mechanisms resulting in global inhibition of NF-κB or AP-1 activity.[25]

A requirement of NCoR for anti-inflammatory effects of PPARγ in macrophages was initially suggested by experiments in which the knockdown of NCoR expression with specific siRNAs resulted in loss of transrepression. Chromatin-immunoprecipitation experiments demonstrated that the treatment of cells with PPARγ agonists prevented the clearance of NCoR corepressor complexes from iNOS target gene in response to an inflammatory stimulus. This effect was found to require ligand-dependent SUMOylation of PPARγ with SUMO1, using PIAS1 as the obligatory SUMO E3 ligase. This modification in turn targeted PPARγ to NCoR corepressor complexes bound to the inflammatory promoters. The recruitment of SUMOylated PPARγ was found to prevent the signal-dependent ubiquitylation and the clearance of the NCoR complex required for full gene activation.[35]

A parallel but distinct transrepression pathway has recently been defined for the LXRs. LXRα and LXRβ are members of the nuclear receptor superfamily that play essential roles in cholesterol and fatty acid homeostasis.[74] Recent studies have characterized LXRs as regulators of innate immunity and inflammation.[55,75] Mice lacking LXRs show an exaggerated response to LPS, and synthetic LXR ligands inhibit the macrophage response to bacterial pathogens and antagonize the induction of a number of pro-inflammatory genes.[54] As in the case of PPARγ-dependent transrepression, LXR transrepression in macrophages and hepatocytes has also been shown to be NCoR-dependent.[76] NCoR corepressor complexes appear to be present on all LPS-responsive genes differentially repressed by these nuclear receptors, and the ability of PPARγ and LXR ligands to transrepress them is correlated with their ability to prevent NCoR clearance. Interestingly, recent studies provide evidence for distinct and parallel pathways that are used in a receptor-specific manner by LXRs and PPARγ based on their conjugation to different SUMO proteins: SUMO1 for PPARγ and SUMO2/3 for LXRs.[77] These studies suggest that SUMOylated PPARγ and LXRs target distinct components of the mechanisms that direct signal-specific clearance of NCoR complexes and are consistent with the finding that LXRs and PPARγ repress overlapping but distinct subset of inflammatory response genes.

14.7 Integrative Functions of Coregulators

SRC3 was initially characterized as a coactivator of nuclear receptors and other signal-dependent transcription factors, including NF-κB.[78] Unexpectedly, it has recently been shown that SRC3$^{-/-}$ mice are markedly susceptible to LPS-induced endotoxin shock due to the increased production of pro-inflammatory cytokines like TNF-alpha, IL-1β and IL-6. Lack of SRC3 significantly increased the proportion of cytokines mRNA that accumulated on polysomes, indicating that in addition to functioning as a coactivator, SRC3 also functions to inhibit cytokine translation, in this case providing a feedback mechanism limiting excessive cytokine production. Intriguingly, while the loss of SRC3 function as a coactivator was compensated for by other proteins, the loss of its function as a translational inhibitor was not.[79]

14.8 Conclusions

Transcriptional coregulators have emerged as required factors for essentially all Pol II-dependent transcription based on their roles in regulating chromatin remodeling, nucleosome architecture and the recruitment of core transcription factors. It is now apparent that these factors are themselves critical targets of signal transduction pathways that control the transcription of genes that underlie reproduction, development, homeostasis and immunity. As illustrated in this chapter, coactivators and corepressors work in a coordinated manner to ensure that genes involved in inflammation and immunity are repressed in the absence of injury or infection, but can be rapidly and highly activated in their presence. Information is now also emerging that coregulators may integrate distinct cellular functions, as provided by the example of SRC3 as a transcriptional coactivator and translational suppressor. Coregulators and the molecular mechanisms that control their interactions with sequence-specific transcription factors thus represent new targets for therapeutic intervention in a variety of disease settings. Drugs that directly or indirectly target coregulators will have complex effects on cellular, physiological and pathological processes due to their influence on multiple sets of genes. Based on our limited knowledge of the full range of biological activities of any coregulator, it is not possible at present to predict whether the ability to stimulate or inhibit coregulator function will translate into a desirable therapeutic profile for a particular disease state. This is clearly an avenue of investigation in which much remains to be learned, and from which much may be gained.

References

1. Akira S, Takeda K, Toll-like receptor signaling, *Nat Rev* **4**:499–511, 2004.
2. McGreal EP, Miller JL, Gordon S, Ligand recognition by antigen presenting cell C-type lectin receptors, *Curr Opin Immunol* **17**:18–24, 2005.
3. Brown GD, Dectin-1: A signalling non-TLR pattern-recognition receptor, *Nat Rev* **6**:33–43, 2006.
4. Creagh EM, O'Neill LA, TLRs, NLRs and RLRs: A trinity of pathogen sensors that co-operate in innate immunity, *Trends Immunol* **27**:352–357, 2006.
5. Karin M, Lawrence T, Nizet V, Innate immunity gone awry: Linking microbial infections to chronic inflammation and cancer, *Cell* **124**:823–835, 2006.
6. Charo IF, Ransohoff RM, The many roles of chemokines and chemokine receptors in inflammation, *New Engl J Med* **354**:610–621, 2006.
7. Li Q, Verma IM, NF-$\kappa\beta$ regulation in the immune system, *Nat Rev* **2**:725–734, 2002.
8. Bonizzi G, Karin M, The two NF-kappaB activation pathways and their role in innate and adaptive immunity, *Trends Immunol* **25**:280–288, 2004.
9. Ghosh S, Karin M, Missing pieces in the NF-kappaB puzzle, *Cell* **109** Suppl:S81–96, 2002.
10. Zhong H, May MJ, Jimi E, *et al.*, The phosphorylation status of nuclear NF-κβ determines its association with CBP/p300 or HDAC-1, *Mol Cell* **9**:625–636, 2002.
11. Wu RC, Qin J, Hashimoto Y, *et al.*, Regulation of SRC-3 (pCIP/ACTR/AIB-1/RAC-3/TRAM-1) coactivator activity by I κβ kinase, *Mol Cell Biol* **22**: 3549–3561, 2002.
12. Na SY, Lee SK, Han SJ, *et al.*, Steroid receptor coactivator-1 interacts with the p50 subunit and coactivates nuclear factor κβ-mediated transactivations, *J Biol Chem* **273**:10831–10834, 1998.

13. Miao F, Li S, Chavez V, *et al.*, Coactivatorassociated arginine methyltransferase-1 enhances nuclear factor-κβ mediated gene transcription through methylation of histone H3 at arginine 17, *Mol Endocrinol (Baltimore, Md)* **20**:1562–1573, 2006.
14. Dechend R, Hirano F, Lehmann K, *et al.*, The Bcl-3 oncoprotein acts as a bridging factor between NF-kappaB/Rel and nuclear co-regulators, *Oncogene* **18**:3316–3323, 1999.
15. Baek SH, Ohgi KA, Rose DW, *et al.*, Exchange of N-CoR corepressor and Tip60 coactivator complexes links gene expression by NF-kappaB and beta-amyloid precursor protein, *Cell* **110**:55–67, 2002.
16. Perkins ND, Post-translational modifications regulating the activity and function of the nuclear factor κβ pathway, *Oncogene* **25**:6717–6730, 2006.
17. Honda K, Taniguchi T, IRFs: Master regulators of signalling by Toll-like receptors and cytosolic pattern-recognition receptors, *Nat Rev* **6**:644–658, 2006.
18. Suhara W, Yoneyama M, Iwamura T, *et al.*, Analyses of virus-induced homomeric and heteromeric protein associations between IRF-3 and coactivator CBP/p300, *J Biochem (Tokyo)* **128**:301–307, 2000.
19. Kamijo R, Harada H, Matsuyama T, *et al.*, Requirement for transcription factor IRF-1 in NO synthase induction in macrophages, *Science* **263**:1612–1615, 1994.
20. Sharma S, tenOever BR, Grandvaux N, *et al.*, Triggering the interferon antiviral response through an IKK-related pathway, *Science* **300**:1148–1151, 2003.
21. Hemmi H, Takeuchi O, Sato S, *et al.*, The roles of two IkappaB kinase related kinases in lipopolysaccharide and double stranded RNA signaling and viral infection, *J Exp Med* **199**:1641–1650, 2004.
22. Honda K, Ohba Y, Yanai H, *et al.*, Spatiotemporal regulation of MyD88-IRF-7 signalling for robust type-I interferon induction, *Nature* **434**:1035–1040, 2005.
23. Takaoka A, Yanai H, Kondo S, *et al.*, Integral role of IRF-5 in the gene induction programme activated by Toll-like receptors, *Nature* **434**:243–249, 2005.
24. Wietek C, Miggin SM, Jefferies CA, *et al.*, Interferon regulatory factor-3-mediated activation of the interferon-sensitive response element by Toll-like receptor (TLR) 4 but not TLR3 requires the p65 subunit of NF-kappa, *J Biol Chem* **278**:50923–50931, 2003.
25. Ogawa S, Lozach J, Benner C, *et al.*, Molecular determinants of crosstalk between nuclear receptors and toll-like receptors, *Cell* **122**:707–721, 2005.
26. Leung TH, Hoffmann A, Baltimore D, One nucleotide in a kappaB site can determine cofactor specificity for NF-kappaB dimmers, *Cell* **118**:453–464, 2004.

27. Jonas BA, Privalsky ML, SMRT and N-CoR corepressors are regulated by distinct kinase signaling pathways, *J Biol Chem* **279**:54676–54686, 2004.
28. Zhang J, Kalkum M, Chait BT, *et al.*, The N-CoR-HDAC3 nuclear receptor corepressor complex inhibits the JNK pathway through the integral subunit GPS2, *Mol cell* **9**:611–623, 2002.
29. Heinzel T, Lavinsky RM, Mullen TM, *et al.*, A complex containing NCoR, mSin3 and histone deacetylase mediates transcriptional repression, *Nature* **387**:43–48, 1997.
30. Perissi V, Aggarwal A, Glass CK, *et al.*, A corepressor/coactivator exchange complex required for transcriptional activation by nuclear receptors and other regulated transcription factors, *Cell* **116**:511–526, 2004.
31. Frasor J, Danes JM, Funk CC, *et al.*, Estrogen down-regulation of the corepressor N-CoR: Mechanism and implications for estrogen derepression of N-CoR-regulated genes, *Proc Nat Acad Sci USA* **102**:13153–13157, 2005.
32. Ogawa S, Lozach J, Jepsen K, *et al.*, A nuclear receptor corepressor transcriptional checkpoint controlling activator protein 1-dependent gene networks required for macrophage activation, *Proc Nat Acad Sci USA* **101**:14461–14466, 2004.
33. Libby P, Inflammation in atherosclerosis, *Nature* **420**:868–874, 2002.
34. Hoberg JE, Yeung F, Mayo MW, SMRT derepression by the IκB kinase α; A prerequisite to NF-κB transcription and survival, *Mol Cell* **16**:245–255, 2004.
35. Pascual G, Fong AL, Ogawa S, *et al.*, A SUMOylation dependent pathway mediates transrepression of inflammatory response genes by PPAR-gamma, *Nature* **437**:759–763, 2005.
36. Hoberg JE, Popko AE, Ramsey CS, *et al.*, $\kappa\beta$kinase alpha-mediated derepression of SMRT potentiates acetylation of RelA/p65 by p300, *Mol Cell Biol* **26**:457–471, 2006.
37. Evans RM, The steroid and thyroid hormone receptor superfamily, *Science* **240**:889–895, 1988.
38. Mangelsdorf DJ, Evans RM, The RXR heterodimers and orphan receptors, *Cell* **83**:841–850, 1995.
39. Chawla A, Boisvert WA, Lee CH, *et al.*, A PPAR γ-LXR-ABCA1 pathway in macrophages is involved in cholesterol efflux and atherogenesis, *Mol Cell* **7**:161–171, 2001.
40. Reichardt HM, Tuckermann JP, Gottlicher M, *et al.*, Repression of inflammatory responses in the absence of DNA binding by the glucocorticoid receptor, *EMBO J* **20**:7168–7173, 2001.
41. De Bosscher K, Vanden Berghe W, Haegeman G, The interplay between the glucocorticoid receptor and nuclear factor-κB or activator protein-1: Molecular mechanisms for gene repression, *Endocr Rev* **24**:488–522, 2003.

42. Auphan N, DiDonato JA, Rosette C, *et al.*, Immunosuppression by glucocorticoids: Inhibition of NF-κB activity through induction of IκB synthesis, *Science* **270**:286–290, 1995.
43. Caelles C, Gonzales-Sancho JM, Munoz A, Nuclear hormone receptor antagonism with AP-1 by inhibition of the JNK pathway, *Genes & Dev* **11**:3351–3364, 1997.
44. Kamei Y, Xu L, Heinzel T, *et al.*, A CBP integrator complex mediates transcriptional activation and AP-1 inhibition by nuclear receptors, *Cell* **85**:403–414, 1996.
45. Sheppard KA, Phelps KM, Williams AJ, *et al.*, Nuclear integration of glucocorticoid receptor and nuclear factor-κB signaling by CREB-binding protein and steroid receptor coactivator-1, *J Biol Chem* **273**:29291–25294, 1998.
46. Kassel O, Schneider S, Heilbock C, *et al.*, A nuclear isoform of the focal adhesion LIM-domain protein Trip6 integrates activating and repressing signals at AP-1- and NF-kappaB-regulated promoters, *Genes Dev* **18**:2518–2528, 2004.
47. Rogatsky I, Zarember KA, Yamamoto KR, Factor recruitment and TIF2/GRIP1 corepressor activity at a collagenase-3 response element that mediates regulation by phorbol esters and hormones, *EMBO J* **20**: 6071–6083, 2001.
48. Nissen RM, Yamamoto KR, The glucocorticoid receptor inhibits NF-κB by interfering with serine-2 phosphorylation of the RNA polymerase II carboxy-terminal domain, *Genes Dev.* **14**:2314–2329, 2000.
49. De Bosscher K, Vanden Berghe W, Vermeulen L, *et al.*, Glucocorticoids repress NF-kappaB-driven genes by disturbing the interaction of p65 with the basal transcription machinery, irrespective of coactivator levels in the cell, *Proc Natl Acad Sci USA* **97**:3919–3924, 2000.
50. McKay LI, Cidlowski JA, Molecular control of immune/inflammatory responses: Interactions between nuclear factor-kappa B and steroid receptor-signaling pathways, *Endocr Rev* **20**:435–459, 1999.
51. Nagpal S, Lu J, Boehm MF, Vitamin D analogs: Mechanism of action and therapeutic applications, *Curr Med Chem* **8**:1661–1679, 2001.
52. Ricote M, Li AC, Willson TM, *et al.*, The peroxisome proliferator-activated receptor-gamma is a negative regulator of macrophage activation, *Nature* **391**:79–82, 1998.
53. Kersten S, Desvergne B, Wahli W, Roles of PPARs in health and disease, *Nature* **405**:421–424, 2000.
54. Castrillo A, Joseph SB, Vaidya SA, *et al.*, Crosstalk between LXR and toll-like receptor signaling mediates bacterial and viral antagonism of cholesterol metabolism, *Mol Cell* **12**:805–816, 2003.
55. Joseph SB, Castrillo A, Laffitte BA, *et al.*, Reciprocal regulation of inflammation and lipid metabolism by liver X receptors, *Nat Med* **9**:213–219, 2003.

56. Joseph SB, Bradley MN, Castrillo A, *et al.*, LXR-dependent gene expression is important for macrophage survival and the innate immune response, *Cell* **119**:299–309, 2004.
57. Forman BM, Tontonoz P, Chen J, *et al.*, 15-Deoxy-D 12,14-prostaglandin J2 is a ligand for the adipocyte determination factor PPARg, *Cell* **83**:803–812, 1995.
58. Janowski BA, Willy PJ, Devi TR, *et al.*, An oxysterol signalling pathway mediated by the nuclear receptor LXRa, *Nature* **383**:728–731, 1996.
59. Weisberg SP, McCann D, Desai M, *et al.*, Obesity is associated with macrophage accumulation in adipose tissue, *J Clin Invest* **112**:1796–1808, 2003.
60. Xu H, Barnes GT, Yang Q, *et al.*, Chronic inflammation in fat plays a crucial role in the development of obesity-related insulin resistance, *J Clin Invest* **112**:1821–1830, 2003.
61. Glass CK, Ogawam S, Combinatorial roles of nuclear receptors in inflammation and immunity, *Nat Rev* **6**:44–55, 2006.
62. Schimmer B, Parker K, eds. *Goodman and Gilman's Pharmacological Basis of Therapeutics*. McGraw Hill, New York. 63. Barnes, P. J., 2006. How corticosteroids control inflammation: Quintiles Prize Lecture 2005, *Brit J Pharmacol* **148**:245–254, 1996.
64. Matyszak MK, Citterio S, Rescigno M, *et al.*, Differential effects of corticosteroids during different stages of dendritic cell maturation, *Euro J Immunol* **30**:1233–1242, 2000.
65. Scheinman RI, Gualberto A, Jewell CM, *et al.*, Characterization of mechanisms involved in transrepression of N-Fkappa B by activated glucocorticoid receptors, *Mol Cell Biol* **15**:943–953, 1995.
66. Brostjan C, Anrather J, Csizmadia V, *et al.*, Glucocorticoid-mediated repression of NFkappaB activity in endothelial cells does not involve induction of IκBα synthesis, *J Biol Chem* **271**:19612–19616, 1996.
67. Mittelstadt PR, Ashwell JD, Inhibition of AP-1 by the glucocorticoid-inducible protein GILZ, *J Biol Chem* **276**:29603–29610, 2001.
68. Imasato A, Desbois-Mouthon C, Han J, *et al.*, Inhibition of p38 MAPK by glucocorticoids via induction of MAPK phosphatase-1 enhances nontypeable Haemophilus influenzae-induced expression of toll-like receptor 2, *J Biol Chem* **277**:47444–47450, 2002.
69. Rosenfeld MG, Lunyak VV, Glass CK, Sensors and signals: A coactivator/corepressor/epigenetic code for integrating signal-dependent programs of transcriptional response, *Genes & Dev* **20**:1405–1428, 2006.
70. Reily MM, Pantoja C, Hu X, *et al.*, The GRIP1:IRF3 interaction as a target for glucocorticoid receptor-mediated immunosuppression, *EMBO J* **25**:108–117, 2006.

71. Ito K, Barnes PJ, Adcock IM, Glucocorticoid receptor recruitment of histone deacetylase 2 inhibits interleukin-1beta-induced histone H4 acetylation on lysines 8 and 12, *Mol Cell Biol* **20**:6891–6903, 2000.
72. Li AC, Glass CK, PPAR- and LXR-dependent pathways controlling lipid metabolism and the development of atherosclerosis, *J Lipid Res* **45**:2161–2173, 2004.
73. Kelly D, Campbell JI, King TP, *et al.*, Commensal anaerobic gut bacteria attenuate inflammation by regulating nuclear-cytoplasmic shuttling of PPAR-gamma and RelA, *Nat Immunol* **5**:104–112, 2004.
74. Zelcer N, Tontonoz P, Liver X receptors as integrators of metabolic and inflammatory signaling, *J Clin Invest* **116**:607–614, 2006.
75. Valledor AF, Hsu LC, Ogawa S, *et al.*, Activation of liver X receptors and retinoid X receptors prevents bacterial-induced macrophage apoptosis, *Proc Nat Acad Sci USA* **101**:17813–17818, 2004.
76. Blaschke F, Takata Y, Caglayan E, *et al.*, A nuclear receptor corepressor-dependent pathway mediates suppression of cytokine-induced C-reactive protein gene expression by liver X receptor, *Cir Res* **99**:e88–99, 2006.
77. Ghisletti S, Huang W, Ogawa S, *et al.*, Parallel SUMOylation-dependent pathways mediate gene- and signal-specific transrepression by LXRs and PPARgamma, *Mol Cell* **25**:57–70, 2007.
78. Chen H, Lin RJ, Schiltz RL, *et al.*, Nuclear receptor coactivator ACTR is a novel histone acetyltransferase and forms a multimeric activation complex with P/CAF and CBP/p300, *Cell* **90**:569–580, 1997.
79. Yu C, York B, Wang S, *et al.*, An essential function of the SRC-3 coactivator in suppression of cytokine mRNA translation and inflammatory response, *Mol Cell* **25**:765–778, 2007.

Chapter 15

Nuclear Receptor Coactivators in the Cardiovascular System

Jianming Xu

In the recent years, breakthroughs made in the field of nuclear hormone receptor and gene transcription have identified many nuclear hormone receptor coactivators. Through interaction with nuclear receptors and modulation of nuclear receptor-mediated gene expression, these coactivators participate in numerous developmental, physiological and pathological processes. In the cardiovascular system, experiments using genetically manipulated mouse models have demonstrated the essential roles of several well established coactivators in the development and function of the cardiovascular system. CBP (CREB-binding protein), p300, PBP (PPAR binding protein) and AIB3 (amplified-in-breast cancer 3) are involved in heart and vascular development. PGC-1α (PPARγcoactivator-1α) is a key regulator of cardiac mitochondrial biogenesis, fatty acid oxidation and ATP production. SRC-1 (steroid receptor coactivator-1) and SRC-3 are required for efficient vesoprotection by estrogen after vascular injury. This chapter reviews the information of these coactivators in the cardiovascular system and discusses possible linkages of these coactivators to cardiovascular diseases.

15.1 Introduction

Nuclear hormone receptors are ligand-inducible transcription factors involved in regulation of numerous developmental and physiological processes. Although the basic concept and classical models regarding the mechanisms of hormonal actions were established many years ago, the breakthroughs made in recent years have suggested that most nuclear hormone receptors do not directly activate or repress the activity of general transcription factors. Instead, they interact with coregulators, namely coactivators and corepressors, to activate or inhibit the transcription of their target genes.[1,2] To date, more than 200 nuclear receptor coregulators have been documented.[1] Most of the coactivators are recruited to the promoter through interaction with nuclear receptors in a hormone binding-dependent manner. These coactivators either serve as adaptor proteins to recruit other components of the coactivator complexes or bear various kinds of enzyme activities required for chromatin remodeling, transcriptional machinery assembly and coactivator complex turnover. Because the cellular concentrations of these coactivators are usually limiting and these coactivators can serve as a platform to regulate gene expression by integrating signals from many signaling transduction pathways, changes in these coactivator levels and/or activities can lead to a significant alteration of gene expression.[2–4] Therefore, emerging evidences from recent studies suggest that coactivators are key regulators of hormone-regulated gene expression.

Through modulating nuclear receptor-mediated gene expression, nuclear receptor coactivators participate in extremely diverse biological processes and their dysfunctions are relevant to various human diseases (refer to other chapters). This chapter will review the information relevant to the important functions of well established nuclear receptor coactivators involved in heart and vascular development, heart energy metabolism and estrogen-mediated vasoprotection.

15.2 Coactivator in Heart and Vascular Development

15.2.1 *CBP and p300*

The CREB-binding protein (CBP) and its closely related family member p300 are common transcriptional coactivators and cointegrators for many transcription factors including nuclear hormone receptors.[2,5] CBP and p300 interact with these transcription factors and function as histone acetyltransferases.[5–7] Therefore, CBP and p300 are critical coactivators that link many transcription factors to chromatin remodeling, general transcription factor assembly and gene expression.

The generation and characterization of CBP mutant mouse models have revealed an important role of CBP in vascular development.

Mutant mice homozygous for a truncated form of CBP exhibit embryonic lethality between E9.5–E10.5. These mutant embryos show defects in primitive hematopoiesis and lack vascular network formation in the yolk sac. When assayed in a culture system, these defects can be partially rescued by adding VEGF.[8] In another independently generated mutant mouse line, the inactivation of CBP function also results in embryonic death between E10.5–E12.5 due to massive hemorrhage caused by defective blood vessel formation in the central nervous system.[9] The number of endothelial cells in CBP null embryos is lower than that in the wild type embryos. The blood vessel structure in neural tissue is also abnormal.[9] These studies demonstrate that CBP plays an essential role in vasculo-angiogenesis.

The mutant mouse models for p300 also demonstrate a crucial role for p300 in heart and vascular development. The p300 null mice exhibit abnormal heart and vascular development and die between E9 and E11.5.[10] At E10.5, p300 null embryos display severe pericardial effusion, significant reduction in ventricle trabeculation and enlargement of heart cavities. Their heart contractions are weaker and less extensive than the heart contractions in wild type embryos. The expression levels of cardiac muscle structural proteins, myosin heavy chain and α-actinin, are much lower in the p300-deficient mouse hearts than in wild type mouse hearts. Furthermore, the yolk sac of p300-deficient embryos is often poorly vascularized.[10]

The mouse model with specific site mutation that abolishes the acetyltransferase activity of p300 further demonstrates that its histone acetyltransferase activity is required for normal heart development.[11] The p300 acetyltransferase-mutant allele behaves in a dominant-negative manner. When it is not expressed, the heterozygous mice are viable. When it is expressed, the heterozygous mice die between E12.5 and the birth time. At E11, the reduction in myocardium thickness of the p300-acetyltransferase mutants becomes apparent and the formation of epicardium for most mutant hearts is delayed. At E12.5, the subepicardial mesenchymal cells have entered the myocardium in wild type but not in mutant hearts. By E13.5, the development of the coronary vessel network in the mutant hearts is delayed or reduced. Between E14.5 and E15.5, ventricular septum closure defects and the underdeveloped or malformed valve leaflets are present. In some of the mutant hearts, blood leakage through ruptures in the ventricular wall is also apparent.[11] These observations indicate that p300 acetyltransferase activity is required for the development of most heart structures and the coronary vasculature.

On the other hand, CBP and p300 may play a role in diseased heart to promote cardiac hypertrophy. It has been shown that CBP or p300 overexpression can induce hypertrophy,[12] and p300 may acetylate and coactivate GATA-4 that mediates changes in gene expression during myocardial hypertrophy and promotes myocyte hypertrophy and decompensated heart failure.[13] Transgenic mice overexpressing p300 in the heart exhibit progressive left ventricular dilation and reduction in systolic function after myocardial infarction, whereas transgenic mice overexpressing p300-acetyltransferase mutant do not show the same phenotype.[14] These results from transgenic mice demonstrate that the acetyltransferase activity of excess p300 may promote left ventricular remodeling after myocardial infarction.

15.2.2 *PBP/TRAP220/DRIP205*

The peroxisome proliferator-activated receptor (PPAR)-binding protein (PBP), also known as TRAP220 or DRIP205, is a key component of the thyroid hormone receptor-associated protein (TRAP) coactivator complex.[15–17] PBP interacts with nuclear hormone receptors, including thyroid hormone receptor (TR), PPAR and vitamin D receptor, in a ligand-binding dependent manner. Through this interaction, these receptors recruit the TRAP coactivator complex to the promoter to potentiate target gene transcription. Analysis of PBP knockout mouse models has revealed an important role of PBP in heart and placental vascular development.[18,19] At embryonic day 9.5 (E9.5) and thereafter, TRAP220 (PBP)-null embryos show severe heart failure manifested by prominent cardiac and large vessel enlargement. The ventricular trabeculae of the TRAP220-null hearts are hypoplastic. At E10.5, the TRAP220-null embryos exhibit both poor trabeculation and thin and dilated ventricular walls. Probably due to the severe heart failure, the heart and the vena cava of TRAP220-null embryos become dilated. Because of the circulation system failure, all TRAP220-null mouse embryos are dead by E12.5.[19] In normal mouse embryos, ventricular trabeculation is first formed at E9.5 and is the major contractile source of the embryonic hearts throughout E13–E14 before the compact layer of the heart wall becomes the major systolic force. Therefore, the poor trabeculation of Trap220-null hearts at E9.5 may cause heart failure with poor ventricular efficiency, resulting in congestive heart failure with thin and stretched heart walls accompanied by dilated major vessels by E10.5.[19]

In another independently generated knockout mouse model, inactivation of PBP also causes defective development of the placental vascular networks.[18] At E11.5, the labyrinth zone in the PBP mutant placenta is poorly developed, with a less distinct network of embryonic capillary blood vessels compared with the labyrinth of wild type embryos. In addition, the three-cell-layer barrier between the fetal and maternal blood in the PBP-null placenta is poorly differentiated. This usually causes poor supply of oxygen and nutrition to the fetus and excretion of carbon dioxide and other metabolic wastes from the fetus.[18] Thus, the defects in placental vascular development may also contribute largely to the lethal phenotype of the PBP-null mouse embryos.

The above findings clearly indicate that PBP is an essential coactivator in cardiac and placental vascular development. Future characterization of the transcription factors that work with PBP and identification of their target genes in these developmental events will provide more insight into understanding the molecular mechanisms by which PBP regulates heart and vascular development.

15.2.3 *AIB3/NCoA-6*

AIB3 (amplified-in-breast cancer 3), also known as ASC-2, RAP250, PRIP, TRBP, NRC and NCoA-6,[20–25] is initially identified from an amplified chromosomal region (20q11–12) in human breast tumor cells.[20] It is subsequently characterized as a strong transcriptional coactivator for many nuclear hormone receptors including RXRα, PPARα, PPARγ, TRβ and TRα.[26] In addition to nuclear receptors, AIB3 is also a coactivator for certain other transcription factors involved in the regulation of cell survival, proliferation, differentiation, and immune responses such as AP-1, serum response factor (SRF), CREB, C/EBPα and NFκB.[24,27,28] These studies suggest that AIB3 may play important roles in multiple signaling pathways regulated by hormones, growth factors and cytokines.

The generation and characterization of the AIB3-deficient mouse models have uncovered the important function of AIB3 in cardiovascular development. Several independent studies have reported that the inactivation of AIB3 caused embryonic lethality during E9.5–E13.5 due to the defective development of the heart and placenta.[26,29–31] Although all the cardiac cell lineages and basic structures are developed in AIB3-null embryos at E10.5, the atrial and ventricular walls are much thinner than that observed in normal hearts. The trabeculae are not only thinner, but also deformed. In addition, a significant portion of endocardium is detached from the myocardium layer, suggesting that the requisite intimate interaction between the endocardium and the myocardium is affected. Further analysis show that the hypoplastic development of AIB3-deficient mouse heart is correlated with a decreased myocardial proliferation. In AIB3-deficient placenta, the fetal blood vessels do not efficiently invade into the labyrinthine trophoblast cell layer and the fetal blood vessel networks in the labyrinth are poorly developed. Although a few vessels are able to grow into the labyrinthine layer in AIB3-deficient placenta, the invaginating chorionic villi are surrounded by excessively thickened trophoblast cell layers, destroying the characteristic three-cell-layered epithelial barrier required for nutrition, oxygen and metabolic waste exchange. Furthermore, the maternal blood sinuses in AIB3-null placenta are often dilated or ruptured and the trophoblasts lining the wall of maternal blood sinuses are loosely attached or even detached.[29] Finally, the blood vessels in the region bordering the spongiotrophoblast and labyrinthine layers are collapsed in Rap250 (AIB3) knockout placentas. The region in close proximity to the collapsed vessels exhibits extensive tissue necrosis, suggesting that the vascular defect causes placental ischemia that may also contribute to the embryonic lethality.[31] These studies demonstrate that AIB3 is an essential coactivator for heart and placental vascular development.

Noticeably, the cardiac and placental phenotype observed in AIB3 knockout embryos is similar to that encountered in PPARγ and PBP knockout mice, suggesting that these genes may function in a linear pathway. Indeed, the transcriptional activity of PPARγ is impaired in cells lacking either AIB3 or PBP.[18,29–31] Therefore, some, if not all, of the developmental defects of AIB3 knockout mice may be a consequence of PPARγ dysfunction.

Interestingly, the hypoplastic morphology of PPARγ-deficient mice can be rescued following the correction of placental developmental problems by using the tetraploid chimera methodology, suggesting that the heart phenotype may be secondary to the placental defect in PPARγ-null mice.[32] Similarly, it is also possible that the heart developmental abnormalities observed in AIB3-null embryos are secondary to the defective development of the placental vascular networks.[29] Intriguingly, the hypoplastic heart development is associated with developmental defects of fetal vessels in the placenta in many mutant mouse models including PPARγ, PBP and AIB3 knockout mice. As we have previously proposed, the reproducibility of this association may indicate a common structural and functional relationship between heart and placental development. From a hemodynamic standpoint, the placenta is a low resistance vascular network. The lack or reduction of this low resistance vasculature system may cause an increase in afterload for the developing heart and, therefore, contribute to the disintegration of myocardial architecture through a mechanical effect.[29] Future studies will be needed to address whether the placental defect is the initial cause of embryonic lethality of AIB3-null mice and how AIB3 is involved in the development of the placental vascular system.

15.2.4 *PGC-1α as a regulator of heart energy metabolism*

The PPARγ coactivator 1α (PGC-1α) interacts with several nuclear receptors including PPARα, PPARγ, RARα, ER, GR and certain other transcription factors including NRF-1, NRF-2 and MEF2.[33] The interaction of PGC-1α with PPARγ causes further recruitments of other coactivators including SRC-1 and CBP/p300 to form coactivator complexes for chromatin remodeling, general transcription factor assembly and transcriptional activation.[33,34] PGC-1α expression is strongly induced by the β-adrenergic receptor/cAMP pathway in adaptive thermogenic tissues such as brown adipose tissue (BAT) and skeletal muscle in response to cold conditions.[33] PGC-1α serves as a coactivator for PPARs, TR, RAR and NRFs in BAT and plays an important role in upregulation of uncoupling protein-1 (UCP-1) and mitochondrial biogenesis.[33] Therefore, PGC-1α enhances uncoupled respiration in the adaptive thermogenic tissues to generate heat from mitochondrial fatty acid oxidation in response to environmental conditions.

In the heart, PGC-1α expression is induced after birth as the heart starts to use mitochondrial fatty acid oxidation as a major energy source.[35] In contrast to BAT where PGC-1α drives uncoupled respiration, PGC-1α promotes coupled respiration in the cardiac myocytes.[35] Cardiac PGC-1α coactivates PPARα, ERRα, NRF and MEF2 to regulate genes involved in mitochondrial biogenesis, fatty acid oxidation, mitochondrial respiration and ATP production.[35–37] Therefore, PGC-1α plays an important role in the regulation of heart energy homeostasis.

Two independent PGC-1α-deficient mouse lines have been generated and characterized.[38,39] The first line of PGC-1α null mice exhibits reduced expression of genes involved in oxidative phosphorylation in skeletal and cardiac muscles and reduced mitochondrial enzymatic activities and ATP levels in the heart. The isolated PGC-1α deficient hearts also have a diminished increase in work output in response to chemical and electrical stimuli. Cardiac dysfunction also appears in aged PGC-1α null mice.[38] The other line of PGC-1α null mice exhibits multisystemic abnormalities indicative of an abnormal energy metabolic phenotype. In this line of mutant mice, the postnatal heart growth is blunted and the cardiac function is moderately reduced due to abnormal control of heart rate.[39] These findings indicate that PGC-1α is required for normal cardiac energy metabolism and function.

On the other hand, overproduction PGC-1α in the heart is harmful. It has been shown that constitutive overexpression of PGC-1α driven by the cardiac α-myosin heavy chain promoter markedly activates mitochondrial biogenesis and leads, ultimately, to heart failure.[35] Stage-specific overexpression of inducible PGC-1α in cardiac myocytes results in a dramatic expansion of mitochondria in neonatal hearts or a modest mitochondrial biogenesis followed by mitochondrial ultrastructural abnormalities and cardiomyopathy in the adult heart.[40] In addition to the loss-of-function and gain-of-function mouse models that link PGC-1α to heart function, in the mouse models with chronic pressure overload, PGC-1α expression is reduced along with the lower expression of PPARα target genes involved in mitochondrial fatty acid oxidation.[37] Taken together, these molecular genetic evidences demonstrate that PGC-1α is a key regulator of heart energy metabolism. However, it is still unknown whether altered expression levels and/or genetic mutations of PGC-1α contribute to the etiology of human cardiac dysfunction.

15.2.5 *SRC-1 and SRC-3 enhance estrogen receptor-mediated vesoprotection*

Although the molecular mechanisms are not fully understood, the vasoprotective effects of estrogen are well recognized. First, the incidence of cardiovascular disease among premenopausal women is significantly less than that among age-matched men and postmenopausal women.[41,42] Furthermore, estrogen increases vasodilatation and inhibits the response to vascular injury.[41,42] It is now generally accepted that estrogen may protect the vascular system through systemic effects, such as decreased serum total cholesterol and alteration of serum lipoproteins. It also protects the vascular system through direct effects on blood vessels by taking rapid nongenomic and longer-term genomic pathways.[41–44] Importantly, both the nongenomic and the genomic cardiovascular effects of estrogen are mediated by estrogen receptor α (ERα) and ERβ.[45,46] Both ERα and ERβ are expressed in vascular smooth muscle cells (VSMCs) and endothelial cells (ECs). They are reported to be involved in estrogen vasoprotection.[41,42,47–52] The protective effects of ERα include the promotion of reendotheliallization, the inhibition of VSMC proliferation and the attenuation of atherosclerotic plaque progression.[42,47] The protective effects of ERβ include the inhibition of VSMC proliferation, the reduction of neointima formation and the improvement of vascular relaxation response.[50,52,53] Interestingly, it is also reported that estrogen may contribute to the vascular healing process and the prevention of restenosis by improving reendothelialization through ERα activation and by decreasing VSMC migration and proliferation through ERβ stimulation.[49]

Similar to other members of the nuclear hormone receptor superfamily, estrogen- and DNA-bound ERs recruit coactivators to the target gene promoter for transcriptional activation.[2,3] The three coactivator members of the p160 SRC family interact with ERs in a ligand-binding dependent manner and recruit general coactivators such as CBP or p300 to the target gene promoter for activation of gene transcription.[2,3] Studies using knockout mouse models have shown that the p160 SRC coactivators play important pleiotropic roles in the regulation of development, somatic growth, reproduction, metabolic homeostasis and breast cancer.[3] Genetic disruption of the SRC-1 gene in mice results in partial hormone resistance and developmental delay in the Purkinje cell development.[54–56] Mice lacking SRC-2 exhibit severe defects in both male and female reproductive functions.[57–59] Genetic deletion of the SRC-3 gene in mice causes growth retardation; delays pubertal development and mammary gland growth; reduces female reproductive function; and makes mice resistant to oncogene and/or carcinogen-induced mammary gland and prostate carcinogenesis.[60–63] These distinct phenotypes observed in different SRC knockout mice indicate that each SRC family member possesses different tissue-specific functions due to either distinct temporal and spatial expression patterns or distinct preferences to different transcription factors. On the other hand, multiple lines of evidence collected from experiments using cultured cells indicate that the p160 SRC family members share redundant functions when expressed in the same cells.[3]

Since SRC family members are characterized as coactivators for ERs, we hypothesized that these coactivators might play important roles in estrogen receptor-mediated vasoprotection.[64,65] To study the roles of SRC-1 and SRC-3 in the estrogen receptor-mediated vesoprotection, our laboratory first examined the expression of these two coactivators in the vascular wall. We discovered that both SRC-1 and SRC-3 were expressed in the VSMCs, ECs and neointima cells.[64,65] Subsequently, we assessed the roles of SRC-1 and SRC-3 in the estrogen-dependent vasoprotection during injury-induced vascular remodeling by applying a common carotid artery ligation model to the knockout mice.[64–66] In this model, female mice were ovariectomized and neointima growth was induced by unilateral ligation of the common carotid artery in the presence or absence of estrogen replacement. Without estrogen treatment, robust neointima growth near the ligation site was induced in WT, SRC-1 null and SRC-3 null mice. With estrogen treatment, the ligation injury-induced neointima growth was almost completely inhibited in WT mice, but was only partially inhibited in SRC-1 null and SRC-3 null mice. Accordingly, the cell proliferation rates in the intimal regions of SRC-1 null and SRC-3 null mice were significantly higher than that observed in WT mice.[64,65] There were no obvious changes in the expression levels of ERα, ERβ and other SRC family members in the vascular walls of SRC-1 null and SRC-3 null mice. This suggests that SRC-1 and SRC-3 are directly involved in the estrogen-induced inhibition of neointima growth.[64,65] These findings indicate that SRC-1 and SRC-3 expressed in the VSMCs, ECs and intimal cells facilitate estrogen/ER-mediated vasoprotection through the inhibition of neointima formation after a vascular injury.

Interestingly, a recent study demonstrates that SRC-3 also serves as a coactivator for myocardin, a major VSMC transcription factor required for VSMC differentiation to the nonproliferative, contractile state.[67] Myocardin uses its C-terminal activation domain to interact with the N-terminal basic helix-loop-helix/Per-ARNT-Sim domain of

SRC-3 for transcriptional activation of VSMC-specific genes such as SM22 and myosin heavy chain. The knockdown of SRC-3 expression by using small interfering RNA attenuates myocardin transcriptional activity in VSMCs. These findings indicate that SRC-3 can also work with non-nuclear hormone receptor transcription factors to exert its vesoprotective effect. Although the possible crosstalk between the role of SRC-3 in estrogen receptor-mediated inhibition of neointima formation and the role of SRC-3 in myocardin-mediated transcriptional activation remains unclear, the finding of SRC-3 and myocardin interaction provides a site of convergence for nuclear hormone receptor-mediated and VSMC-specific gene regulation. It also suggests a possible mechanism for the vascular protective effects of estrogen on vascular injury.[67]

ERs, SRC-1 and SRC-3 are coexpressed in the VSMCs, ECs and intimal cells in the vascular wall, which is consistent with their functional relationships in the estrogen-induced vasoprotective effects.[64,65,68,69] The overlapping expression pattern of multiple ER coactivators also provides an explanation for the partial inhibition of neointima formation by estrogen in either SRC-1 or SRC-3 null mice since SRC-1 and SRC-3 may be functionally redundant. In the future, it would be interesting to know whether SRC-2 also plays a role in the estrogen-dependent vasoprotection and whether the inactivation of two or all three of the SRC family members will completely abolish the inhibitory effect of estrogen on vascular injury-induced neointima development. It would be also very interesting to discern the contribution of individual SRC family members specific to ERα or ERβ-mediated vasoprotection in additional studies using combinatorial mouse models or selective ER modulators.

Intriguingly, despite the fact that many epidemiological and basic studies have demonstrated that estrogen plays an important protective role in the cardiovascular system in women,[41,42,48] the recent "Women's Health Initiative Randomized Controlled Trial (WHIRCT)" has been unable to demonstrate a cardiovascular benefit of hormonal replacement with either estrogen alone or estrogen and progesterone in postmenopausal women.[70,71] The explanation to this controversy may come from the analysis of estrogen signaling components. In our mouse models, estrogen treatment strongly inhibits vascular injury-induced neointima formation in WT mice. Under this circumstance, estrogen is present and ER and SRCs are normally expressed at the time of the vascular injury. However, after SRC-1 or SRC-3 is inactivated in knockout mice, the beneficiary effect of estrogen on vasoprotection during vascular remodeling is reduced. Our findings raise a possibility that some postmenopausal women might have a decrease in the expression or function of ER coactivators such as SRCs. It would be interesting to look into this matter in future studies. In addition, possibilities of reduced ER levels in diseased human coronary arteries and inappropriate timing of estrogen treatment have been suggested to explain the controversial WHIRCT outcome.[72,73]

15.3 Conclusions

The rapid identification and characterization of numerous nuclear receptor coactivators have provided solid molecular basis and physiological evidence for us to recognize the importance of these coactivators in the regulation of normal developmental events and physiological homeostasis as well as in their relevance to many potential diseases such as breast and prostate cancers,[61,62,74,75] metabolic syndromes,[76] and cardiovascular diseases.[33,37,64,65] However, we are just beginning our journey to understanding the entire coregulator biology. Regarding the roles for coactivators in the cardiovascular system, many issues need to be addressed in the future. First, only a few coactivators have been studied in the cardiovascular system and many others will need to be examined. Second, the detail molecular mechanisms and target genes for SRC-1 and SRC-3 in the vascular wall remain to be determined. It is also unclear whether SRC coactivators work with ERα, ERβ or both of them in the vascular wall and whether cell type-specific function of coactivators can explain why estrogen inhibits VSMC proliferation but enhances EC growth. Third, our recent study has demonstrated that SRC-3 plays a crucial role in the repression of cytokine mRNA translation and total cytokine production in the macrophage.[77] Since macrophage-mediated inflammatory plays an important role in the development of arthrosclerosis, it may be useful to investigate whether SRC-3 plays a role in this common cardiovascular disease. Finally, because much of our knowledge regarding the *in vivo* function and relevant pathology of coactivators in the cardiovascular system is obtained from cell culture and animal models, it is unclear at the current stage how much of these findings can be implicated to human and how many coactivators' dysfunction really contributes to the etiology of human cardiovascular and other diseases. Further studies in years to come will definitely provide much more insight into the understanding of the molecular mechanisms, biological functions and pathological roles of nuclear receptor coregulators. Ultimately, we hope to identify some of the coregulators as therapeutic molecular targets for cardiovascular and other diseases.

Acknowledgments

This work was carried out with the support and funding from the National Institutes of Health.

References

1. O'Malley BW, Coregulators: From whence came these "master genes", *Mol Endocrinol* **21**:1009–1013, 2007.
2. McKenna NJ, Lanz RB, O'Malley BW, Nuclear receptor coregulators: Cellular and molecular biology, *Endocr Rev* **20**:321–344, 1999.
3. Xu J, Li Q, Review of the *in vivo* functions of the p160 steroid receptor coactivator family, *Mol Endocrinol* **17**:1681–1692, 2003.
4. Wu RC, Qin J, Yi P, *et al.*, Selective phosphorylations of the SRC-3/AIB1 coactivator integrate genomic reponses to multiple cellular signaling pathways, *Mol Cell* **15**:937–949, 2004.
5. Iyer NG, Ozdag H, Caldas C, p300/CBP and cancer, *Oncogene* **23**:4225–4231, 2004.
6. Bannister AJ, Kouzarides T, The CBP co-activator is a histone acetyltransferase, *Nature* **384**:641–643, 1996.
7. Ogryzko VV, Schiltz RL, Russanova V, *et al.*, The transcriptional coactivators p300 and CBP are histone acetyltransferases, *Cell* **87**:953–959, 1996.
8. Oike Y, Takakura N, Hata A, *et al.*, Mice homozygous for a truncated form of CREB-binding protein exhibit defects in hematopoiesis and vasculo-angiogenesis, *Blood* **93**:2771–2779, 1999.
9. Tanaka Y, Naruse I, Hongo T, *et al.*, Extensive brain hemorrhage and embryonic lethality in a mouse null mutant of CREB-binding protein, *Mech Dev* **95**:133–145, 2000.
10. Yao TP, Oh SP, Fuchs M, *et al.*, Gene dosage-dependent embryonic development and proliferation defects in mice lacking the transcriptional integrator p300, *Cell* **93**:361–372, 1998.
11. Shikama N, Lutz W, Kretzschmar R, *et al.*, Essential function of p300 acetyltransferase activity in heart, lung and small intestine formation, *EMBO J* **22**:5175–5185, 2003.

12. Gusterson RJ, Jazrawi E, Adcock IM, *et al.*, The transcriptional co-activators CREB-binding protein (CBP) and p300 play a critical role in cardiac hypertrophy that is dependent on their histone acetyltransferase activity, *J Biol Chem* **278**:6838–6847, 2003.
13. Yanazume T, Hasegawa K, Morimoto T, *et al.*, Cardiac p300 is involved in myocyte growth with decompensated heart failure, *Mol Cell Biol* **23**:3593–3606, 2003.
14. Miyamoto S, Kawamura T, Morimoto T, *et al.*, Histone acetyltransferase activity of p300 is required for the promotion of left ventricular remodeling after myocardial infarction in adult mice *in vivo*, *Circulation* **113**:679–690, 2006.
15. Rachez C, Lemon BD, Suldan Z, *et al.*, Ligand-dependent transcription activation by nuclear receptors requires the DRIP complex, *Nature* **398**:824–828, 1999.
16. Yuan CX, Ito M, Fondell JD, *et al.*, The TRAP220 component of a thyroid hormone receptor- associated protein (TRAP) coactivator complex interacts directly with nuclear receptors in a ligand-dependent fashion, *Proc Natl Acad Sci USA* **95**:7939–7944, 1998.
17. Zhu Y, Qi C, Jain S, *et al.*, Isolation and characterization of PBP, a protein that interacts with peroxisome proliferator-activated receptor, *J Biol Chem* **272**:25500–25506, 1997.
18. Zhu Y, Qi C, Jia Y, *et al.*, Deletion of PBP/PPARBP, the gene for nuclear receptor coactivator peroxisome proliferator-activated receptor-binding protein, results in embryonic lethality, *J Biol Chem* **275**:14779–14782, 2000.
19. Ito M, Yuan CX, Okano HJ, *et al.*, Involvement of the TRAP220 component of the TRAP/SMCC coactivator complex in embryonic development and thyroid hormone action, *Mol Cell* **5**:683–693, 2000.
20. Guan XY, Meltzer PS, Dalton WS, *et al.*, Identification of cryptic sites of DNA sequence amplification in human breast cancer by chromosome microdissection, *Nat Genet* **8**:155–161, 1994.
21. Lee SK, Anzick SL, Choi JE, *et al.*, A nuclear factor, ASC-2, as a cancer-amplified transcriptional coactivator essential for ligand-dependent transactivation by nuclear receptors *in vivo*, *J Biol Chem* **274**:34283–34293, 1999.
22. Caira F, Antonson P, Pelto-Huikko M, *et al.*, Cloning and characterization of RAP250, a novel nuclear receptor coactivator, *J Biol Chem* **275**:5308–5317, 2000.
23. Zhu Y, Kan L, Qi C, *et al.*, Isolation and characterization of peroxisome proliferator-activated receptor (PPAR) interacting protein (PRIP) as a coactivator for PPAR, *J Biol Chem* **275**:13510–13516, 2000.

24. Ko L, Cardona GR, Chin WW, Thyroid hormone receptor-binding protein, an LXXLL motif-containing protein, functions as a general coactivator, *Proc Natl Acad Sci USA* **97**:6212–6217, 2000.
25. Mahajan MA, Samuels HH, A new family of nuclear receptor coregulators that integrate nuclear receptor signaling through CREB-binding protein, *Mol Cell Biol* **20**:5048–5063, 2000.
26. Mahajan MA, Samuels HH, Nuclear hormone receptor coregulator: Role in hormone action, metabolism, growth, and development, *Endocr Rev* **26**:583–597, 2005.
27. Lee SK, Na SY, Jung SY, *et al.*, Activating protein-1, nuclear factor-kappaB, and serum response factor as novel target molecules of the cancer-amplified transcription coactivator ASC-2, *Mol Endocrinol* **14**:915–925, 2000.
28. Hong S, Lee MY, Cheong J, Functional interaction of transcriptional coactivator ASC-2 and C/EBPalpha in granulocyte differentiation of HL-60 promyelocytic cell, *Biochem Biophys Res Commun* **282**:1257–1262, 2001.
29. Kuang SQ, Liao L, Zhang H, *et al.*, Deletion of the cancer-amplified coactivator AIB3 results in defective placentation and embryonic lethality, *J Biol Chem* **277**:453–456, 2002.
30. Zhu YJ, Crawford SE, Stellmach V, *et al.*, Coactivator PRIP, the peroxisome proliferator-activated receptor-interacting protein, is a modulator of placental, cardiac, hepatic, and embryonic development, *J Biol Chem* **278**:1986–1990, 2003.
31. Antonson P, Schuster GU, Wang L, *et al.*, Inactivation of the nuclear receptor coactivator RAP250 in mice results in placental vascular dysfunction, *Mol Cell Biol* **23**:1260–1268, 2003.
32. Barak Y, Nelson MC, Ong ES, *et al.*, PPAR gamma is required for placental, cardiac, and adipose tissue development, *Mol Cell* **4**:585–595, 1999.
33. Puigserver P, Spiegelman BM, Peroxisome proliferator-activated receptor-gamma coactivator 1 alpha (PGC-1 alpha): Transcriptional coactivator and metabolic regulator, *Endocr Rev* **24**:78–90, 2003.
34. Puigserver P, Adelmant G, Wu Z, *et al.*, Activation of PPARgamma coactivator-1 through transcription factor docking, *Science* **286**:1368–1371, 1999.
35. Lehman JJ, Barger PM, Kovacs A, *et al.*, Peroxisome proliferator-activated receptor gamma coactivator-1 promotes cardiac mitochondrial biogenesis, *J Clin Invest* **106**:847–856, 2000.
36. Huss JM, Kelly DP, Nuclear receptor signaling and cardiac energetics, *Circ Res* **95**:568–578, 2004.
37. Finck BN, Kelly DP, PGC-1 coactivators: Inducible regulators of energy metabolism in health and disease, *J Clin Invest* **116**:615–622, 2006.

38. Arany Z, He H, Lin J, *et al.*, Transcriptional coactivator PGC-1 alpha controls the energy state and contractile function of cardiac muscle, *Cell Metab* **1**:259–271, 2005.
39. Leone TC, Lehman JJ, Finck BN, *et al.*, PGC-1alpha deficiency causes multi-system energy metabolic derangements: Muscle dysfunction, abnormal weight control and hepatic steatosis, *PLoS Biol* **3**:e101, 2005.
40. Russell LK, Mansfield CM, Lehman JJ, *et al.*, Cardiac-specific induction of the transcriptional coactivator peroxisome proliferator-activated receptor gamma coactivator-1alpha promotes mitochondrial biogenesis and reversible cardiomyopathy in a developmental stage-dependent manner, *Circ Res* **94**:525–533, 2004.
41. Mendelsohn ME, Karas RH, The protective effects of estrogen on the cardiovascular system, *N Engl J Med* **340**:1801–1811, 1999.
42. Mendelsohn ME, Karas RH, Molecular and cellular basis of cardiovascular gender differences, *Science* **308**:1583–1587, 2005.
43. Orshal JM, Khalil RA, Gender, sex hormones, and vascular tone, *Am J Physiol Regul Integr Comp Physiol* **286**:R233–249, 2004.
44. Xing D, Miller A, Novak L, *et al.*, Estradiol and progestins differentially modulate leukocyte infiltration after vascular injury, *Circulation* **109**:234–241, 2004.
45. Chen Z, Yuhanna IS, Galcheva-Gargova Z, *et al.*, Estrogen receptor alpha mediates the nongenomic activation of endothelial nitric oxide synthase by estrogen, *J Clin Invest* **103**:401–406, 1999.
46. Bakir S, Mori T, Durand J, *et al.*, Estrogen-induced vasoprotection is estrogen receptor dependent: Evidence from the balloon-injured rat carotid artery model, *Circulation* **101**:2342–2344, 2000.
47. Pare G, Krust A, Karas RH, *et al.*, Estrogen receptor-alpha mediates the protective effects of estrogen against vascular injury, *Circ Res* **90**:1087–1092, 2002.
48. Ouyang P, Michos ED, Karas RH, Hormone replacement therapy and the cardiovascular system lessons learned and unanswered questions, *J Am Coll Cardiol* **47**:1741–1753, 2006.
49. Geraldes P, Sirois MG, Tanguay JF, Specific contribution of estrogen receptors on mitogen-activated protein kinase pathways and vascular cell activation, *Circ Res* **93**:399–405, 2003.
50. Watanabe T, Akishita M, Nakaoka T, *et al.*, Estrogen receptor beta mediates the inhibitory effect of estradiol on vascular smooth muscle cell proliferation, *Cardiovasc Res* **59**:734–744, 2003.
51. Christian RC, Liu PY, Harrington S, *et al.*, Intimal estrogen receptor (ER)beta, but not ERalpha expression, is correlated with coronary calcification and atherosclerosis in pre- and postmenopausal women, *J Clin Endocrinol Metab* **91**:2713–2720, 2006.

52. Schrepfer S, Deuse T, Munzel T, *et al.*, The selective estrogen receptor-beta agonist biochanin A shows vasculoprotective effects without uterotrophic activity, *Menopause* **13**:489–499, 2006.
53. Zhu Y, Bian Z, Lu P, *et al.*, Abnormal vascular function and hypertension in mice deficient in estrogen receptor beta, *Science* **295**:505–508, 2002.
54. Xu J, Qiu Y, DeMayo FJ, *et al.*, Partial hormone resistance in mice with disruption of the steroid receptor coactivator-1 (SRC-1) gene, *Science* **279**:1922–1925, 1998.
55. Nishihara E, Yoshida-Komiya H, Chan CS, *et al.*, SRC-1 null mice exhibit moderate motor dysfunction and delayed development of cerebellar Purkinje cells, *J Neurosci* **23**:213–222, 2003.
56. Weiss RE, Xu J, Ning G, *et al.*, Mice deficient in the steroid receptor co-activator 1 (SRC-1) are resistant to thyroid hormone, *EMBO J* **18**: 1900–1904, 1999.
57. Gehin M, Mark M, Dennefeld C, *et al.*, The function of TIF2/GRIP1 in mouse reproduction is distinct from those of SRC-1 and p/CIP, *Mol Cell Biol* **22**:5923–5937, 2002.
58. Mark M, Yoshida-Komiya H, Gehin M, *et al.*, Partially redundant functions of SRC-1 and TIF2 in postnatal survival and male reproduction, *Proc Natl Acad Sci USA* **101**:4453–4458, 2004.
59. Mukherjee A, Soyal SM, Fernandez-Valdivia R, *et al.*, Steroid receptor coactivator 2 is critical for progesterone-dependent uterine function and mammary morphogenesis in the mouse, *Mol Cell Biol* **26**:6571–6583, 2006.
60. Xu J, Liao L, Ning G, *et al.*, The steroid receptor coactivator SRC-3 (p/CIP/RAC3/AIB1/ACTR/TRAM-1) is required for normal growth, puberty, female reproductive function, and mammary gland development, *Proc Natl Acad Sci USA* **97**:6379–6384, 2000.
61. Kuang SQ, Liao L, Wang S, *et al.*, Mice lacking the amplified in breast cancer 1/steroid receptor coactivator-3 are resistant to chemical carcinogen-induced mammary tumorigenesis, *Cancer Res* **65**:7993–8002, 2005.
62. Kuang SQ, Liao L, Zhang H, *et al.*, AIB1/SRC-3 deficiency affects insulin-like growth factor I signaling pathway and suppresses v-Ha-ras-induced breast cancer initiation and progression in mice, *Cancer Res* **64**:1875–1885, 2004.
63. Chung ACK, Zhou S, Liao L, *et al.*, Genetic ablation of the amplified-in-breast cancer-1 (AIB1) inhibits spontaneous prostate cancer progressio in mice, *Cancer Res* **67**:5965–5975, 2007.
64. Yuan Y, Xu J, Loss-of-function deletion of the steroid receptor coactivaor-1 gene in mice reduces estrogen effect on the vascular injury response, *Arterioscler Thromb Vasc Biol* **27**:1521–1527, 2007.
65. Yuan Y, Liao L, Tulis DA, *et al.*, Steroid receptor coactivator-3 is required for inhibition of neointima formation by estrogen, *Circulation* **105**:2653–2659, 2002.

66. Kumar A, Lindner V, Remodeling with neointima formation in the mouse carotid artery after cessation of blood flow, *Arterioscler Thromb Vasc Biol* **17**:2238–2244, 1997.
67. Li HJ, Haque Z, Lu Q, *et al.*, Steroid receptor coactivator 3 is a coactivator for myocardin, the regulator of smooth muscle transcription and differentiation, *Proc Natl Acad Sci USA* **104**:4065–4070, 2007.
68. Karas RH, Patterson BL, Mendelsohn ME, Human vascular smooth muscle cells contain functional estrogen receptor, *Circulation* **89**:1943–1950, 1994.
69. Venkov CD, Rankin AB, Vaughan DE, Identification of authentic estrogen receptor in cultured endothelial cells. A potential mechanism for steroid hormone regulation of endothelial function, *Circulation* **94**:727–733, 1996.
70. Rossouw JE, Anderson GL, Prentice RL, *et al.*, Risks and benefits of estrogen plus progestin in healthy postmenopausal women: Principal results From the women's health initiative randomized controlled trial, *Jama* **288**:321–333, 2002.
71. Anderson GL, Limacher M, Assaf AR, *et al.*, Effects of conjugated equine estrogen in postmenopausal women with hysterectomy: The women's health initiative randomized controlled trial, *Jama* **291**:1701–1712, 2004.
72. Dubey RK, Imthurn B, Barton M, *et al.*, Vascular consequences of menopause and hormone therapy: Importance of timing of treatment and type of estrogen, *Cardiovasc Res* **66**:295–306, 2005.
73. Losordo DW, Kearney M, Kim EA, *et al.*, Variable expression of the estrogen receptor in normal and atherosclerotic coronary arteries of premenopausal women, *Circulation* **89**:1501–1510, 1994.
74. Anzick SL, Kononen J, Walker RL, *et al.*, AIB1, a steroid receptor coactivator amplified in breast and ovarian cancer, *Science* **277**:965–968, 1997.
75. Chung A, Zhou S, Liao L, *et al.*, Genetic ablation of the amplified-in-breast cancer-1 (AIB1) inhibits spontaneous prostate cancer progression in mice, *Cancer Res*, 2007.
76. Picard F, Gehin M, Annicotte J, *et al.*, SRC-1 and TIF2 control energy balance between white and brown adipose tissues, *Cell* **111**:931–941, 2002.
77. Yu C, York B, Wang S, *et al.*, An essential function of the SRC-3 coactivator in suppression of cytokine mRNA translation and inflammatory response, *Mol Cell* **25**:765–778, 2007.

Chapter 16

Coregulators as Determinants of Selective Receptor Modulator (SRM) Activity

Margaret C. Pace and Carolyn L. Smith

Selective receptor modulators (SRMs) are nuclear receptor ligands which display a subset of the effects of the receptors' cognate hormones. The concept was first elucidated for selective estrogen receptor modulators (SERMs) based on the recognition that the estrogen receptor ligand, tamoxifen exerts estrogen-like and antiestrogen-like effects in a tissue selective manner. In environments favoring interactions with coactivators, tamoxifen exerts greater estrogen receptor agonist effects while corepressor dominant milieus promote the antagonist properties of tamoxifen. In this chapter we summarize the evolution of the SERM concept and review recent data on cellular mechanisms employed by SERM-bound estrogen receptors to achieve cell and tissue selective effects. The SERM concept has been translated to other members of the nuclear receptor superfamily and an overview of SRMs for progesterone, androgen and thyroid hormone receptors is presented.

16.1 Introduction

The nuclear receptor (NR) superfamily comprises a group of structurally-related, ligand-inducible transcription factors that co-ordinate the expression of genes important for regulating a wide variety of physiological and pathological processes. NRs can be divided into three classes: type I receptors include the steroid hormone receptors (SR) that typically translocate to the nucleus upon ligand activation to bind to

DNA as homodimers at specific inverted repeat half-sites called hormone response elements (HRE). Type II receptors include the thyroid hormone (TR) and retinoic acid (RAR) receptors and are typically bound in a ligand-independent manner as homodimers or heterodimers with the retinoid X receptor (RXR) to HREs consisting of direct DNA repeats. Type I and type II receptors may also bind to DNA indirectly through interactions with other non-receptor transcription factors such as Sp1 or AP-1. Type III receptors include the orphan receptors whose structure resembles that of type I and II receptors for which ligands have not been identified.[1] NRs share several conserved functional domains including an N-terminal constitutively-active activation function-1 (AF1), a centrally located DNA-binding and hinge domains and a C-terminal ligand binding domain (LBD) that encompasses the ligand-regulated AF2 domain. The zinc finger-based DNA binding domain is highly conserved among the three NR classes and co-ordinates the binding of specific DNA sequences within the HREs. The AF2 domain is less conserved, although similarities can be observed among receptors that bind similar ligands such as steroids, and the AF1 region is widely variable among the NRs.

Investigating the mechanisms by which small molecule ligands regulate the activities of nuclear receptors has been an area of interest for many decades, and the use of receptor agonists and antagonists has contributed significantly to our understanding of the molecular events involved in nuclear receptor transcriptional activity. In part because of its physiological importance and the availability of agonist and antagonist ligands, estrogen receptor-α (ERα) was used extensively in the early studies of nuclear receptor function. These studies revealed key aspects of estradiol (E2) agonist action, but also demonstrated that the ERα antagonist 4-hydroxytamoxifen (4HT) could stimulate receptor transcriptional activity in a gene- and cell-specific fashion. When these findings on the selectivity of 4HT effects were considered together with the observations that tamoxifen acted like an estrogen in the human skeleton and an antiestrogen in the breast, the concept of tamoxifen as a selective estrogen receptor modulator (SERM) was established. Since then, extensive studies have been performed to elucidate the mechanisms of SERM action *in vitro* and *in vivo*, and successes in that area have spurred efforts to identify new SERMs as well as novel selective receptor modulators (SRMs) for other nuclear receptors. In this chapter, we will discuss the role of nuclear receptor coregulators in determining the biological properties of SERMs, and review recent developments in the identification and utilization of SRMs for other members of the nuclear receptor superfamily.

16.2 Nuclear Receptor Agonist and Antagonist Activities: Regulation by Coactivators and Corepressors

16.2.1 *Nuclear receptor agonists and conformational changes*

Agonist binding induces conformational changes in NRs to generate distinct receptor configurations and functional outcomes. The crystal structure of the LBD of several nuclear receptors has been determined and a common overall arrangement has been identified. For example, the ERα LBD comprises 12 α-helices that form a three-layered anti-parallel sandwich, the center of which creates a hydrophobic ligand binding pocket, and this basic LBD structure is prototypic for the SRs.[2] Agonist binding to the LBD of ERα alters the conformation of the receptor such that helix 12 is precisely positioned against helices 3, 5/6 and 11, burying the ligand within the binding pocket and exposing the coactivator binding groove which allows AF2 to contribute to the receptor's transcriptional activity (Fig. 16.1). The AF1 domain of ERα, and NRs in general, is highly unstructured and there is very little information available on the three-dimensional structure of this region. However, molecular biological studies have shown that the AF1 domain of many of the NRs possess intrinsic transcriptional activity and can interact with diverse components of the transcriptional machinery.[3] The requirement of AF1 and AF2 in mediating transcriptional activity is receptor-, ligand-, promoter- and cell-specific, and in some cases, AF1 and AF2 have synergistic transcriptional activity. Because of this, the full impact of AF1/AF2 interactions on the recruitment of coactivators and corepressors and transcriptional regulation will not be known until full length NRs can be crystallized in the presence of ligands and appropriate coregulators.

16.2.2 *Coactivators and activation of gene expression*

The conformational change in the LBD induced by agonist binding creates a docking site for a diverse group of coactivators that can be recruited to the regulatory regions of target genes and stimulate their

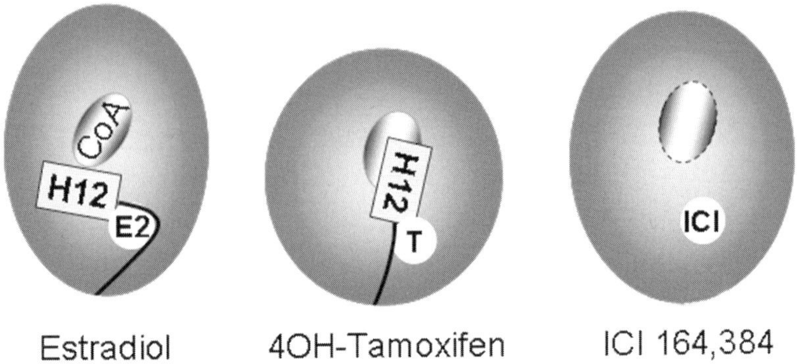

Fig. 16.1. Model of ligand-dependent allosteric modulation of the ER ligand binding domain. When ER is bound to the agonist estradiol (E2), the transactivation helix 12 (H12) adapts a conformation in which it forms a portion of the coactivator binding groove (shown as the surface marked CoA), allowing for coactivator recruitment to the receptor. When ER is bound to the partial antagonist 4-hydroxytamoxifen (T), the position of helix 12 is shifted so that it partially occludes the coactivator binding groove, thus preventing coactivator recruitment. Helix 12 is not visible in crystal structures of ER bound to full antagonists such as ICI 164,384 (ICI), indicating that this compound completely abolishes the interaction of helix 12 with the remainder of the LBD.

transcription.[4] To date, several hundred of these coregulator proteins which include coactivators and corepressors (discussed below) have been identified (listed at http://www.nursa.org). Coactivators bind to NRs through a structural motif called a NR box that consists of five amino acids, L-X-X-L-L, where L is leucine and X is any amino acid.[5] Sequences surrounding the NR box may also influence the specificity and/or strength of the coregulator-NR interaction. Some coactivators, such as the steroid receptor coactivator-1 (SRC-1), also bind to AF1 and synergize the transcriptional activity of the two activation functions. In the classical model, DNA-bound agonist-occupied NRs recruit coactivator proteins to gene regulatory regions. Coactivators do not bind to DNA directly but rather act as bridges between the DNA-bound NR and additional proteins that must be recruited to gene regulatory regions to stimulate gene expression. Several of the coactivators and their recruited accessory factors possess enzymatic activities that either modify other coregulators or chromatin to create a local environment conducive for transcription. Some of these enzymatic activities include the

ATP-dependent chromatin remodeling SWI/SNF complex, histone acetyltransferases (e.g. CBP/p300, SRC-1 and SRC-3), histone methyltransferases (e.g. CARM-1), histone demethylases (e.g. LSD1) and ubiquitin ligases (e.g. E6-AP).[6] Histone modifications relax the nucleosomal structure and allow for the recruitment of the TRAP/DRIP mediator complex which contacts the basal transcriptional machinery and facilitates the assembly of a stable pre-initiation complex. Finally, RNA polymerase II is recruited and gene expression is initiated.

Single-cell real time imaging and chromatin immunoprecipitation (ChIP) studies examining NR/coregulator interactions have been used to characterize the sequential events leading to the initiation of transcription, and both approaches reveal scenarios in which coregulator occupancy of a given promoter is dynamic. Fluorescence recovery after photobleaching (FRAP) experiments showed rapid (i.e. seconds) and transient associations of NRs and coactivators on promoters containing tandem arrays of HREs in response to ligand stimulation.[7] This supports a model in which transcription initiation is the result of stochastic interactions between highly mobile transcription factors and regulatory proteins. In contrast, ChIP studies of the E2-responsive pS2 gene revealed cycles of ERα binding to the promoter that were significantly longer (i.e. minutes to hours) as well as sequential recruitment of distinct complexes of coactivator proteins.[8] To reconcile these divergent data sets, a model has been proposed whereby the promotion of transcription occurs through a specific sequence of longer term productive associations that occur infrequently in the course of the many, rapid unproductive associations between NRs and coregulators.[9] This would allow for a continuous "sampling" of the cell environment to ensure appropriate cellular responses.

16.2.3 *Nuclear receptor antagonists*

As for agonists, antagonists bind to NRs such as ERα and induce conformational changes that are distinct from those induced by agonists. There are two major classes of ER antagonists that show differing pharmacologic profiles. Full antagonists completely block the receptor's transcriptional activity while partial antagonists exhibit a tissue-selective ability to either stimulate or block ER function. The crystal structures of the ERα LBD bound to the full antagonist ICI 182,780 and to selective antagonists including 4HT and raloxifene have been resolved. They offer insight into how these two different classes of antagonist exert their effects on ERα activity.[2] The binding of pure antagonists results in a disordered structure for helix 12 such that it can no longer contact the LBD and contribute to the formation of the coactivator docking site, thus preventing a transcriptionally-productive coactivator-ERα interaction. In addition, full antagonists such as ICI 182,780 employ a range of other molecular mechanisms that contribute to the ability of this class to block ERα-dependent gene expression. These include the immobilization of the receptor within the nuclear matrix, the redistribution of the antagonist-occupied receptor from the nucleus to the cytoplasm and the promotion of rapid receptor degradation by the ubiquitin-proteasome pathway.[10,11] The latter mechanism has led to the term of selective ER down-regulators (SERDs) to describe some full ER antagonists.

In contrast to SERDs, SERMs such as 4HT bind to the LBD of ERα and reorient the position of helix 12 such that it overlaps a portion of the coactivator binding groove and thereby occludes coactivators from binding to the receptor's AF2 domain. Thus, AF2 transcriptional activity is inhibited while AF1 activity remains intact.[12] Because SERMs do not promote cellular redistribution or degradation of the receptor as SERDs do, the receptor's AF1 activity is able to promote transcription in a cell-type and gene-specific manner. The crystal structures of ERα bound to either 4HT or raloxifene indicate that both compounds alter the position of helix 12 in a similar fashion to inhibit coactivator binding and AF2 activity. Yet, the two SERMs yield a distinct profile of biological activities, and this is reflected in molecular studies using peptide interaction assays which demonstrate structural differences between raloxifene-ERα and 4HT-ERα complexes. Thus, each ligand induces a unique receptor conformation that produces a functionally distinct transcriptional profile.[2] This realization prompted investigators to explore a wide range of cellular mechanisms that influence the ability of SERMs to modulate ER activity. This work has broadened our perspective on how cellular signaling pathways can influence multiple aspects of NR function.

16.2.4 *Corepressors and inhibition of gene expression*

Corepressors were originally identified as the proteins that enabled DNA bound, unliganded-TR and RAR to repress transcription and were later shown to bind to antagonist-occupied SRs. The best studied corepressors are NCoR (nuclear receptor corepressor) and SMRT (silencing mediator of retinoic acid and thyroid hormone receptors).[13] Corepressors bind to the LBD of unliganded (apo) TR and RAR through an L/I-X-X-I/V-I motif called a CoRNR box (where L = leucine, I = isoleucine, V = valine and X = any amino acid).[14] The corepressor binding site on NRs overlaps but it is not identical to the coactivator binding groove. Like coactivators, corepressors function as part of a large multi-protein complex, recruiting proteins with enzymatic activities antagonistic to those of coactivator complexes including histone deacetylases (HDACs), the WD-40 repeat proteins transducin-like protein 1 (TBL-1), TBL-1 related protein (TBLR-1), RbAp46/48 and the closely-related G-protein pathway suppressor (GPS2). The HDACs create a hypoacetylated chromatin environment, tightening DNA-histone interactions and restricting access of the transcriptional machinery to DNA, while TBL-1 and TBLR-1 target and anchor the corepressor complex to the promoter.[15,16]

Histones also can be modified by phosphorylation, sumoylation, ubiquitination and methylation and these post-translation events produce a "histone code" that determines spatial and temporal control of gene expression.[17] In particular, changes in histone methylation have been correlated with changes in gene expression. Several histone methyltransferases (HMT) and histone demethylases (HDM) have been identified.[18] Recently, the methylation of specific lysine residues in histone H3 has been shown to reduce the binding of unliganded ERα to the promoters of its target genes and thereby help to maintain genes in an "off" state.[19] Agonist-bound ERs recruit HDMs to target gene promoters, thus leading to the removal of inhibitory methylation and promotion of gene expression.

16.2.5 *Regulation of SRM activity by corepressors*

In the case of receptors for estrogens, progesterone, glucocorticoids and androgens, initial reports suggested that the affinity of corepressor for apo-receptors was relatively poor in comparison to the interaction of corepressors with receptors bound to their respective antagonists; 4HT for ER, RU486 for PR and glucocorticoid receptor (GR), and cyproterone acetate for androgen receptor (AR).[20] Crystallographic studies of ERα bound to 4HT or raloxifene indicated that the antagonist repositions helix 12 of the LBD over the hydrophobic coactivator binding cleft, blocking AF2 activity and preventing coactivator recruitment. The CoRNR box motif of corepressors bind to the same region, and evidence obtained from helix 12 deletion studies supports a model in which helix 12 competes with corepressor for binding to the same site on antagonist-bound receptor.[2] Interestingly, a recent study of ERβ complexed with 4HT indicated that in addition to binding to the ligand binding pocket, 4HT can also bind to the coactivator interaction site to directly antagonize the receptor-coactivator interactions.[21]

With regard to the effects of SERMs on ER function, initial transient transfection studies indicated that exogenous expression of corepressors reduced their partial agonist activity. For instance, in HepG2 cells where 4HT is an agonist, overexpression of exogenous SMRT blocked 4HT-stimulated gene expression.[22] Studies in which inhibitory antibodies to SMRT or NCoR were microinjected into Rat-1 fibroblast cells showed that blocking corepressor activity promoted the agonist activity of 4HT. It also suggested that these corepressors were directly mediating the inhibitory effects of SERMs.[23] Experiments using embryonic fibroblasts derived from NCoR null mice found that NCoR was required for 4HT antagonist activity.[24] Finally, simultaneous siRNA-mediated silencing of SMRT and NCoR expression in MCF-7 cells enabled 4HT to stimulate cell cycle progression and increase the expression of an ER target gene, X-box binding protein 1, consistent with the idea that corepressors blocked the partial agonist activity of SERMs.[25] It is interesting to note, however, that the depletion of SMRT and NCoR did not enable 4HT to stimulate the expression of two other ER target genes, c-Myc and cyclin D1, indicating that the ability of these corepressors to influence the activity of SERM-bound ER was gene specific.

Early ChIP assays were consistent with these studies, showing that treatment of MCF-7 cells with 4HT, but not E2, recruited SMRT and/or NCoR to the cathepsin D and pS2 gene promoters.[26] Subsequent ChIP studies demonstrated that 4HT treatment of MCF-7 cells resulted in the recruitment of ERα and a NCoR/SMRT, HDAC3 and TBL-1 chromatin modifying complex, followed by the recruitment of a NuRD corepressor complex containing HDAC1 and metastasis-associated protein 1/2 (MTA1/2) to the promoter region of the ERα target genes cathepsin D, c-Myc and pS2. Sequential recruitment of the different corepressor complexes coincided with the deacetylation of histones H3 and H4 and the dismissal of RNA polymerase II from the DNA template.[27,28] With these studies, a model was established in which agonists recruit coactivators to the promoters of steroid responsive genes to stimulate transcription while antagonist-bound receptors recruit corepressor complexes to the DNA in order to inhibit gene expression. This model, however, did not explain how SERMs could exert agonist or antagonist effects in a tissue-, cell- and gene specific manner.

16.2.6 *Regulation of SRM activity by coactivators*

The model of corepressors as mediators of SRM antagonist activity did not explain the phenomena of 4HT-mediated transcriptional activation and this led to research examining whether coactivators as well as corepressors could influence the relative agonist/antagonist profile of SRMs in a cell specific manner. Initial studies in HepG2 cells found that transient overexpression of the SRC-1 coactivator significantly enhanced 4HT-stimulated ERα transcriptional activity.[22] Similarly, transient overexpression of SRC-1 enhanced the agonist activity of the selective PR modulator (SPRM) RU486 on a PR target gene in intact HeLa cells.[29] Taken together, these data suggested that high levels of coactivator expression were sufficient to promote the agonist activity of SERMs and other SRMs.

Further evidence of the ability of coactivators to promote SERM agonist activity was obtained in ChIP studies. In MCF-7 and T-47D breast cancer cell lines, 4HT is an ER antagonist and treatment with this drug induced the recruitment of a corepressor complex containing NCoR, SMRT, HDAC2 and HDAC4 to the c-Myc promoter; while in Ishikawa endometrial carcinoma cells, 4HT exposure promoted the recruitment of a coactivator complex containing SRC-1, AIB1 and CBP to the c-Myc promoter.[27] In these experiments, it was the relative expression levels of the SRC-1 coactivator, which is high in endometrial cancer cells and low in MCF-7 breast cancer cells, that correlated with 4HT's agonist activity. Moreover, the changes in SRC-1 levels achieved by overexpression or siRNA-mediated depletion influenced whether 4HT could stimulate ER-dependent gene expression.

16.3 Coregulators as Determinants of SRM Agonist and Antagonist Activities

The ability of coactivators to promote SERM agonist activity and corepressors to inhibit the partial agonist activity of SERMs and render them better antagonists suggested that the relative expression of coactivators *versus* corepressors within a cell, and hence their availability for binding to SERM-occupied receptors was a major determinant of the ability of a SERM to stimulate or repress the transcriptional activity of ERs in that cell.[22] For PR, studies based on *in vitro* transcription assays demonstrated that the agonist activity of the SPRM, RU486 was governed by the relative ratio of coactivator and corepressor supplied to the transcription reactions.[29] These studies support a model in which coactivators and corepressors promote SERM agonist and antagonist activities, respectively, and that the relative expression levels of coregulators in a given cell environment are critical determinants of the biocharacter of the SRM; that is their relative ability to activate or inhibit transcription (Fig. 16.2).

16.3.1 *Relationships between coregulator expression and SERM clinical responses*

16.3.1.1 *Tissue-selective SERM responses*

The recognition of the tissue responses to drugs such as tamoxifen led to the creation of the SERM classification of molecules. In particular, tamoxifen exerts estrogen-like effects in the endometrium and skeleton while in the mammary gland, it is an anti-estrogen that inhibits cell proliferation. In part because of the high number of potential ER coactivators and corepressors, there has not been a comprehensive effort to examine the relative levels of all coactivators and corepressors in normal organ systems with differential responses to SERMs such as tamoxifen. Indeed, most studies to date have concentrated on examining coregulator levels in diseased *versus* normal tissues and this has revealed that coregulator over- or underexpression is frequently implicated in many pathological conditions.[30] As discussed above, high levels of SRC-1 coactivator expression were associated with the ability of tamoxifen to stimulate the expression of select ER target genes such as c-Myc in endometrial cells.[27] However, it is important to note that in these studies, raloxifene did not function as an agonist in endometrial cells even with high levels of SRC-1 expression and not all ER target genes were stimulated by tamoxifen in this environment. This suggests that relationships between SERM-receptor complexes, DNA and coregulators are important determinants of gene expression.

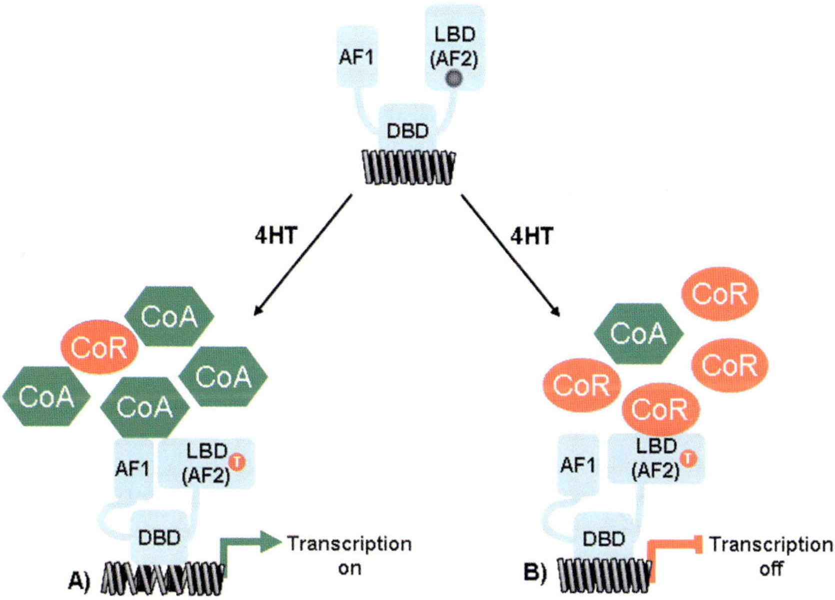

Fig. 16.2. Classical model for coactivator and corepressor regulation of 4HT-bound ERα transcriptional activity. The transcriptional outcome of 4HT binding to ERα is dependent on the relative expression levels of coactivators and corepressors in a given cell. (A) In an environment where coactivator expression is high, the receptor interactions with coactivator are favored and the partial agonist activity of 4HT is apparent, resulting in stimulation of transcription. Note: AF1 activity is required for 4HT to turn on gene expression. (B) When corepressor expression is high relative to coactivators, receptor interactions with corepressors are favored and 4HT functions as a receptor antagonist, turning off transcription.

16.3.2 *Resistance to SERM therapy*

The ability of tamoxifen to inhibit ER transcriptional activity and breast cancer cell growth is intimately involved with its success as an adjuvant therapy for ER-positive breast cancer. Unfortunately, however, some breast cancers exhibit *de novo* resistance to tamoxifen therapy and others will acquire resistance during the course of treatment. In these cases, tamoxifen loses its antagonist properties and may even stimulate tumor growth much like an agonist. Based on models of SERM action derived from *in vitro* studies, early predictions were that alterations in coregulator expression would influence tamoxifen antagonist *versus* agonist properties, and this prompted a number of studies.

Several reports have noted a correlation with elevated SRC-1 expression and resistance to tamoxifen[31,32] but at least one failed to reveal a relationship between tamoxifen resistance (i.e. development of tamoxifen agonism) and increased SRC-1 expression.[33] This raises the non-exclusive possibilities that overexpression of SRC-1 is not obligatory for resistance to this SERM and/or other factors may be required for SRC-1 to exert this effect. For instance, overexpression of the PELP-1 coactivator as well as src kinase signaling to SRC-1 increases tamoxifen agonist activity specifically in endometrial cells.[34,35] In another example, SRC-1 cooperates with another coactivator, PGC-1β to promote tamoxifen agonist activity,[36] and since the latter coactivator's expression is tissue restricted, this is another mechanism that could influence the ability of SRC-1 to promote tamoxifen stimulatory activity in a cell-specific fashion.

Other coactivators have also been associated with poor response to tamoxifen. For instance, in breast cancer patients treated with adjuvant tamoxifen therapy whose tumors expressed high levels of the erbB2 receptor (HER2), high expression levels of the coactivator AIB1 (amplified in breast cancer-1, also known as SRC-3) correlated with significantly lower disease free survival as compared to those patients with low AIB1 expression.[37] This finding has been confirmed by another study that found high AIB1 expression was correlated with early disease recurrence in tamoxifen-treated patients.[38] Therefore, it appears likely that, in addition to the role of AIB1 as an oncogene associated with the onset and progression of breast cancer, increased expression of this coactivator is associated with the development of tamoxifen resistance and poorer patient survival.

Tamoxifen resistance in breast cancer also has been associated with low NCoR expression in ERα-positive tumors. This was first observed in a mouse xenograft model of 4HT-resistant breast cancer and was later observed for a group of tamoxifen-treated breast cancer patients.[23,39] In this population, low tumor levels of NCoR were associated with a significantly shorter relapse-free survival, and together, these reports suggest that low levels of NCoR limit the antagonist properties of tamoxifen. The available data for the SMRT corepressor is mixed. One study found no difference in SMRT mRNA expression levels between 4HT treated and untreated breast tumors,[40] while another report indicates that SMRT mRNA expression is increased in anti-estrogen resistant MCF-7 breast cancer cells.[33] Thus, while there are data consistent with the hypothesis that breast cancer resistance to hormonal therapy is related to decreased corepressor or increased coactivator expression, this is not the case for all studies and this suggests that other factors can influence SRM biocharacter.

16.4 Multiple Cellular Processes Influence SRM Biocharacter

The original model in which the relative expression of coactivators and corepressors determined the biological activity of SERMs was proposed at a time before it was appreciated that there are literally hundreds of coactivator and corepressor molecules. This fact significantly complicates the assessments of whether the relative expression levels of coactivators and corepressors in any given tissue influence SRM activity because of the large number of factors that need to be considered. Moreover, as new coregulators were identified, and additional studies on the functional regulation of coregulators were completed, it became clear that the original model in which relative coregulator expression determined SRM biological activity was incomplete. Indeed, recent studies have revealed that the activity of coregulators are influenced by a large number of intracellular signaling pathways, and thus the cell specificity of SRM agonist *versus* antagonist action results from not only cell- or tissue-specific expression of coregulators, but also the ability of the cell environment to influence the function of the coregulators themselves and their ability to interact with other components of the cellular machinery that regulates the expression of genes, from transcription to protein synthesis.

16.4.1 *Relationships between nuclear receptor expression and response to SRMs*

The original SRM-coregulator hypothesis was based on the idea that distinct ligands induce distinct changes in the conformation of the receptor that enable it to interact with either coactivators or corepressors, and suggested that knowing the identify of the ligand-receptor-coregulator complex would enable an accurate prediction of whether a given SRM would elicit a hormone or anti-hormone-like response. However, recent work by many investigators has demonstrated that the expression and the activity of receptors as well as coregulators can be modified, and this makes the predictions of SRM activity difficult.

Gene expression profiling studies in anti-androgen-resistant LNCaP prostate cancer cells found that the treatment of AR over-expressing cells with the selective AR modulator (SARM) bicalutamide resulted in the concomitant recruitment of AR, NCoR, RNA polymerase II, SRC-1 and the coactivator PCAF to the PSA (prostate specific antigen) promoter, followed by the acetylation of histones H3 and H4 and transcriptional activation.[41] In this instance, it was the relative expression levels of receptor that correlated with the SARM's agonist activity. Similar results were found from profiling isogenic prostate cancer xenograft models whereby a modest increase in AR mRNA and protein expression was the only change consistently correlated with the development of anti-androgen resistance.[41] Increased AR expression altered the recruitment profile of coactivators and corepressors to the promoter region of AR-responsive genes shifting the normal response to bicalutamide from inhibition to stimulation of gene expression. Gene expression profiling studies of androgen-ablation resistant human prostate tumors similarly found increases in AR expression, as well as several genes involved in steroid biosynthesis, indicating the androgen-responsive pathway was reactivated in these tumors.[42]

In the case of ER, results from studies that examine the relationship between receptor expression and the development of resistance to SERM treatment are less clear. One study reported that the acquisition of 4HT resistance was positively correlated with changes in expression of PR and several coactivators, but not ER,[43] while another reported that resistance to anti-estrogen treatment and estrogen-independent growth of breast cancer cells was accounted for by high levels of ER expression driving increased rates of transcription.[44] More clear is the effect of SERM treatment on ER-mediated coregulator expression levels. The treatment of breast cancer cells with 4HT or raloxifene increased the steady-state levels of the coactivators SRC-1 and SRC-3 in an ER-dependent manner, thereby increasing the transcriptional activity of nuclear receptors in the cells and providing a potential mechanism for the tissue-selective biological effects of SERMs.

16.4.2 *Effect of cell signaling on NR-coregulator functional interactions*

In addition to the relative expression levels of NRs and coregulators, post-translational modifications of the proteins that comprise a coregulator complex also affect the transcriptional response to ligand. In general, the activation of SRs either in response to ligand or through activation of second-messenger signaling pathways is accompanied by increased receptor phosphorylation,[45,46] and the increase in transcriptional activity is mediated in part by the alterations in NR-coregulator interactions. For example, the treatment of cells overexpressing SRC-1 with forskolin, an upstream stimulator of protein kinase A (PKA), enabled 4HT to stimulate ERα transcriptional activity,[22] and a recent study revealed that phosphorylation of ERα's serine[305] by PKA alters the orientation of SRC-1 binding to ERα and renders the complex transcriptionally active.[47] Epidermal growth factor-mediated phosphorylation of ERα at serine[118] inhibits the interaction of the 4HT-bound receptor with the corepressor NCoR, resulting in a switch in 4HT action from antagonist to agonist.[23] Similarly, PKA phosphorylation of PR inhibits the association of the corepressors SMRT and NCoR with PR bound to the SPRM RU486, thus promoting the agonist activity of this compound.[48]

For AR, the treatment of the LNCaP prostate cancer cell line with proinflamatory signaling molecules such as interleukin-1β (IL-1β) enables the partial AR antagonists bicalutamide and CPA to stimulate the expression of AR target genes, albeit at a reduced efficiency compared with a pure agonist.[49] This effect also was proposed to reflect an allosteric modulation of the receptor-coactivator interaction resulting in the formation of transcriptionally-active complexes distinct from those formed in the presence of agonist. The phosphorylation of AR by PKA activates receptor-dependent gene expression and inhibits the recruitment of the SMRT corepressor to CPA-bound AR. This further indicates that SR-coregulator interactions are regulated by the signaling environment of the cell.[50]

Recent reports have shown that coactivators and corepressors themselves are substrates for a number of kinases. Phosphorylation of coregulators can influence their ability to interact with NRs as well as their ability to recruit additional coregulator complex components to target gene regulatory regions.[46,51] While these modifications are important for the activity of agonist bound receptors, there is some evidence to suggest that they may also play a role in determining whether SRMs will stimulate or block receptor function. In studies conducted with a MCF-7 breast cancer cell line engineered to over-express the HER2 receptor, tamoxifen functioned as an agonist stimulating cell growth and the expression of several ER target genes.[52] In these cells, tamoxifen treatment was associated with AIB1 phosphorylation and exposure to the HER2 tyrosine kinase inhibitor gefitinib blocked AIB1 phosphorylation and tamoxifen stimulation of gene expression. Moreover, ChIP analyses indicated that exposure of the HER2-overexpressing cells to 4HT resulted in the gefitinib-sensitive recruitment of the coactivators AIB1, CBP and p300 to the pS2 promoter while similar treatment of control MCF-7 cells was associated with recruitment of the NCoR corepressor and HDAC3, regardless of gefitinib treatment.

Phosphorylation pathways may also affect corepressor function. It has been shown that IL-1β stimulation of the MAPK signaling cascade in prostate cancer tumor cells promotes MEKK1-mediated phosphorylation of transforming growth factor β-activated kinase 1 binding protein 2 (TAB2), an adaptor protein and integral component of SARM- and SERM-bound steroid receptor corepressor complexes. This led to the removal of a NCoR-TAB2-HDAC complex from the promoters of steroid responsive genes, and the promotion of the agonist activity of the AR antagonists CPA and bicalutamide as well as the SERM 4HT and the SPRM RU486.[53] Thus, the cell and tissue environment is an important determinant of the functional consequences of NR-coregulator interaction on transcription and the agonist/antagonist activities of SRMs.

Phosphorylation of coregulators may also regulate their activity by influencing their subcellular localization. For example, the phosphorylation of SRC-3 by mitogen activated protein kinases (MAPKs) leads to its translocation to the nucleus and therefore in appropriately stimulated cells, there would be greater levels of this coactivator available to interact with SRM-bound receptors.[54] Correspondingly, the phosphorylation of SMRT and NCoR by various members of the MAPK signal transduction cascade leads to the export of these corepressors out of the nucleus to the cytoplasm and consequently, they are not available to participate in transcriptional repression.[55] In both of these instances, cellular redistribution of coactivators to the nucleus and corepressors to the cytoplasm would presumably create a cellular environment favoring the promotion of the agonist activity of SRMs and transcriptional activation. However, further research is needed to provide support for this mechanism of SRM agonist action.

16.4.3 *SRM regulation of transcription is gene-specific*

Nuclear receptors can bind to DNA directly via perfect and imperfect receptor-specific response elements or indirectly through tethering to non-receptor transcription factors such as Sp1 and AP-1. In addition to ligand-induced alterations in NR conformation, NR binding to DNA also influences receptor ligand affinity and conformation. These effects have functional consequences with regard to the transcriptional response to ligands including SRMs. For example, it has been shown that while the SERMs 4HT and raloxifene inhibit the transcription of promoters containing consensus EREs, ERα or ERβ bound to 4HT increases expression of AP-1-responsive reporter genes in HeLa cells.[56] The studies in the Ishikawa and ECC1 endometrial carcinoma cell lines also showed that the transcriptional response to the SERMs 4HT and raloxifene was variable and dependent on the receptor-DNA association.[27] In genes whose promoters contained a classical ERE such as cathepsin D and EBAG9, both SERMs function as antagonists, whereas 4HT, but not raloxifene, stimulated transcription of the ER-responsive genes c-Myc and IGF-1, whose promoters do not contain a classical ERE. Further examination found that 4HT, but not raloxifene, recruited a coactivator complex including SRC-1, AIB1 and CBP to the c-Myc promoter, which contains an AP-1 site. This suggests that the composition of the promoter is an important determinant of the agonist *versus* antagonist properties of SRMs. These results further highlight the functional differences between 4HT-mediated and raloxifene-mediated transcriptional activity.

In yet another example of the gene-specific nature of SERM activity, microarray analyses reveal the existence of groups of genes that are up-regulated by SERMs and down-regulated by E2, as well as a class of genes that are specifically regulated by SERMs but not by E2.[57] These include genes such as retinoblastoma 1 coiled-coil, the product of which has been identified as a tumor suppressor that could provide an additional benefit of SERM treatment of breast cancer.[57] Although the finding that there are a large number of genes that are up-regulated by agents such as tamoxifen in MCF-7 cells may seem, at first glance, to be

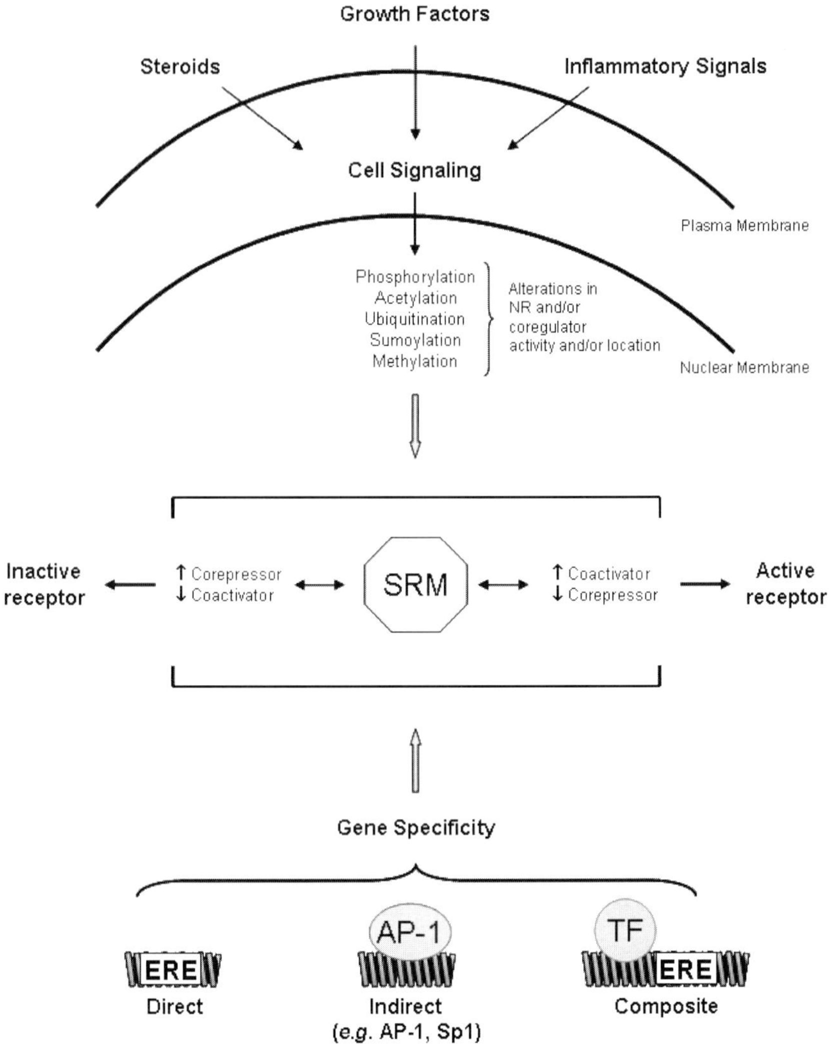

Fig. 16.3. Multiple factors determine how coactivators and corepressors influence the agonist/partial-antagonist properties of SRMs. As in the classical model (Fig. 16.2), the expression level of coactivators and corepressors in a cell influences whether a particular SRM promotes the formation of an active or inactive receptor-coregulator complex, with high corepressor/low coactivator environments favoring inactive receptors and high coactivator/low corepressor contexts promoting gene expression. The ability of coregulators to determine SRM biocharacter is influenced by extracellular stimuli such as steroids (via extranuclear signal transduction), growth factors and inflammatory molecules that

surprising, gene expression profiling studies of E2-treated MCF-7 cells indicate that a majority (70%) of the genes showing regulation are in fact down-regulated by E2,[58] and for many of these, 4HT partially or completely reverses the E2-induced repression. The involvement of coactivators in SERM-mediated reversal remains to be explored in greater detail, while the few studies that have examined the mechanism of E2-dependent transcriptional repression have implicated the corepressor NCoR and its associated HDACs in this process.[53,59] Nonetheless, the findings from these microarray studies indicate that the nature of the target gene is an important determinant of whether a given ligand will up- or down-regulate gene expression (Fig. 16.3). Given our current state of knowledge, it seems likely that cellular signaling events will also influence the ability of gene regulatory regions to determine whether E2 represses and SERMs stimulate gene expression.

16.5 Coregulators Play Diverse Roles in Gene Expression

The initial identification of many coactivators relied on the assumption that agonist-bound receptors will recruit coactivators, and the ability of molecules identified by this approach to enhance agonist-dependent gene expression was used to confirm that such molecules were indeed, coactivators. Likewise, apo-receptors as well as antagonist-bound nuclear receptors were used to isolate putative corepressors and again, functional assays were used to formally establish that such molecules were corepressors. The original SRM hypothesis proposed that the relative levels of coactivators and corepressors within a given cell environment will regulate the ability of SRMs to exert agonist-like or antagonist-like

---

initiate intracellular signaling pathways (shown on top). These signals alter coregulator function and subcellular localization, as well as interactions with nuclear receptors, and can thereby influence the ability of a SRM to stimulate or block the receptor's transcriptional activity. In addition, SRM regulation of transcription is gene-specific and the nature of the DNA binding site (e.g. direct or indirect) and the presence and proximity of additional transcription factors in the gene regulatory regions provides an organizational context that influences whether SRMs can activate gene expression (shown below). It is the combination of these multiple influences that ultimately determines the transcriptional outcome of SRM-nuclear receptor complexes.

effects on gene expression. It is based on the general belief that molecules designated as "coactivators" will stimulate gene expression and that molecules determined to be "corepressors" will inhibit gene expression. However, there are an increasing number of examples where coactivators are involved in the repression of genes and corepressors contribute to the stimulation of gene expression. This unexpected change in the role of these proteins adds an additional layer of complexity and likely contributes to the diverse range of responses that can be elicited by any given ligand-receptor-coregulator complex. Examples of "non-traditional" roles for coactivators and corepressors are given below.

16.5.1 *Repression of gene expression by coactivators*

Although there is abundant evidence that GRIP1 (SRC-2/TIF2) is a coactivator that can stimulate the transcriptional activity of agonist-bound nuclear receptors, several groups have also provided clear evidence that this molecule can repress gene expression. For instance, estradiol-bound ERα represses the expression of the TNFα gene, and this response is dependent on recruitment of GRIP1 to the TNFα promoter.[60] This is also observed for agonist-occupied GR bound to DNA via indirect tethering to AP-1 for genes such as collagenase-3, and it has been noted that the unique surfaces of GRIP1 are important for its repression activity.[61] Taken together, these data indicate that GRIP1 can function as a coactivator or corepressor in a promoter-specific manner.

There are additional examples which demonstrate that the recruitment of coactivators to promoters is not sufficient for activation of gene expression, and may even be involved in repression of the gene's expression. For example, estrogen treatment inhibits the proliferation of MDA-MB-231 cells stably expressing ERα, and this is accompanied by decreased c-Myc expression. ChIP analyses reveal E2-stimulated recruitment of the coactivators Med220 and SRCs, increased histone H4 acetylation and modest recruitment of RNA polymerase II to the c-Myc promoter even though c-myc expression was not stimulated.[62] ERα binds the c-Myc promoter indirectly through interaction with the AP-1 transcription factor and as suggested above, the nature of the receptor-ligand complex with DNA may influence the recruitment of coactivators and their effects on transcription.

16.5.2 *Stimulation of gene expression by corepressors*

The nature of the interaction between the ligand-receptor complex and DNA can substantially influence the ability of corepressors to modulate gene expression. For example, the SMRT corepressor stimulates transcription via unliganded TRα and a negative thyroid hormone response element[63] and N-CoR has been shown to stimulate RAR activity on a DR+1 element.[24] Recent data also indicates that full-length SMRT can stimulate estrogen-dependent ERα transcriptional activity measured on a synthetic reporter in a cell type-dependent manner, and selectively increase the expression of endogenous ERα target genes in breast cancer cells.[64] The stimulation of agonist-dependent transcription by SMRT is not, however, common to all receptors as the agonist-dependent activity of the androgen, thyroid hormone-β and vitamin D receptors is increased in cells depleted of SMRT expression.[64] This indicates that in addition to cell- and gene-specific events that determine coregulator function, there are also receptor-specific interactions that influence the ability of coregulators to positively or negatively regulate gene expression.

Typically, corepressors act in conjunction with histone deacetylases to achieve and maintain a closed chromatin structure associated with transcriptional repression. However, HDAC1 associates with hormone-activated GR on the native MMTV promoter and this correlates with increased expression of this transcriptional unit.[65] Interestingly, HDAC1 itself is acetylated, perhaps by the coactivator p300 and this negatively regulates HDAC1 deacetylase activity. Thus, the hypoacetylated form of HDAC1 associated with agonist-bound GR has relatively high deacetylase activity. Continued activation of the gene results in the acetylation of HDAC1 and consequently a reduction in its deacetylase activity. This suggests that the enzymatic activity of HDAC1 contributes to agonist-induced gene expression.[65] Thus, this scenario provides another example of coactivator (i.e. p300) playing an inhibitory role through reducing the function of HDAC1.

16.5.3 *Ligand regulation of coactivator and corepressor interactions with nuclear receptors*

Although agonists are generally associated with coactivator recruitment and antagonists with corepressor recruitment, this is clearly an oversimplification and examples of antagonist-bound receptors interacting with coactivators have been discussed above. Perhaps less well appreciated is the growing list of examples in which agonist-bound receptors bind to corepressors. This is observed for target genes positively regulated by E2-bound ERα with SMRT.[64] It is also seen for genes whose expression is repressed by agonist-occupied receptors. For example, cyclin G2, which encodes a negative regulator of the cell cycle, is a primary ER target gene in MCF-7 cells that is down-regulated in response to treatment with E2. ChIP analyses reveal the recruitment of ERα, NCoR and HDAC1 to the cyclin G2 promoter in response to treatment with E2, followed by hypoacetylation of chromatin and the dismissal of RNA polymerase II.[59] Regions outside of the ER's LBD were shown to be required for down-regulation indicating a possible novel interaction site between ER and NCoR . In another study, E2 treatment of MCF-7 cells resulted in the recruitment of NCoR to the promoters of BMP7, ABCG2 and BCL3, all of which are E2 down-regulated genes.[53] These findings suggest a role for corepressors in mediating the down-regulation of gene expression by agonist-occupied NRs. Other corepressors, including LCoR, REA, MTA1 and RIP140 also serve to inhibit the transcriptional activity of agonist-bound receptors.[66]

Another way in which corepressors influence the activity of agonist-bound receptors is via competing with coactivators for interaction with the receptor. In LNCaP prostate cancer cells, the SMRT and NCoR corepressors not only mediate antagonist-dependent inhibition of gene expression but also suppress agonist-dependent transcriptional activation. ChIP analyses demonstrate that SMRT and NCoR compete with coactivators for binding to agonist-occupied receptors, and in doing so, suppress gene expression.[21]

The recruitment of corepressors by agonist-occupied NRs, corepressor-mediated transcriptional up-regulation and coactivator-mediated transcriptional down-regulation clearly do not conform to the traditional view that ligand-induced conformational changes determine whether NRs recruit coactivators or corepressors to increase or decrease gene expression, respectively. Thus, these examples collectively provide an interesting perspective on the ability of ligands to regulate receptor interaction with coregulators. Not only do antagonist-bound receptors bind to coactivators and corepressors (i.e. the SERM concept), but this is also true to some extent for receptors in the presence of agonists. This emphasizes that the context of cell environment and target gene plays a critical role in determining the transcriptional output of any given receptor-ligand combination, and suggests that the selective activities of SERMs make use of molecular mechanisms employed by the receptor's cognate ligand to fine tune its transcriptional activity under normal physiological contexts.

16.6 SRMs — Translation Across the NR Superfamily

The ability to define molecular mechanisms that distinguish the ability of estradiol, tamoxifen and raloxifene to regulate ER function has had important implications for drug discovery. Indeed, in demonstrating that different SERMs induce ERs to adopt different structural conformations, and that these conformations enable the receptor to differentially interact with potentially hundreds of coregulators in a gene- and cell-specific manner, it became clear that not only should it be possible to identify a new generation of SERMs such as bazedoxifene, arzoxifene, lasofoxifene and ospemifen, but that the concept of selective receptor modulators should extend to other members of the nuclear receptor superfamily as has been done for PR, AR, thyroid hormone receptor (TR) and proliferating peroxisome activated receptor (PPAR). Thus, in addition to tamoxifen use for breast cancer patients in the treatment and chemoprevention settings, and the use of raloxifene for osteoporosis indications in otherwise healthy individuals, clinical trials are now evaluating selective receptor modulators for a number of other applications. Some of these will be reviewed below.

16.6.1 *Selective progesterone receptor modulators (SPRMs)*

Progesterone regulates mammary gland proliferation and opposes estrogen-mediated mitogenic activity in the uterus and these effects are mediated through the PR. The synthetic PR agonist R5020 and the antagonists mifepristone (RU486) and ZK98299 are currently used for regulating female reproduction and uterine function. However, neither compound is uterine-selective. In contrast, the SPRM, LG120838, suppresses estrogen-mediated uterine and vaginal epithelial proliferation and has low mitogenic activity in breast tissue.[67] Another potentially useful compound, the SPRM asoprisnil, has been shown to inhibit ovulation but not to induce labor, as seen with RU486. It is currently in clinical trials for the treatment of uterine fibroids and endometriosis. Asoprisnil can inhibit the proliferation of uterine leiomyoma cells, but does not affect the growth of cultured myometrial cells demonstrating its ability to act in a cell-specific fashion.[68] Crystallographic studies demonstrate that the PR LBD bound to asoprisnil interacts with peptides derived from the corepressors SMRT and NcoR.[69] In addition, asoprisnil-bound PR interacted with the coactivators AIB1 and TIF2, and this distinguishes this compound from RU486. Thus, as is the case for SERMs and ERs, this SPRM induces a unique conformational change in the PR LBD that enables it to interact with corepressors and coactivators in a manner distinct for other SPRMs, and presumably this is the basis of its unique biological properties.

16.6.2 *Selective androgen receptor modulators (SARMs)*

Androgens (testosterone and dihydrotestosterone) play an important role in male reproductive function, building bone and muscle mass and regulating spermatogenesis, prostate growth and male sexual behavior. The effects of androgens are mediated through binding to the AR and replacement androgen therapy has been used to treat primary osteoporosis, primary and secondary hypogonadism, delayed puberty in boys and muscle wasting. Androgen use, however, is limited by androgen effects in the prostate including concerns about androgens and the development and progression of prostate cancer. Several SARMs, including BMS-564929, LGD2226, JNJ-26146900 and S-4 have androgen-like effects in the muscle and skeleton without inducing prostate growth.[70–73] Peptide interaction assays and X-ray crystallography suggests that LGD2226 and BMS-264929 can induce structural changes in the AR LBD that are distinct from those obtained for AR agonists.[70,71]

16.6.3 *Selective thyroid hormone receptor modulators (STORMs)*

As for the class II receptors, TRs mediates the effects of thyroid hormone which is important for fetal and post-natal development, as well as the maintenance of homeostasis in adults. Excess TR activity has deleterious effects including muscle and bone atrophy, tachycardia and atrial arrhythmia. The development of selective TR modulators (STORMs) for use in tissue-specific interventions is ongoing, but candidates including

GC-1 and KB-141 show TR isoform-specific effects along with decreases in plasma cholesterol and triglyceride levels, and fat loss, without affecting heart or muscle tissue.[74] Interestingly, the nature of the T3 response element influences whether GC-1-bound TR recruits SRC-1 coactivator or corepressor complexes. This suggests that both DNA and ligand influence the structure of the receptor and hence its ability to interact with specific coregulators and regulate gene expression.[75] As for SERMs, this increases the potential for the development of STORMs to be able to mediate tissue- and pathway-selective responses.

16.6.4 *SRMs — More to come*

The similarities in mechanisms of action of nuclear receptors suggest that it will be possible to identify selective receptor modulators for additional members of this superfamily of transcription factors. Indeed, leads for selective modulators of retinoid X receptors (RXRs) such as LG101506 and peroxisome proliferator-activated receptor-γ (PPARγ) such as FK614 support this contention. However, even with the knowledge obtained from many investigators about the mechanisms that enable tissue or gene selective effects to be achieved, screening for new selective modulators still requires a significant investment, in no small part because of the number of factors that can influence SRM biocharacter, and this suggests a bright, if not rapid future for the development of new SRMs.

Acknowledgments

The authors apologize to the many authors whose work was not cited due to space limitations. The authors acknowledge support by NIH grants DK53002 and DK64038.

References

1. Mangelsdorf DJ, Thummel C, Beato M, *et al.*, The nuclear receptor superfamily: The second decade, *Cell* **83**(6):835–839, 1995.
2. Nettles KW, Greene GL, Ligand control of coregulator recruitment to nuclear receptors, *Annu Rev Physiol* **67**:309–333, 2005.
3. Lavery DN, McEwan IJ, Structure and function of steroid receptor AF1 transactivation domains: Induction of active conformations, *Biochem J* **391**(Pt 3):449–464, 2005.

4. Shiau AK, Barstad D, Loria PM, *et al.*, The structural basis of estrogen receptor/coactivator recognition and the antagonism of this interaction by tamoxifen, *Cell* **95**(7):927–937, 1998.
5. Tsai MJ, O'Malley BW, Molecular mechanisms of action of steroid/thyroid receptor superfamily members, *Annu Rev Biochem* **63**:451–486, 1994.
6. Heery DM, Kalkhoven E, Hoare S, *et al.*, A signature motif in transcriptional co-activators mediates binding to nuclear receptors, *Nature* **387**(6634):733–736, 1997.
7. Hager GL, Nagaich AK, Johnson TA, *et al.*, Dynamics of nuclear receptor movement and transcription, *Biochim Biophys Acta* **1677**(1–3):46–51, 2004.
8. Metivier R, Penot G, Hubner MR, *et al.*, Estrogen receptor-alpha directs ordered, cyclical, and combinatorial recruitment of cofactors on a natural target promoter, *Cell* **115**(6):751–763, 2003.
9. Metivier R, Reid G, Gannon F, Transcription in four dimensions: Nuclear receptor-directed initiation of gene expression, *EMBO Rep* **7**(2):161–167, 2006.
10. Htun H, Holth LT, Walker D, *et al.*, Direct visualization of the human estrogen receptor alpha reveals a role for ligand in the nuclear distribution of the receptor, *Mol Biol Cell* **10**(2):471–486, 1999.
11. Long X, Nephew KP, Fulvestrant (ICI 182,780)-dependent interacting proteins mediate immobilization and degradation of estrogen receptor-alpha, *J Biol Chem* **281**(14):9607–9615, 2006.
12. McKenna NJ, Lanz RB, O'Malley BW, Nuclear receptor coregulators: Cellular and molecular biology, *Endocr Rev* **20**(3):321–344, 1999.
13. Privalsky ML, The role of corepressors in transcriptional regulation by nuclear hormone receptors, *Annu Rev Physiol* **66**:315–360, 2004.
14. Hu X, Lazar MA, The CoRNR motif controls the recruitment of corepressors by nuclear hormone receptors, *Nature* **402**(6757):93–96, 1999.
15. Yoon HG, Chan DW, Huang ZQ, *et al.*, Purification and functional characterization of the human N-CoR complex: The roles of HDAC3, TBL1 and TBLR1, *EMBO J* **22**(6):1336–1346, 2003.
16. Zhang J, Kalkum M, Chait BT, *et al.*, The N-CoR-HDAC3 nuclear receptor corepressor complex inhibits the JNK pathway through the integral subunit GPS2, *Mol Cell* **9**(3):611–623, 2002.
17. Strahl BD, Allis CD, The language of covalent histone modifications, *Nature* **403**(6765):41–45, 2000.
18. Martin C, Zhang Y, The diverse functions of histone lysine methylation, *Nat Rev Mol Cell Biol* **6**(11):838–849, 2005.
19. Garcia-Bassets I, Kwon YS, Telese F, *et al.*, Histone methylation-dependent mechanisms impose ligand dependency for gene activation by nuclear receptors, *Cell* **128**(3):505–518, 2007.

20. Jepsen K, Rosenfeld MG, Biological roles and mechanistic actions of co-repressor complexes, *J Cell Sci* **115**(Pt 4):689–698, 2002.
21. Wang Y, Chirgadze NY, Briggs SL, *et al.*, A second binding site for hydroxy-tamoxifen within the coactivator-binding groove of estrogen receptor beta, *Proc Natl Acad Sci USA*. **103**(26):9908–9911, 2006.
22. Smith CL, Nawaz Z, O'Malley BW, Coactivator and corepressor regulation of the agonist/antagonist activity of the mixed antiestrogen, 4-hydroxyta-moxifen, *Mol Endocrinol* **11**(6):657–666, 1997.
23. Lavinsky RM, Jepsen K, Heinzel T, *et al.*, Diverse signaling pathways mod-ulate nuclear receptor recruitment of N-CoR and SMRT complexes, *Proc Natl Acad Sci USA* **95**(6):2920–2925, 1998.
24. Jepsen K, Hermanson O, Onami TM, *et al.*, Combinatorial roles of the nuclear receptor corepressor in transcription and development, *Cell* **102**(6):753–763, 2000.
25. Keeton EK, Brown M, Cell cycle progression stimulated by tamoxifen-bound estrogen receptor-alpha and promoter-specific effects in breast cancer cells deficient in N-CoR and SMRT, *Mol Endocrinol* **19**(6): 1543–1554, 2005.
26. Shang Y, Hu X, DiRenzo J, *et al.*, Cofactor dynamics and sufficiency in estrogen receptor-regulated transcription, *Cell* **103**(6):843–852, 2000.
27. Shang Y, Brown M, Molecular determinants for the tissue specificity of SERMs, *Science* **295**(5564):2465–2468, 2002.
28. Liu XF, Bagchi MK, Recruitment of distinct chromatin-modifying com-plexes by tamoxifen-complexed estrogen receptor at natural target gene promoters *in vivo*, *J Biol Chem* **279**(15):15050–15058, 2004.
29. Liu Z, Auboeuf D, Wong J, *et al.*, Coactivator/corepressor ratios modulate PR-mediated transcription by the selective receptor modulator RU486, *Proc Natl Acad Sci USA* **99**(12):7940–7944, 2002.
30. Lonard DM, Lanz RB, O'Malley BW, Nuclear receptor coregulators and human disease, *Endocr Rev* **28**(5):575–587, 2007.
31. Myers E, Fleming FJ, Crotty TB, *et al.*, Inverse relationship between ER-beta and SRC-1 predicts outcome in endocrine-resistant breast cancer, *Br J Cancer* **91**(9):1687–1693, 2004.
32. Fleming FJ, Hill AD, McDermott EW, *et al.*, Differential recruitment of coregulator proteins steroid receptor coactivator-1 and silencing mediator for retinoid and thyroid receptors to the estrogen receptor-estrogen response element by beta-estradiol and 4-hydroxytamoxifen in human breast cancer, *J Clin Endocrinol Metab* **89**(1):375–383, 2004.
33. Sarvilinna N, Eronen H, Miettinen S, *et al.*, Steroid hormone receptors and coregulators in endocrine-resistant and estrogen-independent breast can-cer cells, *Int J Cancer* **118**(4):832–840, 2006.

34. Vadlamudi RK, Balasenthil S, Broaddus RR, *et al.*, Deregulation of estrogen receptor coactivator proline-, glutamic acid-, and leucine-rich protein-1/modulator of nongenomic activity of estrogen receptor in human endometrial tumors, *J Clin Endocrinol Metab* **89**(12):6130–6138, 2004.
35. Shah YM, Rowan BG, The SRC kinase pathway promotes tamoxifen agonist action in Ishikawa endometrial cells through phosphorylation-dependent stabilization of estrogen receptor (alpha) promoter interaction and elevated steroid receptor coactivator 1 activity, *Mol Endocrinol* **19**(3):732–748, 2005.
36. Kressler D, Hock MB, Kralli A, Coactivators PGC-1beta and SRC-1 interact functionally to promote the agonist activity of the selective estrogen receptor modulator tamoxifen, *J Biol Chem,* 2007.
37. Osborne CK, Bardou V, Hopp TA, *et al.*, Role of the estrogen receptor coactivator AIB1 (SRC-3) and HER-2/neu in tamoxifen resistance in breast cancer, *J Natl Cancer Inst* **95**(5):353–361, 2003.
38. Dihge L, Bendahl PO, Grabau D, *et al.*, Epidermal growth factor receptor (EGFR) and the estrogen receptor modulator amplified in breast cancer (AIB1) for predicting clinical outcome after adjuvant tamoxifen in breast cancer, *Breast Cancer Res Treat*, 2007.
39. Girault I, Lerebours F, Amarir S, *et al.*, Expression analysis of estrogen receptor alpha coregulators in breast carcinoma: Evidence that NCOR1 expression is predictive of the response to tamoxifen, *Clin Cancer Res* **9**(4):1259–1266, 2003.
40. Chan CM, Lykkesfeldt AE, Parker MG, *et al.*, Expression of nuclear receptor interacting proteins TIF-1, SUG-1, receptor interacting protein 140, and corepressor SMRT in tamoxifen-resistant breast cancer, *Clin Cancer Res* **5**(11):3460–3467, 1999.
41. Chen CD, Welsbie DS, Tran C, *et al.*, Molecular determinants of resistance to antiandrogen therapy, *Nat Med* **10**(1):33–39, 2004.
42. Holzbeierlein J, Lal P, LaTulippe E, *et al.*, Gene expression analysis of human prostate carcinoma during hormonal therapy identifies androgen-responsive genes and mechanisms of therapy resistance, *Am J Pathol* **164**(1):217–227, 2004.
43. Scott DJ, Parkes AT, Ponchel F, *et al.*, Changes in expression of steroid receptors, their downstream target genes and their associated co-regulators during the sequential acquisition of tamoxifen resistance *in vitro*, *Int J Oncol* **31**(3):557–565, 2007.
44. Kuske B, Naughton C, Moore K, *et al.*, Endocrine therapy resistance can be associated with high estrogen receptor alpha (ERalpha) expression and reduced ERalpha phosphorylation in breast cancer models, *Endocr Relat Cancer* **13**(4):1121–1133, 2006.

45. Lannigan DA, Estrogen receptor phosphorylation, *Steroids* **68**(1):1–9, 2003.
46. Wu RC, Smith CL, O'Malley BW, Transcriptional regulation by steroid receptor coactivator phosphorylation, *Endocr Rev* **26**(3):393–399, 2005.
47. Zwart W, Griekspoor A, Berno V, *et al.*, PKA-induced resistance to tamoxifen is associated with an altered orientation of ERalpha towards coactivator SRC-1, *EMBO J* **26**(15):3534–3544, 2007.
48. Wagner BL, Norris JD, Knotts TA, *et al.*, The nuclear corepressors NCoR and SMRT are key regulators of both ligand- and 8-bromo-cyclic AMP-dependent transcriptional activity of the human progesterone receptor, *Mol Cell Biol* **18**(3):1369–1378, 1998.
49. Baek SH, Ohgi KA, Nelson CA, *et al.*, Ligand-specific allosteric regulation of coactivator functions of androgen receptor in prostate cancer cells, *Proc Natl Acad Sci USA* **103**(9):3100–3105, 2006.
50. Dotzlaw H, Moehren U, Mink S, *et al.*, The amino terminus of the human AR is target for corepressor action and antihormone agonism, *Mol Endocrinol* **16**(4):661–673, 2002.
51. Smith CL, O'Malley BW, Coregulator function: A key to understanding tissue specificity of selective receptor modulators, *Endocr Rev* **25**(1): 45–71, 2004.
52. Shou J, Massarweh S, Osborne CK, *et al.*, Mechanisms of tamoxifen resistance: Increased estrogen receptor-HER2/neu cross-talk in ER/HER2-positive breast cancer, *J Natl Cancer Inst* **96**(12):926–935, 2004.
53. Zhu P, Baek SH, Bourk EM, *et al.*, Macrophage/Cancer cell interactions mediate hormone resistance by a nuclear receptor derepression pathway, *Cell* **124**(3):615–629, 2006.
54. Amazit L, Pasini L, Szafran AT, *et al.*, Regulation of SRC-3 intercompartmental dynamics by ER and phosphorylation, *Mol Cell Biol,* 2007.
55. Paech K, Webb P, Kuiper GG, *et al.*, Differential ligand activation of estrogen receptors ERalpha and ERbeta at AP1 sites, *Science* **277**(5331): 1508–1510, 1997.
56. Webb P, Nguyen P, Valentine C, *et al.*, The estrogen receptor enhances AP-1 activity by two distinct mechanisms with different requirements for receptor transactivation functions, *Mol Endocrinol* **13**(10):1672–1685, 1999.
57. Frasor J, Stossi F, Danes JM, *et al.*, Selective estrogen receptor modulators: Discrimination of agonistic versus antagonistic activities by gene expression profiling in breast cancer cells, *Cancer Res* **64**(4):1522–1533, 2004.
58. Frasor J, Danes JM, Komm B, *et al.*, Profiling of estrogen up- and down-regulated gene expression in human breast cancer cells: Insights into gene networks and pathways underlying estrogenic control of proliferation and cell phenotype, *Endocrinology* **144**(10):4562–4574, 2003.

59. Likhite VS, Stossi F, Kim K, *et al.*, Kinase-specific phosphorylation of the estrogen receptor changes receptor interactions with ligand, deoxyribonucleic acid, and coregulators associated with alterations in estrogen and tamoxifen activity, *Mol Endocrinol* **20**(12):3120–3132, 2006.
60. Cvoro A, Tzagarakis-Foster C, Tatomer D, *et al.*, Distinct roles of unliganded and liganded estrogen receptors in transcriptional repression, *Mol Cell* **21**(4):555–564, 2006.
61. Rogatsky I, Luecke HF, Leitman DC, *et al.*, Alternate surfaces of transcriptional coregulator GRIP1 function in different glucocorticoid receptor activation and repression contexts, *Proc Natl Acad Sci USA* **99**(26): 16701–16706, 2002.
62. Acevedo ML, Lee KC, Stender JD, *et al.*, Selective recognition of distinct classes of coactivators by a ligand-inducible activation domain, *Mol Cell* **13**(5):725–738, 2004.
63. Berghagen H, Ragnhildstveit E, Krogsrud K, *et al.*, Corepressor SMRT functions as a coactivator for thyroid hormone receptor T3Ralpha from a negative hormone response element, *J Biol Chem* **277**(51):49517–49522, 2002.
64. Peterson TJ, Karmakar S, Pace MC, *et al.*, The silencing mediator of retinoic acid and thyroid hormone receptor (SMRT) corepressor is required for full estrogen receptor {alpha} transcriptional activity, *Mol Cell Biol* **27**(17):5933–5948, 2007.
65. Qiu Y, Zhao Y, Becker M, *et al.*, HDAC1 acetylation is linked to progressive modulation of steroid receptor-induced gene transcription, *Mol Cell* **22**(5): 669–679, 2006.
66. Gurevich I, Flores AM, Aneskievich BJ, Corepressors of agonist-bound nuclear receptors, *Toxicol Appl Pharmacol* 2007.
67. Wardell SE, Edwards DP, Mechanisms controlling agonist and antagonist potential of selective progesterone receptor modulators (SPRMs), *Semin Reprod Med* **23**(1):9–21, 2005.
68. Chen W, Ohara N, Wang J, *et al.*, A novel selective progesterone receptor modulator asoprisnil (J867) inhibits proliferation and induces apoptosis in cultured human uterine leiomyoma cells in the absence of comparable effects on myometrial cells, *J Clin Endocrinol Metab* **91**(4): 1296–1304, 2006.
69. Madauss KP, Grygielko ET, Deng SJ, *et al.*, A structural and *in vitro* characterization of asoprisnil: A selective progesterone receptor modulator, *Mol Endocrinol* **21**(5):1066–1081, 2007.
70. Ostrowski J, Kuhns JE, Lupisella JA, *et al.*, Pharmacological and X-ray structural characterization of a novel selective androgen receptor modulator: Potent hyperanabolic stimulation of skeletal muscle with hypostimulation of prostate in rats, *Endocrinology* **148**(1):4–12, 2007.

71. Miner JN, Chang W, Chapman MS, *et al.*, An orally active selective androgen receptor modulator is efficacious on bone, muscle, and sex function with reduced impact on prostate, *Endocrinology* **148**(1):363–373, 2007.
72. Kearbey JD, Gao W, Narayanan R, *et al.*, Selective Androgen Receptor Modulator (SARM) treatment prevents bone loss and reduces body fat in ovariectomized rats, *Pharm Res* **24**(2):328–335, 2007.
73. Allan G, Lai MT, Sbriscia T, *et al.*, A selective androgen receptor modulator that reduces prostate tumor size and prevents orchidectomy-induced bone loss in rats, *J Steroid Biochem Mol Biol* **103**(1):76–83, 2007.
74. Flamant F, Gauthier K, Samarut J, Thyroid hormones signaling is getting more complex: STORMs are coming, *Mol Endocrinol* **21**(2):321–333, 2007.
75. Gloss B, Giannocco G, Swanson EA, *et al.*, Different configurations of specific thyroid hormone response elements mediate opposite effects of thyroid hormone and GC-1 on gene expression, *Endocrinology* **146**(11): 4926–4933, 2005.

Chapter 17

Coregulators in Toxicology

Joëlle Rüegg, Malin Hedengran-Faulds, Manuela Hase, Ingemar Pongratz and Jan-Åke Gustafsson

The response to a xenobiotic insult is mainly mediated by the bHLH-PAS protein aryl hydrocarbon receptor (AhR) and the nuclear receptors SXR/PXR and CAR. They are therefore called xenosensors. The binding of toxic substances to these receptors results in the transcription of genes encoding enzymes involved in the metabolism and transport of xenobiotics and their metabolites. Coregulators play an important role in the signal transduction of AhR, SXR/PXR, and CAR. Many of these coregulators are also crucial for the signaling of the hormone receptors in the nuclear receptor family. Thus, in the presence of xenobiotics, the recruitment of coactivators to the xenosensors can lead to impaired hormonal signaling. This is one of the mechanisms that may explain endocrine disruption (impaired hormone response after chemical exposure), a phenomenon that has raised attention and concern during recent decades.

17.1 Introduction

All living organisms continuously encounter a changing environment. To survive, they have developed different systems to meet and adapt to challenges posed by the varying conditions, including fluctuating temperatures, variable levels of nutrients and the presence of xenobiotics, for instance, drugs and environmental pollutants.

Chemicals are a part of contemporary life for all members of society. A recent European survey reported that the average European consumer is exposed to up to 10 000 different chemicals on a daily basis. However, the health implications of this exposure have not been characterized for the vast majority of chemicals and there is a lack of knowledge both with respect to whether and how different chemicals cause disease.

Nuclear receptors (NRs) as well as aryl hydrocarbon receptor (AhR) are the main players in activating genes for biological responses to xenobiotic exposure. As coregulators are crucial for the function of both NRs and the AhR, it is evident that they play an important role in the metabolism of toxic substances. However, this is not the only way in which NRs and coregulators are involved in toxic responses. Some xenobiotics can also interfere with hormonal pathways in the organism, a phenomenon called endocrine disruption (ED). In particular, many environmental pollutants are known to affect steroid signaling, which is mediated by a subfamily of NRs. ED has raised attention and concern during recent decades, and might even represent a prominent mechanism by which toxic substances can cause diseases such as cancer. The involvement of coregulators in xenobiotic responses and endocrine disruption is not well understood yet. However, we will here make an attempt to summarize current knowledge in this field.

17.2 Coregulators in Xenobiotic Response

Exposure of the organism to toxic substances induces a wide array of genes whose products are involved in the metabolism of these substances. This process occurs in three phases: phase I includes the oxidation of xenobiotics which leads to activated hydrophobic products, in phase II these products are converted into hydrophilic molecules via conjugation reactions with glutathione, sulfate, or glucuronic acid, and in phase III the conjugates are transported out of the cell by ATP-dependent export pumps (Fig. 17.1).[1]

In mammals, studies have demonstrated that at least three different transcription factor superfamilies are involved in the induction of the enzymes necessary for the xenobiotic response, namely basic-helix-loop-helix/Per-ARNT-Sim (bHLH-PAS) proteins, nuclear receptors (NRs) and basic leucine zipper (bZIP) proteins. In particular, selected members of the bHLH-PAS family of transcription factors are critical mediators of the biological effects of some environmental pollutants like, for example, dioxin. The bHLH-PAS protein aryl

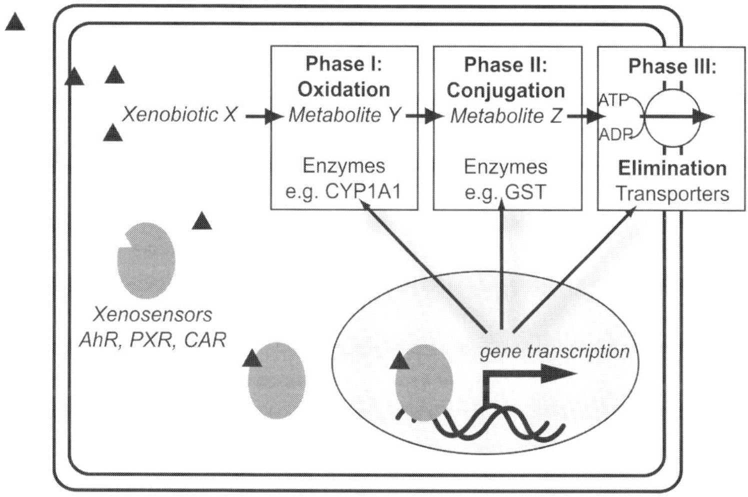

Fig. 17.1. Overview over the xenobiotic response. Xenobiotic substances are eliminated in 3 phases, oxidation, conjugation, and transport out of the cell. Genes needed for these processes are regulated by xenosensors like arylhydrocarbon receptor (AhR), pregnane X receptor (PXR) and constitutive androstane receptor (CAR), which are activated by binding of the xenobiotic substance.

hydrocarbon receptor (AhR), which is a ligand-dependent transcription factor, has been clearly implicated in the biological response to polycyclic aromatic hydrocarbons (PAH) like benzo-[a]-pyrene and 3-methyl-cholanthrene, and toxic compounds which belong to the dioxin family such as 2,3,7,8-tetrachlorodibenzo-p-dioxin (TCDD, also known as Seveso Dioxin). Other transcription factors, such as the NR xenobiotic receptor/rodent pregnane X receptor (SXR/PXR) and constitutive androstane receptor (CAR), can interact with and be activated by xenobiotics, and are hence called xenosensors.[2] Other NRs, like farnesoid X receptor (FXR), liver X receptor (LXR) and perox some proliferator-activated receptors (PPARs), as well as bZIP proteins like Nrf2-Keap1, SREBP, and C/EBP are also involved in the regulation of xenobiotic responses. However, the main functions of these transcription factors are related to other signaling pathways, like regulation of lipid metabolism and modulation of the cellular response to oxidative stress.

17.3 The Aryl Hydrocarbon Receptor — Signaling Pathways, Interaction with Coregulators, and Physiological Functions

During evolution, mammalian organisms have developed the ability to protect themselves to exposure to different water-insoluble, often toxic, compounds by upregulating a battery of enzymes that metabolize xenobiotics into water-soluble derivatives which can be readily excreted. Among the best studied examples is the induction of an array of microsomal enzymes upon cellular exposure to polycyclic aromatic hydrocarbons (PAHs) and halogenated aromatic hydrocarbons (HAHs). Enzymes that are induced by PAHs and HAHs include cytochrome P-450-dependent monooxygenases, notably CYP1A1 and CYP1A2, glutathione S-transferase and NAD(P)H:quinone reductase.[3] In early studies, it was noticed that inducibility of microsomal enzymes by PAHs varied among different mouse strains. For example, C57 and C3H strains were known to be highly responsive to PAHs, whereas no response to these compounds was observed in DBA or AKR strains. Crosses and back-crosses of these strains demonstrated that differences in response to PAHs were mediated by a single autosomal locus, termed *Ah*, for aryl hydrocarbon responsiveness.[4] Later, it was found that HAHs, such as 2,3,7,8-tetrachlorodibenzo-*p*-dioxin (TCDD; dioxin) or 2,3,7,8-tetrachlorodibenzofuran, were much more potent inducers of microsomal enzymes than PAHs. Notably, HAH-dependent effects were also observed in the PAH non-responsive mouse strains. It was predicted that the greater potency of HAHs in induction of xenobiotic-metabolizing enzymes is related to their higher affinity to the putative receptor encoded by *Ah* locus.[4] The existence of this receptor, which today, carries two alternative names, aryl hydrocrbon receptor (AhR) and dioxin receptor, was confirmed by *in vitro* ligand-binding experiments. They revealed the presence of a small pool of high-affinity sites in cytosolic preparations, which specifically and reversibly bound ligand (TCDD) with an estimated K_d of approximately 1 nM.[4] Using photoaffinity labelling and noncovalent immobilization techniques, it was demonstrated that the dioxin receptor is an approximately 100 kDa protein that interacts with other proteins and forms hetero-oligomeric complexes reaching up to 300 kDa in molecular weight. Subsequently, cloning of the dioxin receptor from mouse and human genomic libraries revealed that the receptor shares regions of homology with the bHLH/PAS family proteins.[5] The dioxin receptor is a highly polymorphic protein, with allele variants differing both in ligand-binding activity and in molecular weight. According to recent studies, differences in molecular weights between various isoforms of the receptor are primarily due to differences in the position of the translation termination codon of the dioxin receptor, rather than differential splicing or posttranslational modifications.[3] According to Northern blot analysis, the highest expression levels of the dioxin receptor in humans were observed in the lung and placenta. It was expressed as the lowest in kidney, brain and skeletal muscle. In rats, however, ribonuclease protection assays showed that the receptor was most abundant in the lung, thymus, liver and kidney and less abundant in the heart and spleen.[6]

After cloning of the AhR, more ligands for the receptor have been identified. They can be divided into two major categories, synthetic and natural. Most of the ligands are synthetic and form the above mentioned subgroups, HAHs, including polychlorinated dibenzodioxins (PCDD), dibenzofurans (PCDF) and biphenyls (PCB), and PAHs like 3-methylcholantrene (3MC), benzo[a]pyrene (BaP) and benzoflavones.[7] TCDD, the most potent and toxic AhR ligand, is a congener of PCDD. The natural ligands include several prostaglandins that transiently induce Cyp1A1 activation.[8] The flavonoids genistein and resveratrol, components of green tea and cruciferous vegetable-derived compounds, or their metabolites, are also potent activators of AhR.[7] AhR is further strongly induced by tryptophan-derived products, like indirubin, indigo and 6-formylindolo[3, 2-6]carbazole (FICZ) found in human urine.[9,10]

The AhR is a member of the bHLH-PAS family of proteins. This family includes transcription factors like hypoxia-inducible factors HIF-1α and HIF-2α and the circadian regulatory proteins Clock and Per. In addition, the AhR nuclear translocator (ARNT) proteins form the ARNT sub-family within the bHLH-PAS family of transcription factors.[11] Three different ARNT factors have been identified, ARNT-1, ARNT-2, and ARNT-3 (bMAL). ARNT-1 and ARNT-2 display a high degree of sequence similarity and are often functionally interchangeable.[12] ARNT-1 and ARNT-2 function as general dimerization partners for the AhR, as well as the hypoxia-inducible factors HIF-1α and HIF-2α. In contrast, the third member of this subfamily, bMAL, is selectively recruited by the circadian regulator clock and does not support HIF-1α or AhR function.[13]

bHLH-PAS transcription factors share considerable sequence homology over two conserved structural domains, namely the bHLH domain and the PAS domain.[14] The bHLH domain is the key mediator of dimerization and DNA binding. In particular, specific DNA-binding activity is determined by the basic (b) region, whereas the HLH region serves as a major dimerization interface between bHLH-PAS proteins. The PAS domain is located immediately after the bHLH domain. This domain comprises around 130 amino acids and contains two highly conserved hydrophobic repeats termed PAS A and PAS B domains. The PAS domain is not as well characterized as the bHLH domain but some studies have shown that it provides dimerization specificity between bHLH-PAS proteins. In addition, this domain contains a nuclear export sequence and is involved in regulation of intracellular localization of some bHLH-PAS transcription factors.

Interestingly, the SRC type of coactivators also has bHLH-PAS domains and thus has considerable structural similarities with the bHLH-PAS transcription factors. The SRCs also have biological features in common with the ARNT proteins; both types of factors support the activity of several different transcription factors, are constitutively active, and have a constitutive nuclear localization pattern. In addition, recent studies suggest that ARNT proteins, like the SRC cofactors, can enhance the transcriptional response of selected members of the NR family, namely ERα and in particular ERβ.[15]

The non-activated form of AhR resides in the cytoplasm in complex with the chaperone proteins heat-shock protein (hsp)90, hepatitis B virus protein X-associated protein (XAP)2 and p23 co-chaperone.[16] The hsp90 complex is important for localizing AhR in the cytoplasm, as well as for stabilizing the ligand-binding conformation and protecting the AhR from degradation. XAP2 is thought to further stabilize the AhR-hsp90 complex against ubiquitination and thus against degradation. Finally, p23 is involved in the release of AhR from the hsp90 complex upon ligand-binding.[16] When activated by ligand, the receptor translocates to the nucleus, dissociates from the chaperone complex and binds the AhR dimerisation partner ARNT. The activated AhR/ARNT heterodimer complex binds to regulatory DNA elements, xenobiotic response elements (XREs) or AhREs, and activates the expression of AhR target genes. Finally, AhR is exported out of the nucleus and degraded by the ubiquitin-26S proteasome pathway.[17]

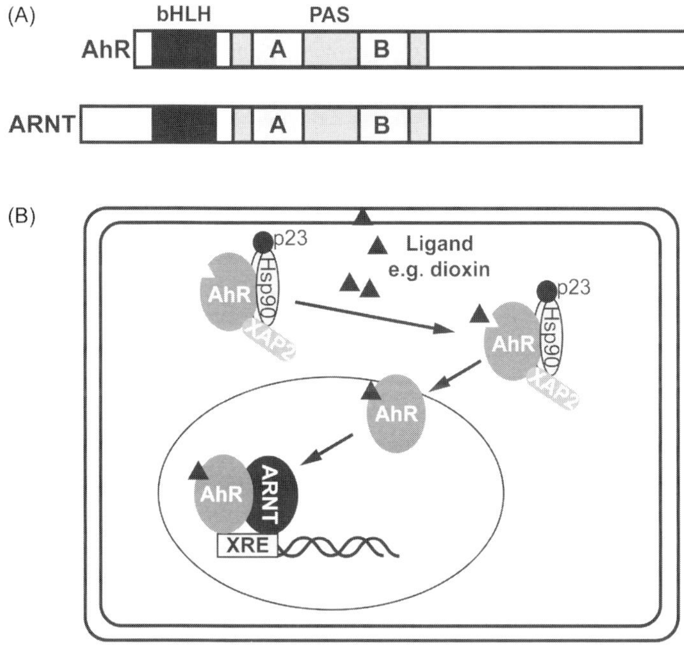

Fig. 17.2. (A) Structure of AhR and ARNT. (B) Signaling pathway of the AhR. The AhR resides in the cytoplasm bound to chaperones and co-chaperones like hsp90 and XAP2. After ligand binding, the receptor dissociates from the chaperone complex, translocates to the nucleus where it binds to ARNT and transcriptional coactivators. Finally, this complex binds to xenobiotic response elements (XREs) in the promoter region of target genes

There are also other mechanisms for AhR signaling (Fig. 17.2). The AhR is known to interact with other transcription factors, for example the Rb protein and NF-κB.[18] There are also reports on ligand-dependent activation of the AhR by phosphorylation,[19] facilitating nuclear translocation of AhR without interaction with ARNT. Furthermore, c-Src is a tyrosine kinase associated with AhR that triggers several phosphorylation cascades upon activation by AhR agonists.

The AhR is involved in cross talk with other signaling pathways, e.g. NF-κB, resulting in suppressed Cyp1A1 expression by inhibited histone acetylation.[18] Furthermore, HIF-1α is known to compete with AhR for ARNT leading to decreased transcriptional activity of AhR. A hypoxic state of the cell thus leads to decreased AhR gene induction. The activity of AhR is also influenced by activation of hormonal pathways. Adrenal corticoids are reported to modulate AhR functions. Furthermore, Cyp1A1 induction by AhR ligands is potentiated in the presence of glucocorticoids.[20] Cortisol treatment of pregnant mice results in pups with increased AhR levels in craniofacial tissues.[21] However, the synthetic glucocorticoid dexamethasone represses AhR levels in mammary fibroblasts of rats.[21] Additionally, thyroid hormones modulate AhR signaling; thyroidectomy protects rats from some of the toxic effects following TCDD-exposure.[22] An important cross talk is the one between AhR and estrogen receptors. 17β-Estradiol (E2) enhances AhR function, e.g. in case of AhR-mediated induction of Cyp1A1 in a mouse ovarian cancer cell line.[23] In breast cancer cells, E2 increases expression of AhR and thereby influences the activity of AhR-regulated genes.[24] The underlying mechanism for this cross talk is not fully understood. A recent study demonstrated that ERα is recruited to the AhR-regulated CYP1A1 promoter after TCDD treatment, and this recruitment is enhanced in the presence of E2.[25] This coincided with increased transcription of CYP1A1 mRNA after treatment with TCDD and E2. On the other hand, when ERα was down regulated using small inhibitory RNA, E2 did not increase AhR transcriptional activity. These findings suggest a mechanism whereby ERα is recruited to AhR-regulated promoters and thereby enhances AhR-mediated gene expression. Not only does ERα affect AhR signaling, but activation of AhR also impacts on ER function. Dioxins have well-established anti-estrogenic properties in the female reproductive tract, like inhibition of estradiol-induced increase in uterine wet weight, as well as decreased levels of ERs and progesterone receptors (PRs) in rodent uterus.[26] The inhibitory action of AhR ligands on ER signaling is in fact the best-studied case of endocrine disruption. The underlying mechanisms are discussed below.

AhR target genes include enzymes involved in metabolism of xenobiotics, mainly phase I drug metabolizing monooxygenases, such as cytochrome P450 1A1 (CYP1A1), CYP1A2, and CYP1B1 but also phase II enzymes like glutathione-S-transferase (GST), UDP-glucuronyltransferase, NADPH-quinone oxidoreductase and xanthinoxidase.[27] Furthermore, AhR activates a multitude of genes other than the drug metabolizing enzymes. A battery of genes involved in growth, differentiation and cellular homeostasis are induced by

TCDD-activated AhR signaling including transforming growth factors (TGFs) α and β, plasminogen activator inhibitor-2 and c-fos and c-jun.[28] Additionally, AhR regulates expression of AhR repressor (AhRR), a novel protein belonging to the bHLH-PAS family. AhRR inhibits AhR function by competing with AhR for dimerization with ARNT and subsequently binds to XREs. The AhRR/ARNT complex is transcriptionally inactive and formation of this complex competes with the transcriptionally active AhR/ARNT complex. Thus, AhR and AhRR form a regulatory circuit in the xenobiotic signaling pathway.[29] The fact that these proteins regulate their own expression via up-regulation of specific repressor molecules designed to silence their own signal transduction is a common property of members of the bHLH-PAS protein family. Another example of this paradigm is the regulation of the expression of the repressor Per by the members of the clock gene family.[30]

Mechanistic studies regarding the regulation of gene expression by the AhR/ARNT complex have mainly been focused on the CYP1A1 gene promoter in response to TCDD. Numerous studies have characterized the CYP1A1 promoter structure. It consists of an enhancer region approximately 1000 bp upstream of the transcription start site that comprises several XREs, and a TATA box region with a TATA box as well as a CCAAT box in proximity to the transcription initiation site.[31] A model has been suggested where TCDD-activated AhR/ARNT binds to XREs located in the enhancer, leading to recruitment of numerous coactivators and transcription factors to both regions, thus forming a large multiprotein complex bridging the two promoter regions. This event leads to nucleosomal disruption over the transcribed region and subsequent mRNA synthesis.[25,31]

An array of coactivators has been shown to physically interact with the AhR, including CBP/p300, the p160 coactivators, BRG1, TRAP220,[31] and nuclear receptor coactivator 4.[32] Using chromatin immunoprecipitation (ChIP) assays, most of these coactivators were confirmed to bind to the promoter region of the CYP1A1 gene. Binding of the p160 coactivators precedes recruitment of CBP/p300, TRAP220 and polymerase II to the promoter.[25,31] The model in Fig. 17.3(B) illustrates coactivator recruitment to the CYP1A1 promoter. Several factors are known to negatively regulate AhR activity besides the repressor AhRR mentioned above. These include corepressors like RIP140[33] and small heterodimer partner (SHP).[34] Additionally, the ARNT interacting

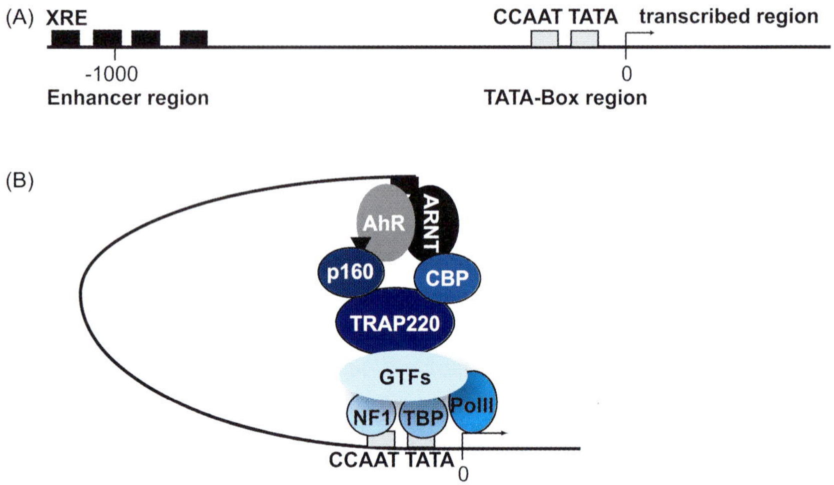

Fig. 17.3. (A): Structure of the CYP1A1 promoter. The enhancer region lies approx. 1000 bp upstream of the transcription start site and contains several XREs. The TATA-box region is located close to the transcription start site and consists of a CCAAT-box in addition to the TATA-box. (B) Activation of the CYP1A1 promoter. The activated AhR/ARNT complex binds to XREs in the enhancer region. This leads to the recruitment of coactivators like CBP and p160 coactivators, of the mediator complex subunit TRAP220, and other transcription factors to both regions, thus forming a large multiprotein complex bridging the two promoter regions. GTF: general transcription factors, NF1: nuclear factor 1, TBP: TATA binding protein.

protein (AINT) is known to bind ARNT and thereby decrease its nuclear localization.[35]

A wide variety of toxic responses to TCDD has been observed in animals, as well as in humans including endocrine disruption, cardiovascular diseases and certain types of cancer.[36] Humans exposed to TCDD suffer from various ailments, like endometriosis and breast cancer.[37] Another exposure effect is hyperinsulinemia, which links AhR signaling to type 2 diabetes.[38] Other significant responses are chloracne, immunosuppression, teratogenic responses and tumor promotion.[39]

Although AhR has been extensively studied, its physiological functions are poorly understood. $AhR^{-/-}$ mice have reduced liver size at birth due to reduced blood supply during fetal development, slowed early growth, reduced female fertility and disturbed immune system.[40] These knock-out mice also show a number of differencies to wild type animals upon TCDD-exposure. Teratogenic responses, like cleft palates, are lost suggesting a role for AhR in this deformity.[41] Other significant responses to TCDD, like wasting syndrome, thymic atrophy and decreased prostate weight are also missing in the $AhR^{-/-}$ mice.[42] Furthermore, AhR-defective liver cells show morphological changes, decreased synthesis of albumin and a slower growth rate than normal cells.[43] On the other hand, transgenic mice with constitutively active AhR develop tumors in the stomach, suggesting a role for AhR in growth and proliferation.[44]

17.4 SXR/PXR and CAR — Interaction with Coregulators and Linkage to Other Metabolic Pathways

Of the nuclear receptor family, rodent pregnane X receptor (PXR) and its human ortholog human steroid and xenobiotic receptor (SXR) as well as constitutive androstane receptor (CAR) are prominent regulators of metabolism of xenobiotics. PXR is an important regulator of expression of genes involved in all phases of drug metabolism and excretion. This transcription factor responds to a large number of ligands, including many endogenous compounds (e.g. pregnanes, bile acids and hormones) as well as pharmaceuticals, such as the anticancer drugs tamoxifen and taxol, and the antibiotic rifampicin.[1] Despite the homology between SXR and PXR, certain xenobiotics have opposite effects on these orthologs, activating PXR but antagonizing SXR.[45] This is an important finding, considering that rodent models are widely used to predict toxicity of a compound in humans.

PXR functionally overlaps with CAR by binding to the same ligands and DNA response elements to control target gene expression.[46] The extent of this overlap depends on a number of factors, such as the availability of endogenous or exogenous ligands as well as potential post-transcriptional regulation of target genes in different tissues. The diversity of drugs and xenobiotics recognized by PXR and CAR increases the protective capacity of the xenobiotic response system.

Both receptors are highly expressed in the liver and intestine where they regulate the expression of drug-metabolizing enzymes and transporters. Through the induction of phase I (cytochrome P450), phase II (conjugating), and phase III (ABC family transporters) metabolizing and detoxifying enzymes, CAR and PXR regulate the metabolism and secretion of endogenous and exogenous molecules.

PXR may control transcriptional events as a heterodimer with the retinoid X receptor-α (RXRα) and, depending on the ligands, RXR serves as a silent or an active partner of PXR.[47] Moreover, by interaction with different coregulators, PXR can activate or repress transcriptional events. By recruitment of the coactivators of the p160 familiy, e.g. steroid receptor coactivator (SRC), PXR mediates upregulation of gene expression. On the other hand the corepressors small heterodimer partner (SHP) and silencing mediator of retinoid and thyroid hormone receptors (SMRT) interact with PXR and repress its activity. By interactions of PXR with these coregulators, the functional repertoire of PXR is expanded. Additionally, the coactivator PGC-1α interacts with and increases the activity of PXR. On the other hand, the recruitment of PGC-1α by ligand-activated PXR impairs the function of hepatocyte nuclear factor 4 (HNF-4).[48] HNF-4 is a key regulator in cholesterol and glucose metabolism. Thus, the response of hepatocytes to foreign and endogenous substances that activate PXR is more than just activation of hepatic enzymes for detoxification and metabolism of those substances. PGC-1α links xenobiotic response to energy metabolism.

In contrast to the ligand dependent activation of PXR, CAR is constitutively active. However, CAR activity can be increased by agonists and decreased by so-called inverse agonists. It is thought that without ligand, the binding between CAR and coactivators is weak. The presence of an agonist strengthens this interaction, whereas an inverse agonist abolishes the binding to coactivators. CAR activity is altered by several classic endocrine-disrupting compounds e.g. metabolites of the pesticide DDT.[49] Coactivators involved in CAR signaling include those of the p160 family, PGC-1α, activating signal cointegrator-2 (ASC-2), TRAP220, PRIC320, and PPAR-binding protein (PBP).[49] Both PBP and the p160 coactivator GRIP-1 are important for the nuclear translocation of CAR, which in turn is essential for the regulation of its transcriptional activity. Like in the case of PXR, competition between CAR and HNF-4 for PGC-1α inhibits HNF-4 function. Additionally, competition for GRIP-1 is also involved in the inhibitory effect of activated CAR on HNF-4 activity.[50] So far, SHP is the only CAR interacting corepressor identified.[51]

More recently, additional roles for CAR and PXR have been discovered. For example, these xenosensors are involved in the homeostasis of cholesterol, bile acids, bilirubin, and other endogenous hydrophobic molecules in the liver. Hereby, CAR and PXR form an intricate regulatory network with other members of the NR superfamily, such as the cholesterol-sensing liver X receptor (LXR) and the bile-acid-activated farnesoid X receptor (FXR).[52]

Farnesoid X receptor (FXR) was identified as a bile acid receptor, and its activation results in the inhibition of hepatic bile acid biosynthesis and increased transport of bile acids from the intestinal lumen to the liver. Jung *et al.*[53] demonstrated that the expression of PXR is regulated by FXR. Here, bile acids activate FXR, which leads to increased transcriptional activation of PXR. This in turn results in the expression of cytochromes P450 and transporter genes involved in the breakdown and elimination of bile acids. In addition, the activation of FXR by high bile acid concentrations leads to increased expression of SHP and thereby inhibits the *de novo* synthesis of bile acids. The combination of these FXR-induced mechanisms leads to an efficient protection of the liver against bile acid induced toxicity.

The overlap of endogenous lipids that activate CAR, PXR, FXR, and LXR suggests a functional connection between these receptors in liver physiology. It appears that they act as generalized steroid and chemical sensors in the liver protecting against chemical perturbation. In addition to their function in the xenobiotic response, both PXR/SXR and CAR play an important role in the metabolism of steroids.[1] As a result, activation of these receptors by xenobiotics can change steroid hormone levels, leading to altered hormone responses and thus endocrine disruption.[54] CAR can alter endocrine signaling by promoting the clearance of molecules such as estrogen, progesterone and thyroid hormone through the regulation of the same enzymes that metabolize drugs. Conversely, PXR and CAR are also influenced by endocrine signals. For example, insulin is able to repress drug metabolism and PXR is activated by pregnanes, progesterone, and glucocorticoids.

17.5 Coregulators in Endocrine Disruption

Many environmental pollutants, but also some natural compounds, are known to interfere with hormonal signaling. There are at least three ways by which these endocrine disruptive chemicals (EDCs) can interfere with hormonal pathways. Firstly, they can bind directly to hormone receptors, either as agonists or as antagonists, and mimic the action of an endogenous hormone or block it. Secondly, they can affect the synthesis, transport, metabolism, and excretion of hormones, thus altering the endogenous hormonal levels. Thirdly, they can interfere with the activity of hormone receptors by changing the availability or the functioning of the receptors.

17.5.1 *Competition for common coactivators between hormone receptors and transcription factors involved in xenobiotic responses*

The biological function of the nuclear hormone receptors is closely connected to the availability of different coactivators. The recruitment of coactivators to another signaling pathway can decrease their availability; thus all coactivators that are shared between steroid receptors and xenobiotic receptors have the potential to be involved in the mechanism of endocrine disruption. However, this mechanism has only been described in a few cases so far. CAR inhibits estrogen receptor (ER) signaling by reducing the intracellular levels of the p160 coactivator GRIP-1 (so called squelching). The activation of CAR leads to reduced transcriptional activity of ERα, and the introduction of additional amounts of GRIP-1 by means of transient transfection into the cells can overcome this inhibition.[55]

In a similar fashion, competition for coactivators has been suggested as a mechanism for the anti-estrogenic effect of different AhR agonists, in particular ligands like the environmental pollutant TCDD.[56] The AhR and ERs share many coactivators, e.g. SRC-1, RIP140, and CBP/p300. Furthermore, overexpression of the transactivation domain of AhR harboring the cofactors binding sites inhibits ERα transcriptional activity.[27] This suggests that recruitment of coactivators to the AhR transcriptional activation domain inhibits the transcriptional activity of ERα and ERβ independently of other regulatory functions of the AhR. More recently, it has been shown that the AhR dimerisation partner AhR nuclear translocator (ARNT) acts as coactivator of both ERα and ERβ.[15] Overexpression of ARNT leads to increased ER activity, and small interference RNA mediated downregulation of ARNT results in an impairment of ER function.[15] In addition, ARNT is recruited to promoters of ER activated genes and activation of AhR by TCDD decreases this recruitment.[15] This suggests a competition between AhR and ER, not only for classical coactivators but also for ARNT. There is indication that ARNT can act as coactivator for other NRs than the ERs (unpublished observations), implying that this mechanism could account for the endocrine disruptive effects of AhR ligands on other hormonal systems as well. These findings suggest that the intracellular levels of different coactivators can dictate the cellular response to different hormones, and that sharing of coactivators may represent an important mechanism that regulates the cellular response to different signals.

17.5.2 *Changes of coregulator levels and activity by endocrine disruptors*

The levels of coregulators can be affected or modulated by additional means such as alterations in mRNA intracellular levels or by changes in the rate of protein degradation. In one case, it has been shown that the endocrine disrupting chemical bisphenol A increases mRNA and protein levels of the mediator complex TRAP220, thereby enhancing ER transcriptional activity (Fig. 17.4). Xenobiotic dependent regulation of degradation of coregulators has not been shown directly yet. However, there is evidence that EDC's may act on proteasome-mediated degradation of certain proteins. TCDD induces ubiquitination and consecutive proteasome-mediated degradation of ERα. A different EDC, phthalic acid, on the other hand, is able to block proteasome-mediated degradation of PXR. Coactivators of the p160 family like GRIP1 and SRC-1 are degraded via the proteasome, thus suggesting that their levels could also be affected by xenobiotic substances.[57]

The activity of coregulators can be changed by phosphorylation; for instance, phosphorylation by the mitogen-activated protein kinase (MAPK) enhances the activity of the coactivator AIB1. MAPK is activated by some xenobiotic short fatty acids e.g. methoxyacetic acid (MAA), which leads to enhanced transcriptional efficacy of nuclear hormone receptors. Thus, it is likely that short fatty acids regulate the activity of AIB1 and possibly other coactivators by inducing phosphorylation.[57]

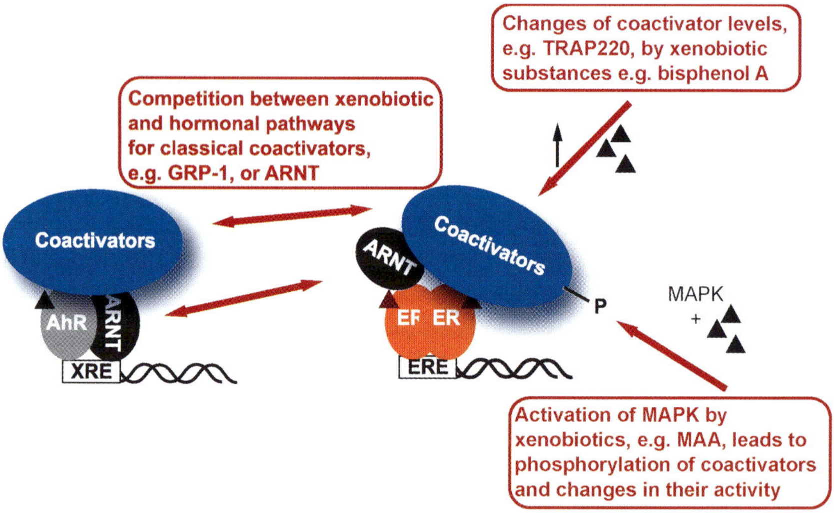

Fig. 17.4. Involvement of coactivators in endocrine disruption. MAA: methoxyacetic acid, MAPK: mitogen-activated kinase.

17.6 Concluding Remarks

In this chapter, we have tried to summarize the events following a toxic insult and mechanisms behind endocrine disruption, with special focus on the involvement of coregulators. Exposure to xenobiotic substances is associated with changes in gene expression, mediated by transcription factors like AhR, PXR, and CAR. Endocrine disruptive substances disturb hormonal pathways and thus lead to changes in gene induction by the respective hormone. In both cases, coregulators play a role in mediating the effects of the foreign compound. However, only little is known about the specific functions of coregulators in toxicology. It is thus evident that their role in mediating toxic effects leading to deleterious consequences, like the development of certain forms of cancer, needs to be investigated in much more depth. Important and alarming is a recent finding that the effects of endocrine disruptive substances can be transmitted to consecutive generations by epigenetic mechanisms. Exposure of mice to the pesticide methoxychlor and the fungicide vinclozolin in the late embryonic or early postnatal period leads to reduced spermatogenic capacity associated with increased spermatogenic cell apoptosis and decreased sperm number and motility in the adult F_1 generation. These effects are most likely promoted by inappropriate activation of AR and ERβ by these chemicals.[57] Importantly, this phenotype was shown to be transferred to the fourth generation without further exposure of the animals.[58] This transgenerational effect was not caused by gene mutation, but rather associated with epigenetic changes, as the DNA-methylation pattern of the affected animals was changed compared to non-treated controls. A number of coactivators display histone methyltransferase and histone acetylase activity and are known to be involved in histone modifications that lead to changes in the histone code. This in turn is correlated with changes in the DNA-methylation pattern. It is thus very likely that coregulators play a crucial role in mediating the epigenetic effect of xenobiotic substances. It will of course be necessary to confirm these findings and clarify the underlying mechanisms, to estimate how important epigenetic transmission is for the damage caused by toxic substances.

References

1. Nakata K, Tanaka Y, Nakano T, *et al.*, Nuclear receptor-mediated transcriptional regulation in Phase I, II, and III xenobiotic metabolizing systems, *Drug Metab Pharmacokinet* **21**:437–457, 2006.
2. Pascussi JM, Gerbal-Chaloin S, Drocourt L, *et al.*, Cross-talk between xenobiotic detoxication and other signalling pathways: Clinical and toxicological consequences, *Xenobiotica* **34**:633–664, 2004.
3. Schmidt JV, Bradfield CA, Ah receptor signaling pathways, *Annu Rev Cell Dev Biol* **12**:55–89, 1996.
4. Poland A, Knutson JC, 2,3,7,8-tetrachlorodibenzo-p-dioxin and related halogenated aromatic hydrocarbons: Examination of the mechanism of toxicity, *Annu Rev Pharmacol Toxicol* **22**:517–554, 1982.
5. Poland A, Bradfield C, A brief review of the Ah locus, *Tohoku J Exp Med*, **168**:83–87, 1992.
6. Wilson CL, Safe S, Mechanisms of ligand-induced aryl hydrocarbon receptor-mediated biochemical and toxic responses, *Toxicol Pathol* **26**:657–671, 1998.
7. Denison MS, Nagy SR, Activation of the aryl hydrocarbon receptor by structurally diverse exogenous and endogenous chemicals, *Annu Rev Pharmacol Toxicol* **43**:309–334, 2003.
8. Seidel SD, Winters GM, Rogers WJ, *et al.*, Activation of the Ah receptor signaling pathway by prostaglandins, *J Biochem Mol Toxicol* **15**:187–196, 2001.
9. Rannug U, Rannug A, Sjoberg U, *et al.*, Structure elucidation of two tryptophan-derived, high affinity Ah receptor ligands, *Chem Biol* **2**:841–845, 1995.

10. Adachi J, Mori Y, Matsui S, *et al.*, Indirubin and indigo are potent aryl hydrocarbon receptor ligands present in human urine, *J Biol Chem* **276**:31475–31478, 2001.
11. Gu YZ, Hogenesch JB, Bradfield CA, The PAS superfamily: Sensors of environmental and developmental signals, *Annu Rev Pharmacol Toxicol* **40**:519–561, 2000.
12. Keith B, Adelman DM, Simon MC, Targeted mutation of the murine aryl-hydrocarbon receptor nuclear translocator 2 (Arnt2) gene reveals partial redundancy with Arnt, *Proc Natl Acad Sci USA* **98**:6692–6697, 2001.
13. Takahata S, Sogawa K, Kobayashi A, *et al.*, Transcriptionally active heterodimer formation of an Arnt-like PAS protein, Arnt3, with HIF-1a, HLF, and clock, *Biochem Biophys Res Commun* **248**:789–794, 1998.
14. Rowlands JC, Gustafsson JA, Aryl hydrocarbon receptor-mediated signal transduction, *Crit Rev Toxicol* **27**:109–134, 1997.
15. Brunnberg S, Pettersson K, Rydin E, *et al.*, The basic helix-loop-helix-PAS protein ARNT functions as a potent coactivator of estrogen receptor-dependent transcription, *Proc Natl Acad Sci USA* **100**:6517–6522, 2003.
16. Petrulis JR, Perdew GH, The role of chaperone proteins in the aryl hydrocarbon receptor core complex, *Chem Biol Interact* **141**:25–40, 2002.
17. Davarinos NA, Pollenz RS, Aryl hydrocarbon receptor imported into the nucleus following ligand binding is rapidly degraded via the cytosplasmic proteasome following nuclear export, *J Biol Chem* **274**:28708–28715, 1999.
18. Puga A, Xia Y, Elferink C, Role of the aryl hydrocarbon receptor in cell cycle regulation, *Chem Biol Interact* **141**:117–130, 2002.
19. Oesch-Bartlomowicz B, Huelster A, Wiss O, *et al.*, Aryl hydrocarbon receptor activation by cAMP vs. dioxin: Divergent signaling pathways, *Proc Natl Acad Sci USA* **102**:9218–9223, 2005.
20. Monostory K, Kohalmy K, Prough RA, *et al.*, The effect of synthetic glucocorticoid, dexamethasone on CYP1A1 inducibility in adult rat and human hepatocytes, *FEBS Lett* **579**:229–235, 2005.
21. Abbott BD, Perdew GH, Buckalew AR, *et al.*, Interactive regulation of Ah and glucocorticoid receptors in the synergistic induction of cleft palate by 2,3,7,8-tetrachlorodibenzo-p-dioxin and hydrocortisone, *Toxicol Appl Pharmacol* **128**:138–150, 1994.
22. Rozman K, Rozman T, Scheufler E, *et al.*, Thyroid hormones modulate the toxicity of 2,3,7,8-tetrachlorodibenzo-p-dioxin (TCDD), *J Toxicol Environ Health* **16**:481–491, 1985.
23. Son DS, Roby KF, Rozman KK, *et al.*, Estradiol enhances and estriol inhibits the expression of CYP1A1 induced by 2,3,7,8-tetrachlorodibenzo-p-dioxin in a mouse ovarian cancer cell line, *Toxicology* **176**:229–243, 2002.

24. Spink DC, Katz BH, Hussain MM, *et al.*, Estrogen regulates Ah responsiveness in MCF-7 breast cancer cells, *Carcinogenesis* **24**:1941–1950, 2003.
25. Matthews J, Wihlen B, Thomsen J, *et al.*, Aryl hydrocarbon receptor-mediated transcription: Ligand-dependent recruitment of estrogen receptor alpha to 2,3,7,8-tetrachlorodibenzo-p-dioxin-responsive promoters, *Mol Cell Biol* **25**:5317–5328, 2005.
26. Safe SH, Modulation of gene expression and endocrine response pathways by 2,3,7,8-tetrachlorodibenzo-p-dioxin and related compounds, *Pharmacol Ther* **67**:247–281, 1995.
27. Reen RK, Cadwallader A, Perdew GH, The subdomains of the transactivation domain of the aryl hydrocarbon receptor (AhR) inhibit AhR and estrogen receptor transcriptional activity, *Arch Biochem Biophys* **408**:93–102, 2002.
28. Nebert DW, Roe AL, Dieter MZ, *et al.*, Role of the aromatic hydrocarbon receptor and [Ah] gene battery in the oxidative stress response, cell cycle control, and apoptosis, *Biochem Pharmacol* **59**:65–85, 2000.
29. Mimura J, Ema M, Sogawa K, Identification of a novel mechanism of regulation of Ah (dioxin) receptor function, *Genes Dev* **13**:20–25, 1999.
30. Dunlap JC, Loros JJ, Liu Y, *et al.*, Eukaryotic circadian systems: Cycles in common, *Genes Cells* **4**:1–10, 1999.
31. Hankinson O, Role of coactivators in transcriptional activation by the aryl hydrocarbon receptor, *Arch Biochem Biophys* **433**:379–386, 2005.
32. Kollara A, Brown TJ, Functional interaction of nuclear receptor coactivator 4 with aryl hydrocarbon receptor, *Biochem Biophys Res Commun* **346**:526–534, 2006.
33. Kumar MB, Tarpey RW, Perdew GH, Differential recruitment of coactivator RIP140 by Ah and estrogen receptors. Absence of a role for LXXLL motifs, *J Biol Chem* **274**:22155–22164, 1999.
34. Klinge CM, Jernigan SC, Risinger KE, *et al.*, Short heterodimer partner (SHP) orphan nuclear receptor inhibits the transcriptional activity of aryl hydrocarbon receptor (AHR)/AHR nuclear translocator (ARNT), *Arch Biochem Biophys* **390**:64–70, 2001.
35. Sadek CM, Jalaguier S, Feeney EP, *et al.*, Isolation and characterization of AINT: A novel ARNT interacting protein expressed during murine embryonic development, *Mech Dev* **97**:13–26, 2000.
36. Bock KW, Kohle C, Ah receptor: Dioxin-mediated toxic responses as hints to deregulated physiologic functions, *Biochem Pharmacol* **72**:393–404, 2006.
37. Warner M, Eskenazi B, Mocarelli P, *et al.*, Serum dioxin concentrations and breast cancer risk in the Seveso Women's Health Study, *Environ Health Perspect* **110**:625–628, 2002.

38. Remillard RB, Bunce NJ, Linking dioxins to diabetes: Epidemiology and biologic plausibility, *Environ Health Perspect* **110**:853–858, 2002.
39. IARC Working Group on the Evaluation of Carcinogenic Risks to Humans: Polychlorinated Dibenzo-Para-Dioxins and Polychlorinated Dibenzofurans. Lyon, France, 4–11 February 1997. IARC Monogr Eval Carcinog Risks Hum **69**:1–631, 1997.
40. Lahvis GP, Bradfield CA, Ahr null alleles: Distinctive or different? *Biochem Pharmacol* **56**:781–787, 1998.
41. Peters JM, Narotsky MG, Elizondo G, *et al.*, Amelioration of TCDD-induced teratogenesis in aryl hydrocarbon receptor (AhR)-null mice, *Toxicol Sci* **47**:86–92, 1999.
42. Lin TM, Ko K, Moore RW, *et al.*, Effects of aryl hydrocarbon receptor null mutation and in utero and lactational 2,3,7,8-tetrachlorodibenzo-p-dioxin exposure on prostate and seminal vesicle development in C57BL/6 mice, *Toxicol Sci* **68**:479–487, 2002.
43. Ma Q, Whitlock JP Jr, The aromatic hydrocarbon receptor modulates the Hepa 1c1c7 cell cycle and differentiated state independently of dioxin, *Mol Cell Biol* **16**:2144–2150, 1996.
44. Andersson P, McGuire J, Rubio C, *et al.*, A constitutively active dioxin/aryl hydrocarbon receptor induces stomach tumors, *Proc Natl Acad Sci USA* **99**:9990–9995, 2002.
45. Tabb MM, Kholodovych V, Grun F, *et al.*, Highly chlorinated PCBs inhibit the human xenobiotic response mediated by the steroid and xenobiotic receptor (SXR), *Environ Health Perspect* **112**:163–169, 2004.
46. Moore LB, Parks DJ, Jones SA, *et al.*, Orphan nuclear receptors constitutive androstane receptor and pregnane X receptor share xenobiotic and steroid ligands, *J Biol Chem* **275**:15122–15127, 2000.
47. Wang K, Mendy AJ, Dai G, *et al.*, Retinoids activate the RXR/SXR-mediated pathway and induce the endogenous CYP3A4 activity in Huh7 human hepatoma cells, *Toxicol Sci* **92**:51–60, 2006.
48. Bhalla S, Ozalp C, Fang S, Ligand-activated pregnane X receptor interferes with HNF-4 signaling by targeting a common coactivator PGC-1alpha. Functional implications in hepatic cholesterol and glucose metabolism, *J Biol Chem* **279**:45139–45147, 2004.
49. Tzameli I, Moore DD, Role reversal: New insights from new ligands for the xenobiotic receptor CAR, *Trends Endocrinol Metab* **12**:7–10, 2001.
50. Miao J, Fang S, Bae Y, *et al.*, Functional inhibitory cross-talk between constitutive androstane receptor and hepatic nuclear factor-4 in hepatic lipid/glucose metabolism is mediated by competition for binding to the DR1 motif and to the common coactivators, GRIP-1 and PGC-1alpha, *J Biol Chem* **281**:14537–14546, 2006.

51. Bae Y, Kemper JK, Kemper B, Repression of CAR-mediated transactivation of CYP2B genes by the orphan nuclear receptor, short heterodimer partner (SHP), *DNA Cell Biol* **23**:81–91, 2004.
52. Handschin C, Meyer UA, Regulatory network of lipid-sensing nuclear receptors: Roles for CAR, PXR, LXR, and FXR, *Arch Biochem Biophys* **433**:387–396, 2005.
53. Jung D, Mangelsdorf DJ, Meyer UA, Pregnane X receptor is a target of farnesoid X receptor, *J Biol Chem* **281**:19081–19091, 2006.
54. Kretschmer XC, Baldwin WS, CAR and PXR: Xenosensors of endocrine disrupters? *Chem Biol Interact* **155**:111–128, 2005.
55. Min G, Kim H, Bae Y, *et al.*, Inhibitory cross-talk between estrogen receptor (ER) and constitutively activated androstane receptor (CAR). CAR inhibits ER-mediated signaling pathway by squelching p160 coactivators, *J Biol Chem* **277**:34626–34633, 2002.
56. Safe S, Wormke M, Inhibitory aryl hydrocarbon receptor-estrogen receptor alpha cross-talk and mechanisms of action, *Chem Res Toxicol* **16**:807–816, 2003.
57. Tabb MM, Blumberg B, New modes of action for endocrine-disrupting chemicals, *Mol Endocrinol* **20**:475–482, 2006.
58. Anway MD, Cupp AS, Uzumcu M, *et al.*, Epigenetic transgenerational actions of endocrine disruptors and male fertility, *Science* **308**:1466–1469, 2005.

Chapter 18

Nuclear Receptor Coactivators Co-ordinate Metabolic Responses to Hormonal and Environmental Stimuli

Ronald M. Evans, Michael Downes, Russell R. Nofsinger, Jun Sonoda and Ruth T. Yu

The p160 and PGC-1 families of nuclear receptor coactivators are important modulators of the metabolic responses by nuclear receptor pathways to hormonal and environmental stimuli. This chapter focuses on the characterization and role of SRC-1, SRC-2, SRC-3, and PGC-1β in the regulation of adipose homeostasis and thermogenesis. Metabolic consequences of the genetic knockout of each of the above coactivators in mice are discussed, along with their involvement in human disease.

18.1 Introduction

Nuclear receptors (NRs) comprise a superfamily of transcriptional factors that serve as broad regulators of development, reproduction, metabolism and immunity. There are 48 human and 49 murine members, ranging from the well characterized steroid hormone and lipophilic vitamin receptors to the relatively enigmatic "orphan" receptors. Expression profiling of the entire nuclear receptor superfamily in 39 tissues from mouse has demonstrated a functional relationship between clustered receptor expression and physiology. Overall, we found that 21 NRs were expressed in all tissues, 17 were present in at least 50% of the tissues examined, and 11 NRs were more restricted in expression. Hierachical unsupervised clustering of tissue distribution profiles revealed a higher-order network of six expression families that links nuclear receptor function to 1) steroidogenesis; 2) reproduction and development; 3) CNS, circadian and basal metabolism; 4) dietary lipid metabolism; 5) energy homeostasis; and 6) bile acid and xenobiotic clearance. This was depicted in our proposed "Nuclear Receptor Ring of Physiology."[1] In a parallel study, the diurnal expression profiles of the NR superfamily in four key metabolic tissues; white and brown adipose tissue, liver, and skeletal muscle were determined, revealing that 45 NRs were expressed, 25 in rhythmic cycles. The dynamic but coordinated expression patterns that were observed with some NRs, when correlated with that of their key target genes, proffer a mechanism for the known cyclic behavior of lipid and glucose metabolism.[2]

Approximately one-third of the NR superfamily responds to metabolic (e.g. steroid and thyroid hormones) and dietary (e.g. Vitamin A) lipophilic ligands and modulates transcription via binding to consensus DNA response elements in the promoters of target genes. While the "active" state (ligand bound receptor) has classically been believed to be the principal mode of action, repression was first observed as a property of the unliganded TR and its oncogenic homolog, v-erbA.[3]

The identification of NR coregulators in 1995 revealed the first glimpse into how the presence or absence of ligand might confer a way to tightly regulate the expression of target genes. Beginning with the steroid receptor coactivator-1 (SRC-1), numerous NR coactivators and corepressors were subsequently cloned and characterized.[4] This duality of opposing co-regulators led to the classic mechanistic explanation of NR action as a ligand-activated switch. In the absence of ligand, some NRs remain in an inactive state, while in others, the LBD adopts a confirmation that facilitates the binding of corepressors and associated histone deacetylases, silencing transcription by "active repression". Conversely, ligand binding induces a conformational change in the receptor-coregulator complex, releasing corepressors in exchange for coactivators and resulting in enhanced target gene expression. This mechanism (Fig. 18.1) allows the balance of coactivators and corepressors to tightly control induction of NR mediated gene expression.[5]

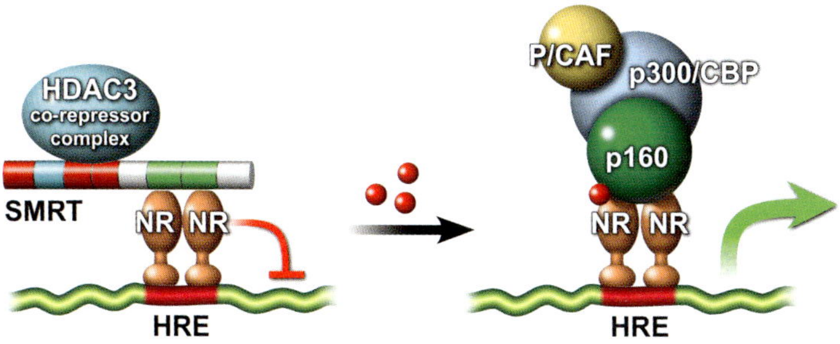

Fig. 18.1. Ligand mediated nuclear receptor corepressor and coactivator complexes. In the absence of ligand, the nuclear receptor homo- or hetero-dimer is associated with corepressor complexes. The corepressors (SMRT/NCoR) recruit histone deactylases through interaction HDAC3. Deacetylation of histone tails leads to chromatin compaction and transcriptional repression. Ligand binding causes release of the corepressor complex and the AF-2- dependent recruitment of a coactivator complex that contains at least the p160 coactivators (P/CIP or SRC-1), CBP/p300, and P/CAF. All of these proteins possess histone acetyltransferase activity that allows chromatin decompactation and gene activation.

In 1998, a new type of NR coactivator family was isolated with the identification of the first member of the family PGC-1alpha.[6] Two additional members termed PGC-1β and PRC were rapidly cloned and characterized to complete the family.[7] The PGC-1 family differed from the p160 family in that it was able to enhance NR mediated transcription independent of their cognate ligands. The gene family is also highly induced in response to thermogenic and metabolic stress. This chapter will focus on the characterization and role of the p160 family of coactivators and PGC-1b in the regulation of adipose and thermogenesis.

18.2 p160 Coactivator Gene Family

The p160 family of coactivators comprise three members (SRC-1, TIF2/SRC-2, ACTR/SRC-3) that were isolated through their interaction with the nuclear receptor family. Numerous studies have now shown that the SRCs potentiate the transcriptional activation of ligand-bound NRs via recruitment of histone modifying enzymes such as the arginine methyltransferase CARM1 and the histone acetyltransferase CBP/p300.[8]

The first p160 member identified, SRC-1, was isolated via its ligand-dependent interaction with the human progesterone receptor[4,9] and has subsequently been demonstrated to interact with many members of the adopted orphan NR subclass (PPARs, LXRs, FXR and HNF4s). Various studies have established that SRC-1 is a key player in NR-dependent regulation of key metabolic pathways that maintain homeostasis.

Soon afterwards, SRC-2 and -3 were cloned, completing the p160 family. SRC-2, originally named transcriptional intermediary factor 2 (TIF2), was isolated on the basis of its estradiol-dependent interaction with the estrogen receptor ligand binding domain.[10] It was shown by a number of studies to be a general NR coactivator through its ability to interact and coactivate both steroid and nonsteroid classes of NRs, including androgen receptor (AR), glucocorticoid receptor (GR), retinoic acid receptor (RAR), thyroid hormone receptor (TR), vitamin D receptor (VDR) and retinoid X receptor (RXR).[10] More recently, constitutive androstane receptor (CAR) function was also shown to be influenced by its interaction with SRC-2.

The importance of the third member of the family, SRC-3 (AIB1/ACTR) was quickly realized due to its original characterization as an unknown gene amplified in human breast carcinomas.[11] Parallel NR coactivator screening studies using liganded ER or RAR directly identified SRC-3 as the third and final member of the p160 family.[8,12] SRC-3 is unique in that it possesses intrinsic histone acetyltransferase activity.[12] Despite their similarities in sequence homolog (Fig. 18.2) and protein-protein interaction domains, there are distinct differences in the physiological roles of each member of the p160 gene family members.

18.3 Metabolic Consequences of Genetic Knockout of p160 Gene Family

One of the most efficient and informative techniques of determining the function of a given gene is to knock it out in mice. Genetic knockout models of all three SRCs have been created and characterized to study their *in vivo* functions. As could be predicted from their broad expression across tissues and their interactions with multiple nuclear receptors, the resulting phenotypes in the knockout mice range from hormone resistance, defects in reproductive organs and general growth inhibition, to metabolic consequences such as altered adipogenesis and

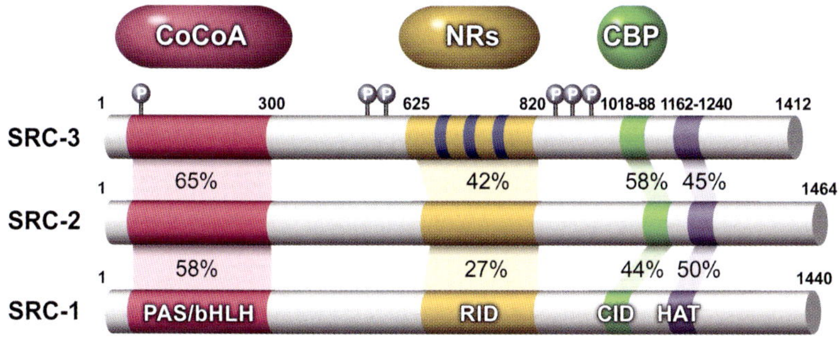

Fig. 18.2. Structural and functional domains of the SRC family members. The similarity and identity of amino acid sequences for full-length human SRC proteins and their specific conserved regions are indicated by color coded regions. PAS, Per/ARNT/Sim homologous domain (Pink); RID domain: receptor interaction domain (yellow); CID domain CBP/p300 interaction domain (green); HAT domain histone acetyltransferase (purple). The *letters within the cylinders* indicate structural domains, and the *percentage similarity is indicated between the domain*. Different factors are color coded to match the domains that they interact with. *Encircled* in blue in the RID domain are the LXXLL α-helix motifs that nuclear receptors interact with in the presence of ligand. Phosphorylation sites are also depicted in the SRC proteins. CoCoA, is a nuclear receptor coactivator which acts through an N-terminal activation domain of p160 coactivators.

thermogenesis. In the next section we will focus on the metabolic consequences observed in the p160 knockout mice (Fig. 18.3).

18.4 Metabolic Phenotypes of SRC-1 Mutant Mice

SRC-1 null mice exhibit a complex phenotype of reduced energy expenditure and increased susceptibility to diet-induced obesity. The mutant mice maintain normal lipolytic activity in white adipose tissue (WAT), but exhibit a reduced thermogenic capacity and decreased fatty acid oxidation in brown adipose tissue (BAT). Consequently, a significant reduction in thermogenic gene expression was observed, including uncoupling protein-1 (Ucp-1), PPARγcoactivator-1 alpha (PGC-1α), and AcetylCoA oxidase (Aox). Metabolic analysis of the mutant mice revealed decreased respiration as well as a primary defect in fatty acid utilization. When these mice were challenged a high-fat diet, a significant

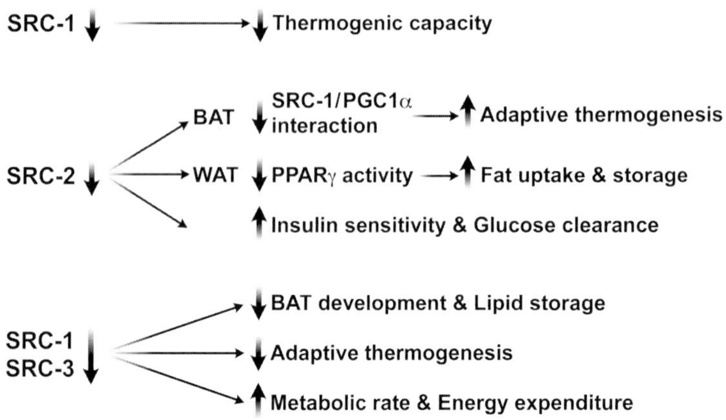

Fig. 18.3. Regulation of adipose and energy metabolism by the members of the p160 family. Summary of metabolic phenotypes observed in single and double SRC mutant mice.

increase in weight gain with a nearly 50% increase in total adiposity was observed. The mechanism by which the decreased gene expression and reduced thermogenic capacity in the $SRC^{-/-}$ BAT occurs is thought to be a result of the reduced activity of PGC1α/PPARγ.[13] Expression profiling of the liver in SRC-$1^{-/-}$ mice identified an increased expression of several important genes involved in glycolysis and glycogen storage such as pyruvate kinase, aldolase 1, 2,3-bisphosphoglycerate mutase and glycogenin 1. In addition, a fatty acid synthesis enzyme (acyl-CoA synthetase long chain family member 4), is increased in the SRC-$1^{-/-}$ mice. Collectively, these observations suggest that a metabolic shift toward increased energy storage occurs in the livers of SRC-$1^{-/-}$ mice via alterations in specific gene expression, resulting in the increased propensity of these mice to high-fat diet-induced obesity.[14,15]

18.5 Metabolic Phenotypes of SRC-2 Knockout Mice

In contrast to the obesity-prone SRC-1 null mice, SRC-2 knockout mice are resistant to obesity and display enhanced adaptive thermogenesis.[13] Instead of the increased energy storage observed in the SRC-$1^{-/-}$ mice, SRC-2 deletion leads to increased energy expenditure. Metabolic analyses of these mice revealed decreased fasting glucose levels and improved insulin sensitivity. Increased lipolytic activity and a reduced potential for fatty acid storage in WAT was observed. In SRC-2 null animals, increased oxygen consumption and enhanced adaptive thermogenesis was observed. In contrast to the repression seen in SRC-1$^{-/-}$ BAT, *Ucp-1*, *Pgc-1*α, and *Aox* are upregulated in the SRC2$^{-/-}$ BAT. These findings suggested that the absence of SRC-2 results in increased energy expenditure as a result of enhanced fatty acid oxidation and uncoupling of respiration.

The disparity in metabolic phenotypes resulting from the loss of SRC-1 compared to the loss of SRC-2 is proposed to result from the preferential activation of target genes in adipose tissue. Genes involved in energy utilization are preferentially activated by PPARγ/SRC-1 and genes involved in energy storage are preferentially activated by PPARγ/SRC-2. Furthermore, SRC-1 and SRC-2 expression are both increased in adipose tissue in response to a high-fat diet, with SRC-2 expression being more robustly induced. In response to a high-fat challenge, the increased TIF2/SRC-1 ratio leads to fat accumulation in WAT and reduced thermogenic potential in BAT. These conditions prohibit increasing energy expenditure to compensate for increased energy intake, thereby promoting the development of obesity and insulin resistance.[13]

Genome-wide expression profiling in SRC-2$^{-/-}$ livers revealed reduced expression of glucose uptake, fatty acid and cholesterol synthesis genes and increased expression of glycolytic and lipolytic genes. As was the case for the SRC-1 knockout, most of the dysregulated genes are known targets of one or more NRs, however, non-NR transcription factors may also contribute to this dysregulation.

18.6 Metabolic Phenotypes of SRC-3 and SRC-1/-3 Genetic Knockout

SRC-3 null mice exhibit a lean phenotype which appears linked to impaired white adipocyte differentiation,[16] suggesting that SRC-3 is a critical regulator in WAT. But overall, the SRC-3$^{-/-}$ mice appear to exhibit fewer metabolic complications than the SRC-1 or -2 mutant mice. Interestingly, the compound SRC-1$^{-/-}$/SRC3$^{-/-}$ mice showed a severe defect in adipose tissue development and energy balance. These double mutant mice exhibit a developmental arrest in BAT, have an increased metabolic rate and are resistanct to diet-induced obesity. This is similar to the SRC-2 phenotype but is contrary to that of SRC-1. The developmental block in adipogenesis was shown to result from the inability of PPARγ to recruit the necessary cofactor complement to direct differentiation of preadipocytes in BAT. The dysregulation of PPARγ was further implicated as a cause of the observed metabolic defects in these mice, where PPARγ targets *Ucp1*, *Ucp2*, *Cd36*, and *Aox* are specifically down-regulated compared to mice or either of the single null animals.[17,18]

Deletion of SRC-1, -2, or -3 yields mice with varying degrees of hormone resistance, reproductive defects, and factor specific metabolic phenotypes.[19] Microarray analysis of hepatic RNA in mice with deletion of specific p160 genes demonstrated that these coactivators play a critical role in regulating the expression of hepatic genes essential for energy homeostasis. Surprisingly, there was little overlap among the gene sets differentially regulated between wild type and the individual null mice. SRC-1 deletion shifts the expression profile toward energy storage, while SRC-2 deletion led to a shift toward energy usage. Although it is highly expressed, SRC-3 deletion did not have a dramatic effect on the transcription profile within the liver.[18] The large number of differentially expressed targets specific to a particular coactivator and the unique defects observed in the individual p160 cofactor null mice clearly demonstrate that SRC-1, -2, and -3 are not functionally redundant *in vivo*. However, *in vitro* data and the dramatic metabolic defects observed in SRC-1$^{+/-}$/SRC-2$^{+/-}$ compound heterozygous and the SRC-1$^{-/-}$/SRC-3$^{-/-}$ compound null mice also indicate that some receptor targets utilize more than one of the p160s for full activation. Thus, energy balance and adipocyte differentiation are controlled by multiple coactivators; each with specific yet partially overlapping roles in lipid homeostasis.[17]

18.7 The PGC-1 Family and the Mitochondria

Mitochondrial dysfunction has been implicated in a number of common pathological conditions including aging, insulin resistance/type II diabetes mellitus, Alzheimer's disease and Parkinson's disease. In addition, germline mutations in mitochondria DNA can cause rare mitochondrial disorders such as mitochondrial cardiac myopathies and the Leigh syndrome. Recently, a family of unique ligand-independent coactivators, PGC-1α (peroxisome proliferator activated receptor gamma coactivator-1 alpha), PGC-1β and PRC (PGC-1-related coactivator), have been identified as master transcriptional regulators of mitochondrial function, providing novel therapeutic opportunities for mitochondria-related disorders. PGC-1α is the founding and the most extensively studied member of the family (reviewed elsewhere in this book). A less characterized member, PGC-1β, has also been established as a central regulator of mitochondrial energy metabolism in oxidative tissues. Here, we will focus on the current knowledge of PGC-1β function and discuss its potential role in human metabolic disease.

18.8 PGC-1β: Identification and Initial Characterization

PGC-1β (also named as PERC or ERRL1) was originally identified based on its sequence similarity to PGC-1α.[20–22] Figure 18.4 shows the primary structure of PGC-1β compared to PGC-1α and PRC. The most conserved regions are in the N-terminal acidic and the C-terminal RNA recognition domains. The N-terminal acidic domain of PGC-1α acts as a docking site for HAT complexes (through SRC-1 and

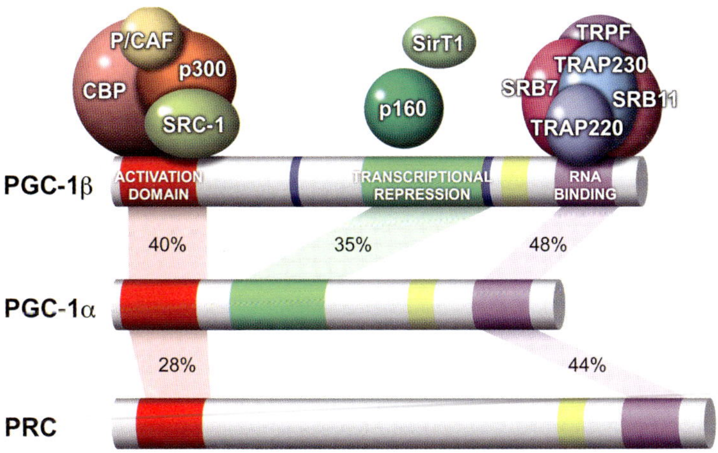

Fig. 18.4. Structure and function of the PGC-1 family coactivators. Sequence homology of PGC-1α, PGC-1β, and PRC. Note that activation domain (red), Arg/Ser-rich domain (yellow), and RNA binding domain (purple) present in all three PGC-1 family members. PGC-1α and PGC-1β share an additional domain of similarity in the central region. PGC-1β binds to the HAT and TRAP/DRIP/Mediator complexes at the amino and carboxyl termini, respectively while SirT1 and p160 bind to the repression domain.

CBP/p300).[23] The C-terminal RNA recognition domain in PGC-1α is essential for the induction of endogenous target gene expression, and is thought to act as a binding site for a chromatin remodeling complex[24] as well as transcription elongation and splicing factors.[25] Although the mechanisms by which PGC-1β activates mRNA synthesis are not yet defined, the structural similarity to PGC-1α suggests that PGC-1β also functions by controlling both transcriptional initiation, transcriptional elongation and pre-mRNA maturation. Like many other nuclear receptor coactivators, all three PGC-1 family members have conserved LXXLL sequences, the signature motif that directly binds to the NR ligand binding domain. However, unlike classical NR coactivators such as p160, CBP/p300 and TRAP220, whose recruitment is ligand-dependent, PGC-1α and PGC-1β bind to and coactivate NRs even in the absence of ligand (Fig. 18.5).[7,26]

PGC-1α was originally identified as a brown adipocyte enriched transcriptional coactivator for PPARγ.[6] When overexpressed by adenovirus-mediated gene transduction, both PGC-1α and PGC-1β induce expression of mitochondrial proteins including Cytochrome C, ATP synthase, CPT-1 and MCAD.[27] Indeed, either PGC-1α or PGC-1β overexpresson results in an increased respiration rate and mitochondrial

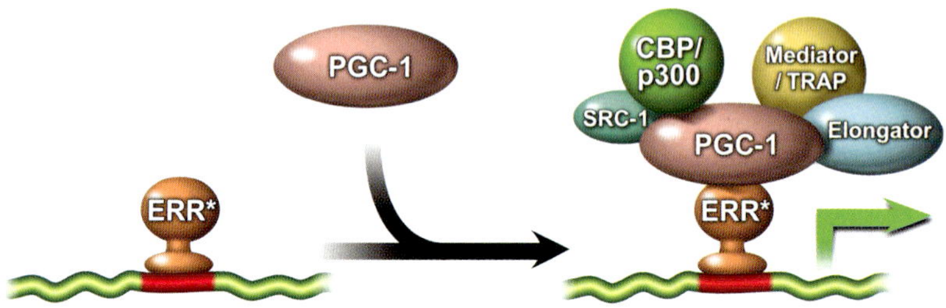

Fig. 18.5. The ligand independent PGC-1 coactivator family: Inducible boosters of gene transcription. The schematic uses generic estrogen related receptor (ERR) as an example of how nuclear receptors dock with the inducible PGC-1 coactivators and recruit protein complexes that activate transcription via either enzymatic modification of chromatin, such as histone acetylation (e.g. by steroid receptor coactivator-1 [SRC-1] or p300), or direct interaction with the transcription initiation machinery (e.g. the thyroid hormone receptor–associated protein/vitamin D receptor–interacting protein [TRAP/DRIP] coactivator complex). The ERR binds cognate NR response elements (ERREs depicted in a red) within the promoter region of the target gene.

density.[28] In addition, adenovirus mediated overexpression of PGC-1α activates genes encoding mitochondrial uncoupling proteins and hepatic gluconeogenic enzymes.[23] These later acitivities are specific to PGC-1α, suggesting that the role of PGC-1β is more selective than that of PGC-1α. The selectivity of PGC-1β function is attributed to its interaction with a more restricted set of transcription factors. In contrast to PGC-1α which can potently activate a number of NR and non-NR transcription factors regulating a variety of metabolic processes,[6,7,26] PGC-1β appears to principally activate the estrogen-related receptors (ERRs).[22,29] Although other putative PGC-1β targets have also been identified, the importance of PGC-1β activity on those factors at physiological conditions.

ERRα was the first orphan nuclear receptor to be identified and it shares a high degree of sequence homology with two estrogen receptors (ERα and β) as well as two other ERR subfamily members (ERRβ and γ).[30,31] The ERRs bind to variants of an extended consensus half site (TCAAGGTCA) that are found in the promoters of many genes involved in mitochondrial function to regulate their expression.[32–36] Adenoviral or transgene-mediated overexpression of PGC-1α, PGC-1β and/or ERRα all result in the coordinated induction of genes involved in mitochondrial FAO and OXPHOS.[33,37,38] The effect of PGC-1α overexpression on a few mitochondrial target genes were shown to be compromised by siRNA-mediated ERRα knock-down or an ERRα-specific inverse agonist, suggesting that ERRα mediates PGC-1α activity, at least in part. Furthermore, mice deficient for PGC-1β or ERRα show a reduction in steady state expression of mitochondrial FAO and OXPHOS genes most profoundly in BAT, which contains the highest density of mitochondria in mammals.[37,39,40] The activation of cytokines such as interferon gamma or interleukin-4 induces PGC-1beta expression and mitochondrial gene expression in macrophages[41,42]; this induction of mitochondrial genes is largely absent in ERRα- or PGC-1β deficient cells.[42] These data collectively show that ERRα is a major transcription factor that mediates induction of mitochondria genes by PGC-1α and PGC-1β.

18.9 Overexpression as a Strategy to Study PGC-1β Function *in vivo*

Transgenic overexpression of PGC-1β in mice has been described in two separate studies. In one study, a ubiquitous CAG promoter was used to drive PGC-1β expression, which resulted in expression in restricted tissues including skeletal muscle, heart, BAT and kidney.[22] Skeletal muscle from the trangenic mice showed an increased expression of an ERRα target gene MCAD. Furthermore, the transgene expression resulted in an increased FAO activity, higher respiration rate and resistance to weight gain. In the second study, PGC-1β expression was driven by the muscle creatine kinase promoter, resulting in expression in skeletal muscle and heart.[43] In addition to MCAD, many other genes involved in oxidative phosphorylation and FAO were overexpressed in the transgenic muscle. Furthermore, the transgenic muscle overexpressed MHC IIX, a marker for oxidative IIX fiber, indicating that PGC-1β overexpression results in fiber-type switching. In contrast to PGC-1α transgenic mice which show an increased mitochondria density but reduced endurance, PGC-1β transgenic mice exhibited a significant increase in Type I endurance muscle fiber.[43–45]

18.10 Use of Genetic Knockout Strategy to Study PGC-1β Function *in vivo*

The physiological function of PGC-1β has been studied by the generation of three independent mouse lines deficient for PGC-1β. In one line, a deletion of exons 3 and 4 results in a hypomorphic mutant protein lacking an internal 110 amino acid stretch that retains a substantial activity toward ERRα, the host cell factor and SREBP-1c in transient transfection assays.[46] The other two groups used a similar targeting strategy, deleting exons 4 and 5, which most likely resulted in expression of non-functional truncated proteins.[39,40] Although there are some differences in targeting strategies, genetic backgrounds, and laboratory conditions, which may account for the minor differences observed between the three PGC-1β KO strains, these mice are all viable, exhibiting dysregulated expression of mitochondrial genes in multiple tissues.

18.11 BAT-Mediated Non-Shivering Thermogeneisis

The expression of PGC-1β is highest in BAT, a mitochondria-rich tissue believed to be responsible for generation of heat during both acute and prolonged cold exposure, particular, human neonates and hibernating mammals.[47] Cold exposure induces a signal from the sympathetic nervous system that activates the adrenegic signaling pathway, resulting in lipolysis and a burst of respiration in brown adipose through UCP-1, a mitochondrial inner-membrane protein responsible for heat production. The importance of lypolysis, FAO and Ucp-1 during acute cold exposure is well exemplified by the thermogenic defects observed in mice deficient for the adipose triglyceride lipase, a fatty acid dehydrogenase and Ucp-1 respectively.[47,48]

Consistent with its high expression in BAT, the loss of PGC-1β results in a dramatic sensitivity to acute cold exposure[39] and mice exhibit hypothermia and morbidity under conditions where WT mice show tolerance. The expression of PGC-1β does not change during cold exposure, but apparently, it is required to sustain the constitutive expression of a number of genes that are involved in mitochondrial FAO and OXPHOS.[39,40] A similar reduction in the mitochondrial FAO and OXPHOS gene expression was observed in ERRα KO mice, indicating that ERRα and PGC-1β function together. ERRα KO also exhibited acute cold sensitivity.[37]

Although PGC-1β KO mice do not tolerate acute cold exposure, they could be gradually acclimated to 4°C.[39,40] Although these two observations seem paradoxical, they also hold true for UCP-1 KO mice that completely lack the thermogenic response to β-adrenergic stimuli in brown adipocytes. In the case of UCP-1 KO mice, it is believed that UCP-1-independent thermogenesis in both skeletal muscle and WAT are acquired during cold acclimation to compensate for the loss of UCP-1.[49,50] Consistent with such non-BAT mediated compensatory mechanism in PGC-1β KO mice, cold acclimated PGC-1β KO mice show a reduced respiratory response to β-adrenergic stimulation compared to WT control mice.[40]

18.12 PGC-1β Function in the Liver

The studies of all three KO lines agree that PGC-1β KO mice accumulate triglycerides (TG) in the liver under certain conditions. Modestly increased hepatic TG levels in chow-fed KO mice observed in two of the lines.[39,46] This phenotype is more pronounced and is associated with hepatomegaly when the KO mice are fed a diet containing hydrogenated vegetable oil (saturated trans-fat) for 24 hours or a diet containing saturated animal fat for 12 weeks, indicating the inability of PGC-1β KO mice to efficiently clear out excess dietary lipid. Based on altered

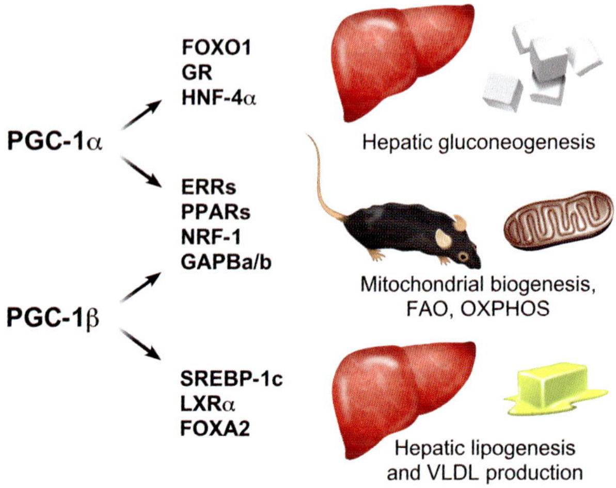

Fig. 18.6. Regulation of hepatic energy metabolism by members of the PGC-1 family. Summary of the different transcription factors whose transcriptional activity is modulated by PGC-1α and β to regulate hepatic function including hepatic gluconeogenesis, mitochondrial biogenesis and hepatic lipogenesis.

mitochondrial gene expression, reduced mitochondrial density, and reduced respiration rate, hepatic lipid accumulation in the KO mice is most likely caused by mitochondrial FAO and OXPHOS defects. In addition, indirect or direct dysregulation of lipogenic SREBP pathway might be involved. In one study, euglycemic-hyperinsulimic clamp showed clear insulin resistance specifically in the KO liver, which might be caused by hepatic triglyceride accumulation. However, none of the PGC-1β KO mouse lines showed clear impairment in the glucose tolerance or insulin tolerance tests, at least in studies using relatively young mice. These studies are summarized along with the effects of PGC-1α in Fig. 18.6.

18.13 Involvement of PGC-1β in Human Diseases

Recent studies have demonstrated a tight association between insulin resistance and mitochondrial dysfunction in skeletal muscle (reviewed in Ref. 51). This association holds true even in non-diabetic lean individuals, suggesting that mitochondrial dysfunction predisposes to intramuscular lipid accumulation and the resulting insulin resistance that ultimately leads to the development of type 2 diabetes. Although the mechanism of mitochondrial dysfunction in insulin resistance is not entirely clear, at least some of the effect appears to be at the level of gene transcription as human diabetic muscles show a decreased expression of mitochondrial genes, perhaps as a consequence of reduced PGC-1α and PGC-1β activity.[33,52] PGC-1α activity is regulated at both the transcriptional and post-translational (e.g. phosphorylation and acetylation) levels. Although virtually nothing is known about post-translational regulation of PGC-1β, a number of studies indicate that aging and insulin resistance reduce PGC-1β expression.[52,53] Furthermore, genetic variations in PGC-1β have been associated with obesity and impaired glucose metabolism in humans.[54,55] Taken together, we believe that PGC-1β is a promising therapeutic target for obesity and type 2 diabetes.

18.14 Perspective

To sense both environmental and metabolic signals, NR coactivators have evolved both ligand independent and ligand dependent mechanisms to control homeostasis. As the PGC-1 gene family can interact with non-liganded NRs, environmental stimuli that change the relative levels of each of these molecules can dictate the activation of target receptors. For this reason, the PGC-1 products are frequently referred to as protein ligands. The p160 gene family, in contrast, is dependent on the availability of the NRs' cognate ligand to modulate the transcriptional activity of NRs. One benefit of this dual system is that secreted hormonal lipid such as steroids can simultaneously and quickly diffuse throughout the body. Thus, steroidal ligands can co-ordinate distributed metabolic programs through bodywide recruitment of p160 proteins. In contrast, because PGC-1 proteins are contained within cells in which they are expressed, they can act in a powerful but more restricted fashion. In conclusion, two NR co-activator systems have evolved that allow the body to be able maintain global regulation of homeostasis while retaining the ability to mount immediate cell specific responses to environmental cues. Exploiting these differences offers numerous opportunities for new classes of drugs to beneficially address the emerging crisis of metabolic disease.

References

1. Bookout AL, Jeong S, Downes M, *et al.*, A nuclear receptor atlas: Anatomical profiling reveals a hierarchican transcriptional network, *Cell*, **126**:789–799, 2006.

2. Yang X, Downes M, Yu RT, *et al.*, Nuclear receptor expression links the circadian clock to metabolism, *Cell* **126**:801–810, 2006.
3. Mangelsdorf DJ, Thummel C, Beato M, *et al.*, The nuclear receptor superfamily: The second decade, *Cell* **83**(6):835–839, 1995.
4. Onate SA, Tsai SY, Tsai MJ, *et al.*, Sequence and characterization of a coactivator for the steroid hormone receptor superfamily, *Science* **270**: 1354–1357, 1995.
5. Shibata H, Spencer TE, Onate SA, *et al.*, Role of co-activators and co-repressors in the mechanism of steroid/thyroid receptor action, *Recent Prog Horm Res* **52**:141–164, 1997.
6. Puigserver P, Wu Z, Park CW, *et al.*, A cold-inducible coactivator of nuclear receptors linked to adaptive thermogenesis, *Cell* **92**(6):829–839, 1998.
7. Finck BN, Kelly DP, PGC-1 coactivators: Inducible regulators of energy metabolism in health and disease, *J Clin Invest* **116**(3):615–622, 2006.
8. Torchia J, Rose DW, Inostroza J, *et al.*, The transcriptional co-activator p/CIP binds CBP and mediates nuclear-receptor function, *Nature* **387**:677–684, 1997.
9. Zhu Y, Qi C, Calandra C, *et al.*, Cloning and identification of mouse steroid receptor coactivator-1 (mSRC-1), as a coactivator of peroxisome proliferator-activated receptor gamma, *Gene Expr* **6**:185–195, 1996.
10. Voegel JJ, Heine MJ, Zechel C, *et al.*, TIF2, a 160 kDa transcriptional mediator for the ligand-dependent activation function AF-2 of nuclear receptors, *EMBO J* **15**:3667–3675, 1996.
11. Guan XY, Xu J, Anzick SL, *et al.*, Hybrid selection of transcribed sequences from microdissected DNA: Isolation of genes within amplified region at 20q11-q13.2 in breast cancer, *Cancer Res* **56**:3446–3450, 1996.
12. Chen H, Lin RJ, Schiltz RL, *et al.*, Nuclear receptor coactivator ACTR is a novel histone acetyltransferase and forms a multimeric activation complex with P/CAF and CBP/p300, *Cell* **90**:569–580, 1997.
13. Picard F, Gehin M, Annicotte J, *et al.*, SRC-1 and TIF2 control energy balance between white and brown adipose tissues, *Cell* **111**:931–941, 2002.
14. Xu J, Qiu Y, DeMayo FJ, *et al.*, Partial hormone resistance in mice with disruption of the steroid receptor coactivator-1 (SRC-1) gene, *Science* **279**:1922–1925, 1998.
15. Weiss RE, Xu J, Ning G, *et al.*, Mice deficient in the steroid receptor co-activator 1 (SRC-1) are resistant to thyroid hormone, *EMBO J* **18**:1900–1904, 1999.
16. Louet JF, Coste A, Amazit L, *et al.*, Oncogenic steroid receptor coactivator-3 is a key regulator of the white adipogenic program, *Proc Natl Acad Sci USA* **103**(47):17868–17873, 2006.

17. Wang Z, Qi C, Krones A, *et al.*, Critical roles of the p160 transcriptional coactivators p/CIP and SRC-1 in energy balance, *Cell Metab* **3**:111–122, 2006.
18. Jeong JW, Kwak I, Lee KY, *et al.* The genomic analysis of the impact of steroid receptor coactivators ablation on hepatic metabolism, *Mol Endocrinol* **20**:1138–1152, 2006.
19. Xu J, Liao L, Ning G, *et al.*, The steroid receptor coactivator SRC-3 (p/CIP/RAC3/AIB1/ACTR/TRAM-1) is required for normal growth, puberty, female reproductive function, and mammary gland development, *Proc Natl Acad Sci USA* **97**:6379–6384, 2000.
20. Lin J, Puigserver P, Donovan J, *et al.*, Peroxisome proliferator-activated receptor gamma coactivator 1beta (PGC-1beta), a novel PGC-1-related transcription coactivator associated with host cell factor, *J Biol Chem* **18**:277(3):1645–1648, 2002.
21. Kressler D, Schreiber SN, Knutti D, Kralli A, The PGC-1-related protein PERC is a selective coactivator of estrogen receptor alpha, *J Biol Chem* **277**(16):13918–13925, 2002.
22. Kamei Y, Ohizumi H, Fujitani Y, *et al.*, PPARgamma coactivator 1beta/ERR ligand 1 is an ERR protein ligand, whose expression induces a high-energy expenditure and antagonizes obesity, *Proc Natl Acad Sci USA* **100**(21):12378–12383, 2003.
23. Puigserver P, Adelmant G, Wu Z, *et al.*, Activation of PPARgamma coactivator-1 through transcription factor docking, *Science* **286**(5443): 1368–1371, 1999.
24. Wallberg AE, Yamamura S, Malik S, *et al.*, Coordination of p300-mediated chromatin remodeling and TRAP/mediator function through coactivator PGC-1alpha, *Mol Cell* **12**(5):1137–1149, 2003.
25. Monsalve M, Wu Z, Adelmant G, *et al.*, Direct coupling of transcription and mRNA processing through the thermogenic coactivator PGC-1, *Mol Cell* **6**(2):307–316, 2000.
26. Lin J, Handschin C, Spiegelman BM, Metabolic control through the PGC-1 family of transcription coactivators, *Cell Metab* **1**(6):361–370, 2005.
27. Lin J, Tarr PT, Yang R, Rhee J, *et al.*, PGC-1{β} in the regulation of hepatic glucose and energy metabolism, *J Biol Chem* **278**: 30843–30848, 2003.
28. St-Pierre J, Lin J, Krauss S, *et al.*, Bioenergetic analysis of peroxisome proliferator-activated receptor {γ} coactivators 1{α} and 1{β} (PGC-1{α} and PGC-1{β}) in muscle cells, *J Biol Chem* **278**:26597–26603, 2003.
29. Hentschke M, Susens U, Borgmeyer U, PGC-1 and PERC, coactivators of the estrogen receptor-related receptor gamma, *Biochem Biophys Res Commun* **299**(5):872–879, 2002.

30. Giguere V, Yang N, Segui P, *et al.*, Identification of a new class of steroid hormone receptors, *Nature* **331**(6151):91–94, 1988.
31. Giguere V, To ERR in the estrogen pathway, *Trends Endocrinol Metab* **13**(5):220–225, 2002.
32. Sladek R, Bader JA, Giguere V, The orphan nuclear receptor estrogen-related receptor alpha is a transcriptional regulator of the human medium-chain acyl coenzyme A dehydrogenase gene, *Mol Cell Biol* (**9**):5400–5409, 1997.
33. Mootha VK, Lindgren CM, Eriksson KF, *et al.*, PGC-1alpha-responsive genes involved in oxidative phosphorylation are coordinately downregulated in human diabetes, *Nat Genet* **34**(3):267–273, 2003.
34. Schreiber SN, Emter R, Hock MB, *et al.*, The estrogen-related receptor alpha (ERRalpha) functions in PPARgamma coactivator 1alpha (PGC-1alpha)-induced mitochondrial biogenesis, *Proc Natl Acad Sci USA* **101**(17):6472–6477, 2004.
35. Dufour CR, Wilson BJ, Huss JM, *et al.*, Genome-wide orchestration of cardiac functions by the orphan nuclear receptors ERRalpha and gamma, *Cell Metab* (**5**):345–356, 2007.
36. Alaynick WA, Kondo RP, Xie W, *et al.*, ERRgamma directs and maintains the transition to oxidative metabolism in the postnatal heart, *Cell Metab* (**1**):13–24, 2007.
37. Villena JA, Hock MB, Chang WY, *et al.*, Orphan nuclear receptor estrogen-related receptor alpha is essential for adaptive thermogenesis, *Proc Natl Acad Sci USA* **104**(4):1418–1423, 2007.
38. Huss JM, Imahashi K, Dufour CR, *et al.*, The nuclear receptor ERRalpha is required for the bioenergetic and functional adaptation to cardiac pressure overload, *Cell Metab* (**1**):25–37, 2007.
39. Sonoda J, Mehl IR, Chong LW, *et al.*, PGC-1beta controls mitochondrial metabolism to modulate circadian activity, adaptive thermogenesis, and hepatic steatosis, *Proc Natl Acad Sci USA* **104**(12):5223–5228, 2007.
40. Lelliott CJ, Medina-Gomez G, Petrovic N, Kis A, *et al.*, Ablation of PGC-1beta results in defective mitochondrial activity, thermogenesis, hepatic function, and cardiac performance, *PLoS Biol* **4**(11):e369, 2006.
41. Vats D, Mukundan L, Odegaard JI, Zhang L, *et al.*, Oxidative metabolism and PGC-1beta attenuate macrophage-mediated inflammation, *Cell Metab* **4**(1):13–24, 2006.
42. Sonoda J, Laganiere J, Mehl IR, Barish GD, *et al.*, Nuclear receptor ERR alpha and coactivator PGC-1 beta are effectors of IFN-gamma-induced host defense, *Genes Dev* **21**(15):1909–1920, 2007.
43. Arany Z, Lebrasseur N, Morris C, *et al.*, The transcriptional coactivator PGC-1beta drives the formation of oxidative type IIX fibers in skeletal muscle, *Cell Metab* **5**(1):35–46, 2007.

44. Lin J, Wu H, Tarr PT, Zhang CY, *et al.*, Transcriptional co-activator PGC-1 alpha drives the formation of slow-twitch muscle fibres, *Nature* **418**(6899): 797–801, 2002.
45. Wende AR, Schaeffer PJ, Parker GJ, *et al.*, A role for the transcriptional coactivator PGC-1alpha in muscle refueling, *J Biol Chem*, 2007.
46. Vianna CR, Huntgeburth M, Coppari R, *et al.*, Hypomorphic mutation of PGC-1beta causes mitochondrial dysfunction and liver insulin resistance, *Cell Metab* (**6**):453–464, 2006.
47. Lowell BB, Spiegelman BM, Towards a molecular understanding of adaptive thermogenesis, *Nature* **404**(6778):652–660, 2000.
48. Haemmerle G, Lass A, Zimmermann R, *et al.*, Defective lipolysis and altered energy metabolism in mice lacking adipose triglyceride lipase, *Science* **312**(5774):734–737, 2006.
49. Golozoubova V, Hohtola E, Matthias A, *et al.*, Only UCP1 can mediate adaptive nonshivering thermogenesis in the cold, *FASEB J* **15**(11): 2048–2050, 2001.
50. Ukropec J, Anunciado RP, Ravussin Y, *et al.*, UCP1-independent thermogenesis in white adipose tissue of cold-acclimated Ucp1$^{-/-}$ mice, *J Biol Chem* **281**(42):31894–31908, 2006.
51. Petersen KF, Shulman GI, Etiology of insulin resistance, *Am J Med* **119**(5 Suppl 1):S10–16, 2006.
52. Patti ME, Butte AJ, Crunkhorn S, *et al.*, Coordinated reduction of genes of oxidative metabolism in humans with insulin resistance and diabetes: Potential role of PGC1 and NRF1, *Proc Natl Acad Sci USA* **100**(14): 8466–8471, 2003.
53. Ling C, Poulsen P, Carlsson E, *et al.*, Multiple environmental and genetic factors influence skeletal muscle PGC-1alpha and PGC-1beta gene expression in twins, *J Clin Invest* **114**(10):1518–1526, 2004.
54. Andersen G, Wegner L, Yanagisawa K, *et al.*, Evidence of an association between genetic variation of the coactivator PGC-1beta and obesity, *J Med Genet* **42**(5):402–407, 2005.
55. Ling C, Wegner L, Andersen G, *et al.*, Impact of the peroxisome proliferator activated receptor-gamma coactivator-1beta (PGC-1beta) Ala203Pro polymorphism on *in vivo* metabolism, PGC-1beta expression and fibre type composition in human skeletal muscle, *Diabetologia* **50**(8):1615–1620, 2007.

Chapter 19

Nuclear Receptor Cofactor Interactions as Targets for New Drug Discovery

Linda L. Grasfeder and Donald P. McDonnell

The classical models of nuclear receptor pharmacology held that agonists functioned by binding to their cognate receptors facilitating their conversion from an inactive form to one that was capable of activating transcription. By extrapolation, it was considered that antagonists functioned by competitively inhibiting agonist binding, freezing the receptor in an inactive state. However, as early as 1967 when the biological actions of the "antiestrogen" tamoxifen were first described, it was clear that this simple model did not adequately describe estrogen receptor pharmacology. Indeed, tamoxifen is now classified as a Selective Estrogen Receptor Modulator (SERM), one of a group of compounds whose relative agonist/antagonist activity differs between cells. Similarly, tissue selective progesterone, androgen and glucococorticoid receptor modulators have also been identified. Significant progress has been made in defining the molecular mechanism(s) by which cells distinguish between agonists and antagonists and how some receptor modulators can manifest their actions in a cell-selective manner. The most important of these are (1) differences in the relative expression level of receptor isoforms or subtypes, (2) the impact which the bound ligand has on the structure of its cognate receptor, and (3) the complement of coactivators and corepressors in a target cell which can interact with the activated receptor. Exploitation of this complexity will lead to the development of novel classes of nuclear receptor modulators with useful therapeutic activities.

19.1 Introduction

In basic models of steroid hormone receptor action, unoccupied receptors are proposed to reside in target cells in an inactive form. Upon agonist binding, the biochemical properties of the steroid receptor (SR) are altered to allow the interaction of a receptor dimer with specific DNA sequences within the promoters of target genes. The DNA bound receptor can then exert either a positive or negative effect on target gene transcription. With respect to the pharmacological actions of SR ligands, these simple models predict that agonists function merely as "switches" that facilitate the conversion of the SRs from an inactive to an active form, whereas antagonists function solely by competitively inhibiting the binding of agonists. Thus, when corrected for affinity, all agonists for a given receptor would be qualitatively the same, as would all compounds classified as antagonists. Not surprisingly, the majority of existing drugs that act on SRs were discovered using simple *in vitro* receptor binding assays with subsequent *in vivo* assays used to distinguish agonists from antagonists. In recent years, however, it has become increasingly clear that there is an unmet medical need for nuclear receptor (NR) interacting drugs that manifest positive activities in a tissue-restricted manner (especially for long-term treatment of chronic diseases). In particular, there are specific needs for (a) a glucocorticoid that exhibits anti-inflammatory actions without causing metabolic disturbances or osteoporosis; (b) an estrogen capable of treating the climacteric symptoms associated with long-term estrogen deprivation but inactive in the breast and uterus; and (c) an androgen that improves lean body mass and bone density without liver toxicity and prostatic hypertrophy. However, within the confines of the classical models of NR pharmacology, it was difficult to understand how absent approaches that relied on differential ligand pharmacokinetics was going to be possible to develop receptor modulators with clinically useful selectivity. Interestingly, a retrospective analysis of the early preclinical studies of the estrogen receptor (ER) modulator, tamoxifen, revealed that it could function as an agonist or an antagonist in a manner that differed between tissues in the same species, and in the same target organ in different species.[1,2] Although these data clearly indicated that the pharmacological definitions "agonist" and "antagonist" did not adequately describe the pharmacology of ER ligands, it took nearly 30 years before the clinical importance of these seminal findings were appreciated. A further 10–15 years of research was needed to define the likely molecular basis for the tissue selectivity of this ER modulator.

Using the insights that have emerged from recent studies, it has been possible to construct a good first draft of the pathways and processes that impinge on and modulate the pharmacological activities of different ER-ligand complexes. Importantly, the information obtained from the study of the molecular pharmacology of ER ligands appears to translate to other receptors. Thus, it is not surprising that we are in the midst of a paradigm shift in NR ligand discovery programs away from classical "grind-and-bind" approaches to the exploitation of the complexity of NR signaling using mechanism-based functionally predictive screens. This review will discuss the seminal findings that have helped to shape our understanding of NR pharmacology and will consider where the field is going and what new types of pharmaceuticals are likely to emerge in the future.

19.1.1 *The "tamoxifen paradox": A clinical observation that reinvigorated interest in nuclear receptors as therapeutic targets*

It was not clear until the early 1990s that it would be possible to develop NR ligands with clinically useful selectivity. However, the findings of a seminal study conducted by Love *et al.* demonstrated that, at least with respect to ER, this goal could be accomplished. Specifically, these investigators examined the impact of adjuvant tamoxifen treatment of two years duration on the skeleton in postmenopausal breast cancer patients.[3] It had been established that longer duration of treatment with tamoxifen had a positive impact on disease-free and overall survival in breast cancer patients. However, given its antiestrogenic properties and the morbidity associated with osteoporosis, it was important to determine the effects of long-term treatment on the skeleton. The surprising result of these studies was that tamoxifen, rather than functioning as an antiestrogen in the lumbar spine, actually significantly increased BMD in a manner similar to that expected of an estrogenic compound.[4,5] Indeed, were it not for the fact that tamoxifen was also shown to exhibit estrogenic activities in the endometrium, it is likely that this drug would have had the utility as an antiresorbtive for use in the treatment and prevention of osteoporosis. This result, referred to as the "tamoxifen paradox", indicated that although this drug can function as an antiestrogen in the breast, it is a significant estrogen in bone. This led to the reclassification of tamoxifen as a selective estrogen receptor modulator (SERM), as opposed to an antiestrogen, to reflect the fact that its relative agonist/antagonist activities manifested in a tissue-dependent manner.[6] The generality of the SERM concept was established in additional studies which demonstrated that another previously described ER antagonist, keoxifene (now called raloxifene), also exhibited a favorable estrogenic activity in bone, but unlike tamoxifen, it functioned as an antiestrogen in the endometrium.[7,8] Not surprisingly, given this selectivity, this SERM has been registered for the treatment and prevention of osteoporosis. It was determined to exhibit comparable activity to tamoxifen, as a breast cancer chemopreventative, in the recently completed STAR trial.[9] Thus, the clinical utility of this drug is likely to expand in the near future. Although tamoxifen and raloxifene exhibit similar pharmacological activities in most target organs, their dramatically different activities in the uterus have served to define the SERM concept. From a practical point of view, these are important findings as they provide the impetus to develop selective nuclear receptor modulators (SNRMs) which, by acting through their targeted receptors, are more clinically useful than classical agonists or antagonists.

19.2 The Molecular Determinants of ER Pharmacology

The identification of SERMs has represented a major advance in the pharmacotherapy of breast cancer and other endocrinopathies. In addition, studies of their molecular mechanism of action has led to the determination that among the most important determinants of NR ligand pharmacology are (a) receptor isoform and subtype expression in target tissues; (b) the impact of ligands on the structure of the receptor; and (c) the ability of differently conformed ligand-receptor complexes to recruit functionally distinct coregulators. A discussion of how these processes influence ER pharmacology will serve to highlight the generality of these concepts.

19.2.1 *The role of receptor subtypes in ER ligand pharmacology*

The biological actions of ER ligands are manifest through high affinity nuclear receptors located within the target cell nuclei. Until relatively recently, it was considered that all the biological actions of ER modulators were manifest through a single receptor that was biochemically identical in all cells. However, the discovery in 1996 of a second ER significantly increased the biological complexity of estrogen.[10] Unexpectedly, Kuiper and coworkers identified a novel receptor cDNA in the prostate of male rats which encoded a protein that bound 17β-estradiol with an affinity equivalent to that of the previously identified "breast/uterus" estrogen receptors.[10] A human homologue of this novel ER, now called ERβ, was subsequently cloned.[11] We now know that the biological action of these two receptor subtypes is not equivalent, and although a discussion of this aspect of ER signaling goes beyond the scope of this review, it is important to note two general themes that have emerged: (1) ERα and ERβ are not functionally redundant, with each being able to regulate a distinct set of biological processes, and[2] in cells where both receptors are expressed, ERβ functions as a trans-dominant inhibitor of ERα signaling.[12–14] It is clear that most of the known receptors have more than one functional isoform or subtype. Not surprisingly, there is a high level of interest in developing small molecules that target these receptor variants as a means of generating more selective drugs.

There has been considerable progress in identifying specific ERβ ligands, and the preliminary studies of these agents *in vivo* have suggested that there are potentially useful clinical outcomes that will emerge by selectively activating each receptor. Two important studies have been published which emphasize this point. Firstly, it has been shown by Harris and colleagues that a specific ERβ agonist, ERB-041, exhibits potent anti-inflammatory activities in rodent models of rheumatoid arthritis and inflammatory bowel disease.[15] These activities can be inhibited by the ER pan-antagonist ICI182,780, confirming the involvement of ERβ. However, these potent anti-inflammatory activities are not as pronounced in animals treated with estradiol, a non-selective ER agonist. This raises the possibility that the anti-inflammatory actions of ERβ ligands may represent a pharmacological activity of this class of compounds that may not occur when the receptor is activated by natural ligands. A second series of ERβ agonists have recently been described that reduce ventral prostate size in rodents.[16] This may relate to the fact that ERβ is a negative regulator of androgen receptor (AR) expression. In discussions at scientific meetings, it appears that the activities of the available ERβ agonists are not equivalent, leading to the suggestion that the existing compounds represent the founding members of a new class of ligands, the selective estrogen receptor beta agonists (SERBAs).[16] Thus, although there is considerable interest in developing novel ERβ agonists, the therapeutic (disease) targets and the profile of the desired drug remains to be determined.

19.2.2 *Ligand induced changes in NR conformation as a determinant of pharmacology*

Initially, it was considered that the selectivity of SERMs like tamoxifen and raloxifene could be explained by differential activation of ERα or ERβ. However, with minor exceptions, the affinity of the SERMs for both receptors has been shown to be equivalent.[17,18] Furthermore, it has been shown that the biological activity of some SERMs can differ among tissues or cells that express the same ER subtype. For instance, using ERα responsive transcription systems reconstituted in a variety of cells, it has been shown that tamoxifen can function in some backgrounds as an antagonist whereas in others, it can manifest partial agonist activity.[19,20] Furthermore, in those cells where tamoxifen manifests partial agonist activity, raloxifene, droloxifene and idoxifene can function as pure ERα antagonists.[21] This suggests that there are differences in the way each ER is able to respond to different SERMs in different cells. Considerable insight into the molecular basis of this selectivity has come from studies aimed at understanding how information flows from the unique chemical structure of each SERM to the transcription apparatus, and how this translates into a different phenotypic response.

According to the classical models of NR action, ERα is a ligand-regulated transcription factor that remains inactive when associated with a heat-shock protein complex until an activating ligand binds. This activity promotes the displacement of the receptor from the inhibitory complex, allows receptor dimerization, and facilitates its interaction with specific estrogen response elements located within target gene promoters. Depending on the cellular and promoter context, the DNA bound promoter can either positively or negatively regulate target gene transcription.[20] In the simplest interpretation of this model, the antiestrogenic activities of SERMs, like tamoxifen, would reflect a displacement of estradiol and the retention of the receptor in an inactive state. Clearly, this was not the correct interpretation. The first departure from this model describing antagonist action came from the studies of Allan *et al.* that used protease digestion to map the conformational changes in the progesterone receptor (PR) which occur upon ligand binding.[22–24] Specifically, these studies demonstrated that agonists and antagonists induce distinct conformational changes in PR upon binding and, importantly, both conformations were distinct from apo-receptor. This implied that antagonists do not merely freeze the receptor in an inactive conformation but facilitate the formation of a uniquely structured complex. A similar series of studies were later performed with ERα and revealed that the overall conformations of the ERα-tamoxifen and ERα-estradiol complexes were different and distinct from apo-receptor.[25,26] Subsequent analysis of the crystal structure of the ERα ligand binding domains complexed with either an agonist or an antagonist confirmed the role of ligands in determining the overall shape of the receptor.[27–29] Furthermore, these latter studies provided insight as to the mechanisms by which agonist-activated receptors physically interacted with nuclear receptor coactivators. Specifically, ligand-induced alterations in the receptor enabled the formation of a hydrophobic cleft that was capable of interacting with the receptor interaction domains of coactivators. This domain is coincident with the previously defined "activation function-2 (AF-2)" of the receptor.[30] In the presence of SERMs, like tamoxifen and raloxifene, this coactivator-binding pocket is not formed, providing a mechanistic explanation for the antagonist activity of these compounds. However, although these structural studies provided an explanation for the agonist activity of molecules, like estradiol, and for the antagonist activity of SERMs, they did not provide insight into the mechanisms underlying the tissue selective agonist activities of different SERM-ER complexes.

19.2.3 *SERMs enable the presentation of specific cofactor interaction surfaces that are required for tissue selective agonist activity*

One interpretation of the results of the assays of receptor conformation was that a specific agonist-induced structure of ER (or any NR) was required for transcriptional activity and that all other conformations were incompatible with activation. This provided a simple model of agonism and antagonism but did not provide an explanation for SERM activity. We proposed that although the canonical coactivator binding pocket was disabled in the SERM-ER complexes, there are additional surfaces presented upon binding these ligands that enable cofactor binding. We further posited that the availability of cofactors that could interact with these surfaces determined the relative agonist/antagonist activities of SERMs. To test this idea, we mapped the protein-protein interaction surfaces on ERα that were presented upon its interaction with different ligands, and then defined the roles of these surfaces in the pharmacological actions of the bound ligand. The complete details of these experiments have been published previously.[21,31–33] In brief, we used combinatorial phage display to screen libraries of random peptides of 15–19 amino acids in length that interacted with purified ERα. We did this in the presence of different ligands. The high-affinity interacting phage identified in this manner were then used in a phage ELISA assay as a means of profiling the surfaces presented on the receptor in the presence of different ligands. The results of one of the most informative assays are presented in Fig. 19.1. One class of peptides, represented by α/β I, interacted with ERα in the presence of estradiol but not other ligands tested. A second class of peptide, represented by α/β V interacted with tamoxifen-activated ERα but not with that activated by estradiol. Neither peptide interacted with the apo-receptor. These important findings confirmed the results of both the protease digestion and crystallography studies, but went further in that they revealed that there are distinct protein-protein interaction surfaces presented on ERα in the presence of estradiol and SERMs. In support of this conclusion, we observed obvious sequence conservation, with distinct consensus motifs apparent, for both the ERα-estradiol and the ERα-tamoxifen interacting peptides. For instance, most, if not all, the peptides that interacted specifically with estradiol-activated ERα contained the sequence LXXLL, a motif commonly found in the receptor interacting surfaces of coactivators. A strong, though unrelated, consensus sequence was also apparent in the peptides that interacted with tamoxifen-activated ERα. We concluded from these results that, in addition to serving as surrogate markers of ER structure, these peptides highlight surfaces on the receptor that are important for ligand pharmacology.

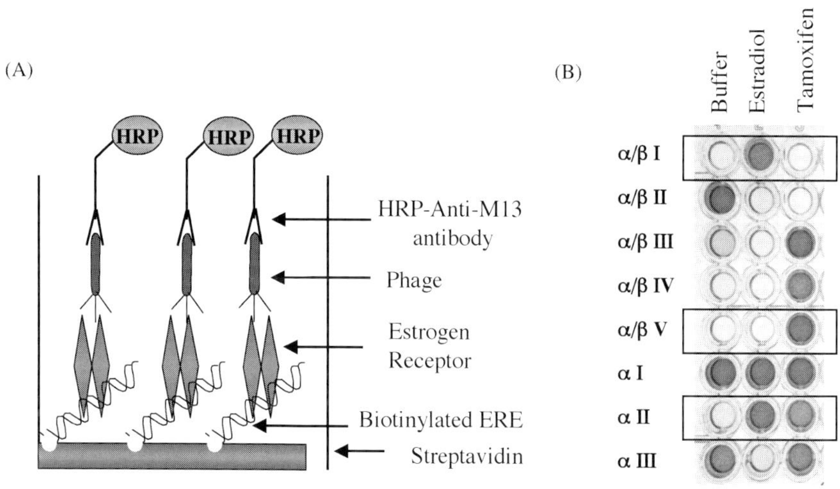

Fig. 19.1. Probing ligand-dependent changes in the conformation of ERα using combinatorial peptide phage display. Bacteriophage expressing peptides which are capable of interacting with ERα in the presence of different ligands were identified and purified as described previously.[32,33] These were then used to probe the conformational changes in the receptor that occurred subsequent to its interaction with different ligands. (A) A diagram of the features of the Phage ELISA assay used for these studies. (B) The ability of selected phage, expressing ERα interacting peptides, to interact with the receptor in the presence of different ligands was assessed using a phage ELISA assay. Those peptides that are discussed specifically in the text are highlighted.

The functionality of the protein-protein interaction surfaces mapped using phage display was tested by assessing the ability of the expressed peptides (α/β I and α/β V) to inhibit the agonist activity of estradiol and tamoxifen in cells where both compounds can function as agonists. Specifically, using a reconstituted ERα-responsive transcription system in HepG2 cells, it was demonstrated that both estradiol and tamoxifen were capable of activating transcription of a C3-luciferase reporter. Interestingly, when the α/β I peptide was expressed in cells, estradiol but not tamoxifen agonist activity was inhibited, whereas the converse was true when the tamoxifen specific peptide α/β V was expressed (Fig. 19.2).[33] We believe that these results indicate that the agonist activity of estradiol and tamoxifen do not occur in the same

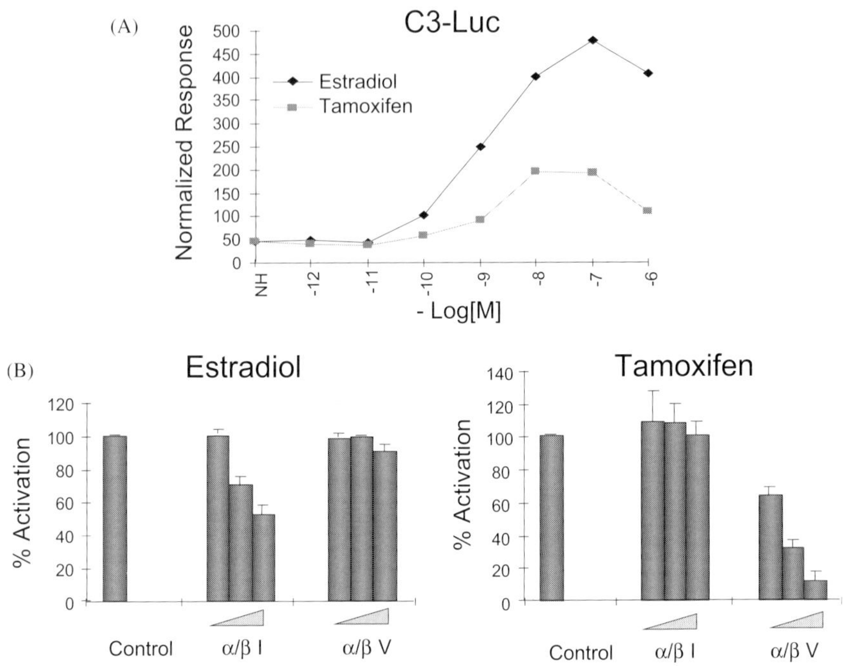

Fig. 19.2. Tamoxifen and estradiol facilitate the presentation of distinct protein-protein interaction surfaces on ERα. (A) HepG2 cells were transfected with the estrogen-responsive C3-Luc reporter gene along with expression vectors for ERα and a normalization plasmid (β-Gal). Cells were treated with either estradiol or tamoxifen as indicated and analyzed for luciferase and β-Gal activity. NH, no hormone. (B) HepG2 cells were transfected as in (A) except that expression vectors for peptide-Gal4 fusions were included as indicated. Control represents the transcriptional activity of estradiol or tamoxifen-activated ERα in the presence of the Gal4-peptide fusion as shown with the resulting transcriptional activity presented as percentage of activation of control. Data are averaged from three independent experiments (each performed in triplicate) with error bars representing SEM.

manner, but rather each compound enables distinct cofactor interactions by presenting different protein-protein interaction surfaces on the receptor. The ER$\alpha/\alpha/\beta$ V complex has not yet been crystallized and thus it has been difficult to assess how "different" the protein-protein interaction surfaces are on the receptor following activation with estradiol or tamoxifen. This is an important issue to resolve as it will enable us to determine if the tamoxifen-ERα complex manifests agonist activity by interacting with a subset of the cofactors with which the ERα-estradiol complex interacts or if the tamoxifen-ERα complex interacts in an ectopic manner with an as yet to be defined cofactor. A working model of our current understanding of some of the basic differences in estradiol and tamoxifen pharmacology is outlined in Fig. 19.3. In the presence of estradiol, a conformational change occurs that induces the formation of the AF-2 coactivator interacting pocket and the subsequent interaction of the receptor with a requisite coactivator (CoA). Upon binding tamoxifen, however, a surface other than AF-2 is presented. This surface is unable to engage a coactivator that utilizes a functional AF-2 and, in most circumstances, this results in tamoxifen functioning as an antagonist. However, in some environments, the surfaces presented on the tamoxifen-activated ERα can engage a cofactor that permits it to manifest agonist activity. Thus, differential expression of the putative coactivator, designated CoA(X), determines the relative agonist/antagonist activity of tamoxifen. Formal proof of this hypothesis will require the identification and functional analysis of proteins that can interact with the tamoxifen-activated receptor and whose knockdown prevents the manifestation of tamoxifen agonist activity.

19.3 Differential Cofactor Interaction Assays as a Means to Identify Novel ER Regulators

Resistance (both *denovo* and acquired) is a significant clinical issue that limits the efficacy and duration of response to tamoxifen in breast cancer.[34] Although resistance is likely to be a multifactoral process, it is clear that the ability of tamoxifen to manifest agonist activity is a key contributing factor. It has been shown in breast cancer cell xenografts propagated in mice that tamoxifen initially functions as an antagonist and then "switches" to functioning as an agonist.[35,36] There were early anecdotal reports of "tamoxifen withdrawal" responses in women who progressed while on drug, a finding that is consistent with that observed in animal studies.[37–39] Additionally, the results of the NSABP-B14 trial revealed that although short-term adjuvant treatment with tamoxifen (<5 years) was clearly beneficial, longer duration of exposure (5–9 years) was associated with an increased risk of breast cancer recurrence.[40,41] This suggested that there was either a selection for cells in the breast that exhibited an innate ability to recognize tamoxifen as an agonist or that some epigenetic event occurred in breast tumor cells that permitted this activity. Given the model we proposed to explain the agonist activity of tamoxifen (Fig. 19.3), we posited that by screening against the presentation of cofactor interaction surfaces on ERα that are apparent upon binding estradiol or tamoxifen, it would be possible to identify a new class of antagonist that would have utility in the treatment of tamoxifen refractory breast cancer. The details of the successful screen that led to

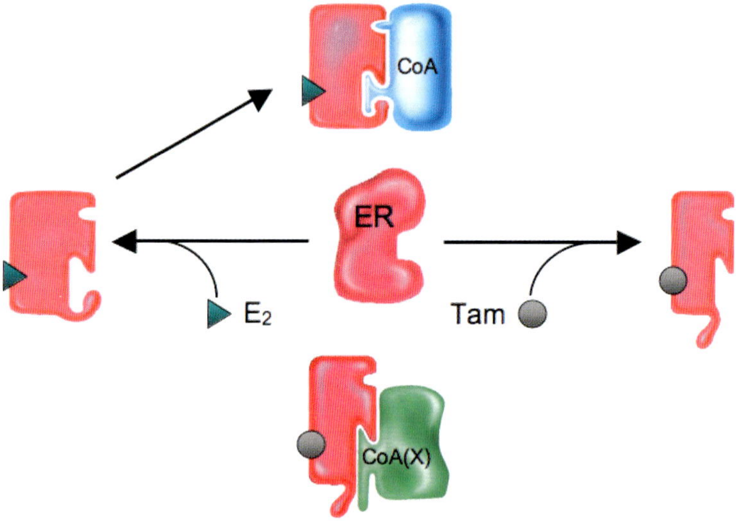

Fig. 19.3. The mechanisms by which estradiol and tamoxifen manifest agonist activity are dissimilar. Using peptide antagonists that inhibit specific protein-protein interactions, it has been possible to show that the mechanisms by which estradiol and tamoxifen manifest agonist activity are dissimilar. Estradiol binding enables ERα to adopt a structure that is compatible with the binding of the p160 class of coactivators. Tamoxifen binding, on the other hand, induces a unique alteration in receptor structure that permits an ectopic interaction of the receptor with an unidentified coactivator. The existence of this coactivator is supported by the fact that peptides of the α/β V class will inhibit tamoxifen, but not estradiol-mediated transcriptional activity when expressed in target cells. Tamoxifen functions as an antagonist in situations where it is unable to facilitate the recruitment of a coactivator to ERα.

the identification of GW5638 (now called IOS974), a compound that inhibits tamoxifen agonist activity *in vitro* and *in vivo*, have been published previously.[35,42] Considering the manner in which it was identified, it is not surprising that we were able to show using peptide profiling and crystallography that this new compound permitted ERα to adopt a novel structure.[21,33,43] This drug is currently being evaluated in the clinic as a second line intervention for patients with tamoxifen refractory disease. The successful outcome of this project highlights the utility of mechanism-based screens in the search for novel NR modulators.

It is clear that ligand-induced alterations in ERα structure are important determinants of cofactor recruitment and downstream biology. A question that has arisen from these studies of ER action is whether or not endogenous ligands exist which manifest SERM activity. In this regard, it has been demonstrated recently by the Mangelsdorf laboratory that the oxysterol metabolite 27-hydroxycholesterol (27HC) exhibits SERM activity *in vivo*.[44] We have confirmed the SERM properties of 27HC in cellular models of estrogen action and demonstrated using peptide profiling that this compound permits ERα to adopt a conformation distinct from other ligands.[45] Interestingly, although 27HC has a relatively low affinity for ERα, it is present in high concentrations within atherosclerotic plaques and is produced at high levels in various tissues by infiltrating macrophages. It remains to be determined if 27HC or similar endogenous SERM-like molecules serve as physiological regulators of ERα (or ERβ), or if they only come into play in pathological conditions.

19.4 Translating Insights from the Study of Serms to Other Nuclear Receptors

One of the unifying themes that has emerged from the study of SERMs is that receptor conformation is a primary determinant of selectivity. This important observation has suggested that it may be possible to develop selective modulators of other nuclear receptors. Among the most advanced programs are those aimed at developing selective progesterone receptor modulators (SPRMs), selective androgen receptor modulators (SARMs) and selective glucocorticoid receptor modulators. A review of the literature, however, indicates that the selective nuclear receptor modulator (SNRM) descriptor is applied loosely and has been used to describe both compounds that achieve selectivity as a result of a unique action at the level of the receptor as well as those that achieve selectivity as a result of differential pharmacokinetics.

19.4.1 *Selective progesterone receptor modulators*

The major clinical uses of progestins to date relate to their ability to oppose estrogen action in the reproductive tract. Thus, progestin-containing medicines have seen widespread use as hormone therapy (HT), contraceptives and as third line therapies in ER-positive breast cancers. However, the currently used progestins have significant side effects creating a niche for molecules with improved therapeutic selectivity. The desired molecules would (a) exhibit antiproliferative activity in the endometrium and in breast tissue; (b) be neutral in the CNS; (c) have no impact on the cardiovascular system; and (d) not negatively impact estrogen action in bone. There is also a high level of interest in developing PR antagonists for the treatment of uterine fibroids and endometriosis.[46] Much of this interest comes from the seminal findings of Sam Yen who first demonstrated that the antiprogestin RU486 could be used as a medical intervention in fibroids.[47,48] However, since this molecule was also a potent antiglucocorticoid, it was unclear as to what activity of RU486 was required for this therapeutic activity.[49] The positive clinical studies that have been published with new and more selective PR modulators have validated PR as an appropriate therapeutic target in this disease. One of the most interesting drugs, asoprisnil (J867), has shown to display agonist, anti-proliferative effects on the endometrium, while functioning as an antagonist of progesterone-driven growth of uterine fibroids.[50–52] Therefore, this molecule exhibits the pharmacological characteristics of a selective progesterone receptor modulator (SPRM). Indeed, in a preliminary study of the molecular basis for its distinct pharmacological actions, it was determined that the PR-asoprisnil complex is capable of binding both coactivator (TIF2) and corepressor (NCoR) peptides with equal affinities in fluorescent binding assays, whereas the corepressor peptide binds with 10-fold greater affinity to RU486-bound PR.[53] Additionally, the crystal structure of PR bound to asoprisnil and an NCoR peptide indicates that the receptor assumes a unique conformation.[53] Thus, the relationship between the structure of the ER-ligand complex and function that explains SERM pharmacology appears to extend to SPRMs.

Finally it is worth mentioning that there are multiple isoforms of PR (PR-A, PR-B, and PR-C), which have distinct biological activities in the female reproductive system. However, since all isoforms are encoded by the same gene, with PR-A having its N-terminal domain truncated, and PR-C additionally missing part of its AF-2 domain; the ligand binding pockets of the receptors are identical. Thus, absent drug-induced allosteric effects on the PR-ligand binding domain that are manifest differently in the context of the three isoforms, it is difficult to see how these receptor isoforms can be exploited differentially. It is possible, however, that they have different cofactor requirements, and that targeting these interaction surfaces may yield drugs with additional selectivity.

19.4.2 *Selective androgen receptor modulators*

There is a high level of interest in developing selective androgen receptor modulators (SARMs) that exhibit androgenic activity in bone and muscle while exhibiting neutral or antagonist activity in the bone.[54–60] Several classes of new molecules have emerged that exhibit these properties in rodent models and are currently being evaluated in the clinic. It is unclear if these compounds are truly functioning as SARMs, or if they are weak/partial agonists that achieve selectivity as a consequence of differential sensitivity of target tissues to agonists. Under normal circumstances, dihydrotestosterone is the androgen that is primarily responsible for AR-dependent prostate growth. Thus, the conversion of testosterone to DHT through the actions of 5α-reductase, amplifies the androgenic stimulus to the prostate. In the presence of a 5α-reductase inhibitor, testosterone has reduced activity in the prostate but retains its anabolic activities. This has led to the suggestion that the selectivity of the currently available SARMs is probably due to the fact that they have sufficient potency/efficacy to mimic the actions of testosterone in bone/muscle but they cannot maximally activate AR in prostate.[61] This would explain why it is relatively easy to develop AR ligands with an improved therapeutic window (prostate versus muscle/bone). One concern, however, from lessons learned from studies of ER pharmacology is that over time the prostatic epithelial cells may adapt and compensate for the reduced androgenic stimulus by increasing the expression of AR, or one of its coactivators, or through a hyperactivation of a signaling pathway that increases cellular sensitivity to androgens. Thus, it would appear to be more useful to develop SARMs that function by facilitating differential cofactor recruitment as opposed to functioning as partial agonists.

Whereas most of the currently available SARMs were discovered in an empirical manner, we have undertaken to develop mechanism-based screens to identify new AR ligands. The goal of these studies was to identify compounds that had different effects on AR structure and to determine if they exhibited SARM-like properties. Specifically, we used AR-interacting peptide probes to select for compounds that altered the structure of the AF-2 pocket in different ways.[62] We reasoned that this might impact the interaction of AR with its cofactors. Using this approach, coupled with modeling and combinatorial peptide phage display, we were able to generate a series of compounds that were either partial agonists or neutral with respect to gene transcription. However, both classes of compounds were fully efficacious as agonists in cell proliferation assays.[62,63] Although clearly not the profile of the desired modulator, this study revealed that as with SERMs, it was possible separate the biological functions of AR using ligands that had a different effect on AR structure. We are currently expanding these efforts with a view to developing an understanding of the relationship between AR structure and function. This is a first step in the rational development of SARMs.

19.4.3 *Selective glucocorticoid receptor modulators*

The anti-inflammatory properties of glucocorticoids have been utilized in the clinics for many years for the treatment of various inflammatory diseases, including asthma, rheumatoid arthritis, and autoimmune diseases. However, the side effect profile of these compounds including hypertension, muscle atrophy, osteoporosis and glucose intolerance, severely limits their long term use. Many of the undesired effects of GR relate to its ability to function as a positive activator of gene transcription while its anti-inflammatory properties are likely to result from its ability to "transrepress" the expression of pro-inflammatory genes regulated by NF-κB and AP-1. Not surprisingly, therefore, the search for selective glucocorticoid receptor modulators (SGRMs) has focused on identifying compounds that permit transpression at the expense of classical transcriptional responses. Significant progress in developing

SGRMs with this favorable pharmacological profile have emerged in recent years. For instance, several novel arylpyrazole compounds have been synthesized that behave different from classical agonists in terms of their activities on target gene transcription, GR nuclear localization, GRE occupancy, and cell proliferation and differentiation.[64–66] While all of these compounds bind GR with strong affinity and selectivity, they have dramatically different effects on receptor–promoter interactions and on gene expression profiles. Thus, as with other SNRMs, there exists a strong relationship between receptor structure and function.

19.5 Development of Novel NR Modulators that Function Outside of the Classical Ligand Binding Pocket

While the classical ligand binding pocket of NRs has been the primary target for pharmacological modulation of NR pathways, there is growing evidence that supports the idea that alternative NR surfaces can be targeted. In this regard, two new approaches appear to hold promise: (a) generation of short peptides that exhibit antagonist activity by inhibiting NR-cofactor interactions; and (b) development of small molecules that bind on the receptor surface and allosterically regulate coactivator function.

19.5.1 *Identification of antagonists that function by directly interfering with NR-cofactor interactions*

As described above, it has been possible to identify short, high affinity, peptides that interact with and inhibit NR transcriptional activity in a specific manner (Fig. 19.2).[31,33,67–69] Whereas these peptides are useful tools with which to probe NR pharmacology *in vitro*, they will require significant modifications and/or formulation to make them useful for studies *in vivo*. However, several groups have made progress in efforts aimed at making these peptides more "drug-like". In one study, it was shown that a macrolactam ring could be used to constrain a peptide from the NR box of GRIP1, a modification that allows the peptide to assume a partial α-helical conformation.[70] This constrained structure exhibited a higher affinity for the TRβ/T$_3$ complex than the uncyclized parent molecule and efficiently blocked the interaction of the receptor-ligand complex with GRIP1. In a similar manner, this group created a library of macrolactam constrained peptides that interacted selectively with ERα, ERβ or TRβ.[71] Other groups created peptide-derived antagonists using both disulfide and thioether bridges to create stabilized, cyclical peptides that are also able to block interaction of the SRC-1 NR box 2 with ERα or ERβ.[72–74] The extent to which these modifications have improved the pharmaceutical properties of the peptide-derived antagonists remains unclear.

Small molecules have several advantages over peptides in terms of drug delivery, stability, and permeability. Not surprisingly, therefore, several groups have attempted to isolate compounds that directly displace cofactor-AF-2 interactions. To this end, a structure-based design approach was used by John Katzenellenbogen and colleagues to identify small molecule inhibitors of NR-coactivator interactions.[75] Using the crystal structure of the LBD of ERα complexed with diethylstilbestrol and a 13 amino acid peptide from GRIP1, they identified key points of contact between the peptide and the receptor and designed compounds using a variety of scaffolds that allowed them to mimic the relevant interactions. Their most successful compound had a pyrimidine core with branched alkyl substituents mimicking the leucines of the coactivator NR-box. This compound was able to displace a coactivator peptide in a fluorescence anistrophy assay, and thus provided the first direct evidence that small molecules can be designed to target the coactivator binding pocket (Rodriguez, 2004).

Using a classical small molecule screen, Guy and Fletterick's groups identified compounds that inhibit TRβ-coactivator interactions.[76,77] The compounds identified in this manner did not compete with T_3 for binding to the LBD of TRβ. Instead, the receptor serves as a surface for catalyzing a reaction that releases an active, unsaturated ketone that covalently binds to the receptor and inhibits coactivator association. These compounds are able to completely block TRβ activity in cell-based transcription assays.[76] A subsequent report describing the crystal structure of TRβ complexed with the active compound, aromatic β-enone 1-(4-hexylphenyl)-prop-2-en-1-one (HPPE), confirmed that it binds within the AF-2 region and led to the identification of the specific residues in this domain with which it interacts.[77] Thus, while the utility of targeting the AF-2 coactivator binding surface was initially demonstrated using peptide antagonists, it is likely that it will be possible to develop small molecules that directly target this surface on the receptor. Drugs that inhibit ER and AR in this manner would have utility in the treatment of breast and prostate cancer where complete inhibition of the receptor is desirable.

While the AF-2 is an attractive alternative target for drug design, there are hints that some NRs display additional surfaces (pockets) that might be amenable to pharmacological manipulation. For instance, peptides have been found that interact with regions of nuclear receptors other than the AF-2 that inhibit their transcriptional activity. One such peptide, αII peptide, was discovered using combinatorial peptide phage display that interacts with ERα in the presence of several different ligands, and inhibits its ability to activate transcription (Fig. 19.1).[33] The crystallization of this peptide in a complex with OHT-bound ERα led to the identification of a previously unidentified protein-protein interaction surface on the β-hairpin face of the LBD, located on the opposite side of the receptor from AF-2 domain.[78] Small molecules that interact with this pocket would be expected to exhibit antagonist activity.

The orphan nuclear receptor Nurr1 has no classical ligand binding pocket, as bulky side chains of the amino acids lining the putative pocket consume all the space of the pocket. Since this receptor appears to be in an active conformation, the question arises as to how the activity of this receptor is regulated. Additionally, the transcriptional activity of this receptor is not affected by LxxLL-containing coactivator. Structural modeling of the receptor identified a potential alternate coactivator binding pocket, and site-directed mutagenesis of the putative region decreased transcriptional activity of the receptor.[79] Pockets of this nature represent potential targets for new drug discovery.

19.5.2 *Identification of NR modulators that function through protein allosterism*

In addition to directly targeting coactivator interaction surfaces, some recently published work from the Fletterick laboratory has validated "allosteric sites" on NRs as targets for drug discovery. Specifically, they identified a novel surface on AR when screening for compounds that dissociate an LxxLL peptide from DHT-bound AR.[80] Using this approach, they identified several molecules which when crystallized with AR were found in a novel pocket that they termed the BF-3 region, a domain distal to AF-2. Interestingly the mutation of various residues within this region altered the activity of the receptor, and furthermore many natural mutations within this region have been identified in patients with prostate cancer or androgen insensitivity syndrome. Clearly, this and other studies have provided the impetus to probe novel ways of manipulating NR activity that may yield drugs that work in place of, or in combination with, current modulators.

19.6 Exploiting Receptor-Cofactor Interactions for New Drug Discovery

The definition of a relationship between overall receptor conformation, differential cofactor recruitment and pharmacological activity has led to the emergence of the "selective nuclear receptor modulator (SNRM) hypothesis". One of the central tenets of this hypothesis is that by screening for small molecules that favor one NR-cofactor interaction at the expense of another, it will be possible to develop compounds that exhibit a high degree of process selectivity.

Although supported by the available data, the SNRM hypothesis remains to be formally proven. It has been shown that the pharmacology of SNRMs (SARMs, SERMs and SPRMs) can be manipulated by overexpression of certain coactivators (i.e. SRC-1/3 or PGC-1α) or inhibition/knockdown of corepressors (NCoR or RTA).[81–83] Thus, there is little doubt that imbalances in cofactor expression, such as it occurs in certain pathological states, affects the pharmacology of NR ligands. However, it remains to be demonstrated that the normal variations in cofactor expression that exist between cells (or tissues) are key determinants of NR ligand pharmacology. This will require the identification of compounds that engender specific NR-cofactor interactions and the subsequent demonstration that these ligands exhibit predictable biology. An example of how this hypothesis applies to ER-ligand pharmacology is shown in Fig. 19.4. To date, attempts to develop SNRMs have generally been performed in an empirical manner with studies of differential NR-cofactor interactions being used in a retrospective manner to provide an explanation for specific pharmacological attributes. However, this field will take a major leap forward when the NR-cofactors involved in specific biological processes are developed and used in mechanism-based screens

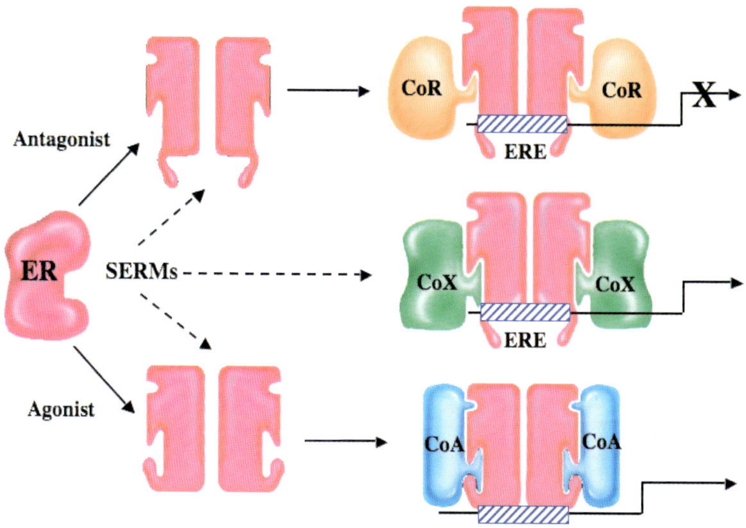

Fig. 19.4. An updated model of ER action may help to explain the activity of ER modulators. Upon binding an agonist or an antagonist, ER undergoes a conformational change that permits its spontaneous dimerization and facilitates the subsequent interaction of the dimer with estrogen response elements (ERE) located within target genes. Two genetically distinct ERs have been identified, ERα and ERβ, which have the potential to form homodimers or heterodimers in cells where both subtypes are expressed. It has recently been determined that different ligands can have different effects on ER structure. The functional consequences of different ligand-induced conformational changes were revealed with the discovery of receptor coactivators (CoA) and corepressors (CoR). Coactivators interact with agonist-activated ER and facilitate transcriptional activation, whereas corepressors interact with antagonist-activated receptor and help to maintain it in a quiescent state. SERMs permit the receptor to adopt a structure that is intermediate between that observed following the binding of agonists or antagonists. Thus, the relative agonist/antagonist activity of SERMs is a reflection of the ability of differently conformed ERα-ligand complexes to engage coactivators or corepressors. It is also possible that some SERMs allow ERα to interact in an ectopic manner with coactivator/corepressor (CoX) that would not interact with ERα in normal physiological circumstances.

for new compounds discovery. It is clear that although NRs are established and well validated targets, there remains a significant opportunity to exploit the complexities of their signaling pathways for new drug discovery.

References

1. Harper MJK, Walpole AL, A new derivative of triphenylethylene: Effect on implantation and mode of action in rats, *J Reprod Fert* **13**:101–119, 1967.
2. Harper MJK, Walpole AL, Contrasting endocrine activities of cis and trans isomers in a series of substituted triphenylethylenes, *Nature* **212**:87–89, 1966.
3. Love RR, Mazess RB, Barden HS, *et al.*, Effects of tamoxifen on bone mineral density in postmenopausal women with breast cancer, *New Engl J Med* **326**:852–856, 1992.
4. Dallenbach-Hellweg G, Schmidt D, Hellberg P, *et al.*, The endometrium in breast cancer patients on tamoxifen, *Arch Gynecol Obstet* **263**:170–177, 2000.
5. Love RR, Wiebe DA, Newcombe PA, *et al.*, Effects of tamoxifen on cardiovascular risk factors in postmenopausal women, *Ann Int Med* **115**:860–864, 1991.
6. McDonnell DP, Seclective estrogen receptor modulators (SERMs): A first step in the development of perfect hormone replacement therapy regimen, *J Soc Gynecol Investig* **7**:S10–S15, 2000.
7. Delmas PD, Bjarnason NH, Mitlak BH, *et al.*, Effects of raloxifene on bone mineral density, serum cholesterol concentrations, and uterine endometrium in postmenopausal women, *N Engl J Med* **337**:1641–1647, 1997.
8. Ettinger B, Black DM, Mitlak BH, *et al.*, Reduction of vertebral risk in postmenopausal women with osteoporosis treated with raloxifene: Results from a 3-year randomized clinical trial. Multiple outcomes of raloxifene evaluation (MORE) investigators, *JAMA* **282**:637–645, 1999.
9. Vogel VG, Costantino JP, Wickerham DL, *et al.*, Effects of tamoxifen vs raloxifene on the risk of developing invasive breast cancer and other disease outcomes, *JAMA* **295**:2727–2741, 2007.
10. Kuiper GGJM, Enmark E, Pelto-Huikko M, *et al.*, Cloning of a novel estrogen receptor expressed in rat prostate and ovary, *Proc Natl Acad Sci USA* **93**:5925–5930, 1996.
11. Mosselman S, Polman J, Dijkema R, ERβ: Identification and characterization of a novel human estrogen receptor, *FEBS Lett* **392**:49–53, 1996.
12. Couse JF, Korach KS, Estrogen receptor null mice: What have we learned and where will they lead us? *Endocrine Reviews* **20**:358–417, 1999.
13. Weihua Z, Saji S, Mäkinen S, *et al.*, Estrogen receptor (ER) β, a modulator of ERα in the uterus, *Proc Natl Acad Sci USA* **97**:5936–5941, 2000.
14. Hall JM, McDonnell DP, The estrogen receptor β-isoform (ERβ) of the human estrogen receptor modulates ERα transcriptional activity and is a key regulator of the cellular response to estrogens and antiestrogens, *Endocrinology* **140**:5566–5578, 1999.

15. Harris HA, Albert LM, Leathurby Y, *et al.*, Evaluation of an estrogen receptor-β agonist in animal models of human disease, *Endocrinology* **144**:4241–4249, 2003.
16. Norman BH, Dodge JA, Richardson TI, *et al.*, Benzopyrans are selective estrogen receptor beta agonists with novel activity in models of benign prostatic hyperplasia, *J Med Chem* **49**:6155–6157, 2006.
17. Kuiper GGJM, Carlsson B, Grandien K, *et al.*, Comparison of the ligand binding specificity and transcript tissue distribution of estrogen receptors α and β, *Endocrinology* **138**:863–870, 1997.
18. Elgort MG, Zou A, Marschke KB, *et al.*, Estrogen and estrogen receptor antagonists stimulate transcription from the human retinoic acid receptor-α-1 promoter via a novel sequence, *Mol Endocrinol* **10**:477–487, 1996.
19. Norris JD, Fan D, McDonnell DP, Identification of the sequences within the human complement 3 promoter required for estrogen responsiveness provides insight into the mechanism of tamoxifen mixed agonist activity, *Mol Endocrinol* **10**:1605–1616, 1996.
20. Tzukerman MT, Esty A, Santiso-Mere D, *et al.*, Human estrogen receptor transactivational capacity is determined by both cellular and promoter context and mediated by two functionally distinct intramolecular regions, *Mol Endocrinol* **8**:21–30, 1994.
21. Wijayaratne AL, Nagel SC, Paige LA, *et al.*, Comparative analyses of the mechanistic differences among antiestrogens, *Endocrinology* **140**:5828–5840, 1999.
22. Allan GF, Leng X, Tsai SY, *et al.*, Hormone and antihormone induce distinct conformational changes which are central to steroid receptor activation, *J Biol Chem* **267**:19513–19520, 1992.
23. Allan GF, Lombardi E, Haynes-Johnson D, *et al.*, Induction of a novel conformation in the progesterone receptor by ZK299 involves a defined region of the carboxyl-terminal tail, *Mol Endocrinol* **10**:1206–1213, 1996.
24. Allan GF, Tsai SY, Tsai M-J, *et al.*, Ligand-dependent conformational changes in the progesterone receptor are necessary for events that follow DNA binding, *Proc Natl Acad Sci USA* **89**:11750–11754, 1992.
25. McDonnell DP, Clemm DL, Hermann T, *et al.*, Analysis of estrogen receptor function *in vitro* reveals three distinct classes of antiestrogens, *Mol Endocrinol* **9**:659–668, 1995.
26. McDonnell DP, Clemm DL, Imhof MO, Definition of the cellular mechanisms which distinguish between hormone and antihormone activated steroid receptors, *Seminars in Cancer Biology* **5**:503–513, 1994.
27. Pike ACW, Brzozowski AM, Hubbard RE, *et al.*, Structure of the ligand-binding domain of oestrogen receptor beta in the presence of a partial agonist and a full antagonist, *EMBO J* **18**:4608–4618, 1999.

28. Shiau AK, Barstad D, Loria PM, *et al.*, The structural basis of estrogen receptor/coactivator recognition and the antagonism of this interaction by tamoxifen, *Cell* **95**:927–937, 1998.
29. Brzozowski AM, Pike AC, Dauter Z, *et al.*, Molecular basis of agonism and antagonism in the oestrogen receptor, *Nature* **389**:753–758, 1997.
30. Webster NJG, Green S, Jin JR, *et al.*, The hormone-binding domains of the estrogen and glucocorticoid receptors contain an inducible transcription activation function, *Cell* **54**:199–207, 1988.
31. Chang C-Y, Norris JD, Gron H, *et al.*, Dissection of the LXXLL nuclear receptor-coactivator interaction motif using combinatorial peptide libraries: Discovery of peptide antagonists of estrogen receptors α and β, *Mol Cell Biol* **19**:8226–8239, 1999.
32. Paige LA, Christensen DJ, Grøn H, *et al.*, Estrogen receptor (ER) modulators each induce distinct conformational changes in ERα and ERβ, *Proc Natl Acad Sci USA* **96**:3999–4004, 1999.
33. Norris JD, Paige LA, Christensen DJ, *et al.*, Peptide antagonists of the human estrogen receptor, *Science* **285**:744–746, 1999.
34. Hu XF, Veroni M, Luise Md, *et al.*, Circumvention of tamoxifen resistance by the pure antiestrogen ICI 182,780, *Int J Cancer* **55**:873–876, 1993.
35. Connor CE, Norris JD, Broadwater G, *et al.*, Circumventing tamoxifen resistance in breast cancers using antiestrogens that induce unique conformational changes in the estrogen receptor, *Cancer Res* **61**:2917–2922, 2001.
36. Gottardis MM, Jordan VC, Development of tamoxifen-stimulated growth of MCF-7 tumors in athymic mice after long-term antiestrogen administration, *Cancer Res* **48**:5183–5187, 1988.
37. Canney PA, Griffiths T, Latief TN, *et al.*, Clinical significance of tamoxifen withdrawal response, *The Lancet* **1**:36, 1987.
38. Belani CP, Pearl P, Whitley NO, *et al.*, Tamoxifen withdrawal response. Report of a case, *Arch Int Med* **149**:449–450, 1989.
39. Legault-Poisson S, Jolivet J, Poisson R, *et al.*, Tamoxifen-induced tumor stumulation and withdrawal response, *Cancer Treat Rep* **63**:1839–1841, 1979.
40. Fisher B, Dignam J, Bryant J, *et al.*, Five versus more than five years of tamoxifen for lymph node-nigative breast cancer: Updated findings from the National Surgical Adjuvant Breast and Bowel Project B-14 randomized trial, *J Natl Cancer Inst* **93**:684–690, 2001.
41. Fisher B, Dignam J, Bryant J, *et al.*, Five versus more than five years of tamoxifen therapy for breast cancer patients with negative lymph nodes and estrogen receptor-positive tumors, *J Natl Cancer Inst* **88**:1529–1542, 1996.
42. Dardes RC, O'Regan RM, Gajdos C, *et al.*, Effects of a new clinically relevant antiestrogen (GW5638) related to tamoxifen on breast and endometrial cancer growth *in vivo*, *Clinical Cancer Research* **8**:1995–2001, 2002.

43. Wu Y-L, Yang X, Ren Z, *et al.*, Structural basis for an unexpected mode of SERM-mediated ER antagonism, *Mol Cell* **18**:413–424, 2005.
44. Umetani M, Domoto H, Gormley A, *et al.*, 27-hydroxycholesterol is an endogenous SERM that inhibits the cardiovascular effects of estrogen, *Nature Med* in press, 2007.
45. DuSell CD, Umetani M, Shaul PW, *et al.*, 27-hydroxycholesterol is an endogenous selective estrogen receptor modulator, *Mol Endocrinol* **22**: in press, 2008.
46. Dixon D, Parrott EC, Segars JH, *et al.*, The second National Institutes of Health International Congress on advances in uterine leiomyoma research: Conference summary and future recommendations, *Fertility and Sterility* **86**:800–806, 2006.
47. Murphy AA, Kettel LM, Morales AJ, *et al.*, Regression of uterine leiomyomata in response to the antiprogesterone RU 486, *J Clin Endocrinol Metab* **76**:513–517, 1993.
48. Murphy AA, Morales AJ, Kettel LM, *et al.*, Regression of uterine leiomyomata to the antiprogesterone RU486: Dose-response effect, *Fertility and Sterility* **64**:187–190, 1995.
49. Wagner BL, Pollio G, Giangrande P, *et al.*, The novel progesterone receptor antagonists RTI 3021-012 and RTI 3021-022 exhibit complex glucocorticoid receptor antagonist activities: Implications for the development of dissociated antiprogestins, *Endocrinology* **140**:1449–1458, 1999.
50. DeManno D, Elger W, Garg R, *et al.*, Asoprisnil (J867): A selective progesterone receptor modulator for gynecological therapy, *Steroids* **68**:1019–1032, 2003.
51. Chwalisz K, Perez MC, DeManno D, *et al.*, Selective progesterone receptor modulator development and use in the treatment of leiomyomata and endometriosis, *Endocrine Reviews* **26**:423–438, 2005.
52. Chwalisz K, Garg R, Brenner RM, *et al.*, Selective progesterone receptor modulators (SPRMs): A novel therapeutic concept in endometriosis, *Annals of the New York Academy of Sciences* **955**:373–388, 2002.
53. Madauss KP, Grygielko ET, Deng S-J, *et al.*, A structural and *in vitro* characterization of asoprisnil: A selective progesterone receptor modulator, *Mol Endocrinol* **21**:1066–1081, 2007.
54. Rosen J, Negro-Vilar A. Novel, non-steroidal, selective androgen receptor modulators (SARMs) with anabolic activity in bone and muscle and improved safety profile, *J Musculoskel Neuron Interact* **2**:222–224, 2002.
55. Hamann LG, Mani NS, Davis RL, *et al.* Discovery of a potent, orally active, nonsteroidal androgen receptor agonist: 4-Ethyl-1,2,3,4-tetrahydro-6-(trifluoromethyl)-8-pyridono[5,6-g]-quinoline (LG121071), *J Med Chem* **42**:210–212, 1999.

56. Sun C, Robl JA, Wang TC, *et al.*, Discovery of potent, orally-active, and muscle-selective androgen receptor modulators based on an N-Aryl-hydroxybicyclohydantoin scaffold, *J Med Chem* **49**:7596–7599, 2006.
57. Edwards JP, Higuchi RI, Winn DT, *et al.*, Nonsteroidal androgen receptor agonists based on 4-(trifluoromethyl)-2H-pyrano[3,2-g]quinolin-2-one, *Bioorganic & Med Chem Lett* **9**:1003–1008, 1999.
58. Marhefka CA, Gao W, Chung K, *et al.*, Design, synthesis, and biological characterization of metabolically stable selective androgen receptor modulators, *J Med Chem* **47**:993–998, 2004.
59. Chen J, Kim J, Dalton JT. Discovery and therapeutic promise of selective androgen receptor modulators, *Mol Interventions* **5**:173–188, 2005.
60. Hwang DJ, Yang J, Xu H, *et al.*, Arlisothiocyanato selective androgen receptor modulators (SARMs) for prostate cancer, *Bioorganic & Medicinal Chemistry* **14**:6525–6538, 2006.
61. Segal S, Narayanan R, Dalton JT, Therapeutic potential of the SARMs: Revisiting the androgen receptor for drug discovery, *Expert Opin Investig Drugs* **15**:377–387, 2006.
62. Kazmin D, Prytkova T, Cook CE, *et al.*, Linking ligand-induced alterations in androgen receptor structure to differential gene expression: A first step in the rational design of selective androgen receptor modulators, *Mol Endocrinol* **20**:1201–1217, 2006.
63. Sathya G, Chang C-Y, Kazmin D, *et al.*, Pharmacological uncoupling of androgen receptor-mediated prostate cancer cell proliferation and prostate-specific antigen secretion, *Cancer Research* **63**:8029–8036, 2003.
64. Coghlan MJ, Jacobson PB, Lane B, *et al.*, A novel antiinflammatory maintains glucocorticoid efficacy with reduced side effects, *Mol Endocrinol* **17**:860–869, 2003.
65. Elmore SW, Pratt JK, Coghlan MJ, *et al.*, Differentiation of *in vitro* transcriptional repression and activation profiles of selective glucocorticoid modulators, *Bioorganic & Medicinal Chemistry Letters* **14**:1721–1727, 2004.
66. Wang JC, Shah N, Pantoja C, *et al.*, Novel arylpyrazole compounds selectively modulate glucocorticoid receptor regulatory activity, *Genes Dev* **20**:689–699, 2006.
67. Hall JM, Chang C-Y, McDonnell DP, Development of peptide antagonists that target estrogen receptor b-coactivator interactions, *Mol Endocrinol* **14**:2010–2023, 2000.
68. Chang C-Y, Abdo J, Hartney T, *et al.*, Development of peptide antagonists for the androgen receptor using combinatorial peptide phage display, *Mol Endocrinol* **19**:2478–2490, 2005.
69. Mettu NB, Stanley TB, Dwyer MA, *et al.*, The nuclear receptor-coactivator interaction surface as a target for peptide antagonists of the peroxisome proliferator activated receptors, *Mol Endocrinol* **21**:2361–2377, 2007.

70. Geistlinger TR, Guy RK, An inhibitor of the interaction of thyroid hormone receptor beta and glucocorticoid interacting protein I, *J Am Chem Soc* **123**:1525–1526, 2001.
71. Geistlinger TR, Guy RK, Novel selective inhibitors of the interaction of individual nuclear hormone receptors with a mutually shared steroid receptor coactivator 2, *J Am Chem Soc* **125**:6852–6853, 2003.
72. Galande AK, Bramlett KS, Trent JO, *et al.*, Potent inhibitors of LXXLL-based protein-protein interactions, *Chem Bio Chem* **6**:1991–1998, 2005.
73. Galande AK, Bramlett KS, Burris TP, *et al.*, Thioether side chain cyclization for helical peptide formation: Inhibitors of estrogen receptor-coactivator interactions, *J Peptide Res* **65**:297–302, 2004.
74. Leduc A-M, Trent JO, Wittliff JL, *et al.*, Helix-stabilized cyclic peptides as selective inhibitors of steroid receptor-coactivator interactions, *Proc Natl Acad Sci USA* **100**:11273–11278, 2003.
75. Rodriguez AL, Tamrazi A, Collins ML, *et al.*, Design, synthesis, and *in vitro* biological evaluation of small molecule inhibitors of estrogen receptor alpha coactivator binding, *J Med Chem* **47**:600–611, 2004.
76. Arnold LA, Estébanez-Perpiña E, Togashi M, *et al.*, Discovery of small molecule inhibitors of the interaction of the thyroid hormone receptor with transcriptional coregulators, *J Biol Chem* **280**:43048–43055, 2005.
77. Arnold L, Estebanez-Perpina E, Togashi M, *et al.*, A high-throughput screening method to identify small molecule inhibitors of thyroid hormone receptor coactivator binding, Science's STKE www.stke.org/cgi/content/full/sigtrans;2006/341/pl3, 2006.
78. Kong EH, Heldring N, Gustafsson J-A, *et al.*, Delineation of a unique protein-protein interaction site on the surface of the estrogen receptor, *Proc Natl Acad Sci USA* **102**:3593–3598, 2005.
79. Volakakis N, Malewicz M, Kadkhodai B, *et al.*, Characterization of the Nurr1 ligand-binding domain co-activator interaction surface, *J Mol Endocrinol* **37**:317–326, 2006.
80. Estebanez-Perpina E, Arnold AA, *et al.*, A surface on the androgen receptor that allosterically regulates coactivator binding, *Proc Natl Acad Sci USA* **104**:16074–16079, 2007.
81. Norris JD, Fan D, Sherk A, *et al.*, A negative coregulator for the human ER, *Mol Endocrinol* **16**:459–468, 2002.
82. Norris JD, Fan D, Stallcup MR, *et al.*, Enhancement of estrogen receptor transcriptionl activity by the coactivator GRIP-1 highlights the role of activation function 2 in determining estrogen receptor pharmacology, *J Biol Chem* **273**:6679–6688, 1998.
83. Smith CL, Nawaz Z, O'Malley BW, Coactivator and corepressor regulation of the agonist/antagonist activity of the mixed antiestrogen, 4-hydroxytamoxifen, *Mol Endocrinol* **11**:657–666, 1997.

Index